Therapeutic Applications of Mesenchymal Stem Cells in Veterinary Medicine

Mudasir Bashir Gugjoo

Therapeutic Applications of Mesenchymal Stem Cells in Veterinary Medicine

Springer

Mudasir Bashir Gugjoo, PhD
Veterinary Clinical Complex,
Faculty of Veterinary Sciences & Animal Husbandry
Sher-e-Kashmir University of Agricultural
Sciences & Technology of Kashmir (SKUAST-K)
Srinagar, Jammu and Kashmir, India

ISBN 978-981-19-3279-3 ISBN 978-981-19-3277-9 (eBook)
https://doi.org/10.1007/978-981-19-3277-9

This Springer imprint is published by the registered company Springer Nature Singapore Pte Ltd.
The registered company address is: 152 Beach Road, #21-01/04 Gateway East, Singapore 189721, Singapore

Preface

Since the inception of human race and idea of cure, concerted efforts have been made to relieve human and animal body sufferings and keep them healthy and productive. The biggest challenge remains the authorized prevention of death of a tissue/organ arising through ageing or disease. To address such concerns, regenerative medicine (a branch of medicine) attempts to effectuate normalization of the deranged physiological processes that occur in ageing or disease. Regenerative medicine essentially employs undifferentiated stem cells, a wonder cell that has the ability to give rise to different cell types and secrete various pro-healing factors. In the last one or two decades huge information has piled upon stem cells: their biological nature and their potential applications as therapeutics and in assisted reproductive technologies. Stem cell research in regenerative medicine has mainly remained focussed towards the human medicine. However, in recent times ethical treatment of animals and their welfare has gained lot more impetus. Besides, animals are being reared for the productive and performance traits and as family members like pets; any ailment in these animals is therefore taken seriously now-a-days. Thus, there is a growing need to impart regenerative medicine to animals for therapeutic purposes to prevent their sufferings and/or enhance their production potential. Since animals are being used as human translational models, such translational studies can become evidence-based medicine for human therapeutics and may become de facto treatment options for animals. Therefore, stem cell research in veterinary science acts as a double-edged weapon: on the one hand it offers treatment option for animals and on the other hand it provides a complementary evidence-based medicine for humans. This forms one of the aspects of the One-Health initiative taken to treat veterinary and human medicine as important tools to prevent diseases.

There are a number of books that focus on stem cell characteristics and their potential therapeutic applications, but all these books focus on human regenerative medicine. Our previous work although provides introduction to the mesenchymal stem cell and summarizes its recent progress in veterinary sciences. A lacuna in the form of detailed systemic approach to the clinical applications of stem cells is felt. Thus, the aim of this book remains to provide a systematic approach towards stem cell clinical applications in relation to the different body organ systems. To fill the lacuna, the current piece is framed to give detailed information on recent progress in stem cell research in relation to the various body systems of different animal species like dog, cat, horse, cattle, sheep and goat. The focus of this book is again on mesenchymal stem cells (MSCs) as they have been extensively investigated as potential therapeutic agents and in enhancing reproductive potential.

The current work offers a systematic approach towards MSCs potential applications in different body organ systems including musculoskeletal, nervous, gastrointestinal, cardiovascular, urinary, reproductive, respiratory and integumentary, besides the special organ like eye and its adnexa. It aims to provide an overview of in vitro basis of MSCs therapeutics for a particular organ/tissue. The in vitro studies translated into in vivo applications are extensively described including their mechanisms like differentiation or paracrine activities. Apart from the experimental studies MSCs clinical trials are discussed to provide evidence-based medicine for veterinary clinicians. Major focus has been put on the available in vivo studies to discuss and present actual feasibility, utility and efficacy of stem cells for a particular ailment of the particular tissue. The limitations in the current literature are highlighted and possible solutions to overcome these limitations are discussed.

Srinagar, Jammu and Kashmir, India Mudasir Bashir Gugjoo

Contents

About the Author

Mudasir Bashir Gugjoo PhD, is an Assistant Professor (senior scale) at the Division of Veterinary Clinical Complex, Faculty of Veterinary Sciences & Animal Husbandry of the Sher-e-Kashmir University of Agricultural Sciences & Technology of Kashmir. He has received his doctorate in mesenchymal stem cell translational studies from the Indian Veterinary Research Institute, Izatnagar. He has various awards to his credit including the best paper presentation, best faculty teacher and best review paper. He has been conferred with Indian Council for Agricultural Research—Junior Research Fellowship (ICAR-JRF) and Senior Research Fellowship (ICAR-SRF) for pursuing his master's and doctoral programmes, respectively. He has attended various faculty improvement programmes and has delivered lectures in conferences and/or workshops as a resource person. His lab focuses on the mesenchymal stem cell culture characteristics, differentiation properties and their therapeutic applications in some common ailments of animals. The research work is being funded by the Science and Engineering Research Board (SERB) of Government of India. He has authored more than 60 publications in different peer-reviewed journals. He has authored a book on stem cells in veterinary sciences. Besides, he has authored or co-authored various book chapters and is a columnist for daily newspapers. His work has been cited in standard books, in addition to numerous peer-reviewed journals. He is a member of various scientific societies and serves on the editorial board of various journals. He is a referee to a number of peer-reviewed journals and also reviews project proposals for the grant of funding.

Abbreviations

AD	Adipose tissue
AF	Amniotic fluid
ALDH1A3	Aldehyde dehydrogenase 1 family member A3
AM	Amniotic membrane
ANG-1	Angiopoietin-1
BALF	Bronchial alveolar lavage fluid
Bcl-2	B-cell lymphoma
BDNF	Brain-derived neurotrophic factor
BM	Bone marrow
CCL	Cranial cruciate ligament
CD	Cluster of differentiation
CNTF	Ciliary neurotrophic factor
Col I	Collagen type I
COL1A1	Collagen type A1
COMP	Cartilage oligomeric matrix protein
CT	Computed tomography
DDFT	Deep digital flexor tendon
DES	Desmin
DKC1	Dyskerin
DMD	Duchenne muscular dystrophy
DNMT1 DNA	Cytosine-5-methyltransferase
DP	Dental pulp
EGF	Epidermal growth factor
Egr1	Early response protein1
EMP2	Epithelial membrane protein 2
End	Endometrium
EPO	Erythropoietin
FGF2	Basic fibroblast growth factor
Gal-1	Galectin-1
GCR	Glucocorticoid receptors
GDNF	Glial cell-derived neurotrophic factor
GFAP	Glial fibrillary acidic protein
HA	Hyaluronic acid

HDAC1	Histone deacetylase 1
HGF	Hepatocyte growth factor
IAD	Inflammatory airway disease
IDO	Indoleamine 2,3-dioxygenase
ILs	Interleukins
INF-γ	Interferon gamma
LDHA	Lactate dehydrogenase A
LE	Limbal epithelium
Li	Liver
M1	Macrophage 1
MCP-1	Monocyte chemoattractant protein-1
MKX	Mohawk
MMP	Matrix metalloproteinase
MRI	Magnetic resonance imaging
MSC	Mesenchymal stem cell
Mu	Muscle
MYF-6	Myogenic Factor-6
MYOD	Myoblast determination protein
MyoD	Myogenic differentiation 1
NF-κB	Nuclear factor-kappa B
NGF	Nerve growth factor
NO	Nitric oxide
NT	Neurotrophin
Oct-4	Octamer-binding transcription factor 4
OE	Olfactory epithelium
Om	Omentum
Ov	Ovary
Pax	Paired box transcriptional factor
PAX6	Paired box protein
PB	Peripheral blood
PCNA	Proliferating cell nuclear antigen
PDGF	Platelet-derived growth factor
Pe	Periosteum
PGE2	Prostaglandin E2
PL	Periodontal ligament
Pl	Placenta
POU5F1	POU domain, class 5, transcription factor 1
PROX1	Prospero homeobox 1
PRP	Platelet-rich plasma
RAO	Recurrent airway obstruction
RUNX2	Runt-related transcription factor 2
SCX	Scleraxis
SEM	Scanning electron microscopy
SLC2A1	Solute carrier family 2 member 1
SOD	Superoxide dismutase

SSEA-1	Stage-specific embryonic antigen 1
Sy	Synovium
TERT	Telomerase reverse transcriptase
TGF-β1	Transforming growth factor
TNF	Tumour necrosis factor
TNMD	Tenomodulin
TSG-6	Tumour necrosis factor stimulated gene-6
TSP-1	Thrombospondin-1
TSPCs	Tendon-specific progenitor cells
TuJ1	Class III beta-tubulin
UCB	Umbilical cord blood
UCT	Umbilical cord tissue
WJ	Wharton's jelly
WNT	Wingless-related integration site

Mesenchymal Stem Cells Therapeutic Applications in Cartilage Regeneration

1

Abstract

Among various joint affections, osteoarthritis is one of the most common joint problems reported in animals. Currently, no effective treatment is available that can treat the condition to the fullest. Mesenchymal stem cells (MSCs) offer a potential tool to address the joint affections. These cells can be harvested from numerous body sources, although the cells derived from musculoskeletal sources appear more suitable. MSCs under cartilaginous-type microenvironment exhibit characteristics of the chondrocytes. This proposes the usefulness of the suitable scaffold/matrices, growth factors or the mechanical factors for the effectiveness of these cells. MSCs in vitro chondrogenesis, however, is affected by various factors and can be correlated with the non-uniform response of MSCs under in vivo system. Numerous in vivo chondrogenic studies, limited in one or the other factors, favour MSCs therapeutic applications although without any guaranteed results. A variable response occurs across the veterinary species with the MSCs transplantation under in vivo studies. The current chapter focuses on the chondrogenic potential of MSCs and the factors that affect their in vitro and in vivo chondrogenesis to provide understanding in the current status and future prospects of the therapy.

Keywords

Cartilage · In vitro · In vivo · Joint affections · Mesenchymal stem cell · Osteoarthritis

M. B. Gugjoo, *Therapeutic Applications of Mesenchymal Stem Cells in Veterinary Medicine*, https://doi.org/10.1007/978-981-19-3277-9_1

1.1 Introduction

Articular cartilage resists compression to offer frictionless movement of the bones in joint (Gugjoo et al. 2019). Cartilage being highly differentiated tissue with indirect blood, lymphatic and nerve supply and a limited number of chondrocytes usually fail to regenerate after an assault (Duarte Campos et al. 2012; Gugjoo et al. 2019). Joint affections in veterinary species range from septic arthritis, degenerative joint diseases/osteoarthritis and auto-immune diseases like rheumatoid arthritis. Assault to the joint may restrict to the cartilage tissue (partial thickness) and its associated subchondral bone (full thickness) (Hunziker 1999; Gugjoo et al. 2016). Partial-thickness defects, comparable to the fissures or clefts of early-stage osteoarthritis (OA), fail to heal spontaneously due to the lack of fibrin clots and the sufficient stem cells (Hunziker 1999). The full-thickness defects are replaced with the mechanically and structurally weaker fibrous tissue that lacks integration with the native tissue. This might arise due to the availability of insufficient reparative stem cell concentration at the lesion site (Hunziker 1999; Tiwary et al. 2014; Gugjoo et al. 2016, 2020). Articular cartilage thickness varies across joints and species (Athanasiou et al. 1995), and in experimental cartilage regenerative medicine, a majority of the studies involve the knee joint. Among various species, horses (1.5–2.0) followed by goats (0.7–1.5) have a comparable knee cartilage thickness to that of humans (2.2–2.5) (Frisbie et al. 2006) and act as suitable human translational model animals. These translational studies can potentially provide proof-of-therapeutics in veterinary medicine.

In clinical veterinary practice, OA is the most common form of arthritis reported, while rheumatoid arthritis remains an uncommon problem as compared to humans (Pedersen et al. 2000; Gershwin 2007). In horses, OA is the common cause of lameness reported in both young as well as the adults (Garcia et al. 2009). This amounts to around 60% of the lameness-related ailments (Clegg and Booth 2000; Caron and Genovese 2003) with 33% of the patients carrying cartilage lesions (Rose 1977). In bovine, more than 90% of the lame bulls are affected by OA (Persson et al. 2007) that potentially makes them infertile (*impotentia coeundi*) (Wolfe 2018). Mostly stifle joint of these bulls remain involved with predilection for medial and lateral condyles, and the patellar groove (Heinola et al. 2006; Persson et al. 2007). In dogs as high as 20–32% (Smith et al. 2006; Lamani et al. 2019) and in humans 21.4% (Barbour et al. 2016) OA cases are reported. Male dogs show more preponderance towards OA and among breeds Labrador retriever seems to be more prone (Lamani et al. 2019).

OA, an active process that initiates due to unknown reasons, is characterized by increase in the production of anti-anabolic, pro-catabolic and pro-inflammatory factors (Loeser 2011). Trauma, abnormal mechanical loading and inflammation may be the predisposing factors for development of OA. Trauma/abnormal mechanical loading may cause microfracture and associated inflammation causes an increase in some enzymatic activity producing wear particles that are being engulfed by macrophages (Wang et al. 2013). Mechanical damage also activates chondrocyte mechanoreceptors triggering the MAPK or NF-κB intracellular signalling pathway and pro-inflammatory and catabolic mediators (Houard et al. 2013; Sanchez-Adams et al. 2014). Notch signalling pathway may also activate in OA, although it is also associated with the inhibition of the OA (Zhu et al. 2021). If the etiological factors continue to act, production of such wear particles may accumilate beyond system's ability to eliminate. This incites inflammation to the extent that resident chondrocytes release degradation enzymes releasing collagen and proteoglycans particles. These particles are engulfed by macrophages that release pro-inflammatory cytokines like IL-1, IL-6 and TNFα that bind to the chondrocyte receptors. This leads to an increased chondrocyte apoptosis and their release of matrix metalloproteinases (MMPs, matrix-degrading enzymes), besides the inhibition of anabolic activities (Stannus et al. 2010; Larsson et al. 2015). In early- and late-stage OA, damaged and inflamed synovium produces reactive oxygen species (ROS), nitrous oxide (NO), prostaglandin E2 and neuropeptides. These factors alter cellular metabolism thereby disturbing the balance of cartilage matrix degradation and restoration (Scanzello and Goldring 2012; Mathiessen and Conaghan 2017). Thus, an imbalance in the cartilage contents arises with higher water content and lower collagen and proteoglycan contents (Buckwalter et al. 2005). This imbalance causes thickening of the subchondral bone and damage to the menisci and ligaments producing osteophytes and enthesophytes (Yang et al. 2013; Wei et al. 2017). In rheumatoid arthritis, an autoimmune inflammatory response damages the synovium producing generalized lesions in the joint and potentially eroding the cartilage. Erosion of cartilage exposes subchondral bone that increases friction thereby eliciting pain upon weight-bearing and/or during movement.

Currently, no therapeutics are available, and both the conservative and surgical management offer palliative treatment. Among cellular therapies, autologous chondrocyte implantation (ACI) makes a majority of the cellular therapies but human stem cell therapy accounts for 15% only, with the potential to replace the ACI (Fraser et al. 2006). In veterinary medicine such a contribution is miniscule. Regenerative medicine employing autologous chondrocytes although show good results but limited availability of autologous chondrocytes, their tendency to senescence and dedifferentiation into fibroblasts make it less feasible (Punwar and Khan 2011). The ageing or damaged chondrocytes due to their declined mitotic and synthetic activities synthesize lower and non-uniform proteoglycan molecules (Adkisson et al. 2001). MSCs due to their wide range of sources, characteristic properties and lack of any ethical or teratogenic effects are being evaluated as replacement options. Unlike chondrocytes, MSCs from non-autologous sources and as ready-to-use therapeutics can be made available due to the absence of major histocompatibility complex (MHC)-II and T-cell co-stimulatory molecules. MSCs or their secretome in the form of soluble factors or exosomes are rich in immunomodulatory/anti-inflammatory and other factors. These factors also help in the rejuvenation of local resident cells, attract distant stem cells and prevent degradation of extracellular molecules by inhibiting matrix metalloproteinases (MMPs) through tissue inhibitor MMPs. MSCs paracrine actions provide immunomodulation (secretion of PGE2, IL-10, TGF-β1, among others), recruitment of other resident cells, growth and proliferation of resident chondrocytes and their matrix secretion (Gugjoo et al.

2020). The immunomodulatory effects may be activated by the dyeing apoptotic MSCs directly or through phagocytosis of dyeing cells (Mancuso et al. 2019). Thus, mesenchymal stem cells (MSCs) or their secretome (exosomes or conditioned medium) is being evaluated in experimental and clinical trials (Lozito and Tuan 2011; Gugjoo et al. 2016, 2017, 2020; Juneau et al. 2016; Mancuso et al. 2019; Gugjoo and Pal 2020; Dar et al. 2021; Mocchi et al. 2021a, b). This chapter focuses on the potentialities and weaknesses of stem cells especially mesenchymal stem cells (MSCs) in relation to cartilage regeneration.

1.2 In Vitro Studies

One of the basic criteria to tag cells as mesenchymal stem cells (MSCs) is their ability to differentiate into the mesodermal lineages of the adipogenic, chondrogenic and osteogenic ones. Such a differentiation potential goes beyond to other types of musculoskeletal tissues as well and is guided by the available microenvironment. As such extensive studies towards the musculoskeletal applications are conducted. The ex vivo chondrogenic differentiation media apes the microenvironment that caters the physiological needs of the stem cells for chondrogenesis. Below are detailed in vitro and in vivo studies that help in understanding the role of MSCs in chondrogenesis and their overall therapeutic effects in cartilage regenerative medicine.

1.2.1 MSCs and Chondrogenesis

MSCs undergo chondrogenic differentiation (characterized mostly through special staining and/or gene expression analysis) under various extrinsic factors when supplemented individually or in combinations. MSCs chondrogenesis is achieved in pellet, micromass and/or 3D culture systems. In the pellet system, chondrogenic differentiation may occur spontaneously due to the local hypoxia that might develop at the centre of the pellet (Bosnakovski et al. 2004). The modifications in the existing pellet system like three-dimensional microwell substrata comprising porous zirconium may potentially improve MSCs chondrogenic potential as reported in equine BM-MSCs (Inui et al. 2019). However, it requires further details for effective utility.

Various cocktail preparations effectively induce chondrogenesis of MSCs harvested from domestic and pet animals like sheep (Chen et al. 2018; Ude et al. 2018; Vivas et al. 2018; Giretova et al. 2019; Dar et al. 2021; Futrega et al. 2021), goat (Somal et al. 2016, 2017; Zhai et al. 2019), dog and cat (Lee et al. 2016; Kim et al. 2017; Iacono et al. 2018; Sasaki et al. 2018; Villatoro et al. 2018; Cabon et al. 2019; Lima et al. 2019), cattle (da Silva et al. 2016; Campos et al. 2017; Cardoso et al. 2017; Chang et al. 2018), buffalo (Sreekumar et al. 2014; Sadeesh et al. 2016; Deng et al. 2018) and horse (Chang et al. 2018; Lara et al. 2017; Deng et al. 2018; Bçrziòð et al. 2018; Bundgaard et al. 2018; Zayed et al. 2018; Cabon et al. 2019; Gale et al. 2019a, b; Taghavi et al. 2020; Chapman et al. 2020; Contentin et al. 2020;

Kafarnik et al. 2020; Kim et al. 2020; Rakic et al. 2020; Fülber et al. 2021; Hotham et al. 2021; Zayed et al. 2021). These studies are detailed as follows:

1.2.2 Effect of Growth Factors and Chemicals

Currently, the cocktail of different growth factors includes transforming growth factor (TGF)-beta, bone morphogenetic protein (BMP) and fibroblast growth factor (FGF), along with the chemicals like dexamethasone, ascorbic acid and transferrin prime MSCs chondrogenic differentiation. These extrinsic growth factors are required in higher concentrations as compared to that of their normal endogenous concentrations (Wang et al. 2010; Gugjoo et al. 2017). A plateau is achieved beyond certain concentrations after which no further improvement in MSCs chondrogenesis occurs.

Among various growth factors, TGF-β3 is most extensively utilized (Goldman and Barabino 2016; Sanjurjo-Rodríguez et al. 2017; Bornes et al. 2018). The chondrogenesis (*collagen II* and *aggrecan* expression) through this growth factor along with dexamethasone may occur as early as 4 days of the sheep BM-MSCs (loaded on hyaluronic acid) chondrogenic priming. Such a chondrogenesis remains comparable to that occurring at day 14 interval (Bornes et al. 2018) and may be achieved through CD markers activation. TGF-β induced chondrogenic differentiation is considered to occur through CD105 or CD44 markers in humans. However, considering the minimal expression of some of the markers like CD105 in sheep AD-MSCs (Martínez-Lorenzo et al. 2009) and BM-MSCs (Mrugala et al. 2008) and their ability to differentiate into chondrogenic lineage, it may not be a sole responsive factor and/or species-wise variations exist in such a chondrogenic induction.

Other growth factors like FGF-2, TGF-β1, BMP-2, BMP-6 and BMP-7 with or without scaffolds variably promote MSCs chondrogenesis (Park et al. 2005; Knippenberg et al. 2006; Zeiter et al. 2009; Goldman and Barabino 2016; Endo et al. 2019b). The addition of serum-free induction medium along with the FGF-2 may further promote dog BM-MSCs chondrogenesis (Endo et al. 2019a, b). As FGF-2 inactivates IGF-1 and TGF-β pathways in human MSCs (Ito et al. 2008), the potential induction mechanism remains to be elucidated. BMP-7 plays contrast roles in MSCs differentiation across the species and tissues. BMP-7 promotes chondrogenesis (Knippenberg et al. 2006) or osteogenesis (Elkhenany et al. 2016) of goat AD-MSCs or adipogenesis in human BM-MSCs (Katja et al. 2007). BMP-2 may promote goat AD-MSCs and BM-MSCs osteogenesis (Knippenberg et al. 2006; Zhang and Jiang 2006) or bovine MSCs chondrogenesis in the presence of the scaffolds like alginate beads (Park et al. 2005; Zeiter et al. 2009). Hydrodynamic loading and BMP-2 application to cattle MSCs promote their osteogenesis (Goldman and Barabino 2016). BMP-2 in contrast to TGF-β1 enhances chondrogenic differentiation (Sox9, collagen type II and aggrecan) of bovine synovial MSCs (Park et al. 2005). Thus, BMP-2 may variably promote differentiation of cattle MSCs on the basis of source and culture conditions. Some of the factors like insulin-like growth factor (IGF)-1 may fail to induce MSCs towards chondrogenic lineage (Endo et al. 2019a, b).

Apart from single growth factor, combined use of these factors may show a variable role in the chondrogenesis. Growth factors like TGF-β3 and BMP-6 may synergistically promote horse BM-MSCs chondrogenesis (Taghavi et al. 2020). Likewise, TGF-β1 along with the growth differentiation factor-5 (GDF-5) promotes dog MSCs chondrogenesis represented by production of glycosaminoglycan and type II collagen without any cartilage hypertrophy (Endo et al. 2019a, b). TGF-β1 along with BMP-2 may also significantly increase glycosaminoglycan production of dog MSCs but simultaneously increases the cartilage hypertrophy (decrease in collagen II) (Endo et al. 2019a, b). Short-term exposure (4 h) of combination of growth factors like BMP-2 and BMP-7 promotes osteogenic differentiation of cattle AD-MSCs, while sole BMP-7 promotes chondrogenic differentiation after long-term (14 days) exposure (Knippenberg et al. 2006). This shows the importance of the microenvironment and thus, under in vivo system, MSCs may behave differently as per the available growth factors. Cattle BM-MSCs under TGF-β and BMP-2 or in combination show upregulated chondrogenic expression although the former has a higher potential than the latter. TGF-β induces simultaneous upregulation of collagen I unlike that of BMP-2 (Zeiter et al. 2009). Furthermore, the incorporation of growth factors like BMP-2 and BMP-6 fails to promote innately lower chondrogenic potential of dog AD-MSCs in comparison to their BM-MSCs (Teunissen et al. 2020). Thus, a variable response arises with the growth factors and the differentiation technique. As in vitro system variation leads to the varied differentiation or expression response of MSCs, it may be expected that MSCs upon in vivo implantation may not show uniform response as multiple factors act simultaneously.

1.2.3 Scaffolds and MSCs Chondrogenesis

Available matrices maintain, activate, guide and dedicate MSCs towards specific lineage. Various natural joint matrices or secretions like synovial fluid, collagen type II, I, chondroitin sulphate, hyaluronic acid (HA) hydrogel, decellularized extracellular matrix of sheep cartilage and others like composite silk extracellular matrix guide chondrogenic differentiation of MSCs derived from veterinary species (Hegewald et al. 2004; Varghese et al. 2008; Toh et al. 2012, 2013; Xue et al. 2012; Yin et al. 2016; Zhang et al. 2018a, b; Zhai et al. 2019; Bordbar et al. 2020). Synthetic scaffolds like polyethylene glycol (PEG) hydrogels, alginate hydrogel microcapsules, porous calcium polyphosphate and polyhydroxybutyrate/chitosan (PCH) porous 3D scaffold facilitate chondrogenic differentiation of MSCs (Hwang et al. 2011; Lee et al. 2015; Santos et al. 2018; Giretova et al. 2019). These cells get activated with these scaffolds through varied receptors as per the scaffold; collagen interacts via integrins (Leitinger and Hohenester 2007) and hyaluronic acid via CD44 (Peach et al. 1993). Synovial fluid may enhance chondrogenic differentiation of horse BM-MSCs (Zayed et al. 2021) as it contains various factors including TGF-β1 (Okazaki et al. 2001), prolactin (Ogueta et al. 2002) and hyaluronic acid (Gupta et al. 2019).

MSCs proteoglycan secretion upon matrices incorporation may not be stimulated to the extent of growth factors (Hegewald et al. 2004), but the combined application of matrices and growth factors tends to show an additive effect (Bosnakovski et al. 2006). Combined application of porous calcium polyphosphate and tri-iodothyronine favours sheep MSCs chondrogenesis (Lee et al. 2015). Application of matrices like PEG and collagen I may waive off the MSCs hypertrophic chondrocyte and lead to hyaline formation (Varghese et al. 2008; Bornes et al. 2016). Matrix-/scaffold-like pentosan polysulphate (PPS) in micromass culture enhances dog MSCs chondrogenesis (increased proteoglycans (PG) deposition) while polysulphated glycosaminoglycans (PSGAG) inhibit their chondrogenesis and promote the fibrocartilage-like phenotype with reduced PG deposition (Bwalya et al. 2017).

An intricacy in design including fibre alignment and a number of layers has a role in determining cellular differentiation (Prins et al. 2016). Cattle BM-MSCs cultured parallel on a single-layer scaffold (poly(ε-caprolactone) follow fibre alignment and give rise to a matrix with better stiffness as compared to that cultured perpendicular. Such a strength is still higher in the multilayered scaffold (Fisher et al. 2015). Scaffold cross-linking like hydrogels of hyaluronic acid and tyramine (HA-Ty) conjugate, and collagen II and GAGs should be designed in a way to ensure MSCs movement and contraction to favour their chondrogenesis without compromising its strength (Vickers et al. 2010; Toh et al. 2012). Thus, culture technique and matrix type and processing influence MSCs chondrogenic differentiation. A recent advancement in tissue engineering is the development of 3D bioprinted scaffolds. In such a technique, scaffold design is developed through the 3D scanning of the tissue to create layer-wise cell-distributed bioscaffolds. In relation to cartilage, the layer-wise scaffolds are being designed with appropriate MSCs positioning. The preliminary reports in goat (Critchley et al. 2020), horse (Mancini et al. 2020) and dog (Zhao et al. 2020) show 3D bioprinted scaffolds have the potential regenerative capacity, although it remains to be evaluated in detail.

1.2.4 Combined Extrinsic Factors and MSCs Chondrogenesis

Mechanical factors play a role in MSCs chondrogenesis. Hydraulic pressure tends to remain neutral while hydrodynamic force effectuates bovine MSCs musculoskeletal (chondro/osteogenic) differentiation (Zeiter et al. 2009; Goldman and Barabino 2016). Combined application of electromagnetic forces and growth factors (TGF-β3) may prevent adverse effects of inflammatory mediators like IL-1β on cattle MSCs chondrogenesis (Ongaro et al. 2012). Growth factors like TGF-β1 play a key role in the mechanical compression of MSCs for chondrogenic differentiation. Simple mechanical compression may inhibit goat BM-MSCs chondrogenesis and tend to enhance their hypertrophy upon shear stress. However, the addition of TGF-β1 leads to better chondrogenesis (Terraciano et al. 2007; Goldman and Barabino 2016). Thus, apart from favouring MSCs chondrogenesis, inhibitory effects may also be alleviated with the combined application of mechanical factors and growth factors. Recommendations of proper walk and motility are therefore a key to the prevention of osteoarthritis. Chondrogenesis may be achieved through rapid chromatin condensation that occurs simultaneously with the fibro-chondrogenic phenotype marker upregulation. Such a finding is confirmed through the blockade of histone methyltransferase EZH2 that usually promotes chromatin condensation and gene silencing (Heo et al. 2015).

1.2.5 Chondrocytes, Hypoxia and MSCs Chondrogenesis

Chondrocytes are superior in chondrogenesis as compared to MSCs (Park et al. 2005; Mauck et al. 2006). Due to their limited number and expansion potential, their alternative in the form of MSCs is being considered (Rackwitz et al. 2014). MSCs tend to promote resident chondrocyte growth and prevent their apoptosis. Equine BM-MSCs-derived nanoparticles promote chondrocyte growth and reduce their apoptosis being induced by the inflammatory environment (Kim et al. 2020; Hotham et al. 2021). These nanoparticles upregulate chondrocyte protein kinase (Akt) phosphorylation in the presence of IL-1β and also reduce the phosphorylation of extracellular signal-regulated kinase 1/2 (ERK1/2) (Kim et al. 2020). MSCs-chondrocyte co-culture may also promote bovine or goat BM-MSCs and AD-MSCs improved matrix secretion and prevent their hypertrophy (McCorry et al. 2016; Arora et al. 2017; McCorry and Bonassar 2017). Thus, MSCs may preserve native chondrocytes and promote regeneration of the lost joint.

Hypoxia may promote cellular differentiation but a variable effect on the viability is seen. Cultured dog AD-MSCs express anti-senescence-related genes including histone deacetylase 1 (HDAC1), cytosine-5-methyltransferase 1 [DNMT1 (DNA)], B-cell lymphoma (Bcl-2, inhibitor of apoptosis), telomerase reverse transcriptase (TERT), lactate dehydrogenase A (LDHA), solute carrier family 2 member 1 (SLC2A1, a glucose transporter) and dyskerin (DKC1, a telomere holoenzyme complex) (Lee et al. 2016). Hypoxia may also enhance the shredding of the microvesicles from the dog AD-MSCs (Iacono et al. 2018), increasing their therapeutic potential. Simultaneously, hypoxia also induces superior chondrogenesis of MSCs exhibited by aggrecan and collagen II mRNA expressions (Zscharnack et al. 2009; Bornes et al. 2015; Lee et al. 2016). Hypoxia may additionally downregulate hypertrophic marker COL10A1 in ovine BM-MSCs and equine synovial MSCs (Bornes et al. 2015; Gale et al. 2019a). Further, hypoxia-induced imbalance in bioactive lipids can be prevented with acid ceramidase in cat and horse MSCs (Simonaro et al. 2013).

1.2.6 Donor Factors Affecting MSCs Chondrogenesis

Quality, quantity and properties of MSCs are donor-dependent involving its age, tissue type, breed and species (Yaneselli et al. 2018; Arnhold et al. 2019; Kamm et al. 2019; Bagge et al. 2020; Hendawy et al. 2021). MSCs from geriatric donors have a limited proliferation potential (Mazzotti et al. 2019; Alicka et al. 2020; Bagge et al. 2020). Aged horse synovial fluid MSCs show reduced proliferation potential with age-related tetraploid cells carrying reduced chondrogenic efficiency. Ultrastructural analysis shows ageing-like changes (including the stressed endoplasmic reticulum, autophagy and mitophagy) in these cells (Mazzotti et al. 2019). Horses above 5 years harbour AD-MSCs of limited proliferation potential (Alicka et al. 2020) with higher chondrogenesis of MSCs derived from young ones as compared to the aged sources (Erickson et al. 2011).

Source also affects MSCs chondrogenesis. Comparing the horse BM- and umbilical cord blood (UCB)-MSCs chondrogenesis in constructs, the former cell construct possesses higher tensile properties and collagen content while the latter constructs show superior compressive properties approaching the native cartilage. It can be concluded that cells from both the sources form neocartilage, but the tissue secreted from BM-MSCs resembles more closely to the native equine articular cartilage (White et al. 2018). Dog AD-MSCs chondrogenic potential remains low as compared to the dog BM-MSCs (Reich et al. 2012) which does not improve with the addition of extrinsic growth factors like BMP-2 and BMP-6 (Teunissen et al. 2020). Among various locations of the adipose tissue, retroperitoneal and subcutaneous offer MSCs with enhanced chondrogenic potency (Arnhold et al. 2019). Equine MSCs from musculoskeletal tissue like synovium and BM may (Gale et al. 2019b) or may not (Zayed et al. 2018) show comparable chondrogenic potential. In relation to the UCB- and UC matrix MSCs, the former cell may be a more favourable option for chondrogenesis (Rakic et al. 2020). In general, MSCs from musculoskeletal sources in dog and equine show superior chondrogenesis (Noel et al. 2008; Malagola et al. 2016; Branly et al. 2017; Sasaki et al. 2018; Contentin et al. 2020; Teunissen et al. 2020). However, equine BM-MSCs chondrogenesis (aggrecan synthesis) in relation to the AD-MSCs and synovial fluid MSCs are also demonstrated to be low (Fülber et al. 2021). As the study employed 3D environment using biomaterials or scaffolds (biocompatible nanoparticles consisting of gold, iron oxide and poly-L-lysine), it may have shown variable response based upon the type of the scaffold and culture technique. Apart from source and species, breed-wise variation may also be seen in MSCs chondrogenesis (Bertolo et al. 2015). Furthermore, depending upon the in vivo priming of MSCs prior to tissue harvest, MSCs may behave accordingly.

1.2.7 Miscellaneous Factors

A major challenge in MSCs culture is to prevent contaminants of the foetal bovine serum, and thus, serum alternatives are being evaluated. One of the options evaluated is the platelet lysate (PL) that improves the growth of the MSCs better than the FBS at 5–10% concentrations (Lima et al. 2019). However, PL fails to enhance horse BM-MSCs and dog AD-MSCs chondrogenic differentiation despite an increase in TGF-β1 (Lima et al. 2019; Chapman et al. 2020).

1.3 Challenges in MSCs Therapy

In general, under in vitro studies MSCs are considered to differentiate into chondrocyte-like cells with an ability to secrete cartilage matrix upon the availability of suitable microenvironment/niche. Such an induction may occur through activation of SOX9 that predisposes chondrogenic lineage and inhibits various hypertrophic chondrocyte pathways like RUNX2 (Zhou et al. 2006), Wnt (Topol et al. 2009), Col10a1 and VEGF-A (Hattori et al. 2010; Leung et al. 2011). Cartilage-specific matrices involving proteoglycans, GAGs and collagen II remain to be confirmed in all the studies (Zayed et al. 2017, 2021). MSCs differentiation mechanism is considered to have little importance under in vivo systems as the cells do not survive for long. The main therapeutic effect is considered to occur through paracrine actions being mediated through the soluble factors and/or microvesicles. There are various limitations that currently remain to be addressed. MSCs undergo hypertrophic cartilage formation and inflammation may not favour their therapeutic efficacy (Fig. 1.1). In horse and sheep osteoarthritis (presence of inflammatory mediators like IL-1b, IL-17 and TNF-α), MSCs from synovium and BM may show reduced chondrogenic (glycosaminoglycan, galactin, aggrecan, collagen II and SOX9 and TGF-β1) expression. Additionally, an increased expression of adhesion molecules with reduced potential to prevent acute inflammation also occurs (Ando et al. 2012;

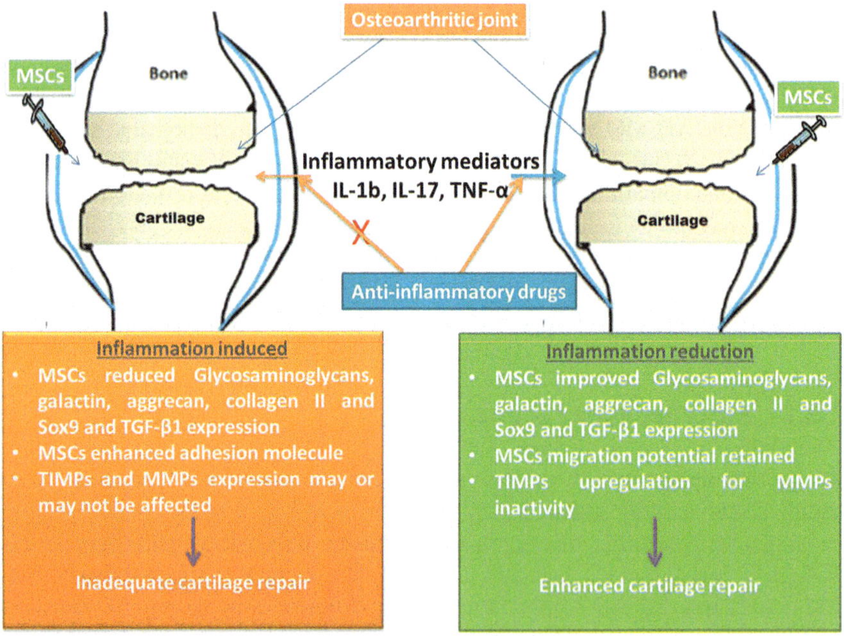

Fig. 1.1 Mesenchymal stem cells (MSCs) intra-articular implantation improves the osteoarthritis. Peak inflammation affects mesenchymal stem cell secretory and migratory potential

Kondo et al. 2013; Zayed et al. 2017, 2021; Reesink et al. 2017; Colbath et al. 2020a, b). This gets further complicated as MSCs from particular sources like synovium are more vulnerable than others like BM-MSCs (Fülber et al. 2016; Zayed et al. 2017). Joint affections like osteoarthritis and osteochondrosis dissecans affect synovial fluid and synovial membrane MSCs and decrease their chondrogenic potential (Fülber et al. 2016). To overcome such problems, tissue engineering techniques may be combined with the gene transfer techniques. One of the studies demonstrates that MSCs transfection with interleukin-1 receptor antagonist (IL-1Ra) blocks IL-1β stimulation to produce IL-1Ra above the basal level (Gabner et al. 2018). Thus, MSCs may not be useful in acute OA and should be avoided or their genetically modified forms may be used for better efficacy.

1.4 In Vivo Studies

Among various stem cells, MSCs contribute the majority of the therapeutics towards the joint affections. The experimental models studied are mostly traumatic involving the osteochondral defects, femoral head osteonecrosis, meniscectomy and anterior cruciate ligament resection although limited enzyme-induced osteoarthritis experimental studies are also conducted. Sheep and goat offer useful animal models for human osteochondral translational studies. The difficulty in arthroscopy to evaluate progress of the condition in these animals is fulfilled in horse models (Wilke et al. 2007; Sah and Ratcliffe 2010; McIlwraith et al. 2011; Korchunjit et al. 2019). Small ruminants show similarity in size and anatomy and lack self-repair as that of the humans. These animals also offer studies on age-related changes (Sah and Ratcliffe 2010). The animals are easily reared and maintained, in addition to being cost-effective as compared to large ruminants and equines (Music et al. 2018). These animal experimental studies have the potential to provide insights in the MSCs regenerative therapeutic potential for veterinary and human joint affections (Gugjoo et al. 2020). Apart from the experimental studies, the clinical applications of MSCs in natural joint affections are also evaluated in dog and horse (Fig. 1.2). MSCs experimental and clinical trials evaluations in animals provides the proof-of-therapeutics for their potential clinical utilization.

The cells are being utilized directly or along with the scaffolds and growth factors. Such a combination of cells, scaffolds and growth factors is being dealt under tissue engineering. Initially, 2D scaffolds were used which are now frequently being replaced by 3D scaffolds for better mimicking of the natural environment. Recent advancements have made it possible to develop in vitro 3D bioprinted tissues (layer-by-layer deposition) based on the actual tissue structure by incorporating the cells like MSCs and scaffolds. Such a technique is thus considered to be a suitable and effective way to cater the growing needs of joint affections (Gugjoo et al. 2016, 2019). Among various 3D bioprinting techniques, bio-extrusion technique seems effective to recreate the layered structures like cartilage (Messaoudi et al. 2021). MSCs, although less effective as compared to chondrocytes in conventional tissue engineering, may better be able to secrete cartilage-specific matrix in 3D bioprinted

Sheep	Goat	Dog	Horse
Experimental Joint defect models in Veterinary species			
I. Mechanical defects 1. Full thickness osteochondral defect 2. Partial chondral defect 3. osteonecrosis of femoral head 4. anterior cruciate ligament excision 5. condyle defect and meniscectomy 6. Meniscectomy **Chemical Induced model** 1. Collagen induced osteoarthritis	**Mechanical defects** 1. Full thickness osteochondral defect 2. Meniscal tear 3. Medial meniscus and anterior cruciate ligament resection	**Mechanical defects** 1. Full thickness osteochondral defect 2. Partial chondral Defect 3. Osteonecrosis 4. osteochondral defects with drilling 5. Meniscectomy	**Mechanical defects** 1. Full thickness osteochondral defect
Clinical affections of the joints			
No study	No study	1. Osteoarthritis 2. Dysplasia 3. Dysplasia & Osteoarthritis	1. Osteoarthritis / Degenerative joint disease

Fig. 1.2 Species-wise details of the joint affection models and clinical conditions undertaken for mesenchymal stem cells (MSCs) applications

cartilage as their differentiation and expression are better guided by the available microenvironment (Gugjoo et al. 2016, 2019). Even combined use of articular cartilage-resident chondroprogenitor cells (ACPCs) and MSCs laden in 3D hydrogel (gelatin methacryloyl (gelMA)-based hydrogels) bioink may offer a potential way to improve cartilage regeneration (Levato et al. 2017), although ACPCs as such may not be as effective as MSCs (Schmidt et al. 2020). As the field is very recent and emerging, studies on in vivo application of 3D bioprinted cartilage currently remain to be undertaken.

In this section, species-wise MSCs experimental and clinical trial studies shall be discussed. It is worth mentioning that MSCs utilized in below given cartilage regenerative studies may necessarily not be fully characterized and their final viability is not defined in many studies.

1.4.1 Sheep Studies

MSCs are evaluated in various experimental studies including chondral defects, osteochondral defects, osteonecrosis of femoral head, anterior cruciate ligament excision and medial meniscectomy, condyle defect and meniscectomy as detailed in Table 1.1. These models are developed mechanically under surgical procedures. One of the models of OA is induced through intraarticular injection of bovine type II collagen (Abdalmula et al. 2017) corresponding to the diffuse rheumatoid arthritis.

Table 1.1 Sheep in vivo MSCs preclinical experimental studies on various joint tissue affections

Study design/study period	Biomaterial used	Evaluation criteria	Overall result	References
Chondral defect model				
Randomized placebo controlled ($n = 28$)/24 weeks	Autologous BM-MSCs (3×10^7)+ beta-tricalcium phosphate (β-TCP)	Macroscopic observation, histological, immune-histochemical Biochemical	Positive outcome	Guo et al. (2004)
Randomized placebo controlled ($n = 3$)/9 weeks	Autologous MSC (1×10^7); transforming growth factor (TGF-β3); Chitosan; fibrin clot.	Macroscopy, histology	Positive outcome	Mrugala et al. (2008)
Randomized placebo controlled ($n = 9$)/12 months	BM-MSCs ($3.3 \times 10^6 \pm 0.4 \times 10^6$), polylactide: polyglycolic acid scaffolds	Biomechanical testing, macroscopic and histological analyses	Positive outcome	Caminal et al. (2014a)
Randomized placebo-controlled comparative study ($n = 12$)/8 weeks	BM-MSCs (2×10^6), amniotic membrane	Gross and histopathology	Positive outcome	Garcia et al. (2015)
Randomized placebo controlled ($n = 15$)/6 months	Xenogeneic AD-MSCs (1×10^6) and collagen/chitosan scaffold	Microscopic and macroscopic analysis	Positive outcome	Zorzi et al. (2015)
Randomized placebo controlled ($n = 12$)/4 weeks	AD-MSCs (1×10^6 cells per cm^2), Collagen membranes	Confocal laser scanning microscopy	Positive outcome	Hindle et al. (2016)
Randomized placebo controlled ($n = 12$)/4 weeks	Allogeneic AD-MSCs (1×10^6/100 μL), Polycaprolactone	Histopathology, qPCR, immunohistochemistry	Positive outcome	Vahedi et al. (2019)
Randomized comparative study ($n = 4$)/4 weeks	Allogeneic BM-MSCs and autologous chondrocyte micropellets	histological analyses	Positive outcome	Futrega et al. (2021)
Osteochondral defect models				
Randomized placebo controlled ($n = 10$)/6 months	Autologous BM-MSCs (4×10^5), collagen I hydrogel constructs	Histopathology	Positive outcome	Zscharnack et al. (2010)
Randomized placebo-controlled comparative study ($n = 28$)/3 months	AD-MSCs differentiated into chondrocytes (1×10^6 cell per mm^3), alginate gel (2%), microfracture	Histopathology, MRI	Positive outcome	Bozkurt et al. (2019)

(continued)

Table 1.1 (continued)

Study design/study period	Biomaterial used	Evaluation criteria	Overall result	References
Randomized placebo controlled ($n = 5$)/6 months	BMP—2 active collagen membrane followed by hydrogel solution (alginate/hyaluronic acid) containing autologous BM-MSCs spheroids (10^7 cells/mL)	Micro-CT, histopathology	Positive outcome	Favreau et al. (2020)
Osteonecrosis of femoral head model				
Randomized placebo controlled ($n = 8$)	Sheep BM-MSCs and human dental stem cells (1×10^6)	Light microscopy	Positive outcome	Feitosa et al. (2010)
Anterior cruciate ligament excision and medial meniscectomy model				
Randomized placebo-controlled comparative study ($n = 16$)/6 weeks	Chondrogenically differentiated MSCs (10×10^6) or undifferentiated MSCs (10×10^6)	Gross, histological and clinical observation	Positive outcome	Al Faqeh et al. (2012)
Randomized placebo-controlled comparative study ($n = 18$)/8 weeks	Autologous BM-MSCs (10×10^6), BM-MNCs after 12 weeks of model creation	Macroscopically and histologically, gene expression levels	Positive outcome	Song et al. (2014)
Randomized placebo-controlled ($n = 24$)/14 weeks after treatment	Allogeneic AD-MSCs (5×10^7, high dose at 3 and 1×10^7 cells, low dose at 6 weeks), hyaluronic acid	Magnetic resonance imaging (MRI), macroscopy, micro-computed tomography, and cartilage-specific staining	Positive outcome	Feng et al. (2018)
Randomized placebo-controlled comparative study ($n = 34$)/27 weeks after treatment	AD-MSCs (2.5×10^7 and 5.0×10^7), hyaluronic acid (2.5 mL)	Histopathology, ELISA, immunoblot assay, immunohistochemistry	Positive outcome	Lv et al. (2018)
Randomized placebo controlled ($n = 18$)/7 weeks	AD-MSCs and BM-MSCs (2×10^6) post 7 weeks of surgery	Gross, histology, electromyography	Positive outcome	Ude et al. (2018)
Condyle defect and meniscectomy				
Randomized placebo controlled ($n = 10$)/12 months	BM-MSCs (1.1×10^7 (in 6-month period study) and 1.2×10^7 (in 12-month period study)	Radiography, MRI, ultrasound, macroscopic and histological analyses	Positive outcome	Caminal et al. (2014a)

Unilateral meniscectomy				
Randomized placebo controlled ($n = 20$)/12 weeks	BM-MSCs + scaffold (Hyaff®-11) and BM concentrate + scaffold (Hyaff®-11) (6×10^6 seeded on scaffold)	Macroscopy, histology, immunohistochemistry, and micro-computed tomography	Positive outcome	Desando et al. (2016)
Randomized placebo controlled ($n = 30$)/6 months	BM-MSCs collagen I scaffold (1×10^6/cm^2)	Macroscopy and histopathology	Positive outcome	Whitehouse et al. (2017)

MSCs mesenchymal stem cells, *BM* bone marrow derived, *AD* adipose tissue derived, *MNCs* mononuclear cells, *BMP* bone morphogenetic protein, *n* total number of animals in all the groups of the single study; Positive outcome: Indicates improved outcome with MSCs transplantation even if only one parameter improves; Study period: Indicates maximum time period for which study is conducted

In all these studies, local intra-articular administration of MSCs is made except for one study that transplants intravenously (Abdalmula et al. 2017). Their localization in the joint and in particular tissue may be made on the basis of magnetic resonance imaging (MRI) (Kaggie et al. 2020) apart form the basic immuno-histochemical localization; however, details of such engraftment are very limited in most of the studies.

1.4.1.1 Chondral Defect Models

In chondral defect models, subchondral bone remains intact while in the case of osteochondral defect models subchondral bone is pierced. In chondral defect models, defect diameter ranges from around 4 mm (Hindle et al. 2016) to 10 mm (Zorzi et al. 2015). Mostly femoral condyles are employed for evaluation of MSCs repair potential with one study using internal groove of the patella (Mrugala et al. 2008). In all these studies, adipose or bone marrow derived MSCs from autologous, allogeneic or xenogeneic sources at variable doses are evaluated (Table 1.1). The animal numbers in these studies are limited ranging from $n = 4$ (Futrega et al. 2021) to $n = 28$ (Guo et al. 2004) with median number around $n = 12$. Further, animal number remains even more limited in an individual group of the study. The study period ranges from 4 weeks (Hindle et al. 2016; Vahedi et al. 2019; Futrega et al. 2021) to 6 months (Guo et al. 2004; Zorzi et al. 2015). All these studies incorporate some kind of the scaffolds/biomaterial as detailed in Table 1.1. The repair potential of MSCs is confirmed through variable evaluation techniques including macroscopic observation, histology, immune-histochemistry, biochemical and confocal laser scanning microscopy.

All these studies demonstrate that MSCs loaded with the scaffolds potentially improve defect repair as compared to the scaffold loaded or control, although a typical hyaline cartilage with typical histology remains to regenerate (Caminal et al. 2014a, b; Zorzi et al. 2015). Defect filling usually occurs with the fibrocartilaginous tissue that might be providing early relief represented by clinical knee-function scores. However, later on, the degenerative changes may ensue (Caminal et al. 2014a, b). There are also studies that show hyaline type of healing with scaffold-loaded MSCs. A typical smooth hyaline cartilage formation resembling adjacent cartilage by 12 weeks occurs post beta tri-calcium phosphate-loaded MSCs. Metachromasia however, remains non-uniform and the transition zone appears to shift towards hypertrophic cartilaginous tissue at this 3-month interval. Further, near the subchondral bone, hypertrophic chondrocytes are calcified through endochondral ossification. However, at 6 months period, the surface of the defect become indistinguishable from the adjacent cartilage although proteoglycan intensity is not sufficient and the chondrocyte arrangement remains a bit irregular (Guo et al. 2004). Similarly, MSCs along with the chitosan powder and TGF-β3 also promote hyaline-like tissue repair that integrates with adjacent cartilage by ninth week post biomaterial transplantation (Mrugala et al. 2008). Polycaprolactone-loaded AD-MSCs transplantation may also promote hyaline-like tissue repair duly characterized for cartilage matrices (collagen II and aggrecan) by fourth week (Vahedi et al. 2019). Thus, MSCs along with the particular scaffold may offer hyaline regeneration, although all the features of typical cartilage remain to be seen and statistically low animal number and short duration of these studies do not give inputs about the long-term results. The cell tracking that essentially helps to understand the role of MSCs and their possible mechanism involved in such a repair remains to be seen. A study that tracks MSCs post transplantation demonstrates that these cells do not achieve rounded chondrocyte appearance by fourth week (Hindle et al. 2016). MSCs as such may potentially act through the paracrine actions. Scaffolds that help in cell growth and proliferation, provide mechanical support, degrade at the pace as that of tissue regeneration and do not create a toxic environment are desired. Collagen, chitosan or hyaluronic acid undergo rapid degradation and offer poor mechanical support for cell proliferation and extracellular matrix deposition. However, one of the studies that employs chitosan also demonstrates better outcome at initial period although growth factor was additionally included (Mrugala et al. 2008). Furthermore, animals additionally pose a challenge of an immediate movement that may further warrant mechanical strength of the scaffold. The mechanical strength should not compromise the cellular movement or growth in the scaffold.

Apart from the fresh MSCs harvested from common adult tissue sources, cryopreserved MSCs or the cryopreserved foetal membrane tissue-derived MSCs are also evaluated. This is essential to establish the data for delayed and/or allogeneic use of MSCs maintained in the tissue or cell bank. Amniotic membrane, irrespective of the fresh or cryopreserved, loaded with MSCs helps in cartilage repair with chondrocytes visible in the defect area (Garcia et al. 2015). To compare MSCs chondrogenic potential against the chondrocytes, MSCs micropellet (prepared in 2% CO_2) appears less effective in chondral defects as compared to the chondrocyte micropellets especially in relation to the hypertrophic cartilage formation. However, typical hyaline-like tissue that could integrate seamlessly fails to generate in either defect (Futrega et al. 2021). One of the main impediments currently against the use of MSCs is hypertrophic cartilage formation. Thus, there is a need to devise tissue-engineered constructs that can mitigate such issue. Besides, the studies that involve large animal numbers and focus on long-term benefits of tissue-engineered constructs are desired. Due to the lack of long-term studies with sufficient animal numbers, and unavailability of detailed data over the fate of MSCs upon transplantation, clinical applications remain distant.

1.4.1.2 Full-Thickness Osteochondral Defect Models

Full-thickness femoral osteochondral defects of variable diameter are evaluated for repair through MSCs derived from bone marrow or adipose tissue (Table 1.1). Simple MSCs (Kaggie et al. 2020), chondrocyte inducted MSCs (Zscharnack et al. 2010; Bozkurt et al. 2019) or MSCs spheroid (Favreau et al. 2020) of variable concentration is transplanted along with the scaffolds like collagen I hydrogel (Zscharnack et al. 2010), alginate microsphere with functionalized chitosan (Bozkurt et al. 2019) or alginate/hyaluronic acid (Favreau et al. 2020). The evaluation period is around 6 months with evaluation conducted on the basis of micro-CT, histopathology and MRI. Superparamagnetic iron oxide (SPIO)-labelled cells in the osteochondral defects may not remain localized under MRI but surrounding tissues like the synovium does (Kaggie et al. 2020).

Like chondral defects, osteochondral defects too improve with the tissue-engineered constructs containing MSCs. There are slight modifications in usual standard procedures in some of the studies. Hydrogel-loaded chondrogenic induced MSCs transplanted by press-fit technique repair defects with hyaline-like tissue characterized by typical columnarization and presence of collagen type II after 6 months. The results, however, vary among individual cases and the design of model is somewhat different from others as two defects (7mm each) in a single condyle are created (Zscharnack et al. 2010). The individual variations are understood considering the immediate mobility of the animal and its stress on the scaffold-cell assembly, besides the physiological differences. To prevent extensive surgical requirement, minimally invasive bio-targeted microsphere-based tissue engineered (TE) construct may be used. These minimally invasive transplanted TE constructs although show non-significant benefits against the cell-scaffold transplantation techniques performed under open surgery. TE constructs promote fibrous tissue alongside the cartilage matrices (Bozkurt et al. 2019). Another modified procedure evaluated is the bi-compartmented implant transplantation. In such a system, one of the compartments (collagen associated with bone morphogenetic protein 2) is aimed at structural support and in promoting the subchondral bone regeneration. The other compartment containing BM-MSCs spheroids may contribute to regeneration. Post 6-month transplantation, the surface of the regenerated articular cartilage remains threefold higher than the control. Osteochondral defect is regenerated with hyaline-like cartilage (Favreau et al. 2020). This process is also being aimed in 3D printing to mimic the actual cartilage structure for effective functioning of the cells. Thus, there is a need to move towards the advanced and more innovative techniques to ensure repair with typical hyaline cartilage.

1.4.1.3 Cruciate Ligament and Medial Meniscectomy Model

Cruciate ligament injury is one of the common injuries seen in human and small animals like dogs. This type of OA is more common than chondral defect-induced OA in animals. Anterior cruciate ligament (ACL) excision and medial meniscal damage develop OA when an animal moves freely. In order to evaluate MSCs therapeutic potential in such cases, various studies are conducted. MSCs transplantation has overall been credited with the improvement in OA and blocking its progression (Table 1.1) (Al faqeh et al. 2012; Song et al. 2014; Feng et al. 2018; Lv et al. 2018; Ude et al. 2018).

In ACL excision and medial meniscectomy model, transplantation of chondrogenic-induced MSCs (CiMSCs) or simple MSCs after 6 weeks of surgical procedure improve International Cartilage Repair Score (ICRS) as compared to the control. Compared to the simple MSCs, CiMSCs implanted joints show non-significant differences in ICRS score but the overall medial meniscal cartilage appears better in CiMSCs-treated joints. Further, the cell-treated joints have good cartilage histology, thickness and quality resembling normal cartilage (Al faqeh et al. 2012). Apart from a single MSCs transplantation, double dose [50 and 10 million along with hyaluronic acid (HA) at 3- and 6-week post-surgery] efficiently blocks OA progression in an 18-week study period. The inflammatory mediators (TNF-α and IL-6) are significantly lower and the cartilage thickness of tibial plateau and femoral epicondyles is significantly increased in biomaterial transplanted joints as compared to others. The SPIO-labelled allogeneic MSCs are located for 14 weeks post cell transplantation mainly in the synovial lining, although not confirmed histologically (Feng et al. 2018). The cellular engraftment to the defect area and their differentiation remains to be confirmed. The possibility of actual cell types to be present in the area for such a period may have to be confirmed as signals from these cells may be transmitted to the resident host cells. The better results with the double dose of cells may be due to the continuous availability of the cells at the defect area and thus, their potential therapeutic benefits.

1.4.1.4 Comparative Studies

BM-MSCs are isolated and cultured from mononuclear cell (MNCs) fraction while AD-MSCs are from stromal vascular fraction (Gugjoo et al. 2020). BM-MSCs (transplanted 12 weeks post-surgery) offer superior results as compared to the MNCs and the control at eighth week. Only partial repair with either cell transplantation may occur; however, MSCs show higher cartilage regenerative effects with lower proteoglycan loss as compared to MNCs. Interestingly, comparable anti-inflammatory effects (inhibition of PGE2, TNF-α and TGF-β in synovial fluid) occur (Song et al. 2014). Similarly, AD-MSCs transplantation blocks OA progression and promotes cartilage regeneration better than the stromal vascular fraction (SVF). Local transplantation of high-dose cells (AD-MSCs @ 50 million) along with 2.5 mL hyaluronic acid (HA) may lead to a significant increase in cartilage thickness of the femoral epicondyle in comparison to the low-dose MSCs (25 million) and hyaluronic acid. Low-dose cells and SVF along with hyaluronic acid (HA) may lead to comparable improvement in cartilage thickness. High-dose cells along with HA additionally inhibits IL-1β and IL-6 expressions in synovial fluid and also prevents hypertrophic cartilage formation while an enhanced SDF-1 expression in synovial fluid may occur with SVF along with HA. Thus, higher of dose cells in the presence of HA may be a better option to prevent osteoarthritis (Lv et al. 2018). As MNCs or SVF contain a small fraction of MSCs, lower efficacy may be expected. As low dose fails to show effective immunomodulation, these cells should be available in some minimal concentrations to offer anti-inflammatory actions in the synovial fluid. In all the above studies, the cell transplantation is avoided in the immediate post-surgical period (ranging from 6 to 12 weeks) to prevent the effect of acute inflammatory mediators over the MSCs properties. Clinically OA becomes obvious when actual joint changes appear and as such cellular transplantation automatically gets delayed. The delay in cell therapy however, should not be to the extent that the osteophytes or enthesophytes form.

All these studies demonstrate the ability of MSCs to prevent OA progression and enhance repair. However, actual hyaline repair remains to be seen especially through extended period studies. Further, cell tracking needs further understanding, and the role of immediate vs. delayed cell transplantation has to be compared to arrive at evidence-based medicine.

1.4.1.5 Meniscectomy Model

Meniscectomy models are created either separately (Desando et al. 2016; Whitehouse et al. 2017) or with full-thickness medial femorotibial defect (Table 1.1) (Caminal et al. 2014a, b). In a bilateral meniscectomy model, transplanted SPIO-labelled MSCs are detectable for 12 weeks and may attach to some of the diseased joint structures (Delling et al. 2015). Transplantation of collagen I-loaded BM-MSCs (@1 million/cm^2) in a meniscal cartilage tear (5 mm × 3 mm) significantly improves repair as compared to the control at 13th week. However, in a later period (6 months), comparable features are visible (Whitehouse et al. 2017). In a unilateral meniscectomy model, BM-MSCs (six million cells) or BM concentrate loaded on the scaffold (Hyaff®-11) show comparable reduction in inflammation of the cartilage, meniscus and synovium and subchondral bone thickness (Desando et al. 2016). In weak joint injuries, discernible osteoarthritic changes may not occur and as such effects of MSCs could not have been obvious (Feng et al. 2018).

In a more extensive combined chronic full-thickness medial femorotibial condyle defect (5 mm) and meniscal tear model, double dose (1.1 and 1.2 million) of BM-MSCs injected at 6- and 12-month intervals after defect creation shows significant improvement. Radiography, MRI, ultrasound, macroscopic and histological analyses show articular cartilage and meniscus regeneration is case-dependent, but statistically significant improvement is found on macroscopic and histological parameters (Caminal et al. 2014a). Thus, MSCs may engraft to joint structures, prevent OA progression and enhance cartilage regeneration. However, numerous factors like inflammation, extent of OA, its duration and movement restrain may all impact MSCs therapeutic effects.

1.4.1.6 Miscellaneous Models

Some of the conditions like femoral condyle osteonecrosis (induced with absolute ethanol), and collagenase II-induced arthropathy have a limited available literature on MSCs therapeutic effects. In the femoral condyle osteonecrosis, transplantation of one million sheep BM-MSCs and human dental stem cells (to the individuals separately after 8 weeks of defect creation) shows better bone/cartilage regeneration on light microscopy (Feitosa et al. 2010).

Collagenase II-induced arthritis resembles more to the immune-mediated arthritis like rheumatoid arthritis. In a collagenase II-induced arthropathy, intravenous transplantation of STRO-3+ mesenchymal progenitor cells (MPCs) significantly reduce clinical signs of lameness, joint pain and swelling as compared to the saline controls. The cell-treated animals show decreased cartilage erosions, synovial stromal cell activation and angiogenesis. A significant reduction (40%) in CD4$^+$ lymphocytes and CD14$^+$ monocytes/macrophages infiltration into the synovial tissues in cell-treated animals may occur. Furthermore, the cell-treated joint has a significant reduction of angiogenesis in synovium, a common problem seen in rheumatoid arthritis (Abdalmula et al. 2017). Limiting recruitment of $CD4^{+}$ lymphocytes and $CD14^{+}$ monocytes/macrophages to the synovial tissues means reduction in arthritis as these cells secrete pro-inflammatory factors like TNF-α that recruits other inflammatory cells. Furthermore, as the study shows comparable Ki-67^{+} cells (proliferation marker) in the control and cell treated, it becomes clear that MPCs show anti-inflammatory action mainly through the prevention of the recruitment of the inflammatory cells. MSCs in general are known to increase angiogenesis via VEGF although in this case a reduction of such a process may occur due to the immune-selected cells (STRO-3+ MPCs) and/or due to the local available environment effectuating this anti-angiogenesis process.

In sheep, irrespective of the model type for OA or rheumatoid arthritis, an overall improved healing with the transplantation of MSCs with a further improvement upon scaffold incorporation occurs. Immunoselection of the MSCs may additionally be a useful option to ensure specific phenotype cells. Scaffold type may have a significant effect on such results as well. In all these studies, however, a hyaline cartilage typical in all features with complete obviation of symptoms remains to be seen. Many cellular/scaffold factors such as source, passage number and concentration, type of scaffold and other factors including resident cells, growth factors also affect the outcome. Further, the case numbers and duration of studies must be considered in order to evaluate the effects in large samples and for long-lasting effects, if any. Above all the clinical utility in sheep remains to be evaluated.

1.4.2 Goat Studies

MSCs therapeutic potential in goat is evaluated in various models like osteochondral defect models of mandible and femur (Jurgens et al. 2013; Pei et al. 2014; Wei et al. 2019), chondral defects, meniscal tear and meniscal and cruciate ligament resection (chronic model) (Table 1.2). The defect size in femoral condyles range from 5 mm (Bekkers et al. 2013) to 10 mm (Wei et al. 2019). A majority of the studies utilize BM-MSCs except two studies that utilize AD-MSCs (Jurgens et al. 2013) and human Wharton's Jelly (Zhang et al. 2018a, b). The cellular concentration ranges from 1 million (Zhang et al. 2018a, b) to 20 million (Port et al. 1996). Matrices/scaffolds like acellular cartilage extracellular matrix (Zhang et al. 2018a, b), β-TCP (Pei et al. 2014), collagen type I/III scaffold (Jurgens et al. 2013), poly-lactic-co-polyglycolic acid scaffold (Zhu et al. 2011), hyaluronan (Murphy et al. 2003) and fibrin clot (Port et al. 1996) are also evaluated. The number of animals vary from as low as 6 (Zhang et al. 2018a, b) to as high 50 (Zhu et al. 2011). The evaluation is made through macroscopic and microscopic scoring, biochemical analysis, histological and immunohistochemical, micro-CT analysis, MRI and gene expression analysis. Like sheep osteochondral defect models, MSCs in goat osteochondral defect models show promising results. In goat, mainly single-step biomaterial transplantation is made (Jurgens et al. 2013; Pei et al. 2014; Wei et al. 2019; Zhang et al. 2020).

A single-step transplantation of collagen I/III scaffolds loaded with the AD-MSCs or SVF in osteochondral defect model of femur (medial condyle and

Table 1.2 Goat in vivo MSCs preclinical experimental studies on various joint tissue affections

Study design/ study period	Biomaterial used	Evaluation criteria	Overall result	References
Osteochondral defect of mandible model				
Randomized placebo controlled ($n = 50$)/6 and 24 weeks	NELL-1 transfect autologous BM-MSCs (3×10^6 cell seeded on PLGA scaffold)	Macroscopic, histology and immunohistochemistry, micro CT	Positive outcome	Zhu et al. (2011)
Osteochondral defect of femur model				
Randomized comparative study ($n = 8$)/ 4 months	Stromal vascular fraction (SVF @ 5×10^6), AD-MSCs (5×10^5), collagen type I/III scaffold	Macroscopy, immunohistochemistry, biomechanical, micro-CT, and biochemistry	Positive outcome	Jurgens et al. (2013)
Randomized placebo-controlled comparative study ($n = 12$)/ 24 weeks	Autologous BM-MSCs, β-TCP, tissue-engineered osteochondral graft (TEO)	Histology and immunohistochemistry, gross morphology	Positive outcome	Pei et al. (2014)
Randomized placebo controlled ($n = 24$)/ 16 weeks	BM-MSCs, porous tantalum (pTa), chondrocytes, collagen membranes (CM)	Radiography, gross observation, and histomorphology	Positive outcome	Wei et al. (2019)
Chondral defect model				
Randomized placebo controlled ($n = 8$)/ 6 months	Chondron (chondrocytes in own matrix) and BM-MSCs or 10% chondron/90% BM-MSC combination (1×10^6 cells/mL)	Macroscopic, Biochemical, Histological Immunohistochemistry	Positive outcome	Bekkers et al. (2013)
Randomized placebo controlled ($n = 18$)/ 6 months	BM-MSCs (1×10^7) after 2 weeks after bone marrow stimulation for 3 consecutive weeks	Macroscopic, histology, biochemical and gene expressions	Positive outcome	Nam et al. (2013)
Randomized placebo controlled ($n = 15$)/ 9 months	Xenogeneic WJ-MSCs, primary cartilage cells, decellularized cartilage extracellular matrix	Gross, MRI, mechanical testing	Positive outcome	Zhang et al. (2020)

(continued)

Table 1.2 (continued)

Study design/ study period	Biomaterial used	Evaluation criteria	Overall result	References
Meniscal tear model				
Randomized placebo-controlled comparative study ($n = 32$)/ 4 months	Autologous BM-MSCs (20×10^6), fibrin clot	Gross, tensile strength and histology	Insignificant outcome	Port et al. (1996)
Medial meniscus and resection of the anterior cruciate ligament model				
Randomized placebo controlled ($n = 24$)/ 26 weeks	BM-MSCs (10×10^6 loaded on hyaluronan)	Histochemistry	Positive outcome	Murphy et al. (2003)

MSCs mesenchymal stem cells, *BM* bone marrow derived, *AD* adipose tissue derived, *WJ* Wharton's Jelly derived, *SVF* stromal vascular fraction, *n* total number of animals in all the groups of the study, *NELL-1* neural epidermal growth factor-like 1, *PLGA* poly-lactic-co-polyglycolic acid; Positive outcome: Indicates improved outcome with MSCs transplantation even if only one parameter improves; Study period: Indicates maximum time period for which study is conducted

trochlear groove) shows that the former biomaterials assembly lead to better regenerative effects than the latter at fourth week post transplantation. Despite better effects as an early stage, a catch-up phenomenon occurs at 4 months. The joints treated with either biomaterial assembly appear to have comparable healing in long run (Jurgens et al. 2013). SVF harbours varied cell types like MSCs, lymphocytes, pericytes, and vascular smooth muscle cells; besides the growth factors, regenerative effects might occur through synergistic effects in a long run. Such a combination of cells in SVF might favour higher differentiation of the cells towards chondrogenic lineage. These mechanisms of the SVF might make the environment conducive and enhance repair in later stages.

In recent times, advanced techniques are being aimed including the bioreactor preparation of tissue-engineered constructs. This helps in developing the tissue-engineered constructs in dynamic and static motion. To compare differences in healing potential, simultaneously prepared osteochondral graft (β-TCP scaffold loaded with BM-MSCs) in a double-chamber stirring bioreactor is evaluated. Transplantation of the former type of the osteochondral graft tends to show better reparative effects (mature cartilage with smooth surface) than the latter that still has better repair than the control at 12th week. Such a potential further improves with the time (24th week) in the cell-treated groups (Pei et al. 2014). As microenvironments including the resident cells also impact MSCs paracrine or differentiation properties, combined use of chondrocytes and MSCs is also evaluated. Wj-MSCs and chondrocytes in equal ratio loaded on to the decellularized cartilage matrix enhance the osteochondral regeneration with high-quality cartilage comparable to the native articular cartilage in terms of its biomechanics and tissue components (Zhang et al. 2020). Furthermore, the combined use of the chondrocytes, BM-MSCs and porous tantalum integrated with 3D collagen-based scaffold too promotes the healing of the large osteochondral defect (10 mm) (Wei et al. 2019). Mechanical stimulation favours differentiation and addition of the chondrocyte creates a conducive environment for MSCs chondrogenic differentiation. In a more innovative way, chondrons (chondrocytes in cartilage territorial matrix) are used along with the MSCs (80%). Such an assembly leads to improved cartilage regeneration as compared to microfracture at 6 months. The joints in either technique show degeneration of the defect opposing and surrounding cartilage (Bekkers et al. 2013). This might be due to the creation of bilateral defects simultaneously and the animal's tendency to shift weight towards better limbs affecting mechanics of the joint. Further, as the MSCs: chondrocyte ratio is higher as compared to the above study, variable results may be expected.

In a single study on mandibular osteochondral defect model (3 mm diameter and 5 mm depth), transplantation of poly-lactic-co-polyglycolic acid loaded with the neural epidermal growth factor-like 1 (NELL-1)-transfected autologous BM-MSCs leads to the rapid and vigorous healing. Although a fibrocartilage forms at a 6-week period, however, complete repair with articular cartilage and subchondral bone occurs in 24 weeks. Contrarily, simple BM-MSCs-loaded poly-lactic-co-polyglycolic acid scaffold although may show complete defect filling but with a fibrocartilage and slightly better tissue (less organized) at 24 weeks (Zhu et al. 2011). NELL-1, being the novel osteo-inductive factor, specifically targets cells that are committed to osteochondral lineage (Gugjoo et al. 2020).

In a single individual study on meniscal tear repaired with vertically oriented suture, BM-MSCs along with the fibrin clot fail to show improved healing comparable to the normal tissue. Healing in fibrin clot-loaded injured areas is comparable to the sutured ones (Port et al. 1996). In the chronic model of meniscal and cruciate ligament resection, transplantation of BM-MSCs loaded in hyaluronan lead to marked regeneration of the medial meniscus with reduced articular cartilage degeneration and subchondral bone sclerosis. Osteophytic remodelling in a cell-treated group at 12 weeks also occurs, but osteoarthritis findings comparable to control at 26 weeks are seen (Murphy et al. 2003). This puts the question mark on the long term beneficial effects of the cell loaded scaffolds.

Thus, MSCs loaded on scaffolds may lead to the improved response at an early period that may further improve with their genetic transfection or through advanced tissue-engineered techniques. However, complete regeneration that maintains its integrity comparable to native tissue remains to be seen in the longer run. In most of these studies, the specific role of MSCs and the scaffolds remains to be underlined. These studies use diverse scaffold/cell assemblies and data on each type is very limited in relation to cell source, concentration, scaffold type and its processing. Additionally, animal number and duration of the studies fall short as there is continuous wear and tear of the joint structure upon motion. Considering the paucity of literature and without any concrete results, further extensive studies are desired. Above all the clinical applications remain to be conducted in this veterinary species.

1.4.3 Dog Studies

1.4.3.1 Experimental Studies

In dog, both experimental and the clinical evaluations of MSCs in various joint affections are conducted. The experimental joint studies involve various models including chondral defects, full thickness osteochondral defects, femoral head osteonecrosis, cruciate ligament resection and medial meniscectomy (Table 1.3).

1.4.3.1.1 Chondral and Osteochondral Defect Models

Chondral defect diameter ranges from 3 mm (Sabry et al. 2018) to 4 mm (Li et al. 2018) and full-thickness osteochondral defects size ranges from 4.2 mm (Yang et al. 2011) to 11 mm (Qiang et al. 2014) (Table 1.3). In all these studies, BM-MSCs (mostly autologous) are transplanted barring a study wherein synovial MSCs are transplanted (Miki et al. 2015). In one of the studies, chondrogenic-induced BM-MSCs (Yang et al. 2011) while in other BMP-7 transfected MSCs (Sun et al. 2020) are transplanted. The cell concentration ranges from 0.5 lakh (Miki et al. 2015) to the 50 million (Qiang et al. 2014; Miki et al. 2015). In addition to the cell transplantation, MSCs microvesicles (MVs) are also evaluated (Sabry et al. 2018). In full-thickness osteochondral defects, the cells are incorporated along with the biomaterials like atelocollagen (Yamazoe et al. 2007), decellularized cancellous bone matrix (Qiang et al. 2014), hyaluronic acid (HA) (Miki et al. 2015; Li et al. 2018), hyaluronic acid and autologous serum (Yamasaki et al. 2015), platelet-rich fibrin (Kazemi et al. 2017), demineralized and decellularized bone and collagen type I (Col I) allograft (Sun et al. 2020). In all these studies, immediate local intra-articular transplantation of biomaterials after defect creation is made except few studies wherein it is delayed by 7 weeks (arthroscopically) (Miki et al. 2015), 1 day and 1 month (El-Tookhy et al. 2008). Chondral defect studies involve variable animal number ranging from $n = 9$ (Sabry et al. 2018) to $n = 24$ (Miki et al. 2015; Li et al. 2018), while in case of osteochondral defects, it ranges from $n = 6$ (Yamazoe et al. 2007; Yang et al. 2011; Qiang et al. 2014; Sun et al. 2020) to $n = 60$ (Kim et al. 2021). In majority of these studies, bilateral joints are employed keeping one side as control. Even one of the studies created multiple defects in a single joint (Li et al. 2018). The studies are evaluated for 1 week (El-Tookhy et al. 2008) up to 28 weeks (Li et al. 2018). The evaluation criterion remains variable incorporating diagnostic techniques/parameters like radiography, micro-CT, single-photon emission computed tomography, gross morphology, biochemical, biomechanical, histopathology, histomorphometric analysis and immunofluorescent staining.

1.4.3.1.2 Chondral Defects

In chondral defects, natural regeneration is very limited. In general, transplantation of the MSCs alone (El-Tookhy et al. 2008) or along with HA (Miki et al. 2015; Li

Table 1.3 Dog in vivo MSCs preclinical experimental studies on various joint tissue affections

Number of animals/ study period	Biomaterial used	Evaluation criteria	Overall result	References
Chondral defect models				
Randomized placebo controlled ($n = 12$)/ 8 weeks	Autologous BM-MSCs ($1.4–1.6 \times 10^6$ cells/ mL) at day 1 and day 30 later)	Morphological and histological analysis	Positive outcome	El-Tookhy et al. (2008)
Randomized placebo controlled ($n = 24$)/ 12 weeks	Autologous Sy-MSCs (0.5×10^5/5×10^6/ 5×10^7 cells + HA (0%, 0.01%, 0.1%, 0.5%) percutaneously at week 7 after defect creation	Macroscopy, histology	Positive outcome	Miki et al. (2015)
Randomized placebo controlled ($n = 24$)/ 28 weeks	BM-MSCs (1×10^7) + hyaluronic acid (HA)	Macroscopy, magnetic resonance imaging (MRI), histopathology, immunohistochemistry	Positive outcome	Li et al. (2018)
Randomized placebo controlled ($n = 9$)/ 6 months	BM-MSCs MVs (5×10^6 cells/mL)	Physical examination and histopathology	Positive outcome	Sabry et al. (2018)
Femoral osteochondral defect models				
Randomized placebo controlled ($n = 6$)/ 10 weeks	BM-MSCs, atelocollagen	Histology and fluorescent microscopy	No improvement than control	Yamazoe et al. (2007)
Randomized placebo controlled/ 6 months	Chondrogenically-induced BM-MSCs and scaffold	Gross morphology, histological, micro-CT biochemical, biomechanical,	Positive outcome	Yang et al. (2011)
Randomized placebo controlled ($n = 12$)/ 6 months	BM-MSCs (5×10^7), decellularized cancellous bone matrix	Gross morphology, radiography, histological, biomechanics	Positive outcome	Qiang et al. (2014)
Randomized placebo controlled ($n = 15$)/ 6 months	BM-MSCs (2.14×10^7), 500 mL of autologous serum suspended in HA	Macroscopy and histology	Positive outcome	Yamasaki et al. (2015)
Randomized placebo controlled ($n = 12$)/ 24 weeks	Autologous BM-MSCs (1×10^6) + platelet-rich fibrin (PRF)	Macroscopic and histopathology	Positive outcome	Kazemi et al. (2017)

(continued)

Table 1.3 (continued)

Number of animals/ study period	Biomaterial used	Evaluation criteria	Overall result	References
Randomized placebo controlled ($n = 6$)/ 12 weeks	Allogeneic BMP-7 transfected BM-MSCs, 200 μL BM-MSCs (5×10^5/ mL), demineralized and decellularized allograft bone-collagen biphasic scaffold	Gross observation and evaluation, histopathology and μCT scanning	Positive outcome	Sun et al. (2020)
Femoral head osteonecrosis of model				
Randomized placebo-controlled comparative ($n = 24$)/ 12 weeks	VEGF 165 transgenic BM-MSCs (2×10^7) or simple BM-MSCs (2×10^7)	Radiography, single-photon emission computed tomography, histopathology, histomorphometric analysis and immunofluorescent staining	Positive outcome	Hang et al. (2012)
Cranial cruciate ligament resection models				
Randomized placebo controlled ($n = 60$)/ 6 months	Xenogeneic Sy-MSCs (2.4×10^6; 4.8×10^6 cells; 9.6×10^6) [4 weeks after crCL transection)	MRI, histopathological and immunohistochemistry	Positive outcome	Kim et al. (2021)
Randomized placebo controlled ($n = 16$)/ 2.5 months	MSCs (2×10^7 cells/ mL), PDGF-MSCs (2×10^7 cells/mL), or HO-1-MSCs (2×10^7 cells/mL) [6 weeks after crCL transection]	Radiography, ELISA	No obvious improvement	Oh et al. (2021)
Medial meniscectomy model				
Randomized placebo controlled ($n = 16$)/ 6 months	BM-MSCs (6×10^6/ mL of scaffold), scaffold [3D printed poly (e-caprolactone)], parathyroid hormone	Histology and immunofluorescent staining	Positive outcome	Zhao et al. (2020)

MSCs mesenchymal stem cells, *BM* bone marrow derived, *Sy* synovium derived, *BMP* bone morphogenetic protein, *PDGF* platelet-derived growth factor, *VEGF* vascular endothelial growth factor, *MVs* microvesicles, *crCL* cranial cruciate ligament, *n* total number of animals in all the groups of the study; Positive outcome: Indicates improved outcome with MSCs transplantation even if only one parameter improves; Study period: Indicates maximum time period for which study is conducted et al. 2018) enhances repair through functional and morphological recovery (El-Tookhy et al. 2008) (Table 1.3). MSCs-MVs transplanted percutaneously using sonography too improve affected joint morphology as well as functionality as early as 45 days (Sabry et al. 2018). A hyaline-like cartilage (neo-cartilage) repair may fill the defect (Li et al. 2018; Sabry et al. 2018). The early (day 1) transplantation of such cells quite effectively regenerates the defect as compared to the delayed (day 30) (El-Tookhy et al. 2008). Such an enhanced effect upon early biomaterial transplantation can be expected from MVs as compared to the cells as inflammatory reaction affects the latter and not the former. In order to evaluate the effective cell-scaffold concentration, different doses of synovial MSCs (@ $0.5 \times 10^5/5 \times 10^6/5 \times 10^7$ cells) and the hyaluronic acid (0, 0.01, 0.1% and 0.5%) have been transplanted. Intermediate dose of five million cells in 0.01% HA significantly improves cartilage regeneration as compared to the others (Miki et al. 2015). Low-dose cells (50,000) might be insufficient and an excessive dose might increase cell apoptosis in the joint due to lack of sufficient nutrients and circulation. HA, part of the cartilage matrix, degrades in osteoarthritis and therefore its supplementation may prevent further degeneration. Additionally, it might provide a microenvironment typical to the cartilage for MSCs paracrine effects. However, it is important to note that the source of the cells and scaffold type included may also affect the outcome.

1.4.3.1.3 Osteochondral Defects

In full-thickness defect models, MSCs along with various scaffolds fail to offer significantly better results at par to the normal joint (Yamazoe et al. 2007; Yang et al. 2011; Qiang et al. 2014; Yamasaki et al. 2015) (Table 1.3). Even one of the studies concluded that it is time and not the cells that heal the osteochondral defects. However, cells with scaffold (HA)-transplanted defects may show early response (3 months) as compared to the control (Yamasaki et al. 2015). Transplantation of allogeneic cancellous extracellular matrix-derived biphasic cell-scaffold construct in femoral condyles (Yang et al. 2011) or femoral head (Qiang et al. 2014) defects also fails to yield any effective repair as compared to the control (Qiang et al. 2014). Allogeneic scaffolds used might undergo early degradation and fail to undergo sufficient vascularization for sustenance of the cells. Besides, the model involves a very large defect (11 mm × 10 mm) that too at the femoral head (Qiang et al. 2014). Contrarily, other studies show improved osteochondral bone stiffness and glycosaminoglycan content as compared to the control cell-scaffold construct (Yang et al. 2011). Transplantation of biphasic scaffold loaded with BMP-7 transfected MSCs in knee joints also completely heals defects by cartilage-like tissue that remains well integrated with the surrounding tissue by twelfth week. As the biphasic scaffold constituted the decellularized and decalcified collagen sponge, BMP-7 transfected MSCs might have mRNA expression as per the available microenvironment with higher tendency towards the osteo- and chondrogenic lineages, leading to the tissue regeneration (Sun et al. 2020). Such studies may be further extended to evaluate the long term results.

Utilizing MSCs along with the atelocollagen gel fail to offer better results in relation to the cartilage, although subchondral bone may be well repaired (Yamazoe et al. 2007). Contrarily, use of MSCs along with the platelet-rich fibrin (PRF, a second-generation platelet concentrate) leads to better osteochondral repair (Kazemi et al. 2017). PRF contains multiple growth factors that may play an important role in chondrogenesis (Lam et al. 2014). Even chondrogenic-induced BM-MSCs along with scaffold too lead to the significant improvement with respect to the gross morphology and histology of the regenerated cartilage (Yang et al. 2011). Thus, it is the combination of the scaffolds and cells with gene transfection that may be incorporated to achieve improvement over the currently available techniques. Incorporation of the scaffolds or grafts needs to be designed to favour cell growth and proliferation, and offer mechanical strength without affecting the cell movement. Therefore, further studies and advancement in techniques are desired to achieve desired results.

1.4.3.1.4 Osteonecrosis

In femoral head osteonecrosis model, VEGF-transfected BM-MSCs transplantation post core decompression leads to the regular arrangement of trabeculae, and bone regeneration becomes obvious with significantly increased capillaries as compared to the simple BM-MSCs by fourth week. Vascularization of BM-MSCs-transplanted defects are comparable to that of the core compression defects (Hang et al. 2012). BM-MSCs transplantation in necrosed areas may fail to express and secrete pro-healing factors and this may be improved through their transduction with vascular growth factors. However, this model may not replicate the actual clinical condition as the subjects are young while osteonecrosis is mostly seen in aged patients. Besides, autologous cells are used in this study which may go against the aged patient that tends to undergo more co-morbidity and the cells have less proliferative and secretory activity. As there are maximum chances of phenotypic changes in MSCs post culture and transfection, it is imperative to study such changes as well.

1.4.3.1.5 Cruciate Ligament Resection Model

In dog, osteochondral defect models were mostly conducted earlier, but due to more frequent clinical problems of cruciate ligament rupture in dog and human, these models are also now-a-days evaluated for MSCs therapeutic efficacy. Compared to the osteochondral defect model, cruciate ligament rupture induced OA takes time and as such immediate MSCs transplantation cannot be made. Osteoarthritis induced with the cranial cruciate ligament (CrCL) transection improves with the intra-articular transplantation of xenogeneic synovial MSCs (2.4 million; 4.8 million; 9.6 million) at fourth week post-surgery. The joint improvement is seen in all the cell-transplanted animals irrespective of the concentration right from the sixth week post injection till 24 weeks of the study. The extracellular matrix (ECM) of regenerated tissue has composition comparable to the naïve articular cartilage (Kim et al. 2021). In another similar type of model, intra-articular transplantation of simple or PDGF and HO-1-transfected MSCs (@ 20 million) has been made at sixth week post-surgical induction. An improvement in symptoms occurs although the radiographic findings and serum MMP-13 levels are comparable to the control at tenth week. Compared to the HO-1, PDGF-transfected MSCs transplantation leads to the better improvement in symptoms (Oh et al. 2021). HO-1 being a stress-responsive enzyme elicited by free radicals and hypoxia, its effect may be more pronounced in acute OA models as against the chronic model like those mentioned above. However, in both the above CrCL transections studies, objective numerical rating scale, quantitative outcome measures (grading of synovitis and osteophytes) are lacking. Besides, natural pathological OA and surgery-induced OA may vary in proteoglycan levels on the articular cartilages. As in surgical OA, hyperactivity enhances repair activity while such a response is lacking in natural pathologies (Liu et al. 2018). Further, the animal number, duration of the studies and overall studies are very limited to be concluded. Thus, there is a big gap in the current literature and clinical conditions in translating the experimental results to the clinical settings.

1.4.3.1.6 Meniscectomy Model

A single available dog medial meniscectomy model utilizes 3D bioprinted scaffold [poly(e-caprolactone)] loaded with BM-MSCs (6×10^6/mL of scaffold) and the parathyroid hormone (PTH). The biomaterial transplantation leads to the better regeneration of the tissue with more similarity to native meniscus as compared to the cells and scaffolds. BM-MSCs along with PTH-loaded 3D scaffold transplantation prevents terminal differentiation markers and also limits degeneration of cartilage matrix (Zhao et al. 2020). PTH may have chondro-protective action, however, due to its short half-life and unknown possible therapeutic mechanisms, it may be premature to attribute efficacy solely to the hormone. Further in this study, chondrogenic activity is demonstrated and not chondro-protective activity. However, MSCs secrete tissue inhibitor matrix metalloproteinases that might limit cartilage matrix degradation. Thus, studies employing gene transfected cells or chondrogenically induced ones along with the scaffold may yield much improved results. However, further extensive studies are desired.

In all the above studies, it is worth mentioning that MSC anti-inflammatory actions, secretory role and differentiation remain to be conclusively confirmed. Further, these studies have employed cells from various sources, passages and concentration along with varied scaffolds. Further, the design of these studies remained variable with limited animal numbers and duration. All the limitations in these studies limit direct translation of the results.

1.4.3.2 Clinical Trials/Studies

Contrary to the small ruminants, dog clinics provide facility of the regenerative medicine involving MSCs as well. MSCs or their conditioned medium (CM) are evaluated in various clinical studies/trials incorporating joints like hip, knee, elbow, carpal, tarsal and shoulder (Table 1.4). Therapeutic benefits of MSCs in these joints are reported either individually or in combination. The studies evaluate MSCs mostly for osteoarthritis. A single study each on dysplasia, arthritis and dysplasia

Table 1.4 Dog in vivo MSCs clinical studies on various joint affections

Clinical case	Number of animals included/ study period/study design	Cell source and dose	Overall result	References
Osteoarthritis of various joints				
Chronic osteoarthritis of coxofemoral joint	Case series (randomized, double-blinded, placebo-controlled trial)/$n = 21$/ 90 days	Autologous AD-MSCs (5×10^6 in 20 dogs) and 4.5×10^6 (in 1 dog)	Positive outcome	Black et al. (2007)
Chronic osteoarthritis of humeroradial joint	Case series (randomized, double-blinded, non-placebo-controlled trial)/ $n = 14$/180 days	Autologous AD-MSCs ($3–5 \times 10^6$)	Positive outcome	Black et al. (2008)
Chronic osteoarthritis of humeroradial joint	Case series (Uncontrolled study)/$n = 4$/ 30 days	Autologous AD-MSCs ($3–5 \times 10^6$) in platelet-rich plasma or hyaluronic acid	Positive outcome	Guercio et al. (2012)
Chronic arthritis of the hip joint	$n = 8$/180	Autologous AD-MSCs (15×10^6)	Positive outcome	Vilar et al. (2013)
Hip osteoarthritis	Randomized comparative clinical trial/ $n = 39$/180 days	Autologous AD-MSCs and PRGF (30×10^6)	Positive outcome	Cuervo et al. (2014)
Chronic arthritis of the hip joint	Case–control (blinded control study)/$n = 15$/ 180 days	Autologous AD-MSCs (15×10^6)	Positive outcome	Vilar et al. (2014)
Osteoarthritis of hip, elbow, stifle, and/or shoulder joints	Case series study (prospective, randomized, masked, and placebo-controlled)/ $n = 82$/60 days	Allogeneic AD-MSCs (12×10^6, cryopreserved with 85.1% viability)	Positive outcome	Harman et al. (2016)
Hip, knee, radiocarpal intercarpal, elbow and ulnar osteoarthritis	Case report/$n = 10$ patients/4 years	Autologous AD-MSCs (3×10^7)	Positive outcome	Dražilov et al. (2018)
Osteoarthritis (hip, knee, elbow and tarsal joints)	Non-randomized, open label and monocentric study/$n = 22$/ 2 years	Allogeneic MSCs (10×10^6, repeated injection after 6 months in 8 dogs or 11 joints)	Positive outcome	Cabon et al. (2019)

(continued)

Table 1.4 (continued)

Clinical case	Number of animals included/study period/study design	Cell source and dose	Overall result	References
Elbow osteoarthritis	Case report/$n = 6$/6 weeks	Xenogeneic (equine derived) chondrogenic-induced PB-MSCs (1×10^6)	Positive outcome	Daems et al. (2019)
Elbow osteoarthritis	Open-label clinical trial ($n = 13$)/6 months	Allogeneic AD-MSCs (2×10^6 cells/kg b wt.)	Positive outcome	Olsen et al. (2019)
Elbow osteoarthritis	Case report/$n = 6$/6 weeks	Double dose of allogeneic AD-MSCs conditioned medium (1 mL) at 2 weeks apart	Positive outcome	Huňáková et al. (2020)
Osteoarthritis of elbow and knee	Case report/$n = 5$/80 days	Double dose of lyosecretome from AD-MSCs (equivalent to 2×10^6 cells)	No observable improvement	Mocchi et al. 2021a, b)
Dysplasia				
Hip dysplasia	Comparative clinical trial/$n = 9$/30 days	Allogeneic AD-MSCs at acupoints or stromal vascular fraction (SVF) ($0.2–0.8 \times 10^6$)	Positive outcome	Marx et al. (2014)
Arthritis and dysplasia				
Elbow dysplasia and elbow osteoarthritis	Case report (uncontrolled study)/$n = 30$/1 year	Allogeneic AD-MSCs ($12 \times 10^6 \pm 3.2 \times 10^6$) + hyaluronic acid (0.5%)	Positive outcome	Kriston-Pál et al. (2017)
Degenerative arthritis of different joints				
Degenerative arthritis	Controlled clinical study/$n = 203$/10 weeks	Allogeneic AD-MSCs	Positive outcome	Shah et al. (2018)

MSCs mesenchymal stem cells, *PB* peripheral blood derived, *AD* adipose tissue derived, *b. wt* body weight, *PRGF* platelet-rich growth factor, *n* total number of animals in all the groups of the study; Positive outcome: Indicates improved outcome with MSCs transplantation even if only one parameter improves; Study period: Indicates maximum time period for which study is conducted

and degenerative arthritis are also evaluated. These studies are performed either as case-controlled (Black et al. 2007, 2008; Vilar et al. 2014; Harman et al. 2016; Shah et al. 2018; Cabon et al. 2019; Daems et al. 2019) or case uncontrolled (Guercio et al. 2012; Kriston-Pál et al. 2017; Dražilov et al. 2018; Olsen et al. 2019). In addition to the direct cellular therapy, cellular products like CM (Huňáková et al. 2020) or lyosecretome (Mocchi et al. 2021a, b) are also evaluated for therapeutic effects. MSCs therapeutic role is additionally compared against PRGF (Cuervo et al. 2014) and SVF (Marx et al. 2014).

1.4.3.2.1 Cell Transplantation, Concentration and Frequency

In majority of the clinical studies, AD-MSCs (autologous or allogeneic) are transplanted locally except one study that transplants xenogeneic chondrogenic induced peripheral blood (PB)-MSCs (Daems et al. 2019). It is worth mentioning that the nomenclature may also vary in some studies as SVF is also claimed as AD-MSCs (Black et al. 2007, 2008). Besides, the characterization of MSCs is not reported in all the studies. MSCs are also transplanted through acupoints (Marx et al. 2014), intravenously (Olsen et al. 2019) or both local and intravenous (Shah et al. 2018). For local transplantation, MSCs concentration ranges from 0.2 million (Marx et al. 2014) to 30 million (Cuervo et al. 2014; Dražilov et al. 2018) with a median range of 10–15 million cells (Vilar et al. 2013, 2014; Harman et al. 2016; Kriston-Pál et al. 2017; Cabon et al. 2019). For parenteral route, higher doses (2×10^6 per kg B. Wt.) are used (Olsen et al. 2019). In addition to the cells, few studies utilize hyaluronic acid along with the cells (Guercio et al. 2012; Kriston-Pál et al. 2017). Mostly, MSCs are transplanted once except in a study wherein second dose of cells is made in some of the individual cases (Shah et al. 2018; Cabon et al. 2019). These transplanted cells are either from single batch (Harman et al. 2016) or from different batches (Cabon et al. 2019). Evaluation period varies ranging from 30 days (Marx et al. 2014) to 1 year (Kriston-Pál et al. 2017). Therapeutic efficacy is evaluated through clinical evaluations like lameness scores, radiography and arthroscopy. A single study additionally makes histopathological analysis in one patient as well (Kriston-Pál et al. 2017). In veterinary medicine clinical evaluation is mainly being conducted through clinical examination, although lack of extensive functional evaluation scales remains a lacuna. Pain (Black et al. 2007, 2008; Guercio et al. 2012; Dražilov et al. 2018) or goniometer (Cuervo et al. 2014; Vilar et al. 2016) as an indicator is also used to assess the range of motion of the treated joint.

1.4.3.2.2 Results

In all these studies, irrespective of the joint or affection type, improvement in clinical condition is reported upon MSCs transplantation. Even one of the studies shows hyaline cartilage regeneration in one of the dogs evaluated histologically (Kriston-Pál et al. 2017). Animals receiving double doses of allogeneic MSCs show significantly higher clinical scores as compared to the single-dose recipients (Cabon et al. 2019). However, not all the individual cases are shown to have a successful and up to the mark outcome (Dražilov et al. 2018; Shah et al. 2018; Cabon et al. 2019). One of the studies reports that around 75% cell recipients have improvement as per the owner assessment at 6 months. At 24 months, $n = 6$ cases out of $n = 22$ show relapse of lameness and $n = 2$ out of that require analgesics (Cabon et al. 2019). One of the extensive detailed study incorporates animals on the basis of OA grades (2–4) and excludes those of grade 1. This study makes intra-articular (IA; $n = 128$) and intravenous injections (IV; $n = 65$) and both IA and IV ($n = 10$). Overall 90% of the dogs below 9 years and 60% of aged dogs show excellent and good improvement in pain and mobility, respectively. Around 12% of dogs fail to show any improvement. Intra-articular cell transplantation is correlated with the excellent improvement in pain and mobility and willingness to play voluntarily in $n = 75$. In dogs receiving

IV therapy, $n = 19$ have excellent or good improvement while in animals receiving both IA and IV MSCs, $n = 9$ have an excellent or good improvement. It is concluded that animals below 9 years of age may have better clinical outcomes especially with the intra-articular MSCs transplantation or with a combination of IA or IV. In relation to intravenous transplantation, further studies are desired as the current study made IV transplantation in case of polyarthritis (Shah et al. 2018). Other study transplanting allogeneic AD-MSCs intravenously reports significant improvement in subjective outcome measures (client-specific outcome measure activity score and its behaviour score) but objective outcome measures (mean peak vertical force and synovial fluid biomarkers) fail to improve in elbow osteoarthritis. However, worth mentioning finding in this study is that NSAIDs are not required based on the client-specific outcome measures (Olsen et al. 2019).

Apart from undifferentiated cells, chondrogenic-induced MSCs may also show reduction in pain and lameness in elbow osteoarthritis based on owner observation. However, improvement in parameters like orthopaedic examination, radiographic findings, synovial fluid, and pressure plate analysis between placebo and chondrogenic-induced MSCs-treated elbows is comparable (Daems et al. 2019). Thus, MSCs or their chondrogenic-induced progenitor cells relapse clinical signs and are therefore promising. However, individual case differences and the appropriate route and dose of the cells are still a matter of discussion. Considering the lack of relapse of pathological lesions especially the osteophytes or enthesophytes, it is imperative to remove them surgically and treat the joint with the cells that may prevent their further formation. Although such types of studies remain to be conducted as above-mentioned experimental studies are conducted mostly as acute models with few as chronic ones but lack details over efficacy of MSCs in relation to the resolve of the lesions.

In recent times, cell-free therapies like MSCs conditioned medium (CM) are being evaluated as an alternative to MSCs. This is possible as MSCs secretome contains the pro-healing factors and as MSCs are considered to have little therapeutic effect through in vivo resident cell differentiation. Transplantation of AD-MSCs conditioned medium (CM) into elbow joint osteoarthritis reduces inflammation. Significant difference in biomolecule (MMP-3, TIMP-1, IL-6 and TNF-α) levels after CM transplantation occurs. The range of motion is very much improved at day 42 post transplantation (Huňáková et al. 2020). CM is considered to be rich in pro-healing growth factors like IGF-1, and in factors that inhibit matrix degradation like tissue inhibitor MMPs. In the above study, such a characterization, however, remains to be performed. Contrarily, AD-MSCs lyosecretome loaded in hyaluronic acid and injected intra-articularly twice at 40-day interval fail to show any significant clinical improvement in elbow or knee joints, although no observable complication is reported. This has been attributed to the co-occurrence of treatment and placebo joints in a single animal making evaluation a complicated matter (Mocchi et al. 2021a, b). Besides, there might have been shifting lameness with the resultant effect on healing.

Comparative studies show beneficial effects of AD-MSCs against the platelet-rich growth factors (PRGF). Functional limitation, range of motion (ROM), owner's and veterinary investigator visual analogue scale (VAS) and patient's quality of life improves from the first month up to 6 months. The outcome is better with the AD-MSCs than PRGF (Cuervo et al. 2014). In contrast, MSCs against the SVF fail to show any significance in dysplastic hips (Marx et al. 2014). In the latter study, the cells are transplanted through acupoints, and effectiveness of MSCs through the route needs further evaluation. Besides, the actual MSCs percent (1–10%) in SVF is also higher than the actual dose of AD-MSCs transplanted (Bourin et al. 2013).

Thus, MSCs transplantation shows mixed clinical response: improvement in pain and lameness but simultaneous improvement on the basis of various diagnostic aids remains limited. These clinical studies mostly employ MSCs as the last resort and thus, any improvement is welcome. However, these studies have some kind of lacunae as not all these studies are blinded and have insufficient patient numbers and additionally lack placebo control. The clinical cases of OA have been of varied joints (involving one or multiple joints) with varied severity and dynamics, diagnosed either through clinical scores or radiography or both.

1.4.4 In Vivo Horse Studies

Horse/donkey due its comparable cartilage thickness and larger joints offer a better translational model animal for the evaluation of therapeutic role of MSCs in human cartilage regenerative medicine. Besides, it also offers the proof of therapeutic principle of MSCs cartilage regeneration potential in veterinary medicine (Gugjoo et al. 2020).

1.4.4.1 Safety of Intra-articular Transplantation of MSCs

Immune reaction to the cells after intra-articular transplantation may variably occur. Intra-articular injection of simple MSCs or chondrogenic-induced allogeneic MSCs or along with other biomaterials like allogeneic plasma does not elicit any significant immune response (Broeckx et al. 2019; Ferris et al. 2014; Van Hecke et al. 2021). Immune response recorded is mild (Ferris et al. 2014; Bertoni et al. 2019), moderate (Pigott et al. 2013) or no response (Broeckx et al. 2018; Colbath et al. 2020a, b; Delco et al. 2020). Some of the studies show some response to the MSCs irrespective of their allogeneic or autologous sources and whether the cells are neonatal cells or adult (Carrade et al. 2011; Ardanaz et al. 2016). One of the studies shows comparable immune response of the cells with that of the placebo (Broeckx et al. 2018) while other shows increased immune response against the BM-MSCs than the foetal membrane-derived MSCs (Bertoni et al. 2019). Others report more specific responses to the allogeneic or xenogeneic cells with both single and repeated doses (Ferris et al. 2014; Magri et al. 2019). Transplantation of chondrogenic-induced MSCs might lack immune response due to the obtunded MHC expression that occurs with the TGF-β (Berglund and Schnabel 2017). Inflammatory environment and MSCs differentiation tend to activate MHC-II expression in the cells and this might lead to immune response. Further, with repeated allogeneic cell transplantations, specific immune responses against MHC antigens may occur as shown under in vitro studies (Pezzanite et al. 2015; Berglund and Schnabel 2017). MSCs during long transportation (36 h) may die and might elicit immune response (Raheja et al. 2011). Even repeated arthrocentesis may additionally induce damage to the joint capsule resulting in reaction. Furthermore, as MSCs are being cultured in FBS, its internalized proteins may also elicit response. This may be prevented by removal of FBS residues through extensive cell washing (Magri et al. 2019). It is therefore important to mention that mere absence of MHC-II in specific cell batches (without priming) does not assure the absence of immune response in horses. Thus, studies that focus on immune-phenotyping of the cells, their tracking and evaluation of immune-mediated responses are highly desired.

1.4.4.2 Experimental Studies

In horse, mostly full-thickness osteochondral defect models are undertaken for MSCs therapeutic evaluations. Joint involved under the studies mostly remains the stifle femoral condyles and trochlea, and to some extent carpal, distal radius, talus of tibiotarsus and metacarpophalangeal joints (Table 1.5). Combined MSCs transplantation and microfracture is also evaluated for healing potential (McIlwraith et al. 2011).

These studies are mostly conducted as chondral defects (Wilke et al. 2007; McIlwraith et al. 2011; Yamada et al. 2013; Goodrich et al. 2016; Dos Santos et al. 2019; Rosa et al. 2021) with few osteochondral defects (Frisbie et al. 2009; Yamasaki et al. 2018; Mancini et al. 2020; Bertoni et al. 2021) and a single study creating osteoarthritis driven with cartilage impact (Delco et al. 2020). The defect diameter in these studies ranges from 10 mm (McIlwraith et al. 2011; Yamada et al. 2013) to 15 mm (Wilke et al. 2007; Frisbie et al. 2009; Goodrich et al. 2016). One of the studies induces synovitis through interleukin-1β (Colbath et al. 2020a, b). MSCs from synovium (Yamasaki et al. 2018), bone marrow (Frisbie et al. 2009; Ferris et al. 2014; Goodrich et al. 2016; Bertoni et al. 2021), adipose tissue (Frisbie et al. 2009; Yamada et al. 2013; Delco et al. 2020) and umbilical cord blood (Bertoni et al. 2021) are evaluated. The concentration ranges from 5 to 6 million (Yamasaki et al. 2018) to the maximum of 20 million (McIlwraith et al. 2011; Ferris et al. 2014). AD-MSCs of specific morphology (Integrin $\alpha 10\beta 1$) are also evaluated for osteoarthritis (Delco et al. 2020). Some of the studies load cells into the scaffolds like hyaluronic acid (McIlwraith et al. 2011; Ferris et al. 2014), autologous platelet-enhanced fibrin scaffolds (Goodrich et al. 2016), self-polymerizing autogenous fibrin vehicle (Wilke et al. 2007) and alginate hydrogel (Dos Santos et al. 2019). The cell-based biomaterials are usually transplanted in a delayed manner after few weeks (Frisbie et al. 2009; Yamasaki et al. 2018; Colbath et al. 2020a, b; Bertoni et al. 2021). The study period ranges from as short as 21 days (Dos Santos et al. 2019) and 9 weeks (Yamasaki et al. 2018) to as high as 24 months (Ferris et al. 2014) post cell transplantation. Therapeutic results are evaluated through variable but relevant techniques like arthroscopy, MRI, gross examination, histology, histomorphometry, immunohistochemistry, gene expression analysis and biochemical examinations.

The transplanted cells localize in the synovial membranes although actual chondral lesions may remain devoid of the cells (Dos Santos et al. 2019). Cell

Table 1.5 Horse in vivo MSCs preclinical experimental studies on various joint tissue affections

Number of animals/ study duration	Study period	Biomaterial used	Evaluation criteria	Overall result	References
Femoral condyles and trochlea model					
Randomized placebo controlled ($n = 6$)/ 8 months		BM-MSCs/fibrinogen (12×10^6 MSC/mL fibrinogen)	Histology, histochemistry, immunohistochemistry, in situ hybridization, matrix biochemical assays	Positive outcome	Wilke et al. (2007)
Randomized placebo controlled comparative study ($n = 10$)/ 12 months		BM-MSCs (20×10^6 MSCs + 22 mg hyaluronan	Radiography, arthroscopy, magnetic resonance imaging, gross, histologic, histomorphometric, immunohistochemical, biochemical examinations	Positive outcome	McIlwraith et al. (2011)
Randomized placebo controlled ($n = 4$)/ 5 months		AD-MSCs (1.35×10^7)	Macroscopic, histopathological and histochemical evaluations	Positive outcome	Yamada et al. (2013)
Randomized placebo controlled ($n = 12$)/ 12 months		BM-MSCs along with autologous platelet enhanced fibrin scaffolds (APEF) (12×10^6)	Arthroscopy, histological examination, MRI, micro-CT, and biomechanical testing	No beneficial effect	Goodrich et al. (2016)
Randomized placebo controlled ($n = 8$)/ 6 months		MSCs (20×10^6 cells/mL of hydrogel) and ACPCs (20×10^6 cells/mL of hydrogel) in 3D printed bioscaffolds [3D printed poly(e-caprolactone) (PCL) and thiol-ene cross-linkable hyaluronic acid/poly(glycidol)]	Microtomography, histology, biomechanical analysis, parenting DNA analysis	Non-significant outcome	Mancini et al. (2020)
Carpal joint model					
Randomized placebo controlled ($n = 24$)/ 70 days		AD-MSCs (16.3×10^6), 10.5×10^6 (BM-MSCs)	Clinical outcome, macroscopy, histopathology, articular cartilage matrix evaluation	No improvement	Frisbie et al. (2009)
Randomized experimental study ($n = 5$)/21 days		Sy-MSCs (10×10^6), microcapsule MSCs (10×10^6)	Histopathological, fluorescence microscopy, flow cytometry	Localization of the cells in the synovial membrane and not lesions	Dos Santos et al. (2019)

(continued)

Table 1.5 (continued)

Number of animals/ study duration	Study period	Biomaterial used	Evaluation criteria	Overall result	References
Radial joint model					
Randomized placebo controlled ($n = 3$)/ 9 weeks		Sy-MSCs (5×10^6)	CT and histology	Positive outcome	Yamasaki et al. (2018)
Randomized placebo-controlled comparative study ($n = 6$)/12 months		BM-MSCs and AD-MSCs + collagen repair patch (each cell @0.5×10^6)	Macroscopy, histochemistry	Positive outcome	González-Fernández et al. (2016)
Talus bone chondral defect in tibiotarsal joint					
Randomized placebo controlled ($n = 8$)/ 6 months		Allogeneic AD- $\alpha 10^{high}$ MSCs (20×10^6)	Synovial fluid examination, arthroscopy, radiography, histology, immunohistochemistry	Positive outcome	Delco et al. (2020)
Randomized placebo controlled ($n = 15$)/ 15 days		Free or alginate hydrogel encapsulated Sy-MSCs (1×10^7)	Physical evaluations, lameness scores and synovial fluid analysis, histological analysis and immunohistochemistry staining	Positive outcome	Rosa et al. (2021)
Metacarpophalangeal joint model					
Randomized placebo-controlled comparative study ($n = 8$)/336 h post cell injection		Allogeneic and autologous BM-MSCs (10×10^6)	Clinical parameters, arthrocentesis, ELISA	Positive outcome with either cell	Colbath et al. (2020a, b)
Randomized case and placebo-controlled comparative study/ $n = 8$/12 weeks		BM-MSCs (1×10^6) and UCB-MSCs (1×10^6) [3 weeks post-surgery]	Clinical, radiography, ultrasonography, MRI, biochemical analysis of synovial fluid, histopathology	Positive outcome, better with BM-MSCs	Bertoni et al. (2021)

MSCs mesenchymal stem cells, *BM* bone marrow derived, *Sy* synovium derived, *AD* adipose tissue derived, *n* total number of animals in all the groups of the study, *UCB* umbilical cord blood derived, *ACPCs* articular cartilage progenitor cells; Positive outcome: Indicates improved outcome with MSCs transplantation even if only one parameter improves; Study period: Indicates maximum time period for which study is conducted transplantation mostly effectuates positive impact on joint affections, although clinically insignificant (Frisbie et al. 2009; Yamasaki et al. 2018). The improvement occurs through prevention of progression of osteoarthritis, reduction in inflammation and enhanced matrix formation, although not necessarily typical to cartilage as one of the studies demonstrates bone formation (Yamada et al. 2013; Goodrich et al. 2016; Delco et al. 2020; Bertoni et al. 2021; Rosa et al. 2021). Better arthroscopic scores are demonstrated as early as day 30 with enhanced tissue formation but typically the fibrous one. At this interval, several defects predominantly contained collagen type II. However, non-significant differences at later periods (8 months) between stem cell treated and control defects occur (Wilke et al. 2007). Insignificant clinical improvement with the cell transplantation also occurs in other study although arthroscopy confirms repair tissue is significantly firm with improved repaired tissue quality (cumulative score of all arthroscopic and gross grading criteria). Further, an immunohistochemical analysis too shows significantly greater levels of aggrecan in repair tissue of cell joints (McIlwraith et al. 2011). The implantation of the cells in scaffold-like alginate hydrogel may limit their immediate migration and shall help in slow and steady release towards the joint space (Dos Santos et al. 2019).

Recent advances make it possible to use MSCs of particular morphology or develop 3D printed constructs for cartilage tissue engineering. Specific morphology AD-MSCs (Integrin $\alpha10^{high}$ MSC) transplantation into the horse talus bone joints significantly limits subchondral bone sclerosis together with limited cartilage fibrillation. The synovial fluid markers signify possible immunomodulatory roles of integrin $\alpha10^{high}$. It is pertinent to mention that integrin $\alpha10^{high}$ is the main collagen type II-binding receptor on chondrocytes and thus might have been possibly effective in reduction of post trauma osteoarthritic changes (Delco et al. 2020). Besides cell type, 3D printed scaffold [poly(ϵ-caprolactone) (PCL) bone anchor reinforcing fibres protruding into the chondral part of the construct over which two layers of a thiol-ene cross-linkable hyaluronic acid/poly(glycidol) hybrid hydrogel (HA-SH/P (AGE-co-G) loaded with articular cartilage progenitor cells (ACPCs in top layer) and MSCs (bottom layer) @ 20 million] are developed. Transplantation of such biomaterial assembly produces stiffer tissue but fails to show any optimal cartilage repair. This is attributed to several factors like early loss of the cells, or an inappropriate hydrogel degradation rate that remain inconsistent with the natural regeneration (Mancini et al. 2020).

One of the studies evaluates the effect of multiple dosing. Transplantation of allogeneic MSCs followed by autologous second dose results in an increase in average scores for gross and histopathological findings amounting to or higher than those of the control (Yamasaki et al. 2018). On the basis of MSCs source (BM- vs AD-MSCs), non-significant differences in defect healing of the carpal joint model occurs. However, cell doses are variable (AD-MSCs, 16.3 million and BM-MSCs, ten million). Even such a healing is insignificant comparable to the control except in relation to the improvement in PGE2 levels (Frisbie et al. 2009).

In general, MSCs appear promising in cartilage defects or joint affections, although further extensive research is desired to shift MSCs to the bedside. As there appears to be a significant effect of scaffold, cell source and dose, it is important to compare such differences through the molecular mechanisms. Besides, MSCs engraftment, paracrine anti- inflammatory/immunomodulatory effects and other pro-healing effects and differentiation features are also desired.

1.4.4.3 Clinical Studies

In horses, various MSCs clinical studies on osteoarthritis or degenerative joint disease of different joints are conducted (Table 1.6). Among various joints evaluated are stifle (femoral condyles) (Raheja et al. 2011), pastern (Spaas et al. 2012; Van Hecke et al. 2021), fetlock (Broeckx et al. 2014, 2019; Magri et al. 2019; Van Hecke et al. 2021) and carpal (Korchunjit et al. 2019). Besides, tibiotarsal joint capsule treated with surgery followed by an intra-articular dose of MSCs is also conducted (Cestari et al. 2021).

1.4.4.3.1 Cell Transplantation, Concentration and Frequency

MSCs remain to be characterized in all the studies with the final percent viability mentioned only in some studies. Mostly single batch cells are transplanted, although different batch cells are also used in one of the studies (Magri et al. 2019). Mainly allogeneic or autologous BM-MSCs, AD-MSCs or PB-MSCs are transplanted. Few studies employ chondrogenic-induced MSCs (Broeckx et al. 2014, 2019; Van Hecke et al. 2021). Some of the studies additionally employ other biomaterials like PRP (Broeckx et al. 2014) or allogeneic plasma (Broeckx et al. 2019) or fibrin glue (Raheja et al. 2011). Transplanted concentrations of MSCs range from 1.4 to 2.5 million (Broeckx et al. 2019; Van Hecke et al. 2021) to 11 million (Magri et al. 2019). The cells are mostly employed as a single dose barring two studies that make a double dose at 13-month (Raheja et al. 2011) and 1-month (Magri et al. 2019) interval after first dose. One of the studies transplants three doses with a minimal gap of 21 days (Van Hecke et al. 2021). The studies are conducted for a period of 6 ± 2 weeks (Van Hecke et al. 2021) to the maximum of 15 months (Raheja et al. 2011) with an average of 1 year (Spaas et al. 2012; Broeckx et al. 2014, 2019). The evaluations are conducted based on the parameters like lameness scores, and the radiography and arthroscopy.

1.4.4.3.2 Results

The studies are reported as monocentric (Spaas et al. 2012; Broeckx et al. 2019) and multicentric (Ferris et al. 2014; Magri et al. 2019). Although multicentric studies prevent bias, there has to be an appropriate and comparable evaluation system. All the above listed equine clinical studies show better clinical outcomes irrespective of the joint type or the affection. Repeated intra-articular (IA) injections within a month's period do not have an edge over single IA. Single dose of the allogeneic cells shows non-significant differences in the improvement of the radiographic osteoarthritis score, osteophyte score, joint space width score and subchondral bone score of metacarpophalangeal joints as compared to that of the repeated doses. Owner-recorded adverse effects comprise around 18% (Magri et al. 2019). The improvements are in the lameness and total clinical scores (Magri et al. 2019)

Table 1.6 Horse in vivo MSCs clinical studies on various joint affections

Study design/duration	Biomaterial used	Overall result	References
Bilateral articular cartilage fissure defects of the medial femoral condyles			
Case report/($n = 1$)/ 15 months	Autologous BM-MSCs (1×10^7 implanted at 90 days and 13 months period) + fibrin glue	Positive outcome	Raheja et al. (2011)
Randomized placebo controlled ($n = 39$)/ 24 months	Autologous BM-MSCs ($15–20 \times 10^6$) + 20 mg of HA	Positive outcome	Ferris et al. (2014)
Bone spavin			
Randomized placebo control comparative study ($n = 16$)/6 months	Autologous AD-MSCs (5×10^6)	Positive outcome	Nicpoń et al. (2013)
Chronic degenerative joint disease of pastern joint			
Case report/$n = 1$/1 year	PB-MSCs (2.5×10^6)	Positive outcome	Spaas et al. (2012)
Degenerative joint disease of fetlock			
Randomized comparative clinical study/12 months	PB-MSCs (6.7×10^3 MSCs/cm^2 cells chondrogen-induced), platelet-rich plasma (200×10^6 platelets)	Positive outcome	Broeckx et al. (2014)
Randomized case controlled study/$n = 75$/ 1 year	Chondrogenic-induced allogeneic MSCs (2×10^6) + allogeneic plasma ($\sim 85 \times 10^6$ platelets)	Positive outcome	Broeckx et al. (2019)
Randomized comparative clinical pilot study/$n = 28$/ 6 months	Allogeneic neonatal UC-MSCs (11×10^6, single and double dose)	Positive outcome	Magri et al. (2019)
Degenerative joint disease of fetlock, pastern and coffin			
Retrospective study/ $n = 10$/6 ± 2 weeks after each injection	Allogeneic chondrogenic-induced allogeneic MSCs ($1.4–2.5 \times 10^6$) + allogeneic plasma 1 mL) three injections (minimal gap between injections is 3 weeks)	Positive outcome	Van Hecke et al. (2021)
Osteochondral fragment-induced osteoarthritis of carpal joint			
Randomized comparative clinical trial/$n = 10$/ 90 days	Allogeneic BM-MSCs (1×10^7)	Positive outcome	Korchunjit et al. (2019)

MSCs mesenchymal stem cells, *PB* peripheral blood derived, *UC* umbilical cord derived, *n* total number of animals in all the groups of the study; Positive outcome: Indicates improved outcome with MSCs transplantation even if only one parameter improves; Study period: Indicates maximum time period for which study is conducted

with animals able to return back to racing (Raheja et al. 2011; Broeckx et al. 2019). However, one of the experimental studies shows increased lameness scores with repeated dosing (Yamasaki et al. 2018). Arthroscopically, MSCs transplantation is associated with the reduction in inflammation and no new osteophyte formation after surgical removal (Raheja et al. 2011; Korchunjit et al. 2019). It is worth mentioning that in some placebo-controlled studies, improvement in clinical scores in 20–25%

of the cases is reported (Broeckx et al. 2019). The differences could have been due to the attendant training program adjustments or to the saline therapeutic effects. However, long-term benefits are mostly seen in the cell-treated cases (Broeckx et al. 2019). Significant improvement in the radiographic scores in the early period is not reported (Spaas et al. 2012; Magri et al. 2019). This may be attributed to the varied OA scores in the cases ranging from mild to severe included in the studies. Further, anatomical changes like osteophytes, joint space narrowing and subchondral bone lysis may not resolve in a shorter period of 6 months to 1 year.

In severe femorotibial joint injuries involving varied structures like meniscus, cartilage or ligament damage, transplantation of MSCs leads to the improvement in 43% of horses that return to the previous level of work and other 33% return to some work, while rest 24% fail to return to work. Animals with meniscal damage may have a higher percentage (75%) of animals that return to some level of work (Ferris et al. 2014). Even bone spavin (early stage) treated intra-articularly with autologous AD-MSCs too lead to the long-term anti-inflammatory effects, promoting joint mechanics up to day 90 as compared to the placebo. However, beyond that non-significant improvement up to 180 days occurs (Nicpoń et al. 2013). In a joint capsule rupture case report, surgical repair together with the intra-articular MSCs transplantation and rehabilitation is correlated with the re-establishment of athletic performance (Cestari et al. 2021).

From the majority of the above studies, it seems that MSCs may offer clinical improvement. Clinical lameness scores improve but may not necessarily be concurrent with the resolve of lesions unless removed surgically. As the studies mostly involve small animal numbers and are of short duration, further large case studies conducted for extended duration are desired to evaluate long-term benefits. Incorporation of the scaffolds may further improve the outcome. HA, PRP may improve cartilage healing by coating the cartilage surface and in localizing the cartilage extracellular matrix among the collagen fibrils and proteoglycans (Maniwa et al. 2001), increasing synovial cell migration (Yoshioka et al. 1997) and above all may localize the extrinsically implanted stem cells. The variabilities in scaffold properties are often seen and need to be evaluated for better outcome. Lack of complete hyaline healing and the long-term evaluation of response in the affected joints remains to be seen.

1.5 Conclusion(s)

MSCs tend to undergo chondrogenesis upon availability of specific microenvironment. However, a single in vitro factor may not mimic the multifactor in vivo system, and thus, results obtained under ex vivo system may not be recapitulated in clinical settings. These results may also change with the source, ageing and/or ailments. In vivo experimental and clinical studies show promising roles of MSCs in chondral, osteochondral, meniscal or other joint affections. Currently, the definitive and standard stem cell therapeutics for joint affections are lacking. The current techniques and their sole application may not offer complete and long-lasting solutions but do appear to resolve pain and lameness to some extent and for some time. The mechanism behind such responses remains to be elucidated in most of the studies. The response may further vary among the species and joints. There is a need for further uniform studies that evaluate other biomaterials like scaffolds, standardize concentration of viable cultured and characterized MSCs, their route and the timing of application. The study design of controlled and uniform one with good animal numbers and for longer duration to act as the evidence-based medicine is desired. Further, comparative studies against the current managemental techniques need to be conducted to arrive at any conclusion.

References

Abdalmula A, Dooley LM, Kaufman C, Washington EA, House JV, Blacklaws BA, Ghosh P, Itescu S, Bailey SR, Kimpton WG (2017) Immunoselected STRO-3+ mesenchymal precursor cells reduce inflammation and improve clinical outcomes in a large animal model of monoarthritis. Stem Cell Res Ther 8:22

Adkisson HD, Gillis MP, Davis EC, Maloney W, Hruska KA (2001) In vitro generation of scaffold independent neocartilage. Clin Orthop Relat Res S391:280–294

Al Faqeh H, Nor Hamdan BM, Chen HC, Aminuddin BS, Ruszymah BH (2012) The potential of intra-articular injection of chondrogenic-induced bone marrow stem cells to retard the progression of osteoarthritis in a sheep model. Exp Gerontol 47:458–464

Alicka M, Kornicka-Garbowska K, Kucharczyk K, Kępska M, Röcken M, Marycz K (2020) Age-dependent impairment of adipose-derived stem cells isolated from horses. Stem Cell Res Ther 11(1):4

Ando W, Heard BJ, Chung M, Nakamura N, Frank CB, Hart DA (2012) Sheep synovial membrane-derived mesenchymal progenitor cells retain the phenotype of the original tissue that was exposed to in-vivo inflammation: evidence for a suppressed chondrogenic differentiation potential of the cells. Inflamm Res 61:599–608

Ardanaz N, Vázquez FJ, Romero A, Remacha AR, Barrachina L, Sanz A, Ranera B, Vitoria A, Albareda J, Prades M, Zaragoza P, Martín-Burriel I, Rodellar C (2016) Inflammatory response to the administration of mesenchymal stem cells in an equine experimental model: effect of autologous, and single and repeat doses of pooled allogeneic cells in healthy joints. BMC Vet Res 12

Arnhold S, Elashry MI, Klymiuk MC, Geburek F (2019) Investigation of stemness and multipotency of equine adipose-derived mesenchymal stem cells (ASCs) from different fat sources in comparison with lipoma. Stem Cell Res Ther 10(1):309

Arora A, Sriram M, Kothari A, Katti DS (2017) Co-culture of infrapatellar fat pad-derived mesenchymal stromal cells and articular chondrocytes in plasma clot for cartilage tissue engineering. Cytotherapy. 19(7):881–894

Athanasiou AK, Agarwal A, Muffoletto A (1995) Biomechanical properties of hip cartilage in experimental animal models. Clin Orthop 316:254–266

Bagge J, MacLeod JN, Berg LC (2020) Cellular proliferation of equine bone marrow- and adipose tissue-derived mesenchymal stem cells decline with increasing donor age. Front Vet Sci 7: 602403

Barbour KE, Helmick CG, Boring M, Zhang X, Lu H, Holt JB (2016) Prevalence of doctor diagnosed arthritis at state and county levels-United States, 2014. MMWR Morb Mortal Wkly Rep 65(19):489–494

Bçrziòð U, Matise-Van Houtana I, Pçtersone I, Dûrîtis I, Òikuïðins S, Bogdanova-Jâtniece A, Kâlis M, Svirskis Đ, Skrastiòa D, Ezerta A et al (2018) Characterisation and *in vivo* safety of canine adipose-derived stem cells. Proc Latvian Acad Sci U S A 72(3):160–171

Bekkers JE, Tsuchida AI, van Rijen MH, Vonk LA, Dhert WJ, Creemers LB, Saris DB (2013) Single-stage cell-based cartilage regeneration using a combination of chondrons and mesenchymal stromal cells: comparison with microfracture. Am J Sports Med 41(9):2158–2166

Berglund AK, Schnabel LV (2017) Allogeneic major histocompatibility complex-mismatched equine bone marrow-derived mesenchymal stem cells are targeted for death by cytotoxic anti-major histocompatibility complex antibodies. Equine Vet J 49(4):539–544

Bertolo A, Steffen F, Malonzo-Marty C, Stoyanov J (2015) Canine mesenchymal stem cell potential and the importance of dog breed: implication for cell-based therapies. Cell Transplant 24(10):1969–1980

Bertoni L, Branly T, Jacquet S, Desancé M, Desquilbet L, Rivory P, Hartmann D-J, Denoix J-M, Audigié F, Galéra P, Demoor M (2019) Intra-articular injection of 2 different dosages of autologous and allogeneic bone marrow- and umbilical cord-derived Mesenchymal stem cells triggers a variable inflammatory response of the fetlock joint on 12 sound experimental horses. Stem Cells Int 2019:9431894

Bertoni L, Jacquet-Guibon S, Branly T, Desancé M, Legendre F, Melin M, Rivory P, Hartmann D-J, Schmutz A, Denoix J-M, Demoor M, Audigié F, Galéra P (2021) Evaluation of allogeneic bone-marrow-derived and umbilical cord blood-derived mesenchymal stem cells to prevent the development of osteoarthritis in an equine model. Int J Mol Sci 22:2499

Black LL, Gaynor J, Gahring D, Adams C, Aron D, Harman S, Gingerich DA, Harman R (2007) Effect of adipose-derived mesenchymal stem and regenerative cells on lameness in dogs with chronic osteoarthritis of the coxofemoral joints: a randomized, double-blinded, multicenter, controlled trial. Vet Ther 8:272–284

Black LL, Gaynor J, Adams C, Dhupa S, Sams AE, Taylor R, Harman S, Gingerich DA, Harman R (2008) Effect of intraarticular injection of autologous adipose-derived mesenchymal stem and regenerative cells on clinical signs of chronic osteoarthritis of the elbow joint in dogs. Vet Ther 9:192–200

Bordbar S, LotfiBakhshaiesh N, Khanmohammadi M, Sayahpour FA, Alini M, Baghaban Eslaminejad M (2020) Production and evaluation of decellularized extracellular matrix hydrogel for cartilage regeneration derived from knee cartilage. J Biomed Mater Res A 108(4):938–946

Bornes TD, Jomha NM, Mulet-Sierra A, Adesida AB (2015) Hypoxic culture of bone marrow-derived mesenchymal stromal stem cells differentially enhances in vitro chondrogenesis within cell-seeded collagen and hyaluronic acid porous scaffolds. Stem Cell Res Ther 6:84

Bornes TD, Jomha NM, Mulet-Sierra A, Adesida AB (2016) Optimal seeding densities for *In Vitro* Chondrogenesis of two- and three-dimensional-isolated and -expanded bone marrow-derived Mesenchymal stromal stem cells within a porous collagen scaffold. Tissue Eng Part C Methods 22(3):208–220

Bornes TD, Adesida AB, Jomha NM (2018) Articular cartilage repair with mesenchymal stem cells after chondrogenic priming: a pilot study. Tissue Eng 24:9–10

Bosnakovski D, Mizuno M, Kim G, Ishiguro T, Okumura M, Iwanaga T, Kadosawa T, Fujinaga T (2004) Chondrogenic differentiation of bovine bone marrow mesenchymal stem cells in pellet cultural system. Exp Hematol 32(5):502–509

Bosnakovski D, Mizuno M, Kim G, Takagi S, Okumura M, Fujinaga T (2006) Chondrogenic differentiation of bovine bone marrow mesenchymal stem cells (MSCs) in different hydrogels: influence of collagen type II extracellular matrix on MSC chondrogenesis. Biotechnol Bioeng 93(6):1152–1163

Bourin P, Bunnell BA, Casteilla L, Dominici M, Katz AJ, March KL, Redl H, Rubin JP, Yoshimura K, Gimble JM (2013) Stromal cells from the adipose tissue-derived stromal vascular fraction and culture expanded adipose tissue-derived stromal/stem cells: a joint statement of the International Federation for Adipose Therapeutics and Science (IFATS) and the International Society for Cellular Therapy (ISCT). Cytotherapy 15(6):641–648

Bozkurt M, Aşık MD, Gürsoy S, Türk M, Karahan S, Gümüşkaya B, Akkaya M, Şimşek ME, Cay N, Doğan M (2019) Autologous stem cell-derived chondrocyte implantation with bio-targeted microspheres for the treatment of osteochondral defects. J Orthop Surg Res 14(1): 394

Branly T, Bertoni L, Contentin R, Rakic R, Gomez-Leduc T, Desancé M, Hervieu M, Legendre F, Jacquet S, Audigié F, Denoix JM, Demoor M, Galéra P (2017) Characterization and use of equine bone marrow mesenchymal stem cells in equine cartilage engineering. Study of their hyaline cartilage forming potential when cultured under hypoxia within a biomaterial in the presence of BMP-2 and TGF-ß1. Stem Cell Rev Rep 13(5):611–630

Broeckx S, Zimmerman M, Crocetti S, Suls M, Marien T, Ferguson SJ, Chiers K, Duchateau L, Franco-Obregón A, Wuertz K et al (2014) Regenerative therapies for equine degenerative joint disease: a preliminary study. PLoS One 9(1):e85917

Broeckx SY, Spaas JH, Chiers K, Duchateau L, Van Hecke L, Van Brantegem L, Dumoulin M, Martens AM, Pille F (2018) Equine allogeneic chondrogenic induced mesenchymal stem cells: a GCP target animal safety and biodistribution study. Res Vet Sci 117:246–254

Broeckx SY, Seys B, Suls M, Vandenberghe A, Marien T, Adriaensen E, Declercq J, Van Hecke L, Braun G, Hellmann K et al (2019) Equine allogeneic chondrogenic induced mesenchymal stem cells are an effective treatment for degenerative joint disease in horses. Stem Cells Dev 28(6): 410–422

Buckwalter JA, Mankin HJ, Grodzinsky AJ (2005) Articular cartilage and osteoarthritis. Instr Course Lect 54:465–480

Bundgaard L, Stensballe A, Elbæk KJ, Berg LC (2018) Mapping of equine mesenchymal stromal cell surface proteomes for identification of specific markers using proteomics and gene expression analysis: an in vitro cross-sectional study. Stem Cell Res Ther 9:288

Bwalya EC, Kim S, Fang J, Wijekoon HMS, Hosoya K, Okumura M (2017) Effects of pentosane polysulfate and polysulfated glycosaminoglycan on chondrogenesis of canine bone marrow-derived mesenchymal stem cells in alginate and micromass culture. J Vet Med Sci 79(7): 1182–1190

Cabon Q, Febre M, Gomez N, Cachon T, Pillard P, Carozzo C, Saulnier N, Robert C, Livet V, Rakic R, Plantier N, Saas P, Maddens S, Viguier E (2019) Long-term safety and efficacy of single or repeated intra-articular injection of allogeneic neonatal mesenchymal stromal cells for managing pain and lameness in moderate to severe canine osteoarthritis without anti-inflammatory pharmacological support: pilot clinical study. Front Vet Sci 6:10

Caminal M, Moll X, Codina D, Rabanal RM, Morist A, Barrachina J, Garcia F, Pla A, Vives J (2014a) Transitory improvement of articular cartilage characteristics after implantation of polylactide: polyglycolic acid (PLGA) scaffolds seeded with autologous mesenchymal stromal cells in a sheep model of critical-sized chondral defect. Biotechnol Lett 36:2143–2153

Caminal M, Fonseca C, Peris D, Moll X, Rabanal RM, Barrachina J, Codina D, García F, Cairó JJ, Gódia F, Pla A, Vives J (2014b) Use of a chronic model of articular cartilage and meniscal injury for the assessment of long-term effects after autologous mesenchymal stromal cell treatment in sheep. N Biotechnol 31(5):492–498

Campos LL, Landim-Alvarenga FC, Ikeda TL, Monteiro BA, Maia L, Freitas-Dell'Aqua CP, De Vita B (2017) Isolation, culture, characterization and cryopreservation of stem cells derived from amniotic mesenchymal layer and umbilical cord tissue of bovine fetuses. Pesquisa Veterinária Brasileira 37(3):278–286

Cardoso TC, Okamura LH, Baptistella JC, Gameiro R, Ferreira HL, Marinho M, Flores EF (2017) Isolation, characterization and immunomodulatory-associated gene transcription of Wharton's jelly-derived multipotent mesenchymal stromal cells at different trimesters of cow pregnancy. Cell Tissue Res 367(2):243–256

Caron JP, Genovese RL (2003) Principals and practices of joint disease treatment. In: Ross MW, Dyson S (eds) Diagnosis and management of lameness in the horse, 1st edn. Saunders, Philadelphia (PA), pp 746–764

Carrade DD, Owens SD, Galuppo LD, Vidal MA, Ferraro GL, Librach F, Buerchler S, Friedman MS, Walker NJ, Borjesson DL (2011) Clinicopathologic findings following intra-articular injection of autologous and allogeneic placentally derived equine mesenchymal stem cells in horses. Cytotherapy 13(4):419–430

Cestari H, Costa e Souza JP, Hussni CA, Rodrigues CA, Watanabe MJ, Garcia Alves AL (2021) Intra-articular mesenchymal stem cell therapy on a joint capsule rupture in a horse—a case report. Res Soc Dev 10(12):e211101220370

Chang L-B, Peng S-Y, Chou C-J, Chen Y-J, Shiu J-S, Tu P-A, Gao S-X, Chen Y-C, Lin T-K, Wu S-C (2018) Therapeutic potential of amniotic fluid stem cells to treat bilateral ovarian dystrophy in dairy cows in a subtropical region. Reprod Domest Anim 53(4):1024–1024

Chapman H-S, Gale AL, Dodson ME, Linardi RL, Ortved KF (2020) Autologous platelet lysate does not enhance chondrogenic differentiation of equine bone marrow-derived mesenchymal stromal cells despite increased TGF-β1 concentration. Stem Cells Dev 29(3):144–155

Chen F, Zhao C, Zhao Y, Li L, Liu S, Zhu Z, Guan W (2018) The biological characteristics of sheep umbilical cord mesenchymal stem cells. Can J Vet Res 82:216–224

Clegg P, Booth R (2000) Drugs used to treat osteoarthritis in the horse. In Practice 22:594–603

Colbath AC, Dow SW, Hopkins LS, Phillips JN, McIlwraith CW, Goodrich LR (2020a) Single and repeated intra-articular injections in the tarsocrural joint with allogeneic and autologous equine bone marrow-derived mesenchymal stem cells are safe, but did not reduce acute inflammation in an experimental interleukin-1β model of synovitis. Equine Vet J 52(4):601–612

Colbath AC, Dow SW, Hopkins LS, Phillips JN, Mcilwraith CW, Goodrich LR (2020b) Allogeneic vs. autologous intra-articular mesenchymal stem cell injection within normal horses: clinical and cytological comparisons suggest safety. Equine Vet J 52(1):144–151

Contentin R, Demoor M, Concari M, Desancé M, Audigié F, Branly T, Galéra P (2020) Comparison of the chondrogenic potential of mesenchymal stem cells derived from bone marrow and umbilical cord blood intended for cartilage tissue engineering. Stem Cell Rev Rep 16(1): 126–143

Critchley S, Sheehy EJ, Cunniffe G, Diaz-Payno P, Carroll SF, Jeon O, Alsberg E, Brama PAJ, Kelly DJ (2020) 3D printing of fibre-reinforced cartilaginous templates for the regeneration of osteochondral defects. Acta Biomater 113:130–143

Cuervo B, Rubio M, Sopena J, Dominguez JM, Vilar J, Morales M, Cugat R, Carrillo JM (2014) Hip osteoarthritis in dogs: a randomized study using mesenchymal stem cells from adipose tissue and plasma rich in growth factors. Int J Mol Sci 15:13437–13460

da Silva CG, Martins CF, Cardoso TC, da Cunha ER, Bessler HC, MacManus CM, Pivato I, Bao SN (2016) Isolation and characterization of Mesenchymal stem cells derived from bovine Wharton's jelly and their potential use in cloning by nuclear transfer. Cienc Rural, Santa Maria 46(10):1830–1837

Daems R, Van Hecke L, Schwarzkopf I, Depuydt E, Broeckx SY, David M, Beerts C, Vandekerckhove P, Spaas JH (2019) A feasibility study on the use of equine chondrogenic induced mesenchymal stem cells as a treatment for natural occurring osteoarthritis in dogs. Stem Cells Int 2019:4587594

Dar ER, Gugjoo MB, Javaid M, Hussain S, Fazili MR, Dhama K, Alqahtani T, Alqahtani AM, Shah RA, Emran TB (2021) Adipose tissue- and bone marrow-derived mesenchymal stem cells from sheep: culture characteristics. Animals 11(8):2153

Delco ML, Goodale M, Talts JF, Pownder SL, Koff MF, Miller AD, Nixon B, Bonassar LJ, Lundgren-Åkerlund E, Fortier LA (2020) Integrin α10β1-selected mesenchymal stem cells mitigate the progression of osteoarthritis in an equine talar impact model. Am J Sports Med 48(3):612–623

Delling U, Brehm W, Metzger M, Ludewig E, Winter K, Jülke H (2015) In vivo tracking and fate of intra-articularly injected superparamagnetic iron oxide particle-labeled multipotent stromal cells in a sheep model of osteoarthritis. Cell Transplant 24:2379–2390

Deng Y, Huang G, Zou L, Nong T, Yang X, Cui J, Wei Y, Yang S, Shi D (2018) Isolation and characterization of buffalo (Bubalus bubalis) amniotic mesenchymal stem cells derived from amnion from the first trimester pregnancy. J Vet Med Sci 80(4):710–719

Desando G, Giavaresi G, Cavallo C, Bartolotti I, Sartoni F, Aldini NN, Martini L, Parrilli A, Mariani E, Fini M, Grigolo B (2016) Autologous bone marrow concentrate in a sheep model of osteoarthritis: new perspectives for cartilage and meniscus repair. Tissue Eng 22(6):608–619

Dos Santos VH, Pfeifer JPH, De Souza JB, De Castro Stievani F, Hussni CA, De Assis Golim M, Deffune E, Garcia Alves AL (2019) Evaluation of alginate hydrogel encapsulated mesenchymal stem cell migration in horses. Res Vet Sci 124:38–45

Dražilov SS, Mrkovački J, Spasovski V, Fazlagić A, Pavlović S, Nikčević G (2018) The use of canine mesenchymal stem cells for the autologous treatment of osteoarthritis. Acta Vet Hung 66(3):376–389

Duarte Campos DF, Drescher W, Rath B, Tingart M, Fischer H (2012) Supporting biomaterials for articular cartilage repair. Cartilage 3(3):205–221

Elkhenany H, Amelse L, Caldwell M, Abdelwahed R, Dhar M (2016) Impact of the source and serial passaging of goat mesenchymal stem cells on osteogenic differentiation potential: implications for bone tissue engineering. J Anim Sci Biotechnol 7:16

El-Tookhy, Abou Elkheir W, Mokbel A, Osman A (2008) Intra-articular injection of autologous mesenchymal stem cells in experimental chondral defects in dogs. Egypt Rheumatol 30(2):1–10

Endo K, Fujita N, Nakagawa T, Nishimura R (2019a) Comparison of the effect of growth factors on chondrogenesis of canine mesenchymal stem cells. J Vet Med Sci 81(8):1211–1218

Endo K, Fujita N, Nakagawa T, Nishimura R (2019b) Effect of fibroblast growth factor-2 and serum on canine mesenchymal stem cell chondrogenesis. Tissue Eng Part A 25(11–12):901–910

Erickson IE, van Veen SC, Sengupta S, Kestle SR, Mauck RL (2011) Cartilage matrix formation by bovine mesenchymal stem cells in three-dimensional culture is age-dependent. Clin Orthop Relat Res 469:2744–2753

Favreau H, Pijnenburg L, Seitlinger J, Fioretti F, Keller L, Scipioni D, Adriaensen H, Kuchler-Bopp S, Ehlinger M, Mainard D, Rosset P, Hua G, Gentile L, Jessel NB (2020) Osteochondral repair combining therapeutics implant with mesenchymal stem cells spheroids. Nanomedicine 29:102253. https://doi.org/10.1016/j.nano.2020.102253

Feitosa MLT, Fadel L, Beltrão-Braga PCB, Wenceslau CV, Kerkis I, Kerkis A, Júnior EHB, Martins JFP, Martins DS, Miglino MA, Ambrósio CE (2010) Successful transplant of mesenchymal stem cells in induced osteonecrosis of the sheep femoral head. Preliminary results. Acta Cir Bras 25(5):416–422

Feng C, Luo X, He N, Xia H, Lv X, Zhang X, Li D, Wang F, He J, Zhang L, Lin X, Lin L, Yin H, He J, Wang J, Cao W, Wang R, Zhou G, Wang W (2018) Efficacy and persistence of allogeneic adipose derived mesenchymal stem cells combined with hyaluronic acid in osteoarthritis after intra-articular injection in a sheep model. Tissue Eng 24(3–4):219–233

Ferris DJ, Frisbie DD, Kisiday JD, McIlwraith CW, Hague BA, Major MD, Schneider RK, Zubrod CJ, Kawcak CE, Goodrich LR (2014) Clinical outcome after intra-articular administration of bone marrow derived mesenchymal stem cells in 33 horses with stifle injury. Vet Surg 43(3): 255–265

Fisher MB, Henning EA, Söegaard N, Bostrom M, Esterhai JL, Mauck RL (2015) Engineering meniscus structure and function via multi-layered mesenchymal stem cell-seeded nanofibrous scaffolds. J Biomech 48(8):1412–1419

Fraser JK, Wulur I, Alfonso Z, Hedrick MH (2006) Fat tissue: an underappreciated source of stem cells for biotechnology. Trends Biotechnol 24(4):150–154

Frisbie DD, Cross MW, McIlwraith CW (2006) A comparative study of articular cartilage thickness in the stifle of animal species used in human pre-clinical studies compared to articular cartilage thickness in the human. Vet Comp Orthop Traumatol 19(3):142–146

Frisbie DD, Kisiday JD, Kawcak CE, Werpy NM, McIlwraith CW (2009) Evaluation of adipose-derived stromal vascular fraction or bone marrow-derived mesenchymal stem cells for treatment of osteoarthritis. J Orthop Res 27:1675–1680

Fülber J, Maria DA, Correia da Silva LCL, Massoco CO, Agreste F, Baccarin RYA (2016) Comparative study of equine mesenchymal stem cells from healthy and injured synovial tissues: an in vitro assessment. Stem Cell Res Ther 7:35

Fülber J, Agreste FR, Seidel SRT, Sotelo EDP, Barbosa ÂP, Michelacci YM, Baccarin RYA (2021) Chondrogenic potential of mesenchymal stem cells from horses using a magnetic 3D cell culture system. World J Stem Cells 13(6):645–658

Futrega K, Music E, Robey PG, Gronthos S, Crawford R, Saifzadeh S, Klein TJ, Doran MR (2021) Characterisation of ovine bone marrow-derived stromal cells (oBMSC) and evaluation of chondrogenically induced micro-pellets for cartilage tissue repair in vivo. Stem Cell Res Ther 12(1):26

Gabner S, Ertl R, Velde K, Renner M, Jenner F, Egerbacher M, Hlavaty J (2018) Cytokine-induced interleukin-1 receptor antagonist protein expression in genetically engineered equine mesenchymal stem cells for osteoarthritis treatment. J Gene Med 20(5):e3021

Gale AL, Mammone RM, Dodson ME, Linardi RL, Ortved KF (2019a) The effect of hypoxia on chondrogenesis of equine synovial membrane-derived and bone marrow-derived mesenchymal stem cells. BMC Vet Res 15(1):201

Gale AL, Linardi RL, McClung G, Mammone RM, Ortved KF (2019b) Comparison of the chondrogenic differentiation potential of equine synovial membrane-derived and bone marrow-derived mesenchymal stem cells. Front Vet Sci 6:178

Garcia RS, Melo UP, Ferreira C, Toscano FS, Cruz GM (2009) Estudoclínico e radiográfico da osteoartritetársicajuvenilempotros da raçamangalargamarchador. Ciên Ani Bras 10(1):254–260

Garcia D, Longo UG, Vaquero J, Forriol F, Loppini M, Khan WS, Denaro V (2015) Amniotic membrane transplant for articular cartilage repair: an experimental study in sheep. Curr Stem Cell Res Ther 10:77–83

Gershwin LJ (2007) Veterinary autoimmunity: autoimmune diseases in domestic animals. Ann NY Acad Sci 1109:109–101

Giretova M, Medvecky L, Petrovova E, Cizkova D, Danko J, Mudronova D et al (2019) Polyhydroxybutyrate/chitosan 3D scaffolds promote in vitro and in vivo chondrogenesis. Appl Biochem Biotechnol 189(2):556–575

Goldman SM, Barabino GA (2016) Hydrodynamic loading in concomitance with exogenous cytokine stimulation modulates differentiation of bovine mesenchymal stem cells towards osteochondral lineages. BMC Biotechnol 16:10

González-Fernández ML, Pérez-Castrillo S, Sánchez-Lázaro JA, Prieto-Fernández JG, López-González ME, Lobato-Pérez S, Colaço BJ, Olivera ER, Villar-Suárez V (2016) Assessment of regeneration in meniscal lesions by use of mesenchymal stem cells derived from equine bone marrow and adipose tissue. Am J Vet Res 77:779–788

Goodrich LR, Chen AC, Werpy NM, Williams AA, Kisiday JD, Su AW, Cory E, Morley PS, McIlwraith CW, Sah RL, Chu CR (2016) Addition of mesenchymal stem cells to autologous platelet-enhanced fibrin scaffolds in chondral defects: does it enhance repair? J Bone Joint Surg 98:23–34

Guercio A, Di Marco P, Casella S, Cannella V, Russotto L, Purpari G, Di Bella S, Piccione G (2012) Production of canine mesenchymal stem cells from adipose tissue and their application in dogs with chronic osteoarthritis of the humeroradial joints. Cell Biol Int 36:189–194

Gugjoo MB, Pal A (2020) Mesenchymal stem cell research in veterinary sciences. Springer Nature, pp 1–337

Gugjoo MB, Amarpal, Sharma GT, Kinjavdekar P, Aithal HP, Pawde AM (2016) Cartilage tissue engineering: role of mesenchymal stem cells along with growth factors and scaffolds. Indian J Med Res 144(3):339–347

Gugjoo MB, Amarpal, Abdelbaset-Ismail A, Aithal HP, Kinjavdekar P, Pawde AM, Kumar GS, Sharma GT (2017) Mesenchymal stem cells with IGF-1 and TGF-β1 in laminin gel for osteochondral defects in rabbits. Biomed Pharmacother 93:1165–1174

Gugjoo MB, Amarpal, Makhdoomi DM, Sharma GT (2019) Equine mesenchymal stem cells: properties, sources, characterization and potential therapeutic applications. J Equine Vet Sci 72:16–27

Gugjoo MB, Amarpal AAI, Aithal HP, Kinjavdekar P, Saikumar GS, Sharma GT (2020) Allogeneic mesenchymal stem cells and growth factors in gel scaffold repair osteochondral defect in rabbit. Regen Med 14. https://doi.org/10.2217/rme-2018-0138

Guo X, Wang C, Zhang Y, Xia R, Hu M, Duan C, Zhao Q, Dong L, Lu J, Qing SY (2004) Repair of large articular cartilage defects with implants of autologous mesenchymal stem cells seeded into beta-tricalcium phosphate in a sheep model. Tissue Eng 10(11–12):1818–1829

Gupta RC, Lall R, Srivastava A, Sinha A (2019) Hyaluronic acid: molecular mechanisms and therapeutic trajectory. Front Vet Sci 6:192

Hang D, Wang Q, Guo C, Chen Z, Yan Z (2012) Treatment of osteonecrosis of the femoral head with VEGF165 transgenic bone marrow mesenchymal stem cells in mongrel dogs. Cells Tissues Organs 195:495–506

Harman R, Carlson K, Gaynor J, Gustafson S, Dhupa S, Clement K, Hoelzler M, McCarthy T, Schwartz P, Adams C (2016) A prospective, randomized, masked, and placebo-controlled efficacy study of intraarticular allogeneic adipose stem cells for the treatment of osteoarthritis in dogs. Front Vet Sci 3:81

Hattori T, Muller C, Gebhard S, Bauer E, Pausch F, Schlund B, Bosl MR, Hess A, Surmann-Schmitt C, von der Mark H et al (2010) SOX9 is a major negative regulator of cartilage vascularization, bone marrow formation and endochondral ossification. Development 137(6): 901–911

Hegewald AA, Ringe J, Bartel J, Krüger I, Notter M, Barnewitz D, Kaps C, Sittinger M (2004) Hyaluronic acid and autologous synovial fluid induce chondrogenic differentiation of equine mesenchymal stem cells: a preliminary study. Tissue Cell 36:431–438

Heinola T, Jukola E, Näkki P, Sukura A (2006) Consequences of hazardous dietary calcium deficiency for fattening bulls. Acta Vet Scand 48:25

Hendawy H, Uemura A, Ma D, Namiki R, Samir H, Ahmed MF, Elfadadny A, El-Husseiny HM, Chieh-Jen C, Tanaka R (2021) Tissue harvesting site effect on the canine adipose stromal vascular fraction quantity and quality. Animals 11:460

Heo S-J, Thorpe SD, Driscoll TP, Duncan RL, Lee DA, Mauck RL (2015) Biophysical regulation of chromatin architecture instills a mechanical memory in Mesenchymal stem cells. Sci Rep 5: 16895

Hindle P, Baily J, Khan N, Biant LC, Simpson AHR, Péault B (2016) Perivascular mesenchymal stem cells in sheep: characterization and autologous transplantation in a model of articular cartilage repair. Stem Cells Dev 25(21):1659–1669

Hotham WE, Thompson C, Szu-Ting L, Henson FMD (2021) The anti-inflammatory effects of equine bone marrow stem cell-derived extracellular vesicles on autologous chondrocytes. Vet Rec Open 8:e22

Houard X, Goldring MB, Berenbaum F (2013) Homeostatic mechanisms in articular cartilage and role of inflammation in osteoarthritis. Curr Rheumatol Rep 15:375–385

Huňáková K, Hluchý M, Špakováb T, Matejová J, Mudroňová D, Kuricová M, Rosocha J, Ledecký V (2020) Study of bilateral elbow joint osteoarthritis treatment using conditioned medium from allogeneic adipose tissue-derived MSCs in Labrador retrievers. Res Vet Sci 132:513–520

Hunziker EB (1999) Biological repair of articular cartilage. Clin Orthop 367:135–146

Hwang NS, Varghese S, Li H, Elisseeff J (2011) Regulation of osteogenic and chondrogenic differentiation of mesenchymal stem cells in PEG-ECM hydrogels. Cell Tissue Res 344:499–509

Iacono E, Pascucci L, Bazzucchi C, Cunto M, Ricci F, Rossi B, Merlo B (2018) Could hypoxia influence basic biological properties and ultrastructural features of adult canine mesenchymal stem/stromal cells? Vet Res Commun 42(4):297–308

Inui T, Haneda S, Sasaki M, Furuoka H, Ito M, Yanagawa M, Hiyama M, Tabata Y, Sasaki N (2019) Enhanced chondrogenic differentiation of equine bone marrow-derived mesenchymal stem cells in zirconia microwell substrata. Res Vet Sci 125:345–350

Ito T, Sawada R, Fujiwara Y, Tsuchiya T (2008) FGF-2 increases osteogenic and chondrogenic differentiation potentials of human mesenchymal stem cells by inactivation of TGF-β signaling. Cytotechnology 56:1

Juneau C, Paine R, Chicas E, Gardner E, Bailey L, McDermott J (2016) Current concepts in treatment of patellofemoral osteochondritis dissecans. Int J Sports Phys Ther 11(6):903–925

Jurgens WJFM, Kroeze RJ, Zandieh-Doulabi B, van Dijk A, Renders GAP, Smit T, van Milligen FJ, Ritt MJ, Helder MN (2013) One-step surgical procedure for the treatment of osteochondral defects with adipose-derived stem cells in a Caprine knee defect: a pilot study. BioRes Open Access 2(4):315–325

Kafarnik C, McClellan A, Dziasko M, Daniels JT, Guest DJ (2020) Canine corneal stromal cells have multipotent mesenchymal stromal cell properties in vitro. Stem Cells Dev 29(7):429–435

Kaggie JD, Markides H, Graves MJ, MacKay J, Houston G, El Haj A, Gilbert F, Henson F (2020) Ultra short echo time MRI of iron-labelled mesenchymal stem cells in an ovine osteochondral defect model. Sci Rep 10:8451

Kamm JL, Parlane NA, Riley CB, Gee EK, Dittmer KE, McIlwraith CW (2019) Blood type and breed-associated differences in cell marker expression on equine bone marrow-derived mesenchymal stem cells including major histocompatibility complex class II antigen expression. PLoS One 14(11):e0225161

Katja N, Michaela E, Jochen R, Bernd F, Rudi M, Thomas H et al (2007) BMP7 promotes adipogenic but not osteo-/chondrogenic differentiation of adult human bone marrow-derived stem cells in high-density micro-mass culture. J Cell Biochem 102(3):626–637

Kazemi D, Asenjan S, Dehdilani K, Parsa H (2017) Canine articular cartilage regeneration using mesenchymal stem cells seeded on platelet rich fibrin. Bone Joint Res 6:98–107

Kim H-R, Lee J, Byeon JS, Gu N-Y, Lee J, Cho I-S, Cha S-H (2017) Extensive characterization of feline intra-abdominal adipose-derived mesenchymal stem cells. J Vet Sci 18(3):299–306

Kim KH, Park TS, Cho B-W, Kim TM (2020) Nanoparticles from equine fetal bone marrow-derived cells enhance the survival of injured chondrocytes. Animals 10:1723

Kim YS, Kim YI, Koh YG (2021) Intra-articular injection of human synovium-derived mesenchymal stem cells in beagles with surgery-induced osteoarthritis. Knee 28:159–168

Knippenberg M, Helder M, Doulabi BZ, Semeins CM, Wuisman PIJM, Klein-Nulend J (2006) Osteogenesis versus chondrogenesis by BMP-2 and BMP-7 in adipose stem cells. Biochem Biophys Res Commun 342:902–908

Kondo M, Yamaoka K, Sonomoto K, Fukuyo S, Oshita K, Okada Y, Tanaka Y (2013) IL-17 inhibits chondrogenic differentiation of human mesenchymal stem cells. PLoS One 8(11): e79463

Korchunjit W, Laikul A, Taylor J, Watchararat K, Ritruechai P, Supokawej A, Wongtawan T (2019) Characterisation and allogeneic transplantation of equine bone marrow derived multipotent mesenchymal stromal cells collected from cadavers. J Equine Vet Sci 73:15–23

Kriston-Pál É, Czibula Á, Gyuris Z, Balka G, Seregi A, Sükösd F, Süth M, Kiss-Tóth E, Haracska L, Uher F, Monostori É (2017) Characterization and therapeutic application of canine adipose mesenchymal stem cells to treat elbow osteoarthritis. Can J Vet Res 81:73–78

Lam J, Lu S, Lee EJ, Trachtenberg JE, Meretoja VV, Dahlin RL, van den Beucken JJJP, Tabata Y, Wong ME, Jansen JA, Mikos AG, Kasper FK (2014) Osteochondral defect repair using bilayered hydrogels encapsulating both chondrogenically and osteogenically pre-differentiated mesenchymal stem cells in a rabbit model. Osteoarthritis Cartilage 22(9):1291–1300

Lamani TSD, Ranganath L, Nagaraja BN, Satyanarayana ML, Jamuna KV, Kamran A (2019) Incidence of osteoarthritis in dogs—a clinical study of 64 patients. Int J Sci Environ Technol 8(3):559–564

Lara E, Velásquez A, Cabezas J, Rivera N, Pacha P, Rodríguez-Alvarez L, Saravia F, Castro FO (2017) Endometritis and in vitro PGE2 challenge modify properties of cattle endometrial mesenchymal stem cells and their transcriptomic profile. Stem Cells Int 2017:4297639

Larsson S, Englund M, Struglics A, Lohmander LS (2015) Interleukin-6 and tumor necrosis factor alpha in synovial fluid are associated with progression of radiographic knee osteoarthritis in subjects with previous meniscectomy. Osteoarthritis Cartilage 23:1906–1914

Lee WD, Hurtig MB, Pilliar RM, Stanford WL, Kandel RA (2015) Engineering of hyaline cartilage with a calcified zone using bone marrow stromal cells. Osteoarthritis Cartilage 23(8):1307–1315

Lee J, Byeon JS, Lee KS, Gu N-Y, Lee GB, Kim H-R, Cho I-S, Cha S-H (2016) Chondrogenic potential and anti-senescence effect of hypoxia on canine adipose mesenchymal stem cells. Vet Res Commun 40(1):1–10

Leitinger B, Hohenester E (2007) Mammalian collagen receptors. Matrix Biol 26(3):146–155

Leung VYL, Gao B, Leung KKH, Melhado IG, Wynn SL, Au TYK, Dung NWF, Lau JYB, Mak ACY, Chan D et al (2011) SOX9 governs differentiation stage-specific gene expression in growth plate chondrocytes via direct concomitant transactivation and repression. PLoS Genet 7(11):e1002356

Levato R, Webb WR, Otto IA, Mensinga A, Zhang Y, van Rijen M, van Weeren R, Khan IM, Malda J (2017) The bio in the ink: cartilage regeneration with bioprintable hydrogels and articular cartilage-derived progenitor cells. Acta Biomater 61:41–53

Li L, Duan X, Fan Z, Chen L, Xing F, Xu Z, Chen Q, Xiang Z (2018) Mesenchymal stem cells in combination with hyaluronic acid for articular cartilage defects. Sci Rep 8:9900

Liu S, Sun Y, Chen T, Tao H, Hu Y, Chen S, Chen J (2018) Time from injury to surgery affects graft maturation following posterior cruciate ligament reconstruction with remnant preservation: a magnetic resonance imaging-based study. Art Ther 34:2846–2854

Lima VP, Tobin GC, de Jesus Pereira MR, Silveira MD, Witz MI, Nardi NB (2019) Chondrogenic effect of liquid and gelled platelet lysate on canine adipose-derived mesenchymal stromal cells. Res Vet Sci 124:393–398

Loeser RF (2011) Aging and osteoarthritis. Curr Opin Rheumatol 23:492–496

Lozito TP, Tuan RS (2011) Mesenchymal stem cells inhibit both endogenous and exogenous MMPs via secreted TIMPs. J Cell Physiol 226:385–396

Lv X, He J, Zhang X, Luo X, He N, Sun Z, Xia H, Liu V, Zhang L, Lin X, Lin L, Yin H, Jiang D, Cao W, Wang R, Zhou G, Wang W (2018) Comparative efficacy of autologous stromal vascular fraction and autologous adipose-derived mesenchymal stem cells combined with hyaluronic acid for the treatment of sheep osteoarthritis. Cell Transplant 27(7):1111–1125

Magri C, Schramme M, Febre M, Cauvin E, Labadie F, Saulnier N, Francois I, Lechartier A, Aebischer D, Moncelet A-S, Maddens S (2019) Comparison of efficacy and safety of single versus repeated intra-articular injection of allogeneic neonatal mesenchymal stem cells for treatment of osteoarthritis of the metacarpophalangeal/metatarsophalangeal joint in horses: a clinical pilot study. PLoS One 14(8):e0221317

Malagola E, Teunissen M, van der Laan LJ, Verstegen MM, Schotanus BA, van Steenbeek FG, Penning LC, van Wolferen ME, Tryfonidou MA, Spee B (2016) Stem characterization and comparison of canine multipotent stromal cells derived from liver and bone marrow. Stem Cells Dev 25(2):139–150

Mancini IAD, Schmidt S, Brommer H, Pouran B, Schäfer S, Tessmar J et al (2020) A composite hydrogel-3D printed thermoplast osteochondral anchor as an example for a zonal approach to cartilage repair: in vivo performance in a long-term equine model. Biofabrication 12(3):035028

Mancuso P, Raman S, Glynn A, Barry F, Murphy JM (2019) Mesenchymal stem cell therapy for osteoarthritis: the critical role of the cell secretome. Front Bioeng Biotechnol 7:9

Maniwa S, Ochi M, Motomura T, Nishikori T, Chen J, Naora H (2001) Effects of hyaluronic acid and basic fibroblast growth factor on motility of chondrocytes and synovial cells in culture. Acta Orthop Scand 72:299–303

Martínez-Lorenzo MJ, Royo-Cañas M, Alegre-Aguarón E, Desportes P, Castiella T, García-Alvarez F, Larrad L (2009) Phenotype and chondrogenic differentiation of mesenchymal cells from adipose tissue of different species. J Orthop Res 27(11):1499–1507

Marx C, Silveira MD, Selbach I, da Silva AS, Gomes de Macedo Braga LM, Camassola M, Nardi NB (2014) Acupoint injection of autologous stromal vascular fraction and allogeneic adipose-derived stem cells to treat hip dysplasia in dogs. Stem Cells Int 2014:391274

Mathiessen A, Conaghan PG (2017) Synovitis in osteoarthritis: current understanding with therapeutic implications. Arthritis Res Ther 19:18–27

Mauck RL, Yuan X, Tuan RS (2006) Chondrogenic differentiation and functional maturation of bovine mesenchymal stem cells in long-term agarose culture. Osteoarthritis Cartilage 14(2): 179–189

Mazzotti E, Teti G, Falconi M, Chiarini F, Barboni B, Mazzotti A, Muttini A (2019) Age-related alterations affecting the chondrogenic differentiation of synovial fluid mesenchymal stromal cells in an equine model. Cells 8(10):1116

McCorry MC, Bonassar LJ (2017) Fiber development and matrix production in tissue engineered menisci using bovine mesenchymal stem cells and fibrochondrocytes. Connect Tissue Res 58(3–4):329–341

McCorry MC, Puetzer JL, Bonassar LJ (2016) Characterization of mesenchymal stem cells and fibrochondrocytes in three-dimensional co-culture: analysis of cell shape, matrix production, and mechanical performance. Stem Cell Res Ther 12(7):39

McIlwraith CW, Frisbie DD, Rodkey WG et al (2011) Evaluation of intraarticular mesenchymal stem cells to augment healing of microfractured chondral defects. Arthroscopy 27:1552–1561

Messaoudi O, Henrionnet C, Bourge K, Loeuille D, Gillet P, Pinzano A (2021) Stem cells and extrusion 3D printing for hyaline cartilage engineering. Cells 10:2

Miki S, Takao M, Miyamoto W, Matsushita T, Kawano H (2015) Intra-articular injection of synovium-derived mesenchymal stem cells with hyaluronic acid can repair articular cartilage defects in a canine model. J Stem Cell Res Ther 5:11

Mocchi M, Bari E, Dotti S, Villa R, Berni P, Conti V, Del Bue M, Squassino GP, Segale L, Ramoni R, Torre ML, Perteghella S, Grolli S (2021a) Canine mesenchymal cell lyosecretome production and safety evaluation after allogenic intraarticular injection in osteoarthritic dogs. Animals 11:3271

Mocchi M, Grolli S, Dotti S, Di Silvestre D, Villa R, Berni P, Conti V, Passignani G, Brambilla F, Bue MD, Catenacci L, Sorrenti M, Segale L, Bari E, Mauri P, Torre ML, Perteghella S (2021b) Equine mesenchymal stem/stromal cells freeze-dried secretome (lyosecretome) for the treatment of musculoskeletal diseases: production process validation and batch release test for clinical use. Pharmaceuticals (Basel). 14(6):553

Mrugala D, Bony C, Neves N, Caillot L, Fabre S, Moukoko D, Jorgensen C, Noël D (2008) Phenotypic and functional characterisation of ovine mesenchymal stem cells: application to a cartilage defect model. Ann Rheum Dis 67:288–295

Murphy JM, Fink DJ, Hunziker EB, Barry FP (2003) Stem cell therapy in a caprine model of osteoarthritis. Arthritis Rheum 48(12):3464–3474

Music E, Futrega K, Doran MR (2018) Sheep as a model for evaluating mesenchymal stem/stromal cell (MSC)-based chondral defect repair. Osteoarthritis Cartilage 26:730–740

Nam HY, Karunanithi P, Poh Loo WC, Naveen SV, Chen HC, Hussin P, Chan L, Kamarul T (2013) The effects of staged intra-articular injection of cultured autologous mesenchymal stromal cells on the repair of damaged cartilage: a pilot study in caprine model. Arthr Res Ther 15:R129

Nicpoń J, Marycz K, Grzesiak J (2013) Therapeutic effect of adipose-derived mesenchymal stem cell injection in horses suffering from bone spavin. Pol J Vet Sci 16(4):753–754

Noel D, Caton D, Roche S, Bony C, Lehmann S, Casteilla L, Jorgensen C, Cousin B (2008) Cell specific differences between human adipose-derived and mesenchymal-stromal cells despite similar differentiation potentials. Exp Cell Res 314:1575–1584

Ogueta S, Muñoz J, Obregon E, Delgado-Baeza E, García-Ruiz JP (2002) Prolactin is a component of the human synovial liquid and modulates the growth and chondrogenic differentiation of bone marrow-derived mesenchymal stem cells. Mol Cell Endocrinol 190(1-2):51–63

Oh J, Son YS, Kim WH, Kwon O-K, Kang B-J (2021) Mesenchymal stem cells genetically engineered to express platelet-derived growth factor and heme oxygenase-1 ameliorate osteoarthritis in a canine model. J Orthop Surg Res 16(1):43

Okazaki R, Sakai A, Uezono Y, Ootsuyama A, Kunugita N, Nakamura T, Norimura T (2001) Sequential changes in transforming growth factor (TGF)-beta1 concentration in synovial fluid and mRNA expression of TGF-beta1 receptors in chondrocytes after immobilization of rabbit knees. J Bone Miner Metab 19(4):228–235

Olsen A, Johnson V, Webb T, Santangelo KS, Dow S, Duerr FM (2019) Evaluation of intravenously delivered allogeneic mesenchymal stem cells for treatment of elbow osteoarthritis in dogs: a pilot study. Vet Comp Orthop Traumatol 32(3):173–181

Ongaro A, Pellati A, Setti S, Masieri FF, Aquila G, Fini M, Caruso A, De Mattei M (2012) Electromagnetic fields counteract IL-1b activity during chondrogenesis of bovine mesenchymal stem cells. J Tissue Eng Regen Med 9(12):E229–E238

Park Y, Sugimoto M, Watrin A, Chiquet M, Hunziker EB (2005) BMP-2 induces the expression of chondrocyte-specific genes bovine synovium-derived progenitor cells cultured in three-dimensional alginate hydrogel. Osteoarthritis Cartilage 13:527–536

Peach RJ, Hollenbaugh D, Stamenkovic I, Aruffo A (1993) Identification of hyaluronic acid binding sites in the extracellular domain of CD44. J Cell Biol 122(1):257–264

Pedersen NC, Morgan JP, Vasseur PB (2000) Canine rheumatoid arthritis. In: Ettinger SJ, Feldman EC (eds) Textbook of veterinary internal medicine: diseases of the dog and cat, vol 1, 5th edn. W.B. Saunders Company, Philadelphia, p 1879

Pei Y, Fan J-J, Zhang X-Q, Zhang Z-Y, Yu M (2014) Repairing the osteochondral defect in goat with the tissue-engineered osteochondral graft preconstructed in a double-chamber stirring bioreactor. BioMed Res Int 2014:219203

Persson Y, Soderquist L, Ekman S (2007) Joint disorder; a contributory cause to reproductive failure in beef bulls? Acta Vet Scand 49:31

Pezzanite LM, Fortier LA, Antczak DF, Cassano JM, Brosnahan MM, Miller D, Schnabel LV (2015) Equine allogeneic bone marrow-derived mesenchymal stromal cells elicit antibody responses in vivo. Stem Cell Res Ther 6:54

Pigott JH, Ishihara A, Wellman ML, Russell DS, Bertone AL (2013) Inflammatory effects of autologous, genetically modified autologous, allogeneic, and xenogeneic mesenchymal stem cells after intra-articular injection in horses. Vet Comp Orthop Traumatol 26(6):453–460

Port J, Jackson DW, Lee TQ, Simon TM (1996) Meniscal repair supplemented with exogenous fibrin clot and autogenous cultured marrow cells in the goat model. Am J Sports Med 24(4): 547–555

Prins H-J, Fernandes H, Rozemuller H, van Blitterswijk C, de Boer J, Martens ACM (2016) Spatial distribution and survival of human and goat mesenchymal stromal cells on hydroxyapatite and β-tricalcium phosphate. J Tissue Eng Reg Med 10:233–244

Punwar S, Khan WS (2011) Mesenchymal stem cells and articular cartilage repair: clinical studies and future direction. Open Orthop J 5(1):296–301

Qiang Y, Yanhong Z, Jiang P, Shibi L, Quanyi G, Xinlong M, Qun X, Baoshan X, Bin Z, Aiyuan W, Li Z, Wengjing X, Chao Z (2014) Xenoimplantation of an extracellular-matrix-derived, biphasic, cell-scaffold construct for repairing a large femoral-head high-load-bearing osteochondral defect in a canine model. Scientific World Journal 2014:127084

Rackwitz L, Djouad F, Janjanin S, Nöth U, Tuan RS (2014) Functional cartilage repair capacity of de-differentiated, chondrocyte- and mesenchymal stem cell-laden hydrogels in vitro. Osteoarthritis Cartilage 22(8):1148–1157

Raheja LF, Galuppo LD, Bowers-Lepore J, Dowd JP, Tablin F, Yelowley CE (2011) Treatment of bilateral medial femoral condyle articular cartilage fissures in a horse using bone marrow-derived multipotent mesenchymal stromal cells. J Equine Vet Sci 31:147–154

Rakic R, Bourdon B, Demoor M, Maddens S, Saulnier N, Galéra P (2020) Differences in the intrinsic chondrogenic potential of equine umbilical cord matrix and cord blood mesenchymal stromal/stem cells for cartilage regeneration. Sci Rep 10(1):12075

Reesink HL, Sutton RM, Shurer CR et al (2017) Galectin-1 and galectin-3 expression in equine mesenchymal stromal cells (MSCs), synovial fibroblasts and chondrocytes, and the effect of inflammation on MSC motility. Stem Cell Res Ther 8:243

Reich CM, Raabe O, Wenisch S, Bridger PS, Kramer M, Arnhold S (2012) Isolation, culture and chondrogenic differentiation of canine adipose tissue- and bone marrow-derived mesenchymal stem cells—a comparative study. Vet Res Commun 36(2):139–148

Rosa VS, Apolonio EVP, Rossi MC, Constantino LV, de Melo TR, Fonseca-Alves CE, Garcia Alves AL (2021) Encapsulated allogeneic synovial membrane mesenchymal stem cells provide better outcomes of chondral lesions in horses. Stem Cell Res Ther. https://doi.org/10.21203/rs.3.rs-587036/v2

Rose R (1977) An analysis of conditions causing lameness in the horse. In: 54th Annual meeting of the Australian Veterinary Association, pp 99–102

Sabry D, Shamaa A, Amer M, El-Tookhy O, Abdallah A, Abd El Hassib DM, Amer E, Elamir A (2018) The effect of mesenchymal stem cell derived microvesicles in repair of femoral chondral defects in dogs. J Musculoskelet Res 21(2):1850006

Sadeesh EM, Shah F, Yadav PS (2016) Differential developmental competence and gene expression patterns in buffalo (Bubalus bubalis) nuclear transfer embryos reconstructed with fetal fibroblasts and amnion mesenchymal stem cells. Cytotechnology 68(5):1827–1848

Sah RL, Ratcliffe A (2010) Translational models for musculoskeletal tissue engineering and regenerative medicine. Tissue Eng Part B Rev 16(1):1–3

Sanchez-Adams J, Leddy HA, McNulty AL, O'Conor CJ, Guilak F (2014) The mechanobiology of articular cartilage: bearing the burden of osteoarthritis. Curr Rheumatol Rep 16:451–460

Sanjurjo-Rodríguez C, Castro-Viñuelas R, Hermida-Gómez T, Fernández-Vázquez T, Fuentes-Boquete IM, de Toro-Santos FJ, Díaz-Prado SM, Blanco-García FJ (2017) Ovine mesenchymal stromal cells: morphologic, phenotypic and functional characterization for osteochondral tissue engineering. PLoS One 12(1):e0171231

Santos VH, Pfeifer JPH, de Souza JB, Milani BHG, de Oliveira RA, Assis MG, Deffune E, Moroz A, Garcia Alves AL (2018) Culture of mesenchymal stem cells derived from equine synovial membrane in alginate hydrogel microcapsules. BMC Vet Res 14:114

Sasaki A, Mizuno M, Ozeki N, Katano H, Otabe K, Tsuji K, Koga H, Mochizuki M, Sekiya I (2018) Canine mesenchymal stem cells from synovium have a higher chondrogenic potential than those from infrapatellar fat pad, adipose tissue, and bone marrow. PLoS One 13(8):e0202922

Scanzello CR, Goldring SR (2012) The role of synovitis in osteoarthritis pathogenesis. Bone 51: 249–257

Schmidt S, Abinzano F, Mensinga A, Teßmar J, Groll J, Malda J, Levato R, Blunk T (2020) Differential production of cartilage ECM in 3D agarose constructs by equine articular cartilage progenitor cells and mesenchymal stromal cells. Int J Mol Sci 21(19):7071

Shah K, Drury T, Roic I, Hansen P, Malin M, Boyd R, Sumer H, Ferguson R (2018) Outcome of allogeneic adult stem cell therapy in dogs suffering from osteoarthritis and other joint defects. Stem Cells Int 2018:7309201

Simonaro CM, Sachot S, Ge Y, He X, DeAngelis VA, Eliyahu E, Leong D, Sun HB, Mason JB, Haskins ME, Richardson DW, Schuchman EH (2013) Acid ceramidase maintains the chondrogenic phenotype of expanded primary chondrocytes and improves the chondrogenic differentiation of bone marrow-derived mesenchymal stem cells. PLoS One 8(4):e62715

Smith GK, Paster ER, Powers MY, Lawler DE, Biery DN, Shofer FS et al (2006) Lifelong diet restriction and radiographic evidence of osteoarthritis of the hip joint in dogs. J Am Vet Med Assoc. 229:690–693

Somal A, Bhat IA, Baiju I, Pandey S, Panda BSK, Thakur N, Sarkar M, Chandra V, Saikumar G, Sharma GT (2016) A comparative study of growth kinetics, in vitro differentiation potential and molecular characterization of fetal adnexa derived caprine mesenchymal stem cells. PLoS One 11(6):e0156821

Somal A, Bhat IA, Baiju I, Singh AP, Panda BSK, Desingu PA, Pandey S, Bharti MK, Pal A, Saikumar G, Chandra V, Sharma GT (2017) Impact of cryopreservation on caprine fetal adnexa derived stem cells and its evaluation for growth kinetics, phenotypic characterization and wound healing potential in xenogenic rat model. J Cell Physiol 232(8):2186–2200

Song F, Tang J, Geng R, Hu H, Zhu C, Cui W, Fan W (2014) Comparison of the efficacy of bone marrow mononuclear cells and bone mesenchymal stem cells in the treatment of osteoarthritis in a sheep model. Int J Clin Exp Pathol 7:1415–1426

Spaas JH, Oosterlinck M, Broeckx S, Dumoulin M, Saunders J, Van Soom A, Pille F, Van de Walle GR (2012) Treatment of equine degenerative joint disease with autologous peripheral blood-derived mesenchymal stem cells: a case report. Vlaams Diergeneeskundig Tijdschrift 81

Sreekumar TR, Ansari MM, Chandra VG, Sharma T (2014) Isolation and characterization of buffalo Wharton's jelly derived mesenchymal stem cells. J Stem Cell Res Ther 4:207

Stannus O, Jones G, Cicuttini F (2010) Circulating levels of IL-6 and TNF-alpha are associated with knee radiographic osteoarthritis and knee cartilage loss in older adults. Osteoarthtitis Cartilage 18:1441–1447

Sun J, Lyu J, Xing F, Chen R, Duan X, Xiang Z (2020) A biphasic, demineralized, and decellularized allograft bone-hydrogel scaffold with a cell-based BMP-7 delivery system for osteochondral defect regeneration. J Biomed Mater Res 108:1909–1921

Taghavi M, Parham A, Raji A (2020) The combination of TGF-β3 and BMP-6 synergistically promotes the chondrogenic differentiation of equine bone marrow-derived mesenchymal stem cells. Int J Pept Res Ther 26:727–735

Terraciano V, Hwang N, Moroni L, Park HB, Zhang Z, Mizrahi J, Seliktar D, Elisseeff J (2007) Differential response of adult and embryonic mesenchymal progenitor cells to mechanical compression in hydrogels. Stem Cells 25:2730–2738

Teunissen M, Verseijden F, Riemers FM, van Osch GJVM, Tryfonidou MA (2020) The lower in vitro chondrogenic potential of canine adipose tissue-derived mesenchymal stromal cells (MSC) compared to bone marrow derived MSC is not improved by BMP-2 or BMP-6. Vet J 105605

Tiwary R, Amarpal, Aithal HP, Kinjavdekar P, Pawde AM, Singh R (2014) Effect of IGF-1 and uncultured autologous bone-marrow-derived mononuclear cells on repair of osteochondral defect in rabbits. Cartilage 5:43–54

Toh WS, Lim TC, Kurisawa M, Spector M (2012) Modulation of mesenchymal stem cell chondrogenesis in a tunable hyaluronic acid hydrogel microenvironment. Biomaterials 33: 3835–3845

Toh WS, Foldager CB, Olsen BR, Spector M (2013) Basement membrane molecule expression attendant to chondrogenesis by nucleus pulposus cells and mesenchymal stem cells. J Orthop Res 31:1136–1143

Topol L, Chen W, Song H, Day TF, Yang Y (2009) Sox9 inhibits Wnt signaling by promoting beta-catenin phosphorylation in the nucleus. J Biol Chem 284(5):3323–3333

Ude CC, Shamsul BS, Ng MH, Chen HC, Ohnmar H, Amaramalar SN, Ruszymah BHI (2018) Long-term evaluation of osteoarthritis sheep knee, treated with TGF-β3 and BMP-6 induced multipotent stem cells. Exp Gerontol 104:43–51

Vahedi P, Jarolmasjed S, Shafaei H, Roshangar L, Rad JS, Ahmadian E (2019) In vivo articular cartilage regeneration through infrapatellar adipose tissue derived stem cell in nanofiber polycaprolactone scaffold. Tissue Cell 57:49–56

Van Hecke L, Magri C, Duchateau L, Beerts C, Geburek F, Suls M, Da Dalt L, Patruno M, Saunders J, Broeckx S, Depuydt E, Spaas JH (2021) Repeated intra-articular administration of equine allogeneic peripheral blood-derived mesenchymal stem cells does not induce a cellular and humoral immune response in horses. Vet Immunol Immunopathol 239:110306

Varghese S, Hwang NS, Canver AC, Theprungsrikul P, Lin DW, Elisseeff J (2008) Chondroitin sulphate based niches for chondrogenic differentiation of mesenchymal stem cells. Matrix Biol 27:12–21

Vickers SM, Gotterbarm T, Spector M (2010) Cross-linking affects cellular condensation and chondrogenesis in type II collagen-GAG scaffolds seeded with bone marrow-derived mesenchymal stem cells. J Orthop Res 28(9):1184–1192

Vilar JM, Morales M, Santana A, Spinella G, Rubio M, Cuervo B, Cugat R, Carrillo JM (2013) Controlled, blinded force platform analysis of the effect of intraarticular injection of autologous adipose-derived mesenchymal stem cells associated to PRGF-Endoret in osteoarthritic dogs. BMC Vet Res 9:131

Vilar JM, Batista M, Morales M, Santana A, Cuervo B, Rubio M, Cugat R, Sopena J, Carrillo JM (2014) Assessment of the effect of intraarticular injection of autologous adipose-derived mesenchymal stem cells in osteoarthritic dogs using a double blinded force platform analysis. BMC Vet Res 10:143

Vilar JM, Cuervo B, Rubio M, Sopena J, Domínguez JM, Santana A, Carrillo JM (2016) Effect of intraarticular inoculation of mesenchymal stem cells in dogs with hip osteoarthritis by means of objective force platform gait analysis: concordance with numeric subjective scoring scales. BMC Vet Res 12:223

Villatoro AJ, Hermida-Prieto M, Fernández V, Fariñas F, Alcoholado C, Rodríguez-García MI, Mariñas-Pardo L, Becerra J (2018) Allogeneic adipose-derived mesenchymal stem cell therapy in dogs with refractory atopic dermatitis: clinical efficacy and safety. Vet Rec 183(21):654

Vivas D, Caminal M, Oliver-Vila I, Vives J (2018) Derivation of multipotent mesenchymal stromal cells from sheep bone marrow. Curr Protoc Stem Cell Biol 44(2B):9.1–9.22

Wang C, Wang Z, Li A, Bai F, Lu J, Xu S, Li D (2010) Repair of segmental bone-defect of goat's tibia using a dynamic perfusion culture tissue engineering bone. J Biomed Mat Res A 92:1145–1153

Wang M, Peng Z, Vasiliev K (2013) Investigation of wear particles generated in human knee joints using atomic force microscopy. Tribol Lett 51:161–170

Wei FY, Lee JK, Wei L, Qu F, Zhang JZ (2017) Correlation of insulin-like growth factor 1 and osteoarthritic cartilage degradation: a spontaneous osteoarthritis in Guinea pig. Eur Rev Med Pharmacol Sci 21(20):4493–4500

Wei X, Liu B, Liu G, Yang F, Cao F, Dou X, Yu W, Wang B, Zheng G, Cheng L, Ma Z, Zhang Y, Yang J, Wang Z, Li J, Cui D, Wang W, Xie H, Li L, Zhang F, Lineaweaver WC, Zhao D (2019) Mesenchymal stem cell-loaded porous tantalum integrated with biomimetic 3D collagen-based scaffold to repair large osteochondral defects in goats. Stem Cell Res Ther 10(1):72

White JL, Walker NJ, Hu JC, Borjesson DL, Athanasiou KA (2018) A comparison of bone marrow and cord blood mesenchymal stem cells for cartilage self-assembly. Tissue Eng Part A 24(15–16):1262–1272

Whitehouse M, Howells NR, Parry M, Austin E, Kafienah W, Brady K, Goodship AE, Eldridge JD, Blom AYW, Hollander AP (2017) Repair of torn avascular meniscal cartilage using undifferentiated autologous mesenchymal stem cells: from in vitro optimization to a first-in-human study. Stem Cells Transl Med 6(4):1237–1248

Wilke MM, Nydam DV, Nixon AJ (2007) Enhanced early chondrogenesis in articular defects following arthroscopic mesenchymal stem cell implantation in an equine model. J Orthop Res 25:913–925

Wolfe DF (2018) Abnormalities of the bull–occurrence, diagnosis and treatment of abnormalities of the bull, including structural soundness. Animal 1–10

Xue JX, Gong YY, Zhou GD, Wei L, Cao LY, Zhang WJ (2012) Chondrogenic differentiation of bone marrow-derived mesenchymal stem cells induced by acellular cartilage sheets. Biomaterials 33(24):5832–5840

Yamada ALM, Carvalho AD, Moroz A, Watanabe MJ, Hussni CA, Rodrigues CA, Garcia Alves AL (2013) Mesenchymal stem cell enhances chondral defects healing in horses. Stem Cell Discov 3(4):218–225

Yamasaki S, Hashimoto Y, Takigami J, Terai S, Mera H, Nakamura H, Wakitani S (2015) Effect of the direct injection of bone marrow mesenchymal stem cells in hyaluronic acid and bone marrow stimulation to treat chondral defects in the canine model. Regen Ther 2:42–48

Yamasaki A, Omura T, Murata D, Kobayashi M, Sunaga T, Kusano K, Ueno Y, Kuramoto T, Hobo S, Misumi K (2018) A pilot study of regenerative therapy by implanting synovium-derived mesenchymal stromal cells in equine osteochondral defect models. J Equine Sci 29(4): 117–122

Yamazoe K, Mishima H, Torigoe K, Iijima H, Watanabe K, Sakai H, Kudo T (2007) Effects of atelocollagen gel containing bone marrow-derived stromal cells on repair of osteochondral defect in a dog. J Vet Med Sci 69(8):835–839

Yaneselli KM, Kuhl CP, Terraciano PB, de Oliveira FS, Pizzato SB, Pazza K, Magrisso AB, Torman V, Rial A, Moreno M, Llambí S, Cirne-Lima E, Maisonnave J (2018) Comparison of the characteristics of canine adipose tissue-derived mesenchymal stem cells extracted from different sites and at different passage numbers. J Vet Sci 19(1):13–20

Yang Q, Peng J, Lu SB, Guo QY, Zhao B, Zhang L, Wang AY, Xu WJ, Xia Q, Ma XL, Hu YC, Xu BS (2011) Evaluation of an extracellular matrix-derived acellular biphasic scaffold/cell construct in the repair of a large articular high-load-bearing osteochondral defect in a canine model. Chin Med J (Engl) 124:3930–3938

Yang CC, Lin CY, Wang HS, Lyu SR (2013) Matrix metalloproteases and tissue inhibitors of metalloproteinases in medial plica and pannuslike tissue contribute to knee osteoarthritis progression. PLoS One 8(11):e79662

Yin H, Wang Y, Sun Z, Sun X, Xu Y, Li P, Meng H, Yu X, Xiao B, Fan T, Wang Y, Xu W, Wang A, Guo Q, Peng J, Lu S (2016) Induction of mesenchymal stem cell chondrogenic differentiation and functional cartilage microtissue formation for in vivo cartilage regeneration by cartilage extracellular matrix-derived particles. Acta Biomater 33:96–109

Yoshioka M, Shimizu C, Harwood FL, Coutts RD, Amiel D (1997) The effects of hyaluronan during the development of osteoarthritis. Osteoarthritis Cartilage 5:251–260

Zayed M, Caniglia C, Misk N, Dhar MS (2017) Donor-matched comparison of chondrogenic potential of equine bone marrow- and synovial fluid-derived mesenchymal stem cells: implications for cartilage tissue regeneration. Front Vet Sci 3:121

Zayed M, Newby S, Misk N, Donnell R, Dhar M (2018) Xenogenic implantation of equine synovial fluid-derived mesenchymal stem cells leads to articular cartilage regeneration. Stem Cells Int 1073705

Zayed M, Adair S, Dhar M (2021) Effects of normal synovial fluid and interferon gamma on chondrogenic capability and immunomodulatory potential respectively on equine mesenchymal stem cells. Int J Mol Sci 22(12):6391

Zeiter S, Lezuo P, Ito K (2009) Effect of TGF-β1, BMP-2 and hydraulic pressure on chondrogenic differentiation of bovine bone marrow mesenchymal stromal cells. Biorheology 46:45–55

Zhai C, Zhang X, Chen J, He J, Fei H, Liu Y, Luo C, Fan W (2019) The effect of cartilage extracellular matrix particle size on the chondrogenic differentiation of bone marrow mesenchymal stem cells. Regen Med 14(7):663–680

Zhang XL, Jiang XQ (2006) Osteoblastic differentiation of goat bone marrow stromal cells after AdBMP-2 mediated transduction in vitro. Shanghai Kou Qiang Yi Xue 15(6):610–613

Zhang X, Zhai C, Fei H, Liu Y, Wang Z, Luo C, Zhang J, Ding Y, Xu T, Fan W (2018a) Composite silk extracellular-matrix scaffolds for enhanced chondrogenesis of mesenchymal stem cells. Tissue Eng Part C Methods 24(11):645–658

Zhang Y, Liu S, Guo W, Wang M, Hao C, Gao S, Zhang X, Li X, Chen M, Jing X, Wang Z, Peng J, Lu S, Guo Q (2018b) Human umbilical cord Wharton's jelly mesenchymal stem cells combined with an acellular cartilage extracellular matrix scaffold improve cartilage repair compared with microfracture in a caprine model. Osteoarthritis Cartilage 30077(18):S1063–S4584

Zhang Y, Hao C, Guo W, Peng X, Wang M, Yang Z, Li X, Zhang X, Chen M, Sui X, Peng J, Lu S, Liu S, Guo Q, Jiang Q (2020) Co-culture of hWJMSCs and pACs in double biomimetic ACECM oriented scaffold enhances mechanical properties and accelerates articular cartilage regeneration in a caprine model. Stem Cell Res Ther 11(1):180

Zhao W, Zou T, Cui H, Lv Y, Gao D, Ruan C, Zhang X, Zhang Y (2020) Parathyroid hormone (1-34) promotes the effects of 3D printed scaffold-seeded bone marrow mesenchymal stem cells on meniscus regeneration. Stem Cell Res Ther 11:328

Zhou G, Zheng Q, Engin F, Munivez E, Chen Y, Sebald E, Krakow D, Lee B (2006) Dominance of SOX9 function over RUNX2 during skeletogenesis. Proc Natl Acad Sci U S A 103(50): 19004–19009

Zhu S, Zhang B, Man C, Ma Y, Hu J (2011) NELL like molecule-1 modified bone marrow mesenchymal stem cells/poly-lactic-co-glycolic acid composite improves repair of large osteochondral defects in mandibular condyle. Osteoarthritis Cartilage 19:743–750

Zhu C, Wu W, Qu X (2021) Mesenchymal stem cells in osteoarthritis therapy: a review. Am J Transl Res 13(2):448–461

Zorzi AR, Amstalden EMI, Plepis AMG, Martins VCA, Ferretti M, Antonioli E, Duarte ASS, Luzo ACM, Miranda JB (2015) Effect of human adipose tissue mesenchymal stem cells on the regeneration of sheep articular cartilage. Int J Mol Sci 16:26813–26831

Zscharnack M, Poesel C, Galle J, Bader A (2009) Low oxygen expansion improves subsequent Chondrogenesis of ovine bone-marrow-derived Mesenchymal stem cells in collagen type I hydrogel. Cells Tissues Organs 190:81–93

Zscharnack M, Hepp P, Richter R, Aigner T, Schulz R, Somerson J, Josten C, Bader A, Marquass B (2010) Repair of chronic osteochondral defects using predifferentiated mesenchymal stem cells in an sheep model. Am J Sports Med 38(9):1857–1869

Mesenchymal Stem Cells Therapeutic Applications in Bone Regeneration

2

Abstract

Bone has good innate healing potential but faces increased morbidity, and in extensive bone damage, infections, or in systemic affections, the said potential is delayed (delayed union) and/or even inhibited (non-union). To boost bone regeneration timely and effectively, regenerative medicine is being evaluated. Mesenchymal stem cells (MSCs) along with scaffolds are being evaluated with both in vitro and in vivo studies demonstrating their potential bone-healing properties. In vitro studies mainly focus on their differentiation capacity towards the osteogenic differentiation lineage through culture incorporated growth factors, matrices/scaffolds and/or mechanical factors. In vivo studies (experimental and clinical) demonstrate tissue-engineered constructs employing MSCs as potential candidates to enhance regeneration of bone defects in various kinds of mechanodynamic models and clinical conditions. MSCs transplantation has mostly been made along with the suitable scaffolds to fill the gap and provide a suitable environment besides maintaining the cells in situ. The current chapter focuses on the bone regenerative potential of MSCs along with the scaffolds in veterinary species.

Keywords

Bone fracture · Mesenchymal stem cell · Osteoporosis · Experimental model · Clinical cases

2.1 Introduction

Bone tends to heal fast due to its innate expanding regeneration potential. Such a natural healing power produces tissue that is comparable to original tissue in its biochemical and biomechanical properties (Taylor et al. 2007). In general, bone

M. B. Gugjoo, *Therapeutic Applications of Mesenchymal Stem Cells in Veterinary Medicine*, https://doi.org/10.1007/978-981-19-3277-9_2

tissue comprises two parts viz. compact cortical and cancellous/spongy trabecular parts. Cortical bone tissue present in long bone shaft forms 80% (majority), while trabecular bone tissue present at ends represents 20% of the bone. In addition to the long bones, vertebrae too comprise trabecular tissue (Tanna 2004; Schindler 2011; Krych and Stuart 2014). The trabecular tissue harbours active bone marrow (BM) containing stem cells like hematopoietic and mesenchymal stem cells. Bone healing is a proliferating type, the remodelling of which is a dynamic and lifelong process (Zaffagnini et al. 2007).

Bone affections form one of the leading causes of disability that decreases man-hour and poses adversity in the quality of life. Numerous affections like infection, extensive bone damage, old age fractures, tumours, osteoporosis and/or metabolic diseases hamper normal bone healing process to ensue (Marycz et al. 2016; Gugjoo et al. 2020a, b). Additionally, some of the bones like tibia have limited innate healing potential as compared to the others like femur (Lovati et al. 2016). Among various causes, fracture accounts for the major bone affections in veterinary species. Fractures in the case of ruminant account for about 10% of the caseload at large animal referral centres (Lozier et al. 2018). In large animals like horse and cattle, fracture management is challenging due to their heavy weight (lack of sufficient strength in implants to bear the weight) and limited control in their movement, especially horses that also burdens contralateral joints as well. This tends to increase the frequency of non-union. In dogs, fracture non-union rate is high, amounting to 3.4% (Millis and Jackson 2003). Bone disorders like osteoporosis and tumours tend to increase steeply, especially in old aged population with poor physical activity and obesity (Amini et al. 2012; Chandran et al. 2017). Fracture fixation of osteoporotic bone poses a challenge to the orthopaedician as the bone undergoes excessive demineralization by the osteoclast cells. This leads to weak osteointegration and finally implant failure (Chandran et al. 2017); thus, an early and effective bone repair is highly desired. Further, patients suffering from degenerative disc disease or spondylolisthesis have a spinal pain. As such spinal fusion surgery to achieve arthrodesis is being performed (O'Brien et al. 2002). However, non-union remains a significant clinical challenge and may lead to added morbidity and require additional surgical interventions (Wheeler et al. 2014).

2.2 Why Mesenchymal Stem Cells for Bone Regeneration

Considering limitations and associated complications in current techniques of fracture management and other bone affections and their impact on bone healing, there is an increased demand for advanced therapeutic options that can cater to the demand in veterinary orthopaedic medicine (Wang et al. 2010). The critical size defects created in the fracture ends make natural bone healing almost impossible. Normally in fracture healing, osteogenesis is governed by the clot formation (acting as a provisional matrix) through which stem cells like mesenchymal stem cells (MSCs) migrate and lead to the osteosynthesis (Gaharwar et al. 2011). It is therefore believed that extrinsic MSCs may enhance bone healing through secretome-containing microvesicles or secretory factors like cytokines, growth factors and extracellular matrix molecules, besides their currently questioned differentiation into the actual bone secreting cells (osteoblasts). The paracrine action modulates endogenous cells to migrate to the area, control inflammation, immuno-modulate, promote angiogenesis and resident cell differentiation. Finally secreted bone tissue is modulated to ensure typical bone regeneration (Benavides-Castellanos et al. 2020; Gugjoo et al. 2020a, b; Re et al. 2021). MSCs may promote early new bone tissue formation and angiogenesis without significant inflammatory cells. MSCs osteogenic differentiation increases the expression levels of osteogenic markers like alkaline phosphatase (ALP), collagen I (Col I*a*2), osteocalcin (OCN), osteopontin (OPN), runt-related transcription factor 2 (RUNX2) and angiogenic genes like vascular endothelial growth factor-A (VEGF-A), angiopoietin (ANG)-1 and ANG-2 (Katagiri et al. 2017; Benavides-Castellanos et al. 2020; Re et al. 2021). Furthermore, their immune-evasive properties and almost nil ethical and teratogenic issues give them an edge over other stem cells like embryonic stem cells or induced pluripotent stem cells (Gugjoo et al. 2018, 2019, 2020a, b). These diverse properties of MSCs make them the potential therapeutic candidates in bone regeneration and are being extensively studied under regenerative medicine (Chandran et al. 2017).

2.3 In Vitro Osteogenesis of MSCs

MSCs may differentiate into osteogenic lineages provided relevant cues elicit from the available environment. In rodents such a differentiation may occur spontaneously, but in higher species like goats and other species it may require specific osteogenic induction (Huang et al. 2013). These species-wise differences may arise due to the imprinting on MSCs by the tissue-specific environmental cues (Seib et al. 2009). Usually in vitro studies employ commercially available medium (containing growth factors among other osteogenic factors) to facilitate MSCs differentiation to osteogenic lineage. However, various other factors including matrices or growth factors, chemicals or cells, in addition to mechanical forces, either alone or in combination, also direct such a differentiation. Furthermore, in vitro culture techniques like 3D spheroids also affect MSCs (like in horse AD-MSCs) bone morphogenesis-inducing factors and differentiation potential (Park et al. 2018). However, subpopulations of MSCs (like equine BM-MSCs and muscle MSCs) sorted with non-equilibrium gravitational field-flow are not created equally and tend to show differences in osteogenic potential (Radtke et al. 2015).

In general, in vitro MSCs osteogenesis is depicted by their ability to lose typical fibroblast morphology, deposit calcium, show higher alkaline phosphatase activity and express markers of both early- and late-phase osteogenesis like RUNX2, collagen I, osterix, osteonectin and osteopontin. The cellular osteogenic differentiation under osteoinduced medium occurs mostly by 3 weeks, although such a time-specific potential may vary depending upon the physiological phase of the donor (Dar et al. 2021) and with extensive passaging (Elkhenany et al. 2016).

2.3.1 Growth Factors and MSCs Osteogenesis

Under the natural body system, growth factors are important to regulate osteogenesis. Advances in developmental biology provide insights into growth and maintenance of bone microenvironment which help in understanding the role of growth factors or other chemicals in MSCs differentiation. Growth factors like basic fibroblast growth factor (FGF), bone morphogenetic protein (BMP)-2, -7 and osteogenic growth peptide (OGP) and chemicals like dexamethasone, ascorbic-acid-2-phosphate, monosodium phosphate, β-glycerolphosphate, 1,25-dihydroxyvitamin-D3 (1,25(OH)2D3) favour MSCs osteogenesis (Jäger et al. 2006; Tjabringa et al. 2008; Zhu et al. 2008; Huang et al. 2013; Kalaszczynska et al. 2013; Randau et al. 2013; Bottagisio et al. 2015; Hu et al. 2015; Elkhenany et al. 2016; Dar et al. 2021).

A short exposure as low as 15-min incubation in growth factors like BMP-2 (10 ng/mL) may be sufficient to promote goat AD-MSCs osteogenic expression (Knippenberg et al. 2006; Zhang and Jiang 2006). In this regard, some of the factors may be more essential than others. Equine BM-MSCs differentiate into osteoblasts (alkaline phosphatase activity) with culture incorporation of the dexamethasone rather than BMP-2. Even dexamethasone-induced osteogenic differentiation decreases the expression of T-Box transcription factor 3 (TBX3), a factor commonly associated with diseases like ulnar-mammary syndrome (Glynn et al. 2013). This is in contrast to the study that induces the osteogenic differentiation of equine BM-MSCs through BMP-2 transfection (Carpenter et al. 2010). BMP-2 transfection induces ovine BM-MSCs osteogenesis (Blum et al. 2004), while culture addition of this factor fails to induce osteogenic matrix deposition (Egermann et al. 2006). Thus, techniques of incorporating osteoinduction factors may have a significant effect on the outcome. Tri-iodothyronine (T3) in the absence of dexamethasone induces terminal osteogenic differentiation in the chondrogenic induced MSCs. Such an exposure increases collagen X but not necessarily other features of the actual hypertrophic chondrocyte. The differential effect of T3 on collagen type II and col X gene expression of chondrogenic induced MSCs and chondrocytes indicates that there exist variable developmental stages (Randau et al. 2013).

It may be noted here that under in vitro conditions, β-glycerolphosphate may not be a suitable phosphate ion source for ovine AD- and BM-MSCs. Sodium dihydrogen phosphate (NaH_2PO_4) appears to improve their ability to secrete alkaline phosphatase and this is in contrast to that of the human MSCs (Kalaszczynska et al. 2013). Further exploration shows that calcium secretion may increase in these ovine cells with the NaH_2PO_4 but with poor osteogenic marker (RUNX2, COL1A1 and osteonectin) expression (Bottagisio et al. 2015). Thus, ovine MSCs may be immune to normally used chemicals for osteogenic induction and mineralization status may be sometimes misleading to confirm the actual osteogenesis of the MSCs. Therefore, further in-depth understanding in the area is required especially in relation to species-wise regulation of osteogenesis, osteoinducing chemicals/growth factors and their technique of application.

2.3.2 Scaffold and MSCs Osteogenesis

Scaffolds resembling bone extracellular matrices are able to activate the body's physiological mechanisms for bone matrix deposition and its integration to the innate bony tissues. As such teh scaffolds are being evaluated for bone tissue engineering. These scaffolds regulate MSCs differentiation towards osteogenesis although a complete understanding in the mechanisms involved remains to be understood (Pereira-Junior et al. 2013; Bwalya et al. 2017; Das et al. 2017; Fang et al. 2017; Trindade et al. 2017). Various scaffolds favour osteogenesis of MSCs derived from veterinary species. These include silica-coated bioactive ceramic, hydroxyapatite and hyaluronic acid, starch and poly-caprolactone (SPCL), porous tantalum, zinc silicate mineral, oxidized graphene, biodegradable calcium-deficient hydroxyapatite (CdHA) tubular bacterial cellulose (BC-TS) composites and synthesized methacrylate-endcapped caprolactone and nano-bioactive glass ceramic-coated poly(L-lactic acid) (PLLA) scaffolds (Ivirico et al. 2009; Nair et al. 2009a, b; Hwang et al. 2011; Favi et al. 2014; Rodrigues et al. 2014; Elkhenany et al. 2015; Mahdavi et al. 2017; Bageshlooyafshar et al. 2019; Wei et al. 2019). Hormones like tri-iodothyronine (T3) add-on to the porous calcium polyphosphate may further promote ovine BM-MSCs osteogenesis but simultaneously the chondrogenic expression as well (Lee et al. 2015).

These scaffolds should be bio-active; however, coating or adding bio-active agents on bio-inert may also increase MSCs differentiation potential. Coating of osteoinductive matrix like hydroxyapatite (HA) over the bio-inert hydrogels like poly ethylene glycol (PEG) enhances goat BM-MSCs osteogenesis in the osteogenic induction medium (Hwang et al. 2011). This osteogenic induction potential of HA is further improved with the silica coating (Nair et al. 2009a, b), that improves irreversible adsorption of serum proteins on MSCs. This is attributed to high negative-surface-surface charge elicited by silanols (Si-OH) (Lobel and Hench 1998). Coating of nano-bioactive glass ceramic and zinc silicate mineral onto the poly(L-lactic acid) (PLLA) improves osteogenic environment for equine AD-MSCs depicted by their significant increase in ALP activity, calcium content and bone-related osteogenic gene expression (Mahdavi et al. 2017; Bageshlooyafshar et al. 2019). Calcium phosphate cements prepared with sintering does not change pH and ionic concentrations while non-sintered calcium phosphate does that has an adverse effect on the stem cells (Zou et al. 2012).

Some of the scaffolds may show little potential to inhibit MSCs chondrogenesis like alginate as compared to the others like Polysulphated glycosaminoglycan (PSAG). Canine BM-MSCs cultured in alginate beads may suppress SOX9 but still show chondrogenic phenotype (express *collagen type II* and *aggrecan*). PSAG shows dose reversal effect on hypertrophic chondrocyte formation. Pentosan polysulphate (PPS) also inhibits hypertrophic chondrocyte formation and thereby MSCs osteogenesis (Bwalya et al. 2017). PPS is a novel inhibitor of canine osteoclastogenesis and mediates this action through suppression of key transcription factors such as nuclear factor of activated T-cells cytoplasmic 1 (NFATc1) and c-Fos (Wijekoon et al. 2018).

Scaffold properties like fibre-cell orientation and alignment, pore size and density, biodegradability, cross-linking and concentration govern MSCs differentiation and their osteosynthesis. The scaffold design is affected by the technique of fabrication that affects MSCs growth and differentiation. Some of the scaffolds may allow cellular growth and differentiation at specific concentrations. Among various scaffolds like electrochemically assembled collagen, monomeric collagen, fibrilized rat tail collagen and fibrin glue, electrochemically aggregated collagen gives the highest cell viability and best osteogenic differentiation (Nino-Fong et al. 2013). Likewise, goat MSCs show good growth and osteogenic differentiation over 70% and 100% poly(CLMA-co-HEA) that have comparable pore size distribution, interconnectivity and behaviour. However, 70% may be superior over 100% concentration as it allows adequate hydrophilicity to enhance the transport of nutrients. Further increase in scaffold hydration may suppress osteogenesis of goat BM-MSCs as lower alkaline phosphatase activity exists with greater scaffold hydration (Ivirico et al. 2009). Small pore-sized hydroxyapatite may be better able to promote the goat BM-MSCs osteogenesis as compared to the dense and large size hydroxyapatite particles (Fischer et al. 2004). On tricalcium phosphate (TCP), goat BM-MSCs remain interconnected and aligned while the same cells remain scattered on hydroxyapatite (Prins et al. 2016).

Biodegradability of scaffold has a direct bearing on cell viability as the change in pH may lead to adjacent cellular senescence (Liu 2011; Johnson et al. 2012). Among various chitosan synthesis techniques, freeze gelation and particle leaching techniques appear effective to effectuate goat BM-MSCs osteogenesis (García Cruz et al. 2010). Crosslinking of natural scaffold like small intestine submucosa and goat acellular lung scaffold with quercetin and nano-hydroxyapatite favour osteogenesis of goat BM-MSCs (Li et al. 2006; Gupta et al. 2017). Demineralized bone matrix promotes better dog MSCs proliferation and osteogenesis as compared to the mineralized bone matrix (Jo et al. 2017). Thus, a cell-matrix/scaffold interaction has a specific role in determining the MSCs differentiation and is open for insights into the mechanisms involved. There has to be insights into the cell surface receptors and associated pathways that respond to the signals arising from the microenvironment.

2.3.3 Mechanical Factors

MSCs osteogenesis may be effectuated through mechanical forces (Versari et al. 2007). Pulsatile fluid flow and extracorporeal shockwave (ESW) therapy favour goat AD-MSCs (Knippenberg et al. 2005) and BM-MSCs (Tan et al. 2017) osteogenesis, respectively. BM-MSCs osteogenesis might possibly occur through adenosine release regulation and A2B receptor activation (Tan et al. 2017). However, an equine study shows differential actions of BM-MSCs on exposure to ESW. A higher production of alkaline phosphatase but with a simultaneous decrease in expression of VEGF and TGF-β of the cells occurs (Colbath et al. 2020).

2.3.4 Tissue Source and MSCs Osteogenesis

In vitro studies demonstrate that MSCs quality and quantity varies with the source (Hendawy et al. 2021). The differentiation potential including osteogenesis varies with tissue source and species. Some sources in a particular species harbour MSCs with a higher preponderance towards osteogenesis. Equine BM-MSCs carry higher osteogenic (Cbfa1, Osx and Omd and staining for ALP activity and calcium deposition) potential as compared to the AD-MSCs and UC-MSCs (Toupadakis et al. 2010). Other similar type equine studies also demonstrate a higher osteogenic potential of BM-MSCs as compared to the periosteal MSCs, AD-MSCs and muscle MSCs (Nino-Fong et al. 2013). Dog AD-MSCs and UCB-MSCs show greater in vitro osteogenesis (degree of Alizarin Red S staining and ALP levels) than BM-MSCs and WJ-MSCs (Kang et al. 2012). Such differences need further transcriptional and translational studies to comprehend the reasons behind such variations in order to utilize MSCs effectively in clinical conditions. One of the reasons may the internal priming of the cells under in vivo systems.

2.3.5 Miscellaneous Factors and MSCs Osteogenesis

Ailments like osteoporosis and equine metabolic syndrome (EMS) impair MSCs osteogenesis. Reduced goat BM-MSCs osteogenic potential in osteoporosis may be improved with specific scaffolds like beta tri-calcium phosphate (β-TCP) (Cao et al. 2012). In EMS, impairments in mitochondrial biogenesis and function affect AD-MSCs osteogenesis (Marycz et al. 2016). Contrary to normal bone regeneration, in osteosarcoma, commonly seen in dogs, there is a need for aberrant proliferation inhibition of MSCs. Dog BM-MSCs culture implantation of rhBMP-2 inhibits the latter's proliferation and osteoinduction and might occur through apoptosis and tumour suppression mediated by p53 (Rici et al. 2012). This occurs contrary to that reported commonly. Aged dogs harbour BM-MSCs with reduced osteogenesis (Volk et al. 2012). Platelet-rich plasma is synergistic with goat BM-MSCs growth but inhibits their osteogenesis (Cheng et al. 2007). Thus, addressal of actual problem may help in enhancing osteogenesis of MSCs.

2.4 Fracture Healing (FH) Model Study

To evaluate MSCs role in fracture healing, FH model (mixing and coagulating MSCs and blood cells) demonstrates that MSCs survive longer in the coagulating blood (hematoma). In normoxia culture conditions, equine MSCs survive for at least 12 h, while the frequency of immune cells is quite reduced. In contrast, hypoxia-cultured MSCs survive for 72 h with an enhanced expression of fracture healing-relevant genes. The immune cell frequency is better under the normoxia culture than the hypoxia (Pfeiffenberger et al. 2019). Thus, MSCs appear to be crucial during the initial phase of fracture healing and its surrounding environment determines its efficacy.

MSCs from various sources and as per the induction agents vary in pathways towards the osteogenesis. Apart from the above-mentioned pathways, other pathways also accentuate MSCs osteogenesis. Functional assay and hierarchical clustering for tracking protein dynamic change demonstrate that osteogenesis of dog BM-MSCs requires the presence of Wnt, transforming growth factor (TGF)-β and bone-morphogenetic protein (BMP) signalling contrary to that of the DP-MSCs that mainly rely on BMP signalling (Nantavisai et al. 2020). Osteogenesis of goat BM-MSCs is mediated through p38 mitogen-activated protein kinase (MAPK) pathway while p44/42 MAPK pathway is essential in goat AD-MSCs (Elkhenany et al. 2016). Osteogenic inductors also affect the pathway involved. Vitamin D3 activates polyamine metabolism of goat AD-MSCs for osteogenesis. During the course, gene expression of the polyamine-modulated transcription factor-1 (PMF-1) and spermidine/spermine N (1)-acetyltransferase (SSAT) gets activated (Tjabringa et al. 2008). These polyamines stimulate tyrosine phosphorylation by tyrosine kinases, mitogen-activated protein kinases (MAPKs) and c-myc and increase the expression of the transcription factors c-fos and c-jun depending on the type of polyamine (Seiler and Raul 2005). Cattle BM-MSCs osteogenesis with osteogenic growth peptide may be mediated through endogenous nitric oxide synthase (NOS) pathway (Zhu et al. 2008). Thus, there are varied pathways in MSCs osteogenesis depending upon the source of cells and osteoinduction agents.

2.5 In Vivo Osteogenic Studies

Numerous in vivo osteogenic tissue engineering studies involving MSCs along with scaffolds are being conducted in veterinary species involving sheep, goat, dog and horse. Sheep and goat are utilized as translational models for humans while the dogs and horses are made subjects for experimental as well as clinical studies. Translational studies on dogs and horses, in addition to the sheep and goats potentially provide MSCs proof-of-therapeutic principle for orthopaedic regenerative medicine. The clinical studies demonstrate the potential feasibility, utility and efficacy of MSCs in bone fractures in veterinary species. In bone regenerative medicine, MSCs are essentially supported by scaffolds to maintain their in situ availability, prevent their death and enhance their smooth and slow release and the function (Fig. 2.1). Various conventional scaffolds are being evaluated; however, recent 3D bioprinted scaffolds seem to carry an edge in harnessing stem cell potential for bone healing. These 3D bioprinted scaffolds are designed to replicate actual bone structure scanned from CT or MRI of patients (Szivek et al. 2018). This field is in infancy but growing and in vivo experimental or clinical studies on such scaffolds are currently very much lacking. The below given studies are discussed to evaluate the bone regeneration potential of the MSCs with or without scaffolds.

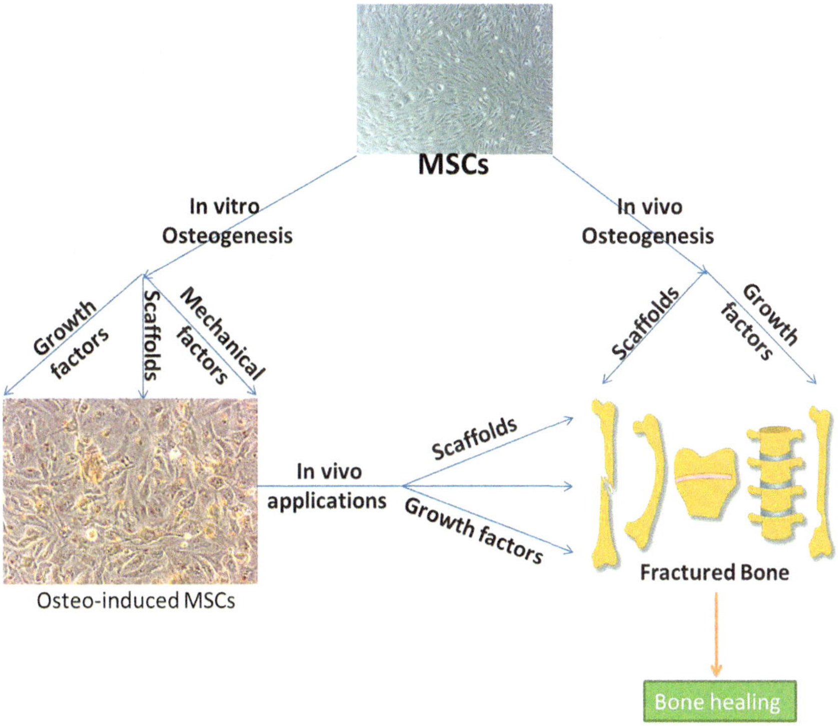

Fig. 2.1 Bone tissue engineering involving mesenchymal stem cells (MSCs), scaffolds and growth factors to enhance osteogenesis. Ex vivo osteoinduced MSCs also contribute to the osteosynthesis in one tissue engineering

2.5.1 Sheep MSCs Osteogenic Studies

MSCs osteogenic potential is being evaluated in various bone defect models including tibia, metatarsus, cranium, spinal fusion, femur, sinus, mandible and parietal bones (Table 2.1). These studies are mostly conducted as placebo-controlled experimental studies comparing osteogenic potential of cell/scaffold assembly against the scaffold and/or control. It is worth mentioning that MSCs characterization is not well elaborated in many of these studies.

2.5.1.1 Tibial Defect Model

Segmental tibial defects ranging from 3.0 to 3.5 cm are mostly evaluated with one of the studies performing trephined holes (8 mm diameter and 4 mm depth) (Lovati et al. 2016). These defects are stabilized with the plates (Niemeyer et al. 2010; Berner et al. 2015), interlocking nails (Field et al. 2011), external skeletal fixation (Kon et al. 2000; Fernandes et al. 2014; Herten et al. 2019) or combination of plate and external skeletal fixation (Giannoni et al. 2008). These studies are mostly randomized placebo control and involve varying number of animals ranging from

Table 2.1 Sheep in vivo MSCs preclinical experimental studies on various bone defects or spinal fusions

Study design/period	Biomaterial used and cell dose	Evaluation criteria	Overall result	References
Tibial bone defect models				
Randomized placebo controlled ($n = 4$)/2 months	Autologous BM-MSCs ($0.5–1.0 \times 10^8$ in fibrinogen), hydroxyapatite ceramic (HAC)	Radiography, histology, indentation assay	Improved outcome	Kon et al. (2000)
Randomized placebo controlled ($n = 45$)/24 weeks	Autologous BM-MSCs ($70–100 \times 10^6$), silicon-stabilized tricalcium phosphate	Micro-radiography, histology and vascular density (VD)	Improved outcome	Giannoni et al. (2008)
Randomized placebo-controlled comparative study ($n = 21$)/6 months	Human BM-MSCs and sheep BM-MSCs (2×10^7), mineralized collagen scaffold (matrices)	Radiology and histology	Improved outcome	Niemeyer et al. (2010)
Randomized placebo controlled ($n = 24$)/9 months	BM-MSCs (225×10^6), collagen and biphasic calcium phosphate ceramic	Radiology, CT, biomechanical, histology	Improved outcome	Field et al. (2011)
Randomized placebo-controlled comparative study/ ($n = 24$)/12 weeks	BM-MSCs (35×10^6), scaffold (polycaprolactone + β-tricalcium phosphate)	Radiology, microcomputed tomography (micro-CT), biomechanical testing, histology	Improved outcome	Berner et al. (2013)
Randomized placebo controlled ($n = 8$)/18 weeks	Autologous BM-MSCs (6×10^7) allogeneic graft	Radiography, CT and histology	Improved outcome	Fernandes et al. (2014)
Randomized placebo controlled/($n = 24$)/12 months	Allogeneic BM-MSCs (100×10^6), PCL/HA (4 weeks after defect surgery)	Biomechanical testing and micro-CT, histology and immune-histochemistry, scanning electron microscopy	Improved outcome	Berner et al. (2015)
Randomized placebo controlled ($n = 10$)/4 months	Hydroxyapatite and autologous osteo-differentiated BM-MSCs (2×10^6 seeded on scaffold)	Macroscopy, radiography, micro-CT and histology	Improved outcome	Lovati et al. (2016)
Randomized placebo controlled (n = 10)/6 months	Autologous UCB-MSCs (2×10^7), hydroxyapatite	Histomorphometric	Improved outcome	Herten et al. (2019)
Spinal fusion models				
Randomized placebo controlled ($n = 24$)/6 months	Bone marrow (selective cell retention technology), tricalcium phosphate	Micro-CT, manual palpation and histology	Improved outcome	Gupta et al. (2007)

Randomized placebo controlled ($n = 18$)/3 months	Allogeneic MSCs (1×10^6), gelfoam sponge, pentosan polysulphate (PPS)	Radiology and histology	Improved outcome	Goldschlager et al. (2010)
Randomized placebo controlled ($n = 30$)/3 months	Allogeneic BM-MSCs (5×10^6 & 10×10^6), hydroxyapatite/tricalcium phosphate (HA/TCP)	CT, functional radiography, histomorphological, biomechanical analysis	Improved outcome	Goldschlager et al. (2011)
Randomized placebo controlled ($n = 34$)/6 months	Osteo-induced BM-MSCs (50×10^6), scaffolds, mineral scaffold	Clinical observation, radiography, CT, histology, histomorphometry	Improved outcome	Cuenca-López et al. (2014)
Randomized placebo controlled ($n = 6$)/12 weeks	BM-MSCs ($5–6 \times 10^7$), hydroxyapatite, tricalcium phosphate	Radiography, manual palpation, histology, scanning electron microscopy	Improved outcome	Shamsul et al. (2014)
Randomized placebo controlled ($n = 36$)/16 weeks	Allogeneic BM-MSCs (2.5×10^6; 6.25×10^6; 12.5×10^6), scaffold (hydroxyapatite, tricalcium phosphate and type I collagen)	Plain radiography, CT, histopathology, histomorphometric and biomechanical competency	Improved outcome	Wheeler et al. (2016)
Metatarsal bone defect models				
Randomized placebo controlled ($n = 21$)/6 months	Autologous BM-MSCs ($8.28 \pm 1.32 \times 10^6$), coral pieces ($138 \pm 22$)	Radiology, histology and CT	Improved outcome	Viateau et al. (2007)
Randomized placebo controlled ($n = 10$)/6 months	BM-MSCs ($7.5 \pm 1.2 \times 10^6$), Acropora coral scaffold	Micro-CT and histology	Improved outcome	Manassero et al. (2013)
Femoral bone defect models				
Randomized placebo controlled ($n = 12$)/16 weeks	β-TCP, BM-MSC-derived osteoblasts (2×10^6 loaded on β-TCP)	Radiography, histology	Improved outcome	Li et al. (2014)
Randomized placebo controlled ($n = 4$)/6 months	AD-MSCs (2×10^6), bio-printed polybutylene terephthalate coated with micron-sized tricalcium phosphate particles	Radiographs and micro-CT scanning	Improved outcome	Szivek et al. (2018)
Parietal bone defect model				
Randomized placebo controlled ($n = 8$)/18 weeks	Autologous BM-MSCs and calcium alginate	Gross, histological and computerized tomography (CT) scan	Improved outcome	Shang et al. (2001)

(continued)

Table 2.1 (continued)

Study design/period	Biomaterial used and cell dose	Evaluation criteria	Overall result	References
Randomized placebo controlled ($n = 5$)/12 weeks	Autologous BM-MSCs/ polyetherketoneketone (PEKK) 3D scaffolds	Micro-computed tomography imaging, histology	No observed improvement	Adamzyk et al. (2016)
Mandibular bone defect model				
Randomized placebo-controlled comparative study ($n = 15$)/32 weeks	Autologous BM-MSCs ($25–30 \times 10^6$) and autologous serum scaffold	Computed and micro-CT	Improved outcome	Gallego et al. (2015)
Sinus augmentation model				
Randomized placebo-controlled comparative study ($n = 16$)/16 weeks	BM-MSCs, bone marrow aspirate, bovine bone mineral (BBM)	Histological and histomorphometrical, CT	Improved outcome	Ardjomandi et al. (2015)
Osteoporotic bone defect model				
Randomized placebo controlled ($n = 8$)/10 months	AD-MSCs, strontium hydroxyapatite (SrHA)	Histological, histomorphometry, micro-CT and elemental analysis	Improved outcome	Chandran et al. (2017)

MSCs mesenchymal stem cells, *AD* adipose tissue derived, *BM* bone marrow derived, *UCB* Umbilical cord blood derived, *PCL/HA* polycaprolactone-hydroxyapatite, *β-TCP* beta tricalcium phosphate, *n* total number of animals in all the groups of the study; Improved outcome: Indicates improved outcome with MSCs transplantation even if only one parameter improves; Study period: Indicates maximum time period for which study is conducted

$n = 04$ (Kon et al. 2000) to $n = 45$ (Giannoni et al. 2008). Among all the sources, allogenic or autologous bone marrow-derived MSCs (BM-MSCs) are utilized with concentrations varying from as low as 20 million (Niemeyer et al. 2010) to as high as 225 million (Field et al. 2011). Various scaffolds including synthetic ones like hydroxyapatite ceramic (HAC) (Kon et al. 2000; Lovati et al. 2016), silicon-stabilized tricalcium phosphate (Giannoni et al. 2008), mineralized collagen scaffold (Niemeyer et al. 2010), collagen and biphasic calcium phosphate ceramic (Field et al. 2011), polycaprolactone and β-tricalcium phosphate (Berner et al. 2013), polycaprolactone-hydroxyapatite scaffold (PCL/HA) (Berner et al. 2015) and natural allogeneic (Fernandes et al. 2014) or autologous (Berner et al. 2013; Berner et al. 2015) grafts are utilized (Table 2.1). The study period varies from 2 months (Kon et al. 2000) to 9 months (Field et al. 2011), although during the course of the study healing evaluations are made at different intervals. Evaluations are made through variable diagnostic methods involving micro-radiography, histology and vascular density (VD) indentations assays, computed tomography (CT), biomechanical, micro-CT, immunohistochemistry and scanning electron microscopy (SEM). Among these diagnostic methods, radiography and histology remain basic evaluations.

In general, combined transplantation of cell and scaffold assembly improves tibial healing irrespective of the source (whether allogeneic or autologous) as compared to the scaffolds alone or control. The cell-cortical allograft assembly ensures bone repair through union at the interface between the host bone and the allograft. Bone defects transplanted with such tissue-engineered constructs show early repair tissue remodelling (18 weeks). The marrow cavity re-establishes with conspicuous inter-trabecular spaces duly filled with marrow. These animals observed for 4–5 years remain symptomless and walk properly (Fernandes et al. 2014). Although the study shows effective strategy for bone healing, however, further studies on MSCs anti-inflammatory and other roles remain to be evaluated. This shall help to elucidate the early and effective regeneration. Apart from the allograft, synthetic scaffolds like cell-loaded porous hydroxyapatite (HA) ceramic (HAC) cylinder lead to the bone formation both within the internal macropore space and on the scaffold surface while in cell-free implants bone formation mostly limits to the outer surface confirming cell-induced osteoinduction (Kon et al. 2000). Further, scaffolds like silicon-stabilized tricalcium phosphate (TCP)-loaded cells improve vascular density significantly greater in the periphery than in the central areas of the constructs (Giannoni et al. 2008). This could be attributed to the higher density of scaffold-loaded cells in the peripheral areas or these cells additively stimulate resident osteoprogenitor cells within the skeletal tissues of the resection ends via a paracrine loop. Comparatively better survival of the loaded cells within the particular area might lead to the faster vascularization in the area than the internal regions. A modification of this preloaded cell scaffold tissue-engineered construct is the delayed transplantation of MSCs loaded onto the scaffold through a minimally invasive injection method. In such case, allogeneic BM-MSCs sheet loaded on the polycaprolactone (80% weight) and hydroxyapatite (20% weight) (mPCL-HA) assembly shows healing potential comparable to the autologous graft after a delay of 4 weeks (Berner et al. 2015). Delayed transplantation may avoid inflammatory environment of the defects and may additionally have some waiver from the toxic fragments of the defect. Besides, the invasiveness is avoided to prevent loss of hematoma and vascular damages where repositioning is not required.

Considering the immune-based reactions that might arise upon cell implantation, BM-MSCs from autologous, allogeneic or xenogeneic sources are evaluated in bone healing (Field et al. 2011; Berner et al. 2013). Autologous and allogeneic BM-MSCs show comparable bone formation that is better than cell-free scaffold, evaluated through radiography, micro-CT and histology (Field et al. 2011; Berner et al. 2013). In one of the studies, biomechanical strength of repaired bone in cell-loaded scaffold and cell-free scaffold remains comparable at 3 months (Berner et al. 2013). As the latter study employs geriatric sheep (7–8 years old) that additionally undergoes periosteal removal (unlike that reported for other studies) might possibly compromise biomechanical strength in the repaired tissue. Considering the xenogeneic MSCs potential against the allogeneic MSCs, radiology and histology reveal significantly better and early bone formation with the latter cell-loaded matrices as compared to the former. Although xenogeneic (human) BM-MSCs do not evoke any local or systemic reaction (Niemeyer et al. 2010), their survival might not last long or activation of surface markers are not stimulated desirably. Xenogeneic cells may not adequately respond to the local and systemic osteoinductive factors in sheep. As these tissue-engineered constructs contain scaffolds alongside, its impact on prevention of cell exposure to the local environment remains to be seen. Also the cell characterization from either source or the in vivo anti-inflammatory role of MSCs remains to be evaluated. Thus, to evaluate possible differences in the healing potential on the basis of the cell source, further studies are desired.

In a non-segmental defect model of tibia and femur, dynamically cultured (bioreactor) osteogenic-induced BM-MSCs lead to the better bone healing in tibia while comparable bone formation among HA-based grafts and static or dynamic cultured grafts is demonstrated for femur. It can be concluded that tibia has a lower bone repair capability as compared to femur, dynamically cultured BM-MSCs loaded on hydroxyapatite scaffold may lead to better bone deposition and osteointegration in disadvantaged anatomical sites (Lovati et al. 2016). However, further studies in this regard are highly desired to provide evidence-based medicine.

2.5.1.2 Spinal Fusion Models

Spinal fusion as compared to the bone defect models involves a complex system as it is aimed at both orthotopic ossification around the decorticated processes as well as the heterotopic ossification to bridge the gap. This heterotopic bridging additionally poses a challenge of neovascularization (Cuenca-López et al. 2014). MSCs are being evaluated in sheep spinal fusion models of lumbar (Gupta et al. 2007; Cuenca-López et al. 2014; Shamsul et al. 2014; Wheeler et al. 2016) and cervical (Goldschlager et al. 2010, 2011) vertebrae. In lumbar spinal fusion models, MSCs role in posterolateral fusion models mostly at the level of L4/L5 (various lumbar vertebrae involved) with a single study of lumbar interbody spine fusion (Wheeler et al. 2016) is evaluated. In the cervical spine model, anterior cervical discectomy

(Goldschlager et al. 2010) and cervical interbody fusion (Goldschlager et al. 2011) are created to evaluate MSCs role. In the case of anterior cervical discectomy, the aim is to achieve cartilaginous tissue to maintain dynamic motion of cervical vertebrae. These studies are conducted as randomized placebo control trials involving varied animal numbers. In lumbar spinal fusion models, animal numbers range from $n = 4$ (Shamsul et al. 2014) to $n = 36$ (Wheeler et al. 2016), in cervical spinal fusion model $n = 30$ (Goldschlager et al. 2011) and in anterior cervical discectomy $n = 18$ was utilized (Goldschlager et al. 2010).

All these studies utilize BM-MSCs with a single study using commercially available MSCs (Mesoblast Ltd.) (Goldschlager et al. 2010). These cells are implanted at various concentrations ranging from 1 million (Goldschlager et al. 2010) to 50 million (Cuenca-López et al. 2014). Alongside the cells, various scaffolds like TCP (Gupta et al. 2007), gelfoam sponge, pentosan polysulphate (PPS) (Goldschlager et al. 2010), HA (Cuenca-López et al. 2014), hydroxyapatite/tricalcium phosphate (HA/TCP) (Goldschlager et al. 2011; Wheeler et al. 2016) and type I collagen (Wheeler et al. 2016) are transplanted. Besides, osteoinduced BM-MSCs seeded on the TCP/HA are also evaluated (Shamsul et al. 2014). The evaluation of bone formation is variably done through radiology, micro-CT, manual palpation, CT, functional radiography, SEM, biomechanical analysis and histomorphometry. These evaluations are conducted for a period ranging from 3 months (Goldschlager et al. 2010, 2011; Shamsul et al. 2014) to 6 months (Cuenca-López et al. 2014).

In lumbar spinal fusion models, variable outcome occurs with MSCs or osteogenic-induced cells loaded onto the scaffolds. MSCs may promote spinal fusion rate (33%) as compared to the autograft (25%), bone marrow (8%) and TCP (0%) (Gupta et al. 2007). Osteoinduced BM-MSCs loaded on HA scaffold induce greater (35%) lumbar fusion rates than mineral scaffold (22%) but lower against the auto and allograft (70%) based on histology and CT. Histologically, repair tissue in cell laden scaffold transplanted models have better bone pattern but statistically non-significant than those transplanted with HA. The greater fusion rate in autologous and allograft transplanted defects correlates with the higher quantity of bone formation (Cuenca-López et al. 2014). However, addition of TCP to HA and loading osteoinduced cells onto it may help in achieving the superior lumbar intertransverse fusion, albeit still lower than the autograft (Shamsul et al. 2014). Most of these studies show bone autografts are better than cell-based mineralized constructs for posterolateral lumbar fusion but cent percent success rate remains to be achieved. Mineralized scaffolds may not be suitable as the scaffold interferes with the callus formation.

In a spinal interbody fusion model, transplantation of MSCs (2.5 million MPCs, 6.25 million mesenchymal progenitor cells, MPCs, or 12.5 million MPCs) loaded on to the scaffold (comprising hydroxyapatite, tricalcium phosphate and type I bovine collagen) achieves fusion rates similar or better than the autograft (iliac crest, AG) in 16-week study period. Higher MSCs doses fail to show any significant effect in the healing potential over the lower cell concentrations. Radiography and manual palpation reveal that more than 75% of cell-treated and 65% of AG-treated animals achieve rigid fusion. CT fusion scores show all animals in either group are fused at 16 weeks, contrary to the X-ray scores that indicate only 67% of the AG-treated animals are fused. Histomorphometry analysis shows equivalent connectivity and bony fusion for both the groups (Wheeler et al. 2016). Thus, interbody spinal fusion may be achieved with improved results using MSCs-loaded scaffold. Further investigations with better advanced regenerative techniques of MSCs transplantation may offer efficient bone formation.

In cervical body fusion (C3-C4), allogeneic cell laden scaffold-treated defects show continuous bony bridging in 75% (9/12) cases, while autologous graft treated ones have 16.7% (1/6) and HA/TCP have 33% (2/6). New bone density is 121% higher in cell-treated as compared to HA/TCP and 128% as compared to autologous groups. Functional radiography reveals significantly reduced macro motion between C3 and C4 (Goldschlager et al. 2011). In anterior cervical discectomy, new bone is observed in 75% (9/12) and 91.6% (11/12) animals in gelfoam sponge only and MSCs/gelfoam-treated groups, respectively. To make movement possible, additional implantation of pentosan polysulphate (PPS) restricts tissue repair to the cartilaginous matrix within the interbody cages as compared to other groups. Only 8% (1/12) in the PPS group have new bone formation (Goldschlager et al. 2010). Since the study is aimed to develop assembly that could maintain mobility of cervical vertebrae through cartilaginous tissue development, PPS potentially prevents bone formation (Bwalya et al. 2017).

In general, spinal fusion may be enhanced with the cell-scaffold assembly; however, the type of the scaffold has a definitive role to play. Mineralized scaffold per se may not be effective unless modified. Besides, the differences in fusion rates across studies put emphasis on the use of instrumented posterolateral fusion as with these techniques good fusion rates even to the level of 100% occur (Wheeler et al. 2014, 2016). Additionally, scaffold properties are important in effectuating the desired results. Some of the scaffolds prevent chondrocyte hypertrophy and thus, are desired in the cervical spine where movement is required. Thus, to develop the typical and all weather tissue-engineered construct, model conceptualization should be followed by selection of the scaffolds and the cells. This should be followed by the evaluation of individual roles played by each biomaterial and in combination at different concentrations and at intervals in order to provide evidence based medicine.

2.5.1.3 Metatarsal Bone Defects

Two randomized control studies are available on metatarsal bone defect models involving animal size of $n = 21$ (Viateau et al. 2007) and $n = 10$ (Manassero et al. 2013). A bone defect size of 2.5 cm created (Viateau et al. 2007; Manassero et al. 2013) is stabilized by plating. A variable concentration of 8+ million (Viateau et al. 2007) and 7+ million (Manassero et al. 2013) BM-MSCs are implanted along with coral scaffolds. In the former study, MSCs loaded onto the scaffold are maintained for 10 days under in vitro culture before in vivo transplantation and are surrounded and lined by the vascularized membrane (Viateau et al. 2007). Coral scaffold supports growth of the MSCs under in vitro cultures. Bone healing is evaluated through radiography, CT and histology for a period of 6 months. The coral loaded cell assembly leads to the two-fold increase in bone formation as compared to the coral scaffolds (Manassero et al. 2013). However, autograft outperforms cell-loaded coral scaffolds in bone formation. Thus, the autografts perform better than the external use of the MSCs/scaffolds. It is worth mentioning that the bone volume in the centre of the cell-scaffold assembly is 16 fold higher while only 1.4- and 1.7-fold increase in the proximal and distal ends is seen. The coral granules degrade at a moderate rate through physicochemical dissolution and osteoclast-like cell mediated degradation and around 5% calcium is localized at the site (Viateau et al. 2007). Variable osteogenesis in the individual animals in the cell-scaffold-treated groups is seen and possibly occurs due to the animal or cell-scaffold-related differences. This includes an individual's innate osteogenic potential, inflammatory environment and mechanical loading of the test limb. Further, cell culture technique, cell-seeding procedure, transplantation technique and the surrounding recipient tissue bed interactions have bearing on such outcomes. However, further insights employing different tissue engineering techniques can provide details for effective utilization.

2.5.1.4 Femoral Bone Defect Models

In sheep, two types of femoral defects are created; one without support (Li et al. 2014) and other stabilized by interlocking nails (Szivek et al. 2018). In the former study, perfusion bioreactor prepared tissue-engineered bone (comprising two million osteoinduced BM-MSCs loaded on porous β-tri-calcium phosphate), and in the latter, two million AD-MSCs loaded on the 3D printed scaffold coated with micron-sized TCP particles are transplanted. The variable animal size is used with the former study employing $n = 16$ and the latter $n = 4$. Bone healing evaluation is conducted through radiography, histology for a period of 4 months (Li et al. 2014) or through radiography, histology, micro-CT for a period of 6 months (Szivek et al. 2018). Osteoinduced BM-MSCs loaded on scaffold successfully repairs bone defects at 16 weeks. The healed bone tissue volume in cell scaffold is sufficient as compared to scaffold only. New bone tissue formation is isochronous with TCP scaffold absorption that occurs through creeping substitution (Li et al. 2014). In traditional system, the cells only grow at the edge of the scaffold, while perfusion bioreactor continuously supplies nutrients to the porous scaffold and removes cell metabolic products and thus, prevents cell apoptosis in the scaffold and their uniform growth (Davisson et al. 2002). In 3D scaffold-loaded AD-MSCs implanted defects, bone formation occurs on the lateral, anterior, medial and posterior-medial aspects. Near muscle attachment area (posterior-lateral aspect), bone formation is absent. At 3-month period, 15.5% more cortical bone deposition around the scaffold circumference while the 6-month time point shows 40.9% more bone deposition within scaffold pores. Lower than complete bridging and apparent non-union in the scaffold only defects demonstrate MSCs potential role in effective bone healing (Szivek et al. 2018). 3D printed scaffolds mimic the architecture of the actual bone and implanting progenitor cells alongside can prove a game changer in the field of bone tissue engineering.

2.5.1.5 Cranial Bone Defect Models

MSCs role in parietal bone defect healing is evaluated in two studies with variable defect of 20 cm (Shang et al. 2001) and 1 cm (four defects per bone) (Adamzyk et al. 2016) treated with the autologous BM-MSCs on calcium alginate and autologous BM-MSCs/or pre-differentiated BM-MSCs on polyetherketoneketone (PEKK) 3D scaffolds, respectively. Variable animal numbers $n = 8$ (Shang et al. 2001) and $n = 5$ (Adamzyk et al. 2016) are employed. Healing is evaluated through gross examination, histology, CT (Shang et al. 2001) or micro-CT and histology (Adamzyk et al. 2016) for a period of 18 weeks and 12 weeks, respectively. Implantation of MSCs/scaffold leads to the bone formation as early as 6 weeks post-implantation unlike that in control defects. The engineered bone tissue becomes more mature at 18 weeks. CT scan reveals almost complete repair of the defect with MSCs/scaffold at 18 weeks (Shang et al. 2001). In another study, a prominent ingrowth from the adjacent bone into the interconnected channels of the scaffolds is observed in all the scaffold groups. Non-significant differences between cell-free, MSC-seeded or osteogenically pre-differentiated MSC-seeded scaffold implantations are detected with respect to the volume of newly formed bone (Adamzyk et al. 2016). Such a contrasting report on the healing potential of MSCs/scaffold might be due to the differences of the model type (defect size), technique of implantation and type of scaffold used. Furthermore, formation of blood clots and fibrous capsule, in addition to the inflammatory reaction might have overwhelmed the CT signals imparted by the scaffold and cells (Haberstroh et al. 2010; Lyons et al. 2010; McCarty et al. 2010).

A single study on mandibular bone defect (30 mm) stabilized by titanium plate and transplanted with the autologous BM-MSCs (25–30 million) along with autologous serum scaffold or only autologous serum (control) is conducted. Complete bone healing occurs in 25% (1/4) cases at 12 weeks and 75% (3/4) at 32 weeks in treatment groups but limited bone healing in control defects occurs only at 32 weeks. In the case of cell/scaffold group, two other sheep present ossification in the areas adjacent to local mandibular bone while one sheep fails to show bone regeneration radiographically. On micro-CT, the control group shows significantly lesser bone mineral density while non-significant differences are observed between native bone and newly formed bone in experimental groups. Histologically, most of the experimental animals show higher levels of graft ossification with more maturity and ossification than the control (Gallego et al. 2015).

2.5.1.6 Osteoporotic Bone Model

In an osteoporotic (ovariectomy-induced) bone defect model (12 mm × 4 mm defect), strontium hydroxyapatite (cSrHA) along with osteoinduced MSCs enhances osteogenesis with mature lamellar bone formation as compared to bare HA, SrHA and tissue-engineered HA-implanted groups. Histomorphometry reveals improved osteogenesis at par with material resorption, as cSrHA/cell implanted group exhibits highest regeneration ratio. Density histograms from micro-CT further signify enhanced osteointegrative ability of cSrHA/cell implants (Chandran et al. 2017). Thus, tissue-engineered constructs employing scaffolds along with stem cells appear promising in regeneration of bone defects even in underlying pathological conditions.

In all these studies, improved outcome with the cell-scaffold transplantation occurs, although variably. This may depend upon the model type, cell culture technique, source, and dose, scaffold type and its design. Besides, inflammatory reaction and formation of blood clots and fibrous capsule, if any may also have bearing on the outcome. Mostly MSCs offer some role to enhance bone formation, although such mechanisms remain to be understood. The cell survival and its engraftment together with the mechanisms involved remain to be evaluated in detail. Scaffold degradation and effect of its degradation products besides, the type and time of degradation are still an open field. In all these studies, mostly a limited number of animals are employed with limited period evaluation and thus, may be undertaken for detailed understanding.

2.5.2 Goat Experimental Bone Defect Models

Like sheep, MSCs osteogenic potential is evaluated in bone defect models in goats. Defects in various bones like tibia, femur and iliac crest are created (Table 2.2). Besides, tissue-engineered constructs employing MSCs are also evaluated in goat models like cranial suture distraction (Shen et al. 2006), maxillary sinus augmentation (Zou et al. 2012; Zhao et al. 2017) and mandibular distraction osteogenesis (El Hadidi et al. 2016).

2.5.2.1 Tibial Bone Defect Models

MSCs osteogenic role is evaluated in two types of tibial defect models viz. full thickness critical size defect (Dai et al. 2005; Liu et al. 2010; Wang et al. 2010; Chu et al. 2018) and non-critical size unicortical defects (Vertenten et al. 2009; Lippens et al. 2010) (Table 2.2). In full-thickness model, a variable tibial defect size is created ranging from 2.5 cm (Liu et al. 2010) to 3.0 cm (Wang et al. 2010; Chu et al. 2018), while in unicortical model, multiple uniform 6-mm defects are created (Vertenten et al. 2009; Lippens et al. 2010). Full-thickness defects are stabilized through bone plating. The animal numbers are variable in segmental full-thickness tibial defects ranging from $n = 6$ (Wang et al. 2010) to $n = 32$ (Liu et al. 2010). In all these studies, BM-MSCs are implanted along with various kinds of scaffolds like biphasic calcined bone (Dai et al. 2005), nano-hydroxyapatite/collagen/poly (L-lactic acid) PLLA/chitin fibres (nHACP/CF) (Liu et al. 2010), β-TCP (Vertenten et al. 2009; Wang et al. 2010; Chu et al. 2018), methacrylate-encapped poly (D,L-lactide-Co-έ-caprolactone (Vertenten et al. 2009) and Pluronic® F127 hydrogel, tubular sintered hydroxy apatite (Lippens et al. 2010). Even BMP-2 transfected BM-MSCs are also evaluated for healing potential (Dai et al. 2005). Bone healing evaluation is made through the radiography, micro-CT, histology, biomechanical examination and immunohistochemistry, although radiography and histology are consistently performed. The evaluation period is variable ranging from 8 weeks (Liu et al. 2010) to 26 weeks (Dai

Table 2.2 Goat in vivo MSCs preclinical experimental studies on various bone defects or distractions

Study design/ period	Biomaterial used	Evaluation criteria	Overall result	References
Tibial bone defect models				
Randomized placebo controlled comparative study ($n = 26$)/ 26 weeks	Biphasic calcined bone (BCB), BMP-2 transfected BM-MSCs, simple BM-MSCs (2×10^8)	Histology, biomechanical examination	Improved outcome	Dai et al. (2005)
Randomized placebo controlled ($n = 8$)/ 12 weeks	BM-MSCs, β-TCP, methacrylate-encapped poly (D,L-lactide-co-έ-caprolactone	Radiography, histology, histomorphology and immunohistochemistry	Improved outcome	Vertenten et al. (2009)
Randomized placebo controlled ($n = 8$)/8 weeks	Pluronic® F127 hydrogel, tubular sintered hydroxy apatite and autologous BM-MSCs (4×10^5)	Clinical, radiology and histology	Improved outcome	Lippens et al. (2010)
Randomized placebo controlled comparative study ($n = 32$)/ 8 weeks	Autologous BM-MSCs (2×10^6), nano-hydroxyapatite/ collagen/poly (L-lactic acid) PLLA/ chitin fibres (nHACP/ CF)	Radiography, histology and biomechanics	Improved outcome	Liu et al. (2010)
Randomized placebo controlled comparative study ($n = 6$)/ 24 weeks	Autologous BM-MSCs (2×10^6/ mL), β-TCP	Radiography, histology and micro-CT	Improved outcome	Wang et al. (2010)
Randomized placebo controlled ($n = 12$)/ 6 months	BM-MSCs, β-TCP [prepared trough novel stem cell screen-enrich-combine (-biomaterials) circulating system (SECCS)]	Radiography, computed tomography (CT), micro-CT, histology, and histomorphology	Improved outcome	Chu et al. (2018)
Femoral bone defect models				
Randomized placebo controlled comparative study) ($n = 23$)/ 16 weeks	Bone morphogenetic protein-2 (BMP-2)- or β-galactosidase (β-gal)-transfected BM-MSCs (1×10^7), β-TCP	Radiography, bone mineral density, histology and biomechanical testing	Improved outcome	Tang et al. (2007)

(continued)

Table 2.2 (continued)

Study design/ period	Biomaterial used	Evaluation criteria	Overall result	References
Randomized placebo controlled ($n = 6$)/ 4 months	BM-MSCs, triphasic ceramic coated hydroxyapatite (HASi) & 1×10^5/ cm^2 of HASi	Radiography, micro-CT, histology, histomorphometry, SEM, plasma spectrometry	Improved outcome	Nair et al. (2008)
Randomized placebo controlled ($n = 16$)/ 16 weeks	Autologous BM-MSCs (oestrogen-deficient donor), β-TCP	Radiography, CT, micro-CT and histomorphometry	Improved outcome	Cao et al. (2012)
Randomized placebo controlled ($n = 12$)/ 16 weeks	Bone marrow aspirate (BMA), Nano-scale self-assembly peptide (SAP) modified demineralized bone matrix (DBM)	Radiography, CT and 3D image processing techniques	Improved outcome	Li et al. (2015)
Iliac crest defect models				
Randomized placebo controlled ($n = 10$)/ 12 weeks	Allogeneic BM-MSCs (2×10^6), bicalcium phosphate (BCP)	Histomorphometry, florescent microscopy, immunohistochemistry	Improved outcome	Loozen et al. (2015)
Cranial suture distraction models				
Randomized placebo controlled ($n = 10$)/ 4 weeks	MSCs (5×10^9)	SEM and histology	Improved outcome	Shen et al. (2006)
Maxillary sinus models				
Randomized placebo controlled ($n = 9$)/ 3 months	BM-MSCs (2×10^7), calcium phosphate cement(CPC)	Micro-CT, histology and histomorphometry	Improved outcome	Zou et al. (2012)
Randomized placebo controlled comparative study ($n = 18$)/ 3 months	DT-MSCs, calcium phosphate cement (CPC)	Computed tomography (CT), histology and histomorphometric analysis	Improved outcome	Zhao et al. (2017)
Mandibular distraction osteogenesis models				
Randomized placebo controlled ($n = 12$)/ 30 days	BM-MSCs (1.5×10^6 two injections at day 10 and 20 after surgery)	Energy-dispersive X-ray, scanning electron microscopy (SEM) and histology	Improved outcome	El Hadidi et al. (2016)

Table 2.2 (continued)

MSCs mesenchymal stem cells, *AD* adipose tissue derived, *BM* bone marrow derived, *BMP* bone morphogenetic protein, *DT* deciduous teeth derived, *βTCP* β-tricalcium phosphate, *n* total number of animals in all the groups of the study; Improved outcome: Indicates improved outcome with MSCs transplantation even if only one parameter improves; Study period: Indicates maximum time period for which study is conducted

et al. 2005) in full-thickness models and 8 weeks (Lippens et al. 2010) to 12 weeks (Vertenten et al. 2009) in unicortical defect models.

In full-thickness bone defect models, an overall improved and early bone formation occurs with the tissue-engineered (TE) constructs employing MSCs as compared to the control. The TE constructs offer superior and early bone formation as compared to the scaffolds (Dai et al. 2005; Liu et al. 2010) with comparability to that of the autograft (Liu et al. 2010). Transplantation of biphasic calcined bone (BCB) with 40% TCP has a slow rate of absorption and by 26th week 24% is absorbed. Such a BM-MSC-loaded scaffold enhances bone repair but with individual animal variations (Dai et al. 2005). This could possibly be due to the larger bone defect (2.6 cm), limited blood supply in tibia and possible removal of periosteum in the study. The cell culture technique may also affect the MSCs bone formation potential as one of the studies demonstrates that segmental bone defect repair by dynamic perfusion reactor cultured tissue-engineered bone outweighs static cultured ones (Wang et al. 2010). Furthermore, a novel method [stem cell screen-enrich-combine (−biomaterials) circulating system (SECCS)] prepared BM-MSCs/β-TCP construct improves the recanalization of the medullary cavity with effective cortical remodelling as compared to the pure porous β-TCP in critical size tibia defect (30 mm). Such a technique compounds appx. 81% of the MSCs in BM to the porous β-TCP and the cell viability is well maintained (Chu et al. 2018).

The osteogenic-induced cells loaded on to the polymer (methacrylate-endcapped poly(D,L-lactide-co-e-caprolactone) show good survival and proliferation, while only limited proliferation occurs on α-TCP under in vitro conditions. However, under in vivo studies, synthetic scaffolds fail to offer good osteoconductive properties and defects in control are filled early and dense than the scaffold/cell-loaded defects (Vertenten et al. 2009). This could possibly occur due to the lack of sufficient porosity in the polymer as in the previous study of same authors effective healing is recorded wherein porogen is used (Vertenten et al. 2008). In the current study, gelatin used as a porogen might negatively influence cell viability (Tielens et al. 2007). Contrarily, cell-hydrogel plus cultiSpher-S carriers (crosslinked by gelatin) enhances early (2 and 4 weeks) bone repair in defects in comparison to the scaffold only or control. This shows osteoinduced cells might secrete bone matrix. As hydrogel-based TE construct-loaded defects show osteoblast phenotype at 4 weeks period and such cells are lacking in the hydrogel only, it indicates these osteoinduced cells recruit osteoblasts cells. HA carrier-loaded cells, however, fail to give rise to new bone formation (Lippens et al. 2010). HA that remains non-degradable up to 8 weeks could hamper the natural bone regeneration. Further, HA, a common and often considered better osteoinduction scaffold, induced MSCs osteogenesis under in vitro settings is favoured by lower porosity and is independent of pore size contrary to that of the in vivo studies (Baroli 2009). Thus, there has to be a proper technique of tissue engineering for effective bone repair.

From all these studies, cells, scaffold, culture method, species and model (including stabilization technique) have an important role to determine outcome with the tissue-engineered constructs. Scaffold must ape natural bone to favour early and effective bone regeneration. This includes its composition (bone comprises 60–70% nano-hydroxyapatite crystals, while 30-40% is collagen and polysaccharide), mechanical properties, degradation products that make microenvironment conducive, and effectively allows transfer of nutrients through well-established porosity (Liu et al. 2010).

2.5.2.2 Femoral Bone Defect Models

MSCs osteogenesis is evaluated in various types of randomized case-controlled femoral models including the femoral head osteonecrosis (Tang et al. 2007), segmental bone defects in healthy (Nair et al. 2009b; Li et al. 2015) or the osteoporotic (Cao et al. 2012) animals. Osteonecrosis model is created with the ligation of lateral medial circumflex arteries and delivery of liquid nitrogen to the femoral head. Femoral segmental bone defect (2 cm) is stabilized with the interlocking nail (ILN). Osteoporotic bone defect model is created through oestrogen deficiency followed by induction of 0.8 cm wide and deep cylindrical defects. Mostly BM-MSCs are utilized except one study that employs BM aspirate (Li et al. 2015). These cells are implanted along with the β-TCP (Tang et al. 2007; Cao et al. 2012), triphasic ceramic-coated hydroxyapatite (HASi) (Nair et al. 2009b) and nano-scale self-assembly peptide (SAP) modified demineralized bone matrix (DBM) (Li et al. 2015). Variable animal sizes are made subjects; in osteonecrosis $n = 23$ (Tang et al. 2007), segmental defect model $n = 6$ (Nair et al. 2009b) and $n = 12$ (Li et al. 2015), osteoporotic bone defect model $n = 16$ (Cao et al. 2012). Bone formation is evaluated through radiography, bone mineral density, histology, micro-CT, SEM, couple plasma spectrometry and biomechanical testing, although variably for a period of 2–4 months.

In case of femoral head osteonecrosis, MSCs transfected with BMP-2 and β-gal genes along with TCP are compared. Transplantation of BMP-2 transfected cells loaded on TCP in the femoral head leads to the bone tissue formation with normal density and surface architecture, while transplantation of β-gal transfected cells loaded on TCP shows low density and irregular surface of femoral heads. Histologically, lamellar bone is formed in BMP-2 cell-treated and residual fibrous tissue in β-gal cell-treated. The newly remodelled lamellar bone formed in BMP-2 group shows significantly higher bone density as compared to the β-gal. The maximum compressive strength and Young's modulus of repaired tissue in the BMP-2-treated cell group are similar to normal bone and significantly higher as compared to β-gal cell-treated. In a non-treated group, femoral heads collapsed at 16 weeks (Tang et al. 2007). Thus, transfection of MSCs with BMP-2 enhances their osteogenic potential. It is worth mentioning that such techniques need further safety evaluations.

In unicortical segmental bone defect models, cell-scaffold assembly enhances bone formation. Good osteointegration and osteoconduction are observed in cell-HASi (porous bioactive triphasic ceramic-coated hydroxyapatite) treated bone defects. Cell-HASi implantation leads to faster and better performance evident by lamellar bone organization of newly formed bone with the degradation of material. In the HASi only group, immature interwoven bony bridges are still intermingled with scattered small remnants of material. In peripheral areas, no observable difference in cell-free and cell-loaded scaffolds is recorded (Nair et al. 2009b). Incorporation of different ingredients in the scaffold in order to resemble normal bone may improve bone healing. Implantation of self-assembly peptide/demineralized bone matrix (SAP/DBM) along with the cells leads to the significantly higher volume of new bone as compared to DBM/MSCs. DBM may provide mechanical strength and stability while SAP may provide extracellular matrix (ECM) that promotes cell and growth factor retention in the nano-fibre mesh and amino acids electrostatic forces. Such an ECM favours cellular growth and proliferation and the sustained release of the growth factors/cytokines secreted by them (Li et al. 2015).

Apart from the normal individual, tissue-engineered scaffolds along with MSCs may have a role in bone regeneration in osteoporotic animals. MSCs harvested from osteoporotic animals have lower proliferation and osteogenic potential. However, combined use of such cells and scaffolds like β-TCP leads to the effective bone formation in osteoporotic femur defects (24-month ovariectomized goats) at 16 weeks period (Cao et al. 2012). In this study, healing evaluated specifically at 16 weeks is still incomplete; it is therefore imperative to evaluate healing is enhanced or not at other periods as well. Besides, TE constructs' ability to heal should be evaluated through the prism of inflammation/immune-modulation, matrix secretion ability, histological characterization and ultimately the remodelling.

2.5.2.3 Maxillary Sinus Models

Apart from the routine defect models corresponding to the fracture, bone augmentation or elevation procedures are also improved with the tissue-engineered constructs. A significant enhancement in bone formation occurs with the tissue-engineered constructs. In maxillary sinus augmentation, BM-MSCs (Zou et al. 2012) or deciduous teeth MSCs (DT-MSCs) (Zhao et al. 2017) loaded with calcium phosphate cement (CPC) promotes early bone formation and mineralization as compared to the control or scaffold. The regenerated tissue effectively maintains volume and height of augmented maxillary sinus with a significant osteointegration as compared to CPC alone or control (Zou et al. 2012). Even new bone tissue in cell-scaffold transplanted defects has significantly higher GFP and OCN in elevated sinuses as compared to the autogenous bone group (Zhao et al. 2017). These studies clearly show that tissue engineering may prove boon in orthopaedic affections, however, right scaffold and cell in right concentrations besides, the method of implantation need to be standardized. Furthermore, the regeneration with the ideal bone tissue in these models remains to be seen and is currently attributed to the lack of understanding in the mechanisms involved. In mandibular distraction osteogenesis, transplantation of MSCs improves bone healing both qualitatively and quantitatively, a limiting factor often reported for simple distraction osteogenesis (El Hadidi et al. 2016). MSCs modulates including inflammation, enhances progenitor cell recruitment and prevents their apoptosis, besides, the enhancement of the matrix secretion and prevention of its degradation (Gugjoo et al. 2020a, b).

In general, goat osteogenesis is improved with bone tissue engineering although with some reservations. Bone and its model, scaffold design and material properties, cell source and its culture and concentration all have bearing on the outcome. Refinement in such factors is desired and the major factor in all these studies is the limited animal size. Above all the clinical applications of all these tissue-engineered constructs remain to be conducted.

2.5.3 Dog Experimental Bone Defect Models

In dogs, tissue-engineered constructs employing MSCs are evaluated in models of various bones like femur (Bruder et al. 1998; Arinzeh et al. 2003), radius (Jang et al. 2008; Kim et al. 2018), ulna (Itoi et al. 2016) and mandible (Jafarian et al. 2008; Abe et al. 2020) (Table 2.3).

2.5.3.1 Femoral Bone Defect Model

In femoral bone defect, diaphyseal critical size defect (21 mm) is transplanted with autologous BM-MSCs (Bruder et al. 1998) or allogeneic BM-MSCs (Arinzeh et al. 2003) along with hydroxyapatite (65%) and beta-tricalcium phosphate ceramic (35%), respectively. Animal numbers incorporated in these studies are $n = 15$ (Bruder et al. 1998) and $n = 12$ (Arinzeh et al. 2003). Evaluation of bone formation is made through histology and morphometry and radiography for a period of 16 weeks.

Irrespective of the cell source (autologous or allogeneic), radiography reveals MSCs along with the scaffold-transplanted critical-sized defects establish callus at the interface between the host bone and the implants and around the implant from fourth week onwards. Histological and morphometric analyses demonstrate woven and lamellar bone fills the pores of the implants at 16 weeks. The amount of bone is significantly greater in the pores of cell-loaded implants than that of cell-free implants. The size and shape are comparable to the resected native bone segment. Callus does not develop around the cortex of the host bone or around the defect in any of the specimens in the scaffold or control for up to 16 weeks (Bruder et al. 1998; Arinzeh et al. 2003). Furthermore, the allogeneic cells remain localized up to 4 weeks in the bone tissue and 8 weeks in the connective tissue (Arinzeh et al. 2003). Thus, MSCs along with scaffold appear quite promising in bone formation and may be kept available as off-the-shelf products.

2.5.3.2 Radial Bone Defect Model

MSCs from UCB and AD are implanted along with the β-TCP covered with poly L-lactide-co-ε-caprolactone (Jang et al. 2008) or demineralized bone matrix (DBM),

Table 2.3 Dog in vivo MSCs preclinical experimental studies on various bone defects

Animal size	Study period	Biomaterial used and cell dose	Evaluation criteria	Overall result	References
Femoral bone defect models					
Randomized placebo controlled (n = 15)/16 weeks		Autologous BM-MSCs, porous ceramic cylinder containing hydroxyapatite (65%) and beta-tricalcium phosphate ceramic (35%)	Histology and morphometry, radiography	Improved outcome	Bruder et al. (1998)
Randomized placebo-controlled comparative study (n = 12)/16 weeks		Allogeneic BM-MSCs, beta tricalcium phosphate (β-TCP)	Histology and radiography	Improved outcome	Arinzeh et al. (2003)
Radial bone defect model					
Randomized placebo controlled (n = 6)/12 weeks		Allogeneic UCB-MSCs (1×10^6), beta tricalcium phosphate covered with β-TCP/poly L-lactide-co-ε-caprolactone composite	Radiography, histology	Improved outcome	Jang et al. (2008)
Randomized placebo controlled (n = 12)/12 weeks		BMP-7 overexpressing AD-MSCs, demineralized bone matrix (DBM), poly ε-caprolactone/β-tricalcium phosphate scaffolds	Micro-CT for bone imaging, histology	Improved outcome	Kim et al. (2018)
Ulnar bone defect model					
Randomized placebo controlled comparative study (n = 9)/12 weeks		BM-MSCs + rBMP (*E. coli*) [560 μg/10^7 cells, 560 μg/10^5 cells, 560 μg/0 cells, 140 μg/10^7, 140 μg/10^5, 140 μg/0]	Radiography, micro CT, histology	Improved outcome	Itoi et al. (2016)
Mandibular bone defect model					
Randomized placebo controlled comparative study (n = 4)/6 weeks		Biphasic hydroxyapatite/-tricalcium phosphate (HA/TCP), natural bovine bone mineral (NBM), BM-MSCs (5×10^5 seeded on scaffolds)	Histomorphometric analysis (immunohistochemistry)	Improved outcome	Jafarian et al. (2008)
Randomized controlled/ 3 months		Allogeneic BM-MSCs (1×10^7), carbonate hydroxyapatite	Radiography and histopathology	Improved outcome	Abe et al. (2020)

MSCs mesenchymal stem cells, *UCB* umbilical cord derived, *BM* bone marrow derived, *rBMP* recombinant bone morphogenetic protein, *DT* deciduous teeth derived, *βTCP* β-tricalcium phosphate, *n* total number of animals in all the groups of the study; Improved outcome: Indicates improved outcome with MSCs transplantation even if only one parameter improves; Study period: Indicates maximum time period for which study is conducted poly ε-caprolactone/β-TCP (Kim et al. 2018), respectively, in induced radius critical size defect (1.5 cm). Animals $n = 6$ (Jang et al. 2008) and $n = 12$ (Kim et al. 2018) are made subjects of the study. Bone formation is evaluated through radiography, histology and micro-CT after a period of 12 weeks.

Radiography reveals that UCB-MSCs/β-TCP-treated defects have continuity between implant and host bone at $n = 3$ interfaces while only $n = 1$ in case of control by 12 weeks. Bone formation occurs around β-TCP in longitudinal sections of implant in both cell and control groups. Histomorphometric analysis reveals significantly increased bone formation in a cell-treated group at 12 weeks after implantation (Jang et al. 2008). Transfection of MSCs with BMP-7 further enhances such a bone formation potential of TE constructs. Newly formed bone and neovascularization is enhanced with BMP-7-MSCs sheets with or without the demineralized bone matrix (DBM) as compared to the control. This may be due to MSCs osteoconduction and osteoinduction activities. BMP-7 may improve osteogenesis of both transplanted and resident MSCs; however, the former cells may mostly function as delivery vehicles (Kim et al. 2018). Thus, further modifications in cells besides the scaffold may promote early and improved bone formation.

2.5.3.3 Mandibular Bone Defect Models

Jaw bone defect models are useful to study healing of the cleft of palate or lip. In one such model, allogeneic BM-MSCs may not survive at defect areas beyond 3 months (5.7%). However, cells along with the carbonate hydroxyapatite enhance bone defect regeneration by that time (Abe et al. 2020). Comparing the type of scaffold, HA/TCP loaded with MSCs may prove superior as compared to the demineralized bone matrix. However, both these scaffolds have an erratic absorption (Jafarian et al. 2008). Thus, MSCs may not show repair process through actual differentiation rather via paracrine actions and such an activity becomes prominent with the particular scaffold.

2.5.3.4 Ulnar Bone Defect Model

A single study demonstrates the role of various concentrations of MSCs along with the rBMP on critical size ulnar bone defect repair. An enhanced bone formation characterized by improved bone mineral density and reduced non-uniformity of the regenerated bone is induced by the higher cell density. Higher cell and growth factor concentration in general improves healing. Micro-CT confirms bone mineral density is improved in the groups that receive cells (10^7 or 10^5) along with the 560 μg of rBMP. BMP is involved in both osteogenic and osteoclastic activities. The former action occurs at the lower doses, while the latter action occurs with the higher doses. The study demonstrates that the higher doses of the BMP have to be integrated with the higher cell concentration for effective bone formation as higher BMP concentration may additionally stimulate MSCs osteogenesis (Itoi et al. 2016). This is contrary to the sheep interbody spinal fusion study where MSCs concentration (higher vs lower) shows a non-significant effect on osteogenesis (Wheeler et al. 2016). Such differences could possibly be due to the differences in the model type, besides the biomaterials used and species involved. This could also be due to the combined use of growth factor in the present study that might have effectuated MSCs enhanced osteogenesis.

2.5.4 Dog In Vivo Clinical Studies

MSCs osteogenic potential under clinical conditions is evaluated in complicated non-union radial fracture case reports (Jang et al. 2008; Lee et al. 2009; Song et al. 2017) and in enhancing tibial tuberosity healing after tibial advancement procedures (dos Santos et al. 2018) or tibial fracture (Franco et al. 2021) (Table 2.4). In these studies, bone ends are stabilized with the implants especially the bone plate except a single study that uses external skeletal fixation (Song et al. 2017).

The cells are implanted along with the scaffolds like hydroxyapatite and chitosan (Lee et al. 2009) and cortical allograft (Jang et al. 2008) or cells are cultured as sheets expressing BMP-7 (10,000 cells/cm^2) (Song et al. 2017). The cells employed are derived either from adipose tissue (Lee et al. 2009; Song et al. 2017; Franco et al.

Table 2.4 Dog in vivo MSCs clinical studies on various bone affections

Study design/ period	Biomaterial	Route	Overall result	References
Non-union radial fracture				
Case report ($n = 1$)/ 8 weeks	Umbilical cord MSCs (1×10^7 repeated after 1 month), cortical allograft	Local implantation	Improved outcome	Jang et al. (2008)
Case report ($n = 1$)/90 days	Autologous AD-MSCs (3.2×10^7), hydroxyapatite and chitosan	Local implantation + compression plate	Improved outcome	Lee et al. (2009)
Case report ($n = 1$)/ 9 weeks	Bone morphogenetic protein-7 AD-MSC sheet (1×10^4 cells/cm^2)	Local implantation + locking plate	Improved outcome	Song et al. (2017)
Tibial tuberosity osteotomy				
Case controlled clinical trial ($n = 9$)/ 120 days	MSCs (1.5×10^6)	Local application	Improved outcome	dos Santos et al. (2018)
Tibial fracture				
Case controlled clinical trial ($n = 36$)/ 120 days	AD-MSCs (3×10^6)	Local implantation + minimally invasive plate osteosynthesis	Improved outcome	Franco et al. (2021)

MSCs mesenchymal stem cells, *UCB* umbilical cord derived, *BM* bone marrow derived, *rBMP* recombinant bone morphogenetic protein, *DT* deciduous teeth derived, *βTCP* β-tricalcium phosphate, *n* total number of animals in all the groups of the study; Improved outcome: Indicates improved outcome with MSCs transplantation even if only one parameter improves; Study period: Indicates maximum time period for which study is conducted

2021) or umbilical cord (Jang et al. 2008). The variable cell concentration used ranges from 10 million (Jang et al. 2008) to 32 million (Lee et al. 2009). Evaluation of bone formation and healing is observed through radiography and clinical lameness scores for a period of 2–3 months.

In all these cases, bone formation and subsequent healing are reported. In non-union cases of radial fracture, cell transplantation in an already allograft implanted case enhances fracture healing as early as 6 weeks. An increased radiopacity at the junction of radius and allograft bone with decreased radiolucent gap is demonstrated that disappears by 2 months. The allograft bone lysis and instability evident by ninth week are succored by the transplanted stem cells that enhance early healing (Jang et al. 2008). Another case report of non-union radial and ulnar fracture (sclerotic ends) heals within 3 months post AD-MSCs (32 million) loaded HA-chitosan implantation duly stabilized by plating (Lee et al. 2009). Similarly, transplantation of the AD-MSCs sheets expressing BMP-7 in an oligotrophic non-union radial fracture (distal 1/3) stabilized by plating ensures healing by ninth week. The possible cause for non-union remains unknown but is possibly due to the biological reasons as fracture ends are stable (Song et al. 2017). In complete non-articular tibia fracture stabilized with the minimally invasive plate osteosynthesis technique, local transplantation of AD-MSCs (three million) significantly reduces bone union time (28.5 days) as compared to the cell-free fractures (70.3 days). However, it is worth mentioning that both these groups show skeletally immature bone tissue ranging from 56% (cell free group) to 60% (cell-transplanted group) (Franco et al. 2021).

In tibial tuberosity osteotomy study, cases treated with MSCs (1.5 million cells) ($n = 4$) show significantly higher ossification (36.45% greater) than the non-cell-treated cases ($n = 5$) at 30 days period. However, the recovery period is comparable to that of the control (dos Santos et al. 2018). This could be due to the non-conducive environment, particularly the local inflammation that might affect MSCs (directly transplanted without any scaffold) effectiveness. Further, limited cell concentrations might show limited therapeutic effects. Thus, MSCs may offer better results provided that the cell transplantation is appropriate.

In general, these clinical studies offer potential evidence to use MSCs and/or scaffolds for early and effective bone healing although complete healing may not occur in every case. Some of the lacunas in these above-detailed studies include lack of control group, although the non-union cases are automatically the case-controlled studies. To develop evidence-based medicine, there is a need to undertake such studies in a greater and extensive manner involving multicentric or unicentric but with greater animal numbers and diverse cases.

2.5.5 Horse In Vivo Experimental Studies

In horses, the lower bone fracture models are created and evaluated for MSCs bone healing potential (Table 2.5). The metacarpal (McDuffee et al. 2012; Tsuzuki et al. 2014) and metatarsal (Tsuzuki et al. 2014) bone defect (1 cm) models, and an

Table 2.5 Horse in vivo MSCs preclinical experimental studies on various bone affections

Number of animals	Study period	Biomaterial used and cell dose	Evaluation criteria	Overall result	References
Splint bone defect models					
Randomized placebo controlled ($n = 10$)/12 weeks		Periosteum-derived osteoprogenitor cells (osteogenic lineage differentiated), fibrin glue and 20×10^6	Radiography and histology	Improved outcome	McDuffee et al. (2012)
Randomized placebo-controlled comparative study ($n = 6$)/16 weeks		Autologous BM-MSCs, BMP-2 and gelatin β-tricalcium phosphate sponges	Radiography, CT and histology	Improved outcome	Tsuzuki et al. (2014)
Pastern joint arthrodesis model					
Randomized placebo controlled ($n = 6$)		BM-MSC, gelatin/β-tricalcium phosphate and BMP-2 & 10×10^6	Radiography, histology, computed tomography	Improved outcome	Seo et al. (2014)

MSCs mesenchymal stem cells, *BM* bone marrow derived, *BMP* recombinant bone morphogenetic protein, *n* total number of animals in all the groups of the study; Improved outcome: Indicates improved outcome with MSCs transplantation even if only one parameter improves; Study period: Indicates maximum time period for which study is conducted

arthrodesis pastern joint model are evaluated (Seo et al. 2014). These models are conducted as randomized control studies in which BM-MSCs along with gelatin/β-tricalcium phosphate and BMP-2 (Seo et al. 2014; Tsuzuki et al. 2014) or only osteodifferentiated periosteum MSCs along with fibrin glue (McDuffee et al. 2012) are used. The case numbers $n = 6$ (Seo et al. 2014; Tsuzuki et al. 2014) to $n = 10$ (McDuffee et al. 2012) are employed. The bone formation is evaluated through radiography, histology and CT for a period of 12–16 weeks.

In the metacarpal segmental ostectomy model, osteoprogenitor cells along with the fibrin glue fail to enhance bone formation as compared to that of the fibrin glue treatment by 12th week. The possible reasons for lack of any superiority or significant differences over the contralateral control bone defects are the cellular migration towards the control defects, use of osteoprogenitor cells and the fibrin glue. Besides, an inflammatory environment might affect the efficacy of the cells or the cells are not typical osteoprogenitor rather fibroblasts as defects are filled with the fibrous tissue (McDuffee et al. 2012). Thus, characterized cell population for transplantation should be made a priority. In contrast, MSCs along with the BMP-2 and gelatin β-TCP sponges lead to enhanced bone formation in 1-cm splint bone defects ($n = 6$) by 16th week (Tsuzuki et al. 2014). Thus, cell transfection and particular scaffold may promote resistant bone defects.

In case of proximal interphalangeal (pastern) joint arthrodesis (surgical cartilage damage) ($n = 6$), gelatin/β-tricalcium phosphate (GT) sponges loaded with MSCs and BMP-2 and stabilized by minimally invasive locking compression plate (LCP) lead to a significant enhancement of arthrodesis in comparison to the simple use of LCP after 16 weeks (Seo et al. 2014).

In horses, the bone tissue engineering studies employing MSCs are limited but show their potential therapeutic efficacy. Again, like studies in other species, there is a need to develop tissue-engineered constructs that ensure safety and efficacy.

2.5.6 Horse In Vivo Clinical Studies

There are very limited studies in the horse that evaluate TE constructs employing MSCs for bone healing (Table 2.6). A clinical study on phalanx digitalis distalis fracture (fixed with orthopaedic screw) demonstrates that implantation of plasma protein adhesive/MSCs biofilm helps in hoof regeneration and complete bone healing occurs at 12th week (Marycz et al. 2012). Orthopaedic screw implantation causes inflammation and induces tension as a reaction against the presence of a foreign body. Combined use of MSCs and fibrin glue thin biofilm might limit inflammation to enhance early healing. In clinical studies of subchondral bone cyst, AD-MSCs (Golonka et al. 2018) or resveratrol-treated AD-MSCs (Golonka et al. 2020) loaded in alginate hydrogel and applied to the debrided cyst area remodels the defect and replaces it with the bone tissue at 6 months. In these cases, 87.75% ($n = 7/8$) have clinical improvement at 6 months and 100% by the end of the year (Golonka et al. 2018). To improve the efficacy of MSCs in bone cyst, resveratrol-treated AD-MSCs loaded on the 3D alginate hydrogel are evaluated. The horse affected with bone cyst treated surgically followed by transplantation of these tissue-engineered scaffolds return back to normal work within 6 months of the treatment without recurrence of the lameness (Golonka et al. 2020). Resveratrol checks cell apoptosis and is considered to carry potential to enhance the survival of

Table 2.6 Horse in vivo MSCs clinical studies on various bone affections

Study design/period	Biomaterial used	Overall result	References
Phalanx digitalis distalis fracture			
Case report ($n = 1$)/ 12 weeks after surgery	Orthopaedic screw wrapped/MSCs biofilm (6×10^6 cells in plasma protein adhesive biofilm)	Improved outcome	Marycz et al. (2012)
Femoral condyle subchondral bone cyst			
Case report ($n = 6$)/ 12 months after surgery	Autologous AD-MSCs, sodium alginate hydrogel	Improved outcome	Golonka et al. (2018)
Case report ($n = 1$)/ 6 months after surgery	SIRT+MSCs (1×10^6 in sodium alginate hydrogel)	Improved outcome	Golonka et al. (2020)

MSCs mesenchymal stem cells, *AD* adipose tissue derived, *n* total number of animals in all the groups of the study; Improved outcome: Indicates improved outcome with MSCs transplantation even if only one parameter improves; Study period: Indicates maximum time period for which study is conducted

MSCs. Resveratrol activates the Sirtuin-1 (SIRT1) in MSCs that suppresses NF-κB signalling (Bhullar and Hubbard 2015) and inactivates hypoxia-inducible factor (HIF)-1α (Sun et al. 2018). SIRT1+ MSCs also protects bone loss by deacetylation of FOXO3a and inhibition of oxidative stress (Sun et al. 2018). Thus, MSCs appear as potential candidates in bone tissue engineering, although the availability of insignificant number of studies and cases fail to provide evidence-based medicine.

2.6 Conclusion(s)

Osteogenic differentiation is a prerequisite for the cells to be named as MSCs. The osteogenic lineage of MSCs can be effectuated through various growth factors and matrices/scaffolds besides the mechanical factors. During osteogenesis both early- as well as late-phase osteogenic markers are expressed, although the pathways involved vary with the source of cells (species and tissue source) and osteogenic agents like growth factors, matrices or hormones. Under in vivo systems, MSCs are mostly transplanted with the grafts/scaffolds/implants/matrices for better survival and efficacy. The cellular osteogenic property is guided by the scaffold type, design and technique of preparation. Mineralized scaffolds/implants may not be effective unless treated with the cell compliable biomaterials. In vivo studies (experimental and clinical) demonstrate that tissue-engineered constructs employing MSCs significantly enhance repair in bone defects. Although it may not amount to that of the autologous bone grafts, in some cases it improves to such an extent. In the case of spinal fusion, TE constructs improve fusion although different with different bone densities being affected by other factors like stabilization and scaffold involved. Even in osteoporosis, such bone healing also improves. Overall interspecies differences, model type, scaffold and its design, cell culture techniques and cell source also affect such outcomes.

As BM-MSCs have mostly been utilized, osteogenic potential of MSCs from other sources may comparatively be evaluated. The cells have been implanted at variable concentrations and along with varied scaffolds, the impact of which must be evaluated comparatively. Further, the small animal number and short duration studies are incomplete in determining the long-term results especially in large veterinary species like horses and cattle. Above all the clinical trials that are very miniscule remain to be conducted to evaluate both safety and efficacy. Few available case reports in dog and horse although show good prospects of the bone tissue engineering using MSCs but are limited enough to provide evidence-based medicine. Furthermore, studies on bone tumour affections remain to be conducted.

References

Abe T, Sumi K, Kunimatsu R, Oki N, Tsuka Y, Awada T, Nakajima K, Sugiyama M, Tanimoto K (2020) Bone regeneration in a canine model of artificial jaw cleft using bone marrow–derived mesenchymal stem cells and carbonate hydroxyapatite carrier. Cleft Palate Craniofac J 57(2): 208–217

Adamzyk C, Kachel P, Hoss M, Gremse F, Modabber A, Hölzle F, Tolba R, Neuss S, Lethaus B (2016) Bone tissue engineering using polyetherketoneketone scaffolds combined with autologous mesenchymal stem cells in a sheep calvarial defect model. J Craniomaxillofac Surg 44(8): 985–994. https://doi.org/10.1016/j.jcms.2016.04.012

Amini AR, Laurencin CT, Nukavarapu SP (2012) Bone tissue engineering: recent advances and challenges. Crit Rev BioMed Eng 40(5):363–408

Ardjomandi N, Duttenhoefer F, Xavier S, Oshima T, Kuenz A, Sauerbier S (2015) In vivo comparison of hard tissue regeneration with sheep mesenchymal stem cells processed with either the FICOLL method or the BMAC method. J Craniomaxillofac Surg 43(7):1177–1183

Arinzeh TL, Peter SJ, Archambault MP, van den Bos C, Gordon S, Kraus K, Smith A, Kadiyala S (2003) Allogeneic mesenchymal stem cells regenerate bone in a critical-sized canine segmental defect. J Bone Joint Surg Am 85-A:1927–1935

Bageshlooyafshar B, Isangcheshmeh AM, Vakilian S et al (2019) Zinc silicate mineral-coated scaffold improved in vitro osteogenic differentiation of equine adipose-derived mesenchymal stem cells. Res Vet Sci 124:444–451

Baroli B (2009) From natural bone grafts to tissue engineering therapeutics: brainstorming on pharmaceutical formulative requirements and challenges. J Pharm Sci 98:1317

Benavides-Castellanos MP, Garzón-Orjuela N, Linero I (2020) Effectiveness of mesenchymal stem cell-conditioned medium in bone regeneration in animal and human models: a systematic review and meta-analysis. Cell Regen 9:5

Berner A, Reichert JC, Woodruff MA, Saifzadeh S, Morris AJ, Epari DR, Nerlich M, Schuetz MA, Hutmacher DW (2013) Autologous vs. allogenic mesenchymal progenitor cells for the reconstruction of critical sized segmental tibial bone defects in aged sheep. Acta Biomater 9(8): 7874–7884

Berner A, Henkel J, Woodruff MA, Steck R, Nerlich M, Schuetz MA, Hutmacher DW (2015) Delayed minimally invasive injection of allogenic bone marrow stromal cell sheets regenerates large bone defects in a sheep preclinical animal model. Stem Cells Transl Med 4:503–512

Bhullar KS, Hubbard BP (2015) Lifespan and healthspan extension by resveratrol. Biochim Biophys Acta 1852(6):1209–1218

Blum R, Alini M, Evans C, Schneider E, Egermann M (2004) The effect of AdBMP-2 on ovine osteoblasts and BMSC including a comparison to human and murine cells. Eur Cell Mater 7:48

Bottagisio M, Lovati AB, Lopa S, Moretti M (2015) Osteogenic differentiation of human and ovine bone marrow stromal cells in response to b-Glycerophosphate and monosodium phosphate. Cell Reprogram 17(4):235–242

Bruder SP, Kraus KH, Goldberg VM, Kadiyala S (1998) The effect of implants loaded with autologous mesenchymal stem cells on the healing of canine segmental bone defects. J Bone Joint Surg Am 80:985–996

Bwalya EC, Kim S, Fang J, Wijekoon HMS, Hosoya K, Okumura M (2017) Effects of pentosan polysulfate and polysulfated glycosaminoglycan on chondrogenesis of canine bone marrow-derived mesenchymal stem cells in alginate and micromass culture. J Vet Med Sci 79(7): 1182–1190

Cao L, Liu G, Gan Y, Fan Q, Yang F, Zhang X, Tang T, Dai K (2012) The use of autologous enriched bone marrow MSCs to enhance osteoporotic bone defect repair in long-term estrogen deficient goats. Biomaterials 33(20):5076–5084

Carpenter RS, Goodrich LR, Frisbie DD, Kisiday JD, Carbone B, McIlwraith CW, Centeno CJ, Hidaka C (2010) Osteoblastic differentiation of human and equine adult bone marrow-derived mesenchymal stem cells when BMP-2 or BMP-7 homodimer genetic modification is compared to BMP-2/7 heterodimer genetic modification in the presence and absence of dexamethasone. J Orthop Res 28(10):1330–1337

Chandran S, Shenoy SJ, Babu SS, Nairm RP, Varma HK, John A (2017) Strontium hydroxyapatite scaffolds engineered with stem cells aid osteointegration and osteogenesis in osteoporotic sheep model. Colloids Surf B Biointerfaces 163:346–354

Cheng W, Jin D, Zhao Y (2007) Effect of platelet-rich plasma on proliferation and osteogenic differentiation of bone marrow stem cells in China goats. Zhongguo Xiu Fu Chong Jian Wai Ke Za Zhi 21(4):386–389

Chu W, Gan Y, Zhuang Y, Wang X, Zhao J, Tang T, Dai K (2018) Mesenchymal stem cells and porous β-tricalcium phosphate composites prepared through stem cell screen-enrich-combine (-biomaterials) circulating system for the repair of critical size bone defects in goat tibia. Stem Cell Res Ther 9(1):157

Colbath AC, Kisiday JD, Phillips JN, Goodrich LR (2020) An extracorporeal shockwave promote osteogenesis of equine bone marrow-derived mesenchymal stem cells in vitro? Stem Cells Dev 29(2):110–118

Cuenca-López MD, Andrades JA, Gómez S, Zamora-Navas P, Guerado E, Rubio N, Blanco J, Becerra J (2014) Evaluation of posterolateral lumbar fusion in sheep using mineral scaffolds seeded with cultured bone marrow cells. Int J Mol Sci 15(12):23359–23376

Dai KR, Xu XL, Tang TT, Zhu ZA, Yu CF, Lou JR, Zhang XL (2005) Repairing of goat tibial bone defects with BMP-2 gene-modified tissue-engineered bone. Calcif Tissue Int 77:55–61

Dar ER, Gugjoo MB, Javaid M, Hussain S, Fazili MR, Dhama K, Alqahtani T, Ahqahtani AM, Shah RA, Emran TB (2021) Adipose tissue- and bone marrow-derived mesenchymal stem cells from sheep: culture characteristics. Animals (Basel) 11(8):2153

Das K, Mili B, Madhusoodan AP, Saxena AC, Kumar A, Singh P, Vermae MR, Sarkara M, Bag S (2017) Proliferation of canine bone marrow derived mesenchymal stem cells on different nanomaterial based thin film scaffolds. Tissue Cell 49(2):270–274

Davisson T, Sah RL, Ratcliffe A (2002) Perfusion increases cell content and matrix synthesis in chondrocyte three-dimensional cultures. Tissue Eng 8:807–816

dos Santos CR, da Rocha Filgueiras R, Malard PF, da Cunha Barreto-Vianna AR, Nogueira K, da Silva Leite C, Mendes de Lima EM (2018) Mesenchymal stem cells in osteotomy repair after tibial tuberosity advancement in dogs with cranial cruciate ligament injury. J Exp Orthop 5:17

Egermann M, Lill CA, Griesbeck K, Evans CH, Robbins PD, Schneider E, Baltzer AW (2006) Effect of BMP-2 gene transfer on bone healing in sheep. Gene Ther 13:1290–1299

El Hadidi YN, El Kassaby M, El Fatah Ahmed SA, Khamis N (2016) The effect of mesenchymal stem cells application on the distracted bone microstructure. (An experimental study). J Oral Maxillofac Surg 74(7):1463.e1–1463.e11

Elkhenany H, Amelse L, Lafont A, Bourdo S, Caldwell M, Neilsen N, Dervishi E, Derek O, Biris AS, Anderson D, Dhar M (2015) Graphene supports in vitro proliferation and osteogenic differentiation of goat adult mesenchymal stem cells: potential for bone tissue engineering. J Appl Toxicol 35(4):367–374

Elkhenany H, Amelse L, Caldwell M, Abdelwahed R, Dhar M (2016) Impact of the source and serial passaging of goat mesenchymal stem cells on osteogenic differentiation potential: implications for bone tissue engineering. J Anim Sci Biotechnol 7:16

Fang J, Wei Y, Teng X, Zhao S, Hua J (2017) Immortalization of canine adipose-derived mesenchymal stem cells and their seminiferous tubule transplantation. J Cell Biochem 119(4): 3663–3670

Favi P, Benson R, Dhar M, Neilsen N (2014) Cell proliferation and osteogenic differentiation of equine mesenchymal stem cells on biodegradable biomimetic tubular bacterial cellulose composites. In: 40th Annual Northeast Bioengineering Conference (NEBEC). https://doi.org/10.1109/NEBEC.2014.6972787

Fernandes MBC, Guimarães JAM, Casadom PL, Cavalcantim AS, Gonçalves NN, Ambrósio CE, Rodrigues F, Pinto ACF, Miglino MA, Duarte MEL (2014) The effect of bone allografts combined with bone marrow stromal cells on the healing of segmental bone defects in a sheep model. BMC Vet Res 10:36

Field JR, McGee M, Stanley R, Ruthenbeck G, Papadimitrakis T, Zannettino A, Gronthos S, Itescu S (2011) The efficacy of allogeneic mesenchymal precursor cells for the repair of an sheep tibial segmental defect. Vet Comp Orthop Traumatol 24:113–121

Fischer EM, Layrolle P, Van Blitterswijk CA, De Bruijn JD (2004) Bone formation by mesenchymal progenitor cells cultured on dense and microporous hydroxyapatite particles. Tissue Eng 9(6):1179–1188

Franco GG, Minto BW, Dreibi RM, Costa Junior JS, Dias LGGG (2021) Percutaneous application of allogeneic adipose-derived mesenchymal stem cell in dogs submitted to minimally invasive plate osteosynthesis of the tibia. Acta Cir Bras 36(2):e360206

Gaharwar AK, Schexnailder PJ, Schmidt G (2011) Nanocomposite polymer biomaterials for tissue repair of bone and cartilage: a material science perspective. Nanomaterials handbook, vol 24. Taylor and Francis Group, LLC

Gallego L, Perez-Basterrechea M, Garcıa-Consuegra L, Alvarez-Viejo M, Megas J, Novoa A, Costilla S, Meana A, Junquera L (2015) Repair of segmental mandibular bone defects in sheep using bone marrow stromal cells and autologous serum scaffold: a pilot study. J Clin Periodontol 42:1143–1151

García Cruz DM, Gomes M, Reis RL, Moratal D, Salmerón-Sánchez M, Gómez Ribelles JL, Mano JF (2010) Differentiation of mesenchymal stem cells in chitosan scaffolds with double micro and macroporosity. J Biomed Mat Res Part A 95A:1182–1193

Giannoni P, Mastrogiacomo M, Alini M, Pearce SG, Corsi A, Santolini F, Muraglia A, Bianco P, Cancedda R (2008) Regeneration of large bone defects in sheep using bone marrow stromal cells. J Tissue Eng Regen Med 2:253–262

Glynn ERA, Londono AS, Zinn SA, Hoagland TA, Govoni KE (2013) Culture conditions for equine bone marrow mesenchymal stem cells and expression of key transcription factors during their differentiation into osteoblasts. J Anim Sci Biotechnol 4(1):40

Goldschlager T, Ghosh P, Zannettino A, Gronthos S, Rosenfeld JV, Itescu S, Jenkin G (2010) Cervical motion preservation using mesenchymal progenitor cells and pentosan polysulfate, a novel chondrogenic agent: preliminary study in an sheep model. Neurosurg Focus 28(6):E4

Goldschlager T, Jeffrey V, Rosenfeld JV, Ghosh P, Itescu S, Blecher C, McLean C, Jenkin G (2011) Cervical interbody fusion is enhanced by allogeneic mesenchymal precursor cells in an sheep model. Spine 36(8):615–662

Golonka P, Szklarz M, Kusz M, Marędziak M, Irwin Houston JM, Marycz K (2018) Subchondral bone cyst surgical treatment using the application of stem progenitor cells combined with alginate hydrogel in small joints in horses. Pol J Vet Sci 21(2):307–316

Golonka P, Kornicka-Garbowska K, Marycz K (2020) SIRT1+ adipose derived mesenchymal stromal stem cells (ASCs) suspended in alginate hydrogel for the treatment of subchondral bone cyst in medial femoral condyle in the horse. Clinical report. Stem Cell Rev Rep 16:1328–1334

Gugjoo MB, Amarpal, Chandra V, Wani MY, Dhama K, Sharma GT (2018) Mesenchymal stem cell research in veterinary medicine. Curr Stem Cell Res Ther 13(8):645–657

Gugjoo MB, Amarpal, Makhdoomi DM, Sharma GT (2019) Equine mesenchymal stem cells: properties, sources, characterization and potential therapeutic applications. J Equine Vet Sci 72:16–27

Gugjoo MB, Amarpal, Fazili MR, Shah RA, Mir MS, Sharma GT (2020a) Goat mesenchymal stem cell basic research and potential applications. Small Rum Res 183:106045

Gugjoo MB, Hussain S, Amarpal, Shah RA, Dhama K (2020b) Mesenchymal stem cell-mediated immuno-modulatory and anti-inflammatory mechanisms in immune and allergic disorders. Recent Pat Inflamm Allergy Drug Discov 14:3–14

Gupta MC, Theerajunyaporn T, Maitra S, Schmidt MB, Holy CE, Kadiyala S, Bruder SP (2007) Efficacy of mesenchymal stem cell enriched grafts in an sheep posterolateral lumbar spine model. Spine 32(7):720–726

Gupta SK, Kumar R, Mishra NC (2017) Influence of quercetin and nanohydroxyapatite modifications of decellularized goat-lung scaffold for bone regeneration. Mat Sci Eng C 71: 919–928

Haberstroh K, Ritter K, Kuschnierz J, Bormann K-H, Kaps C, Carvalho C, Mülhaupt R, Sittinger M, Gellrich N-C (2010) Bone repair by cell-seeded 3D-bioplotted composite scaffolds made of collagen treated tricalciumphosphate or tricalciumphosphate-chitosancollagen hydrogel or PLGA in ovine critical-sized calvarial defects. J Biomed Mater Res B Appl Biomater 93: 520–530

Hendawy H, Uemura A, Ma D, Namiki R, Samir H, Ahmed MF, Elfadadny A, El-Husseiny HM, Chieh-Jen C, Tanaka R (2021) Tissue harvesting site effect on the canine adipose stromal vascular fraction quantity and quality. Animals 11:460

Herten M, Zilkens C, Thorey F, Tassemeier T, Lensing-Höhn S, Fischer JC, Sager M, Krauspe R, Jäger M (2019) Biomechanical stability and osteogenesis in a tibial bone defect treated by autologous ovine cord blood cells—a pilot study. Molecules 24:295

Hu P, Pu Y, Li X, Zhu Z, Zhao Y, Guan W, Ma Y (2015) Isolation, *in vitro* culture and identification of a new type of mesenchymal stem cell derived from fetal bovine lung tissues. Mol Med Rep 12:3331–3338

Huang Y-Z, Cai J-Q, Lv F-J, Xie H-L, Yang Z-M, Huang Y-C, Deng L (2013) Species variation in the spontaneous calcification of bone marrow-derived mesenchymal stem cells. Cytotherapy 15: 323–329

Hwang NS, Varghese S, Li H, Elisseeff J (2011) Regulation of osteogenic and chondrogenic differentiation of mesenchymal stem cells in PEG-ECM hydrogels. Cell Tissue Res 344:499–509

Itoi T, Harada Y, Irie H, Sakamoto M, Tamura K, Yogo T, Soeta S, Amasaki H, Hara Y, Tagawa M (2016) Escherichia coli-derived recombinant human bone morphogenetic protein-2 combined with bone marrow-derived mesenchymal stromal cells improves bone regeneration in canine segmental ulnar defects. BMC Vet Res 12:201

Ivirico JL, Salmerón-Sánchez M, Ribelles JL, Pradas MM, Soria JM, Gomes ME, Reis RL, Mano JF (2009) Proliferation and differentiation of goat bone marrow stromal cells in 3D scaffolds with tunable hydrophilicity. J Biomed Mat Res B Appl Biomater 91(1):277–286

Jafarian M, Eslaminejad MB, Khojasteh A, Abbas FM, Dehghan MM, Hassanizadeh R, Houshmand B (2008) Marrow -derived mesenchymal stem cells-directed bone regeneration in the dog mandible: a comparison between biphasic calcium phosphate and natural bone mineral. Oral Surg Oral Med Oral Pathol Oral Radiol Endod 105:e14–e24

Jäger M, Bachmann R, Scharfstädt A, Krauspe R (2006) Ovine cord blood accommodates multipotent mesenchymal progenitor cells. In Vivo 20:205–214

Jang BJ, Byeon YE, Lim JH, Ryu HH, Kim WH, Koyama Y, Kikuchi M, Kang KS, Kweon OK (2008) Implantation of canine umbilical cord blood-derived mesenchymal stem cells mixed with beta-tricalcium phosphate enhances osteogenesis in bone defect model dogs. J Vet Sci 9:387–393

Jo K, Kim Y, Lee SH, Yoon YS, Kim WH, Kweon O-K (2017) Effect of canine cortical bone demineralization on osteogenic differentiation of adipose- derived mesenchymal stromal cells. Heliyon 3(8):e00383

Johnson I, Perchy D, Liu H (2012) In vitro evaluation of the surface effects on magnesium-yttrium alloy degradation and mesenchymal stem cell adhesion. J Biomed Mat Res Part A 100A:477–485

Kalaszczynska I, Ruminski S, Platek AE, Bissenik I, Zakrzewski P, Noszczyk M, Lewandowska-Szumiel M (2013) Substantial differences between human and ovine mesenchymal stem cells in response to osteogenic media: how to explain and how to manage? Biores Open Access 2(5): 356–363

Kang BJ, Ryu HH, Park SS, Koyama Y, Kikuchi M, Woo HM, Kim WH, Kweon OK (2012) Comparing the osteogenic potential of canine mesenchymal stem cells derived from adipose tissues, bone marrow, umbilical cord blood, and Wharton's jelly for treating bone defects. J Vet Sci 13:299–310

Katagiri W, Kawai T, Osugi M, Sugimura-Wakayama Y, Sakaguchi K, Kojima T et al (2017) Angiogenesis in newly regenerated bone by secretomes of human mesenchymal stem cells. Maxillofac Plast Reconstr Surg 39(8):1–8

Kim Y, Kang B-J, Kim WH, Yun H-s, Kweon O-k (2018) Evaluation of mesenchymal stem cell sheets overexpressing BMP-7 in canine critical-sized bone defects. Int J Mol Sci 19:2073

Knippenberg M, Helder M, Doulabi BZ, Semeins CM, Wuisman PIJM, Klein-Nulend J (2005) Adipose tissue-derived mesenchymal stem cells acquire bone cell-like responsiveness to fluid shear stress on osteogenic stimulation. Tissue Eng 11(11/12):1780–1787

Knippenberg M, Helder M, Doulabi BZ, Semeins CM, Wuisman PIJM, Klein-Nulend J (2006) Osteogenesis versus chondrogenesis by BMP-2 and BMP-7 in adipose stem cells. Biochem Biophys Res Commun 342:902–908

Kon E, Muraglia A, Corsi A, Bianco P, Marcacci M, Martin I, Boyde A, Ruspantini I, Chistolini P, Rocca M, Giardino R, Cancedda R, Quarto R (2000) Autologous bone marrow stromal cells loaded onto porous hydroxyapatite ceramic accelerate bone repair in critical-size defects of sheep long bones. J Biomed Mater Res 49(3):328–337

Krych AJ, Stuart MJ (2014) Advances in articular cartilage defect management. ICRS Focus Meeting the Knee. July 03–04, 2014. FIFA Auditorium Sonnenberg, Zurich, Switzerland

Lee HB, Chung YS, Heo SY, Kim NS (2009) Augmentation of bone healing of nonunion fracture using stem cell based tissue engineering in a dog: a case report. Vet Med 54(4):198–203

Lee H, Kwon JC, Kim JH, Kim KC, Cho MS, Yang YS, Oh W, Choi SJ, Seo ES, Lee SJ, Wang TJ, Hong YM (2015) The effect of umbilical cord blood derived mesenchymal stem cells in monocrotaline-induced pulmonary artery hypertension rats. J Korean Med Sci 30(5):576–585

Li H, Ji T, Xu LQ, Hu YJ, Li SY, Zhang CP (2006) Study on the osteogenesis ability of co-culturing bone marrow stromal cells (BMSCs) and small intestinal submucosa. Shanghai Kou Qiang Yi Xue 15(2):167–171

Li D, Li M, Liu P, Zhang Y, Lu J, Lin J (2014) Tissue-engineered bone constructed in a bioreactor for repairing critical-sized bone defects in sheep. Int Orthop 38(11):2399–2406

Li Z, Hou T, Moyuan D, Luo F, Wu X, Xing J, Chang Z, Xu J (2015) The osteogenetic efficacy of goat bone marrow-enriched self-assembly peptide/demineralized bone matrix *in vitro* and *in vivo*. Tissue Eng 21(7–8):1398

Lippens E, Vertenten G, Gironès J, Declercq H, Saunders J, Luyten J, Duchateau L, Schacht E, Vlaminck L, Gasthuys F, Cornelissen M (2010) Evaluation of bone regeneration with an injectable, in situ polymerizable Pluronic F127 hydrogel derivative combined with autologous mesenchymal stem cells in a goat tibia defect model. Tissue Eng Part A 16(2):617–627

Liu H (2011) The effects of surface and biomolecules on magnesium degradation and mesenchymal stem cell adhesion. J Biomed Mat Res Part A 99A:249–260

Liu X, Li X, Fan Y, Zhang G, Li D, Dong W, Sha Z, Yu X, Feng Q, Cui F, Watari F (2010) Repairing goat tibia segmental bone defect using scaffold cultured with mesenchymal stem cells. J Biomed Mat Res Part B Appl Biomater 94:44–52

Lobel KD, Hench LL (1998) In vitro adsorption and activity of enzymes on reaction layers of bioactive glass substrates. J Biomed Mater Res 39:575–579

Loozen LD, van der Helm YJM, Oner FC, Dhert WJA, Kruyt MC, Alblas J (2015) Bone morphogenetic protein-2 nonviral gene therapy in a goat iliac crest model for bone formation. Tissue Eng 21(9):10

Lovati AB, Lopa S, Recordati C, Talò G, Turrisi C, Bottagisio M, Losa M, Scanziani E, Moretti M (2016) In vivo bone formation within engineered hydroxyapatite scaffolds in a sheep model. Calcif Tissue Int 99(2):209–223

Lozier JW, Niehaus AJ, Muir A, Lakritz J (2018) Short- and long-term success of transfixation pin casts used to stabilize long bone fractures in ruminants. Can Vet J 59(6):635–641

Lyons FG, Al-Munajjed AA, Kieran SM, Toner ME, Murphy CM, Duffy GP, O'Brien FJ (2010) The healing of bony defects by cell-free collagen-based scaffolds compared to stem cell-seeded tissue engineered constructs. Biomaterials 31:9232–9243

Mahdavi FS, Salehi A, Seyedjafari E, Mohammadi-Sangcheshmeh A, Ardeshirylajimi A (2017) Bioactive glass ceramic nanoparticles-coated poly(L-lactic acid) scaffold improved osteogenic differentiation of adipose stem cells in equine. Tissue Cell 49(5):565–572

Manassero M, Viateau V, Deschepper M, Oudina K, Logeart-Avramoglou D, Petite H, Bensidhoum M (2013) Bone regeneration in sheep using acropora coral, a natural resorbable scaffold, and autologous mesenchymal stem cells. Tissue Eng 19(13–14):1554–1563

Marycz K, Grzesiak J, Wrzeszcz K, Golonka P (2012) Adipose stem cell combined with plasma-based implant bone tissue differentiation in vitro and in a horse with a phalanx digitalis distalis fracture: a case report. Vet Med 57(11):610–617

Marycz K, Kornicka K, Marezdziak M, Golonka P, Nicpo J (2016) Equine metabolic syndrome impairs adipose stem cells osteogenic differentiation by predominance of autophagy over selective mitophagy. J Cell Mol Med 20(12):2384–2404

McCarty RC, Xian CJ, Gronthos S, Zannettino ACW, Foster BK (2010) Application of autologous bone marrow derived mesenchymal stem cells to an ovine model of growth plate cartilage injury. Open Orthop J 4:204–210

McDuffee LA, Pack L, Lores M, Wright GM, Esparza-Gonzalez B, Masaoud E (2012) Osteoprogenitor cell therapy in an equine fracture model. Vet Surg 41(7):773–783

Millis DL, Jackson AM (2003) Delayed union, nonunion, and malunions. 1849–1861. In: Slatter D (ed) Textbook of small animal surgery, 3rd edn. W.R, Saunders, Philadelphia, p 2714

Nair MB, Suresh Babu S, Varma HK, John A (2008) A triphasic ceramic-coated porous hydroxy-apatite for tissue engineering application. Acta Biomater 4:173–181

Nair MB, Varma HK, John A (2009a) Triphasic ceramic coated hydroxyapatite as a niche for goat stem cell-derived osteoblasts for bone regeneration and repair. J Mater Sci Mater Med 20: S251–S258

Nair MB, Varma HK, Menon KV, Shenoy SJ, John A (2009b) Tissue regeneration and repair of goat segmental femur defect with bioactive triphasic ceramic-coated hydroxyapatite scaffold. Biomed Mater Res 91A:855–865

Nantavisai S, Pisitkun T, Osathanon T, Pavasant P, Kalpravidh C, Dhitavat S, Makjaroen J, Sawangmake C (2020) Systems biology analysis of osteogenic differentiation behaviour by canine mesenchymal stem cells derived from bone marrow and dental pulp. Sci Rep 10:20703

Niemeyer P, Schönberger TS, Hahn J, Kasten P, Fellenberg J, Suedkamp N, Mehlhorn AT, Milz S, Pearce S (2010) Xenogenic transplantation of human mesenchymal stem cells in a critical size defect of the sheep tibia for bone regeneration. Tissue Eng 16(1):33–43

Nino-Fong R, McDuffee LA, Esparza Gonzalez BP, Kumar MR, Merschrod EFS, Poduska KM (2013) Scaffold effects on osteogenic differentiation of equine mesenchymal stem cells: an in vitro comparative study. Macromol Biosci 13:348–355

O'Brien F, Taylor D, Lee T (2002) An improved labelling technique for monitoring microcrack growth in compact bone. J Biomech 35:523–526

Park MJ, Lee J, Byeon JS, Jeong D-U, Gu N-Y, Cho I-S, Cha S-H (2018) Effects of three-dimensional spheroid culture on equine mesenchymal stem cell plasticity. Vet Res Commun 42(3):171–181

Pereira-Junior OCM, Rahal SCJ, Lima-Neto F, da Cruz Landim-Alvarenga F, Monteiro FOB (2013) In vitro evaluation of three different biomaterials as scaffolds for canine mesenchymal stem cells. Acta Cir Bras 28(5):353–360

Pfeiffenberger M, Bartsch J, Hoff P, Ponomarev I, Barnewitz D, Thöne-Reineke C, Buttgereit F, Gaber T, Lang A (2019) Hypoxia and mesenchymal stromal cells as key drivers of initial fracture healing in an equine *in vitro* fracture hematoma model. PLoS One 14(4):e0214276

Prins H-J, Fernandes H, Rozemuller H, van Blitterswijk C, de Boer J, Martens ACM (2016) Spatial distribution and survival of human and goat mesenchymal stromal cells on hydroxyapatite and β-tricalcium phosphate. J Tissue Eng Regen Med 10:233–244

Radtke CL, Nino-Fong R, Rodriguez-Lecompte JC, Esparza Gonzalez BP, Stryhn H, McDuffee LA (2015) Osteogenic potential of sorted equine mesenchymal stem cell subpopulations. Can J Vet Res 79(2):101–108

Randau TM, Schildberg FA, Alini M, Wimmer MD, Haddouti E-M, Gravius S, Ito KJ, Stoddart M (2013) The effect of dexamethasone and triiodothyronine on terminal differentiation of primary bovine chondrocytes and chondrogenically differentiated mesenchymal stem cells. PLoS One 8(8):e72973

Re F, Gabusi E, Manferdini C, Russo D, Lisignoli G (2021) Bone regeneration improves with mesenchymal stem cell derived extracellular vesicles (EVs) combined with scaffolds: a systematic review. Biology (Basel) 10(7):579

Rici REG, Alcântara D, Fratini P, Wenceslau CV, Ambrósio CE, Miglino MA, Maria DA (2012) Mesenchymal stem cells with rhBMP-2 inhibits the growth of canine osteosarcoma cells. BMC Vet Res 8:17

Rodrigues MT, Leonor IB, Gröen N, Viegas CA, Dias IR, Caridade SG, Mano JF, Gomes ME, Reis RL (2014) Bone marrow stromal cells on a 3D bioactive fiber mesh undergo osteogenic differentiation in the absence of osteogenic media supplements: the effect of silanol groups. Acta Biomater 10(10):4175–4185

Schindler O (2011) Current concepts of articular cartilage repair. Acta Orthop Belg 77:709–726

Seib FP, Franke M, Jing D, Werner C, Bornhäuser M (2009) Endogenous bone morphogenetic proteins in human bone marrow-derived multipotent mesenchymal stromal cells. Eur J Cell Biol 88:257–271

Seiler N, Raul F (2005) Polyamines and apoptosis. J Cell Mol Med 9(3):623–642

Seo J-P, Yamaga T, Tsuzuki N, Yamada K, Haneda S, Furuoka H, Tabata Y, Sasaki N (2014) Minimally invasive proximal interphalangeal joint arthrodesis using a locking compression plate and tissue engineering in horses: a pilot study. Can Vet J 55:1050–1056

Shamsul BS, Tan KK, Chen HC, Aminuddin BS, Ruszymah BHI (2014) Posterolateral spinal fusion with osteogenesis induced BMSC seeded TCP/HA in a sheep model. Tissue Cell 46:152–158

Shang Q, Wang Z, Liu W, Shi Y, Cui L, Cao Y (2001) Tissue-engineered bone repair of sheep cranial defects with autologous bone marrow stromal cells. J Craniofac Surg 12(6):586–593

Shen Q-R, Zou S-J, Chen U-X, Xou H-X, Hu J (2006) Effect of mesenchymal stem cells transplantation on cranial suture distraction osteogenesis in growing goats. West China J Stomatol 24(2):153–155

Song J, Kim Y, Kweon O-K, Kang B-J (2017) Use of stem-cell sheets expressing bone morphogenetic protein-7 in the management of a nonunion radial fracture in a toy poodle. J Vet Sci 18(4): 555–558

Sun W, Qiao W, Zhou B, Hu Z, Yan Q, Wu J et al (2018) Overexpression of Sirt1 in mesenchymal stem cellsprotects against bone loss in mice by FOXO3a deacetylation and oxidative stress inhibition. Metab Clin Exp 88:61–71

Szivek JA, Gonzales DA, Wojtanowski AM, Martinez MA, Smith JL (2018) Mesenchymal stem cell seeded, biomimetic 3D printed scaffolds induce complete bridging of femoral critical sized defects. J Biomed Mater Res Part B 107:242–252

Tan L, Zhao B, Ge FT, Sun DH, Yu T (2017) Shockwaves inhibit chondrogenic differentiation of human mesenchymal stem cells in association with adenosine and A2B receptors. Sci Rep 7: 1437

Tang TT, Lu B, Yue B, Xie XH, Xie YZ, Dai KR, Lu JX, Lou JR (2007) Treatment of osteonecrosis of the femoral head with hBMP-2-gene-modified tissue-engineered bone in goats. J Bone Joint Surg Br 89(1):127–129

Tanna S (2004) Osteoarthritis – opportunities to address pharmaceutical gaps. In: Priority medicines for Europe and the world. A public health approach to innovation. P.6.121, 25

Taylor SE, Smith RKW, Clegg PD (2007) Mesenchymal stem cell therapy in equine musculoskeletal disease: scientific fact or clinical fiction? Equine Vet J 39:172–180

Tielens S, Declercq H, Gorski T, Lippens E, Schacht E, Cornelissen M (2007) Gelatin-based microcarriers as embryonic stem cell delivery system in bone tissue engineering: an in-vitro study. Biomacromolecules 8:825

Tjabringa GS, Zandieh-Doulabi B, Helder MN, Knippenberg M, Wuisman PIJM, Klein-Nulend J (2008) The polymine spermine regulates osteogenic differentiation in adipose stem cells. J Cell Mol Med 12(5A):1710–1717

Toupadakis CA, Wong A, Genetos DC et al (2010) Comparison of the osteogenic potential of equine mesenchymal stem cells from bone marrow, adipose tissue, umbilical cord blood, and umbilical cord tissue. Am J Vet Res 71:1237–1245

Trindade AB, Therrien J, Garcia JM, Smith LC (2017) Mesenchymal-like stem cells in canine ovary show high differentiation potential. Cell Prolif 50(6):e12391

Tsuzuki N, Nakao S, Seo J, Yamada K, Haneda S, Furuoka H, Tabata Y, Sasaki N (2014) Effect of biodegradable gelatin β-tri calcium phosphate sponges containing mesenchymal stem cells and bone morphogenetic protein-2 on equine bone defect. J Equine Vet Sci 34(7):903–910

Versari S, Villa A, Helder MN, Doulabi BZ, van Loon J, Bradamante S (2007) Effects of gravity on proliferation and differentiation of adipose tissue-derived stem cells. J Gravit Physiol 14(1): P127–P128

Vertenten G, Vlaminck L, Gorski T, Schreurs E, van den Broeck W, Duchateau L, Schacht E, Gasthuys F (2008) Evaluation of an injectable, photopolymerizable threedimensional scaffold based on D,L-lactide and epsiloncaprolactone in a tibial goat model. J Mater Sci Mater Med 19: 2761

Vertenten G, Lippens E, Gironès J, Gorski T, Declercq H, Saunders J, Van den Broeck W, Chiers K, Duchateau L, Schacht E, Cornelissen M, Gasthuys F, Vlaminck L (2009) Evaluation of an injectable, photopolymerizable, and three-dimensional scaffold based on methacrylate-endcapped poly(D,L-lactide-co-epsilon-caprolactone) combined with autologous mesenchymal stem cells in a goat tibial unicortical defect model. Tissue Eng Part A 15(7):1501–1511

Viateau V, Guillemin G, Bousson V, Oudina K, Hannouche D, Sedel L, Logeart-Avramoglou D, Petite H (2007) Long-bone critical-size defects treated with tissue-engineered grafts: a study on sheep. J Orthop Res 25(6):741–749

Volk SW, Wang Y, Hankenson KD (2012) Effects of donor characteristics and ex vivo expansion on canine mesenchymal stem cell properties: implications for MSC-based therapies. Cell Transplant 21(10):2189–2200

Wang C, Wang Z, Li A, Bai F, Lu J, Xu S, Li D (2010) Repair of segmental bone-defect of goat's tibia using a dynamic perfusion culture tissue engineering bone. J Biomed Mater Res A 92: 1145–1153

Wei X, Liu B, Liu G, Yang F, Cao F, Dou X (2019) Mesenchymal stem cell-loaded porous tantalum integrated with biomimetic 3D collagen-based scaffold to repair large osteochondral defects in goats. Stem Cell Res Ther 10:72

Wheeler DL, Lane JM, Seim H, Puttlitz C, Itescu S, Turner A (2014) Allogeneic mesenchymal progenitor cells for posterolateral lumbar spine fusion in sheep. Spine J 14:435–444

Wheeler DL, Fredericks DC, Dryer RF, Bae HW (2016) Allogeneic mesenchymal precursor cells (MPCs) combined with an osteoconductive scaffold to promote lumbar interbody spine fusion in an sheep model. Spine J 16:389–399

Wijekoon HMS, Bwalya EC, Fang J, Kim S, Hosoya K, Okumura M (2018) Inhibitory effects of sodium pentosan polysulfate on formation and function of osteoclasts derived from canine bone marrow. BMC Vet Res 14:152

Zaffagnini S, Giordano G, Vascellari A, Bruni D, Neri MP, Iacono F et al (2007) Arthroscopic collagen meniscus implant results at 6 to 8 years follow up. Knee Surg Sports Traumatol Arthrosc 15(2):175–183

Zhang XL, Jiang XQ (2006) Osteoblastic differentiation of goat bone marrow stromal cells after AdBMP-2 mediated transduction in vitro. Shanghai Kou Qiang Yi Xue 15(6):610–613

Zhao W, Lu JY, Hao YM, Cao CH, Zou DR (2017) Maxillary sinus floor elevation with a tissue-engineered bone composite of deciduous tooth stem cells and calcium phosphate cement in goats. J Tissue Eng Regen Med 11(1):66–76

Zhu HB, Guo DZ, Yang SJ, Zhang YH, Wang H, Guo HT, Zhang Y, Cheng DC (2008) Osteogenic actions of the osteogenic growth peptide on bovine marrow mesenchymal stromal cells in culture. Vet Med 53(9):501–509

Zou D, Guo L, Lu J, Zhang X, Wei J, Liu C, Zhang Z, Jiang X (2012) Engineering of bone using porous calcium phosphate cement and bone marrow stromal cells for maxillary sinus augmentation with simultaneous implant placement in goats. Tissue Eng 18(13,14):1464–1478

Mesenchymal Stem Cell Therapeutic Applications in Muscle, Tendon and Ligament Regeneration

3

Abstract

Regeneration of tendon, ligament and muscle remains a challenge. As such regenerative medicine employing mesenchymal stem cells (MSCs) is being employed. MSCs have properties similar to that of the tendon-specific stem cells with a potential to heal the tendinopathies. Therapeutic applications of tendon stem cells are hampered by its limited availability and lack of effective isolation and characterization techniques. As an alternative, mesenchymal stem cells (MSCs) are being employed that have a wide range of sources and standardized procedures with specific properties to heal the tissue. MSCs ex vivo studies in relation to the tendon, ligament and muscle differentiation are very limited in veterinary species. In vivo MSCs therapeutic potential for tendon and ligament affections are being evaluated in sheep, goat, dog and horse among various veterinary species. The former two species are mainly being used as translational experimental models contrary to the latter two species wherein both the experimental and the clinical evaluations of MSCs are being conducted. Dog clinical studies evaluate MSCs therapeutic potential mainly in the gastrocnemius and supraspinatus tendon and cruciate ligament injuries. In horses, MSCs therapeutic potential in clinical affections of the superficial digital flexor tendon and suspensory ligaments are performed. All these studies show a favourable outcome with the MSCs applications in tendon injuries; however, the standardized therapy remains to be established. The current chapter focuses on the MSCs properties in relation to the tendon-specific stem cells and their in vivo applications for tendon, muscle and ligament regeneration.

Keywords

Tendon, Muscle, Ligament · Mesenchymal stem cells · Tendon-specific stem cells · In vitro · In vivo

M. B. Gugjoo, *Therapeutic Applications of Mesenchymal Stem Cells in Veterinary Medicine*, https://doi.org/10.1007/978-981-19-3277-9_3

3.1 Introduction

Tendon and ligament are the two specialized structures that connect muscle to the bone, and bone to bone of the joint in an individual, respectively. Both tendon and ligament share histological properties and can preserve elastic energy and are destined to withstand high tension (Im and Kim 2020). The extracellular matrix (ECM) comprises collagen (especially type I and to some extent type III), proteoglycans and glycosaminoglycans to support tension and bear weight (Costa-Almeida et al. 2019). There are limited nucleated cells including the stem cells in these tissues. Animal tendon and ligament are more prone to tears from overstrain due to their typical location and function (Tyrnenopoulou et al. 2016). Horses often undergo superficial digital flexor tendinopathy and suspensory ligament desmitis, comprising one of the most common causes of lameness in this species (Kasashima et al. 2004). Rupture of cranial cruciate ligament (crCL) in dogs is the most common stifle joint ailment and one of the most common reasons for the hind limb lameness (Johnson et al. 1994). Cattle tend to get affected mostly in medial patellar ligament. Tendinopathies are either acute or chronic (degenerative) (Schneider et al. 2017) with the former condition developing immediately after trauma and the latter condition arising through recurrent over activity and sometimes with other disorders. Chronic tendinopathies develop over long periods (>3 months) and are frequently associated with an unresolved inflammatory condition leading to an impaired performance (Kaux and Crielaard 2013). Ligament, in addition to the injuries, undergoes chondrification/ossification (enthesophyte) and/or degeneration in chronic osteoarthritis (Schulze-Tanzil 2019).

Tendon/ligaments, although, are hypo-vascular and -cellular in nature, but these structures still have intrinsic and extrinsic healing mechanisms that act synergistically (Kajikawa et al. 2007). The former mechanism seems limited in efficacy as compared to the latter (Sharma and Maffulli 2005). These mechanisms fall short in the case of extensive affections and in systemic pathologies as these tissues have an overall limited potential to self-heal (Shojaee and Parham 2019). Their pathologies are further aggravated by an increase in temperature upon loading and such thermal reactions lead to the imperfect metabolism or death of tendinocytes (Pacini et al. 2007; Alsook et al. 2015). Being multi-fibrillar structure, any damage in collagen fibril initiates an inflammatory response consequent upon the release of proteolytic enzymes (Lombana et al. 2015). Acute affections like inflammation and without any tissue loss are mostly managed by conservative methods, while the chronic tendinopathies involving structural damage or tissue loss, surgical repair through suturing or regenerative medicine involves auto- or allografting. The damaged tendon/ligament may recover but with an altered function, and thus, poses a substantial re-injury risk (Violini et al. 2009; Brandão et al. 2018). In horses, re-injury rate goes up to 43% upon pre-injury loading with the current standard treatments (Roth et al. 2017). Despite advances in therapeutics, an effective functional outcome remains to be achieved (Chen et al. 2009). Thus, regenerative medicine employing stem cells is being evaluated for effective and long-lasting solution of the tendon/ligament affections. The main aim remains to develop quality tissue that is free of adhesions similar to pre-injury levels (Longo et al. 2010; Snedeker and Foolen 2017).

3.2 In Vitro Studies

3.2.1 Tendon Stem Cells Vs. Non-tendon MSCs

The intrinsic healing may occur through the tendon cells including the stem cells. Tendon or ligament harbours stem cells that show some of the characteristics comparable to that of the non-tendon and non-ligament mesenchymal stem cells (MSCs). These cells are popularly known as tendon stem/progenitor cells (TSPCs)/ tenogenic stem cells (Bi et al. 2007; Zhang and Wang 2010a, b; Alsook et al. 2015; Lee et al. 2017). TSPCs constitute nearly 1–4% of the tendon-nucleated cells (Bi et al. 2007) and are frequently obtained as a heterogeneous population (Costa-Almeida et al. 2019) depending upon the tissue area. The peritenon cells seem more reactive and differentiate towards the myofibroblast phenotype and thus have a greater potential of scarring and adhesion formation during tendon healing (Cadby et al. 2014). In horse deep digital flexor tendon (DDFT) fibrocartilaginous and tendinous zones harbour TSPCs. These cells have fibroblast morphology and express CD90 and CD29 similar to that of the MSCs but lack CD44; with comparable basal ability to express tenogenic, osteogenic or chondrogenic markers. However, fibrocartilaginous TSPCs are largely restricted to chondrogenic differentiation, while tendinous zone TSPCs differentiate into both osteogenic and chondrogenic lineages. Both the subpopulations may be weak in adipogenesis (Quam et al. 2021). Horse TSPCs have also been grouped from endotenon of superficial digital flexor into three clones on the basis of their ability to differentiate viz., (a) trilineage differentiation and tenogenic; (b) tendon, cartilage and bone without adipogenic and (c) tendon, cartilage and adipose without the osteogenic. Cells able to differentiate into all the four lineages show the highest expression of scleraxis and mohawk upon tenogenesis (Rajpar and Barrett 2020). Ligament (equine suspensory ligament)-derived stem cells too show MSCs like characteristics including fibroblast morphology, expression of CD90, CD73 and CD105 and lacking CD45 expression (Fig. 3.1). These cells differentiate into trilineage (adipogenic, chondrogenic and osteogenic) and express pluripotency markers like octamer-binding transcription factor 4 (oct4) and stage-specific embryonic antigen 1 (SSEA-1). Among these cells, 15% cells are positive for Class III beta-tubulin (TUJ-1), but only a few cells express glial fibrillary acidic protein (GFAP) (Alsook et al. 2015). Thus, tendon stem cells show some of the features like self-replication, surface markers and the basal ability to differentiate into various lineages comparable to the non-tendon MSCs (Zelzer et al. 2014; Im and Kim 2020). However, these cells as per the location in the tendon-like peritenon (paratenon and epitenon) or endotenon exhibit differential expression, are not as effective in trilineage differentiation as MSCs and also may not express all the surface markers like CD44 (Cadby et al. 2014; Williamson et al.

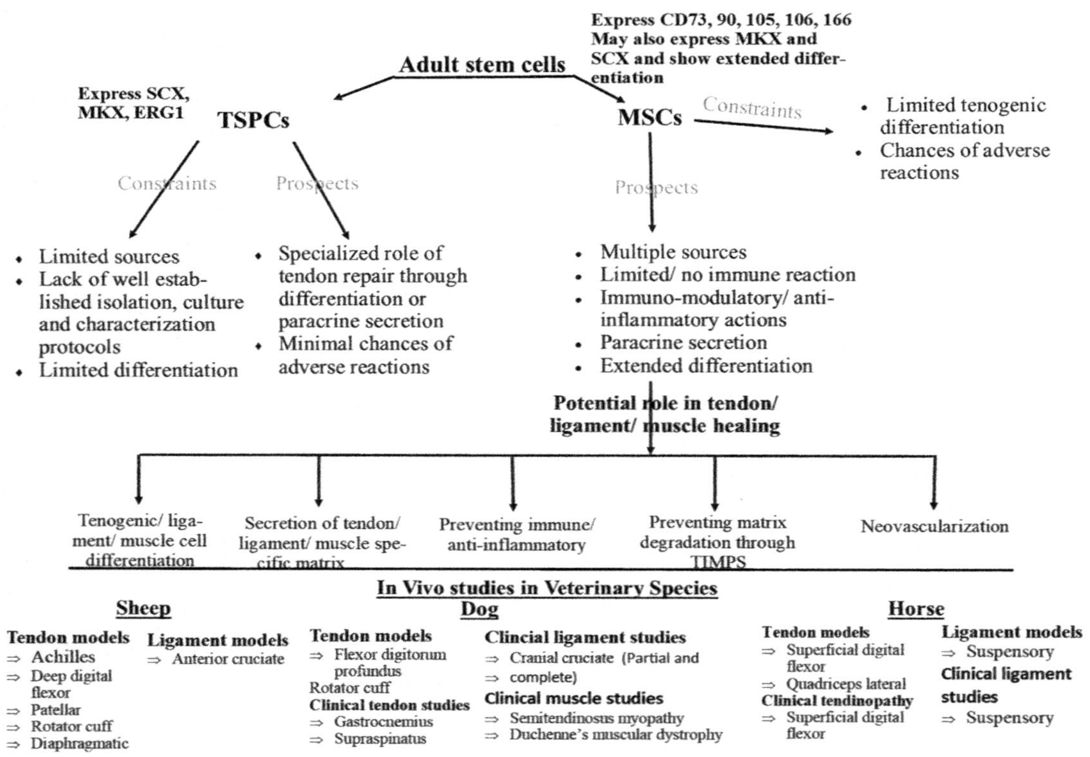

Fig. 3.1 Comparative overview of the tendon specific progenitor cells (TSPCs) and mesenchymal stem cells (MSCs) for their tendon or ligament regenerative medicine. The image also shows the in vivo studies conducted in veterinary species with the MSCs

2015; Rajpar and Barrett 2020; Quam et al. 2021). All these factors along with the limited sources limit their utilization for therapeutic applications.

3.2.2 Tenogenic Differentiation

MSCs tenogenic differentiation is being confirmed through their morphology (stretching and spindle shape) and expression of tenogenic markers (Javanshir et al. 2020) (Fig. 3.1). The definitive consensus on tenogenic differentiation markers is lacking (Tempfer et al. 2009; Nourissat et al. 2015); however, some of the markers like Scleraxis (SCX), Mohawk (MKX) and early response protein (Egr1) are considered to be specific. SCX, a helix–loop–helix transcription factor, is expressed early in tendon development. It promotes the transcription of genes like type 1 collagen a1/a2, aggrecan, and Tnmd in TSPCs (Zelzer et al. 2014). Type 1 collagen expression is additionally regulated by MKX, although its deficiency has a limited impact on the cell number, but collagen fibre size and diameter is reduced. It promotes SCX expression through TGF-β2 promoter and tends to inhibit proliferation and trilineage differentiation potential of MSCs. SCX initiates tenogenic differentiation, while MKX helps in maturation and maintenance of collagen fibres (Tempfer et al. 2009; Mienaltowski et al. 2013; Tan et al. 2013). Egr1 promotes SCX expression and thereby enhances collagen a1/a2 expression. Thus, Erg1 is mainly involved in maintaining tensile strength of tendon and ligament (Tempfer et al. 2009; Nourissat et al. 2015). Tendon MSCs (Yang et al. 2016) and non-tendon MSCs (Yamasaki et al. 2018; Korchunjit et al. 2019; Javanshir et al. 2020) express tendon-specific matrices but unlike TSPCs does not undergo spontaneous tenogenic differentiation (Javanshir et al. 2020). For effective and result-oriented tenogenesis, various extrinsic factors like growth factors/interleukins, matrices and mechanical factors are employed.

3.2.3 Growth Factors and MSCs/TSPCs Tenogenesis

In veterinary species, details on MSCs tenogenic differentiation is limited as compared to the other lineages. Various growth factors including BMP-12 (Violini et al. 2009; Gulati et al. 2013; Mohanty et al. 2014), BMP-13 (Shojaee and Parham 2019), TGF-β3 (Gomiero et al. 2016; Roth et al. 2018, 2019), IGF-1 (Durgam et al. 2012), platelet-derived growth factor (PDGF), PDGF-BB, endothelial growth factor (EGF), fibroblast growth factor (FGF) (Mohanty et al. 2014), growth differentiation factor-6 (GDF-6) (Javanshir et al. 2020), interleukins like interleukin [(IL)-6, IL-1β], platelet-rich plasma (PRP) and prostaglandin E2 (PGE2) (Mishra et al. 2009; Zhang et al. 2010) are proposed to be the important mediators to promote tenogenesis of the stem cells. BMP-12 induces the expression of tenomodulin (TNMD, transmembrane protein expressed by mature tenocytes regulating their proliferation and matrix organization) and decorin (DCN, small cellular or peri-cellular matrix proteoglycan that regulates collagen fibrils assembling and biomechanical properties) in equine

BM-MSCs (Violini et al. 2009) and AF-MSCs (Gulati et al. 2013) and additionally Mkx, Col I and SCX in equine UCB-MSCs (Mohanty et al. 2014) and equine foetal BM-MSCs (Oliva et al. 2021). However, it is not necessary that BMP-12 always induces tenogenic expression of equine MSCs as other factors like extracellular matrix topography may additionally impact their gene expression (Roth et al. 2018). Similarly, PDGF-BB and GDF-6 may induce the expression of MKX and COLI in equine AD-MSCs at day 14, while at the said period non-significant expression of SCX comparable to control occurs (Javanshir et al. 2020). Non-significant expression of SCX may be due to its early expression that decreases as the time passes or there might be differences in MSCs based on the species or tissue type. Further, variable concentration (10 ng/mL to the 50 ng/mL) of BMP-12 may also affect MSCs tenogenic differentiation.

Some of the individual growth factors like TGF-β3, IGF-1 and EGF-2 may be unable to induce tenogenic differentiation; however, their combined use may successfully induce it. TGF-β3 along with bFGF2 upregulates tenogenic expression contrary to the combination of bFGF2-IGF-1 and bFGF2-EGF-2 that fails (Gomiero et al. 2016). The combined use of growth factors (TGF-β1 and IGF-1) and tenocyte spent media induces dog AD-MSCs tenogenic expression of Col I/III, DCN, TNMD, β1-Integrin, MAPKinase pathway (Shc, Erk1/2) and SCX (Schneider et al. 2011). TGF-β3-mediated horse AD-MSCs tenogenesis may occur through activation of Smad 2/3 with nuclear translocation, and this tenogenic expression may further be improved through the inhibition of the Rho/ROCK (Melzer et al. 2021). Growth factors that effectuate each other's tenogenic pathways may induce MSCs tenogenic differentiation.

Mechanical loading also plays a key role in tendon repair. The mechanical forces get translated into biochemical signals through mechano-transduction-capable molecules activating key cellular processes (Galloway et al. 2013; Raabe et al. 2013; Lavagnino et al. 2015). Moderate but not heavy loading patterns induce anabolic processes, and thus, tendon or ligament homeostasis (Heinemeier and Kjaer 2011; Heinemeier et al. 2012; Galloway et al. 2013). IGF-1 may induce tendon acellular matrix-loaded TSPC's tenogenic differentiation but not in BM-MSCs (Durgam et al. 2012).

3.2.4 Matrices/Scaffolds and MSCs Tenogenesis

MSCs co-culture in the tendon tissue matrix improves their tenogenic potential. Tendon tissue enhances equine BM-MSCs tenogenic differentiation (increased expression of DCN, TNMD, tenascin-C and collagen type I) with an ability to retain tenocyte-like phenotype (Lovati et al. 2012; Roth et al. 2018). Dog MSCs loaded onto the tendon slices and aligned between the collagen fibres of the tendon slices show higher TNMD and MMP13 expression and higher Col I expression in the composite at 14th day (Omae et al. 2009). This tendon tissue-induced tenogenic differentiation of MSCs may occur via an integrin/TGF-β crosstalk-mediated mechanism (Wang et al. 2020). Even an engineered electrospun poly (lactide-co-glycolide) (PLGA) fleeces with highly aligned fibres (ha-PLGA) mimic tendon extracellular matrix and enhance ovine amniotic epithelial stem cell tenogenic differentiation (Russo et al. 2020). Joint inflammatory environment (cranial cruciate ligament rupture and medial patella luxation) might enhance synoviocytes multipotency exhibiting markers (CD44 and CD90) comparable to MSCs. Such an enhanced multipotency remains more prominent in cranial cruciate ligament rupture as compared to the medial patellar luxation (Wijekoon et al. 2017). This shows the tendon harbours cells with stem cell properties. Ancillary options like laser (660 nm) therapy along with growth factors like bFGF2 and TGF-β3 may also significantly enhance MSCs tenogenic expression including tenascin c (Gomiero et al. 2016). Further, SCX introduction into MSCs may significantly decrease their SOX-9 (cartilage marker) and osteogenic potential expression (Zelzer et al. 2014). This may additionally inhibit chondrogenic metaplasia of tendons/ligaments to prevent enthesophyte formation (Tempfer et al. 2009; Mienaltowski et al. 2013; Tan et al. 2013).

Thus, further insights are desired to compare various cocktails of growth factors, scaffolds and mechanical stimulation for effective tenogenesis of MSCs. This shall open Pandora's Box and help in understanding the intricate balance of homeostasis of tendon/ligaments and its extrinsic regulation.

3.2.5 Miscellaneous Factors Affecting MSCs Tenogenesis

Age affects stem cell characteristics including that of the TSPCs and MSCs which show limited proliferation and differentiation potential and may be considered as one of the reasons for reduced tendon repair in aged patients (Zhou et al. 2010). Additionally, cell source may also affect tenogenesis as skeletal muscle stem cells show higher expression for tendon-specific matrices (TNMD, Col I and III), while the BM-MSCs express only some of the tendon-specific genes (Col I and Col III) (Sassoon et al. 2012). This demands that pathways involved in such a differentiation of stem cells need to be understood.

3.3 Muscular Disorders

Muscle injuries are common in horse and athletic dogs (Lee et al. 2012), and muscular dystrophy involving Duchenne muscular dystrophy (DMD) is one of the most common and severe forms of muscular dystrophy seen in humans and animals. DMD is an X-linked genetic disorder involving dystrophin genes. It leads to the dystrophin deficiency causing breakdown of the structural integrity of myofibres, resulting in progressive myofibre necrosis, fibroblast proliferation and growth of fibrous tissue and fat (Nakamura and Takeda 2011). Muscle carries inherent repair potential through stem cells that activate even after the exercise in horses. Activation in the form of increased expression of IL-6, paired box transcriptional factor (Pax) 7, myogenic differentiation 1 (MyoD), myogenin, proliferating cell nuclear antigen

(PCNA), insulin-like growth factor (IGF)-I and hepatocyte growth factor (HGF) mRNA occurs (Kawai et al. 2013; Čamernik et al. 2019). These stem cells are subpopulations of mesenchymal stromal cells including Pax-7-positive satellite cells, fibro-/adipogenic progenitors and PW1-positive interstitial cells, among others (Čamernik et al. 2019). Muscle-derived stem cells mainly express CD13, CD10, Sca-1 and CD56 to some extent (Young et al. 1999). Further, these cells lack expression of hematopoietic marker (CD45) and undergo differentiation into various lineages like chondrogenic, adipogenic, osteogenic and neural (Wu et al. 2010). However, standard muscle stem cell characterization is still debatable, and in veterinary science literature on such cells is even more limited. Despite the availability of such cells, complete functional recovery of muscle remains a challenge as fibrosis along with atrophy sets in after severe injury (Milner et al. 2018; El-Habta et al. 2018). Currently, there is no effective treatment for the repair of muscle disorders, and thus regenerative medicine is being considered. Due to the lack of effective standardized protocols to harvest (single-fibre isolation and whole muscle enzymatic digestion) and isolate muscle-derived stem cells, regenerative medicine mainly employs MSCs from other sources (Nitahara-Kasahara et al. 2012; Carlo et al. 2019). These cells are also being considered for ex vivo meat production; however, the current protocol is quite uneconomical with additional lacunas in flavour of the cultured meat burger (Gugjoo et al. 2018; Choi et al. 2021).

Currently, limitations exist in applying tendon/ligament/muscle stem cells due to the lack of standardized isolation and characterization protocols, in addition to their limited availability and reduced proliferation potential with extensive passaging (Costa-Almeida et al. 2019). To develop an alternative source, MSCs from sources other than tendon, muscle and ligament are evaluated in an ex vivo system and utilized under in vivo system with improved results. Mesenchymal stromal cells are not completely characterized in all the individual studies as per International Society for Cellular Therapy (ISCT) criteria and a consensus on the definitive therapeutic mechanisms in these musculoskeletal tissues remain to be developed. Further, MSCs specialized properties specific to tendon/ ligament/ muscle under ex vivo system and their potential therapeutic mechanisms under in vivo environments remain a challenge for evidence-based medicine.

3.3.1 In Vitro Muscle Differentiation Studies

Various in vitro studies demonstrate MSCs ability to differentiate into the muscle-like cells. MSCs derived from sheep endometrium (Letouzey et al. 2015), goat skeletal muscle (Tripathi et al. 2010), cat adipose tissue (Kono et al. 2014), dog adipose tissue (Vieira et al. 2010; Oh et al. 2011), equine endometrium (Rink et al. 2017) and peripheral blood (Patruno et al. 2017) differentiate into muscle cell-like cells. Even equine AD-MSCs derived iPSCs may also undergo muscle differentiation (Lee et al. 2016). Different cocktails of chemicals (5-Aza-2′-deoxycytidine (5-Aza), myoblast-secreted factor galectin-1 (Gal-1)/myoblast culture medium SkGM-2 BulletKit and bezafibrate) effectuate MSCs muscle-like differentiation

(Oh et al. 2011; Ramirez-Espinosa et al. 2015; Okamura et al. 2017). However, complete and effective myogenic differentiation cocktail is yet to be developed. Tat-MyoD remains ineffective to induce myogenic differentiation of equine PB-MSCs and the same can be achieved through culture addition of the conditioned medium from C2C12 cells (Patruno et al. 2017). Variability in gene expression of myogenic differentiated cells with different cocktails is reported. 5-Aza downregulates expression of muscle regulatory factors like myoblast determination protein (MYOD) and upregulates myogenic factor-6 (MYF6), MYF-5 and DES in cattle BM-MSCs. Contrarily, Gal-1 and SkGM-2 BulletKit downregulate early MRF (MYF5), while MYOD and late MRF (DES) mRNA upregulate. Further, concentration of the chemicals also affects differentiation and cell viability. Higher concentration (100 μM) induces higher differentiation rates but are simultaneously toxic to the cattle BM-MSCs, while lower (10 μM) concentration induces gradual myogenic differentiation without any toxicity (Okamura et al. 2017). Goat BM-MSCs co-cultured with muscle-derived cells (MDCs) too exhibit myocyte-like features (Kulesza et al. 2016), correlating the effect of the microenvironment. Thus, MSCs may undergo myogenic-like cell differentiation; however, their typical features for actual myocytes remain debatable. Besides, the in vivo studies on these myocyte-like cells remain to be conducted.

3.4 MSCs Therapeutic Effects

MSCs may offer therapeutic effects either by direct differentiation into tenogenic cells or through release of trophic factors (Yoon et al. 2012; Gugjoo et al. 2018, 2019, 2020). Under in vivo system, MSCs modulate tendon inflammatory response by enhancing regenerative M2 macrophages, inflammatory inhibitors (prostaglandin reductase 1), and proteins involved in tendon formation. At the injury site, mononuclear cell infiltration is hampered while infiltration of the progenitor cells to the damaged site enhances. An enhanced matrix secretion may be induced while simultaneously inhibiting its degradation through release of tissue inhibitor matrix metalloproteinases (TIMPs). Incorporation of growth factors like recombinant bone morphogenetic protein-12 (rBMP-12) may further activate M2 macrophages through IL-6 secretion and M2 effector MMP-12 (involved in matrix remodelling). Such a combination further encourages angiogenesis and the cell migration (Gelberman et al. 2017).

3.5 In Vivo MSCs Studies

In veterinary species, in vivo studies are evaluated mostly in sheep, dog and horse. Sheep is utilized as a human translational model, while dog and horse are made as subjects for the MSCs therapeutic evaluation in both experimental and clinical studies. This diverse nature of studies helps to get insights into different aspects of the MSCs applications and becomes the cornerstone of MSCs evidence-based medicine. Sheep and dog rotator cuff models are considered as a suitable translational model for humans. Comparing the architecture of rotator cuff muscle of various animals (cows, pigs and sheep and dog), muscle parameters of dogs are closer to humans than that of others. However, limitation with canine models is that of the repair failure due to direct weight bearing (Mathewson et al. 2014). The below given session mainly focuses on MSCs in vivo studies in tendon, ligament and muscle regenerative medicine in veterinary species.

3.5.1 Route of Implantation

MSCs are transplanted via intralesional, intravenous and regional perfusion routes in tendon or ligament affections. The cellular tracking is done through histology, MRI or nuclear medicine (Dudhia et al. 2015). In equine naturally occurring tendon or ligament injuries, intralesional route favours the highest cell homing followed by the regional perfusion and at lowest level by the intravenous route (cell 'home in' towards lungs). Sheep MSCs engraft efficiently to the Achilles tendon post local intralesional implantation (Lacitignola et al. 2014). However, intralesional implantation of these cells lacks uniform distribution throughout the entire length of tendon, its muscle and attaching bone, besides, the cell loss may be considerable (Smith et al. 2011). The cells show uneven distribution preferably to the synovial membrane in clusters and the tendon sheath (Khan et al. 2013). Such a distribution pattern may also be affected by lesion and available local microenvironment.

3.5.2 Sheep Studies

3.5.2.1 Tendon

MSCs therapeutic potential is evaluated in various sheep tendon models including Achilles, deep digital flexor, patellar, rotator cuff muscle and diaphragmatic tendons (Table 3.1 and Fig. 3.1). Evaluation of MSCs in varied types of models helps to evaluate their therapeutic potential in diverse tendinopathies.

Achilles tendonitis model is created with collagenase and treated at delayed (2 weeks later) interval to ensure complete tendonitis and/ or to ape chronic condition (Crovace et al. 2008; Lacitignola et al. 2014). These studies are randomized with one of the studies injecting BM-MSCs in fibrin glue (Lacitignola et al. 2014), while another study compares BM-MSCs against the BM-MNCs loaded with the fibrin glue (Crovace et al. 2008). Variable cell concentrations are transplanted with an evaluation made through histomorphology, immunohistochemistry and cartilage expression analysis for a period of 6 weeks (Lacitignola et al. 2014) to 8 weeks (Crovace et al. 2008).

The cell engraftment remains observable up to 6 weeks (Lacitignola et al. 2014). These studies show enhanced tendon repair with the cell transplantation as compared to the control. MSCs/fibrin glue implantation improves tendon architecture with a typical matrix exhibiting higher Col I and cartilage oligomeric matrix protein

Table 3.1 Sheep in vivo MSCs preclinical experimental studies on various tendinopathy and ligament affections

Study design/period	Biomaterial used cell dose	Evaluation criteria	Overall result	References
Tendonitis of Achilles tendon				
Randomized placebo-controlled comparative study ($n = 30$)/8 weeks	BM-MSCs ($1.1 \times 10^6 \pm 1.4 \times 10^5$), BMMNCs ($101 \times 10^6 \pm 26 \times 10^6$) and fibrin glue after 2 weeks	Histomorphology, immunohistochemistry	Improved outcome	Crovace et al. (2008)
Randomized controlled study ($n = 9$)/6 weeks	Allogeneic BM-MSC (6×10^6), fibrin glue after 2 weeks	Evaluation for morphology, expression analysis	Improved outcome	Lacitignola et al. (2014)
Deep digital flexor tendonitis				
Randomized placebo-controlled study ($n = 18$)/120 days	Autologous PB-MSCs (10×10^6), PRP (1 mL of $882 \pm 199 \times 10^3$/mL), hyaluronic acid (only left limb lesions treated while contralateral limb served as control)	Clinical, ultrasonography, histology, immunohistochemistry expression	Improved outcome	Martinello et al. (2013)
Randomized placebo-controlled study ($n = 50$)/24 weeks	Magnetic iron-oxide nanoparticles (MION) labelled autologous BM-MSCs (5×10^6 implanted 2 weeks after lesion)	MRI, histology and flow cytometry	Insignificant result	Khan et al. (2018)
Randomized placebo-controlled study ($n = 24$)/24 weeks	Autologous S-MSCs (5×10^6), implanted 2 weeks after lesion	MRI, histology	Insignificant result	Khan et al. (2020)
Patellar tendon retraction model				
Randomized controlled comparative study ($n = 10$)/12 weeks	BM-MNCs, demineralized bone matrix	Force plate analysis, CT, histology	Improved outcome	Thangarajah et al. (2016)
Rotator cuff injury model				
Randomized placebo-controlled study ($n = 24$)/3 months	BM-MSCs, muscle MSCs	Muscle biopsies and muscle testing, MRI	Improved outcome	Coleman et al. (2011)
Diaphragmatic defect model				
Randomized placebo-controlled study ($n = 9$)/12 months	AF-MSCs	Histology	Improved outcome	Fuchs et al. (2004)

(continued)

Table 3.1 (continued)

Study design/period	Biomaterial used cell dose	Evaluation criteria	Overall result	References
Randomized placebo-controlled study ($n = 27$)/14 months	Amniotic MSCs (5×10^6 cells/mL), Teflon and acellular bioprosthetic implants (collagen I hydrogel placed in between acellular human dermis)	Tensile strength measurements, histology, and quantitative extracellular matrix analysis	Improved outcome	Turner et al. (2011)
Common calcaneal tendon model				
Randomized placebo-controlled study ($n = 16$)/8 weeks	Stromal vascular fraction (derived from 4 gm subcutaneous fat) after 2 weeks	Histology, immunohistochemistry, PCR, biomechanical testing, ultrasonography	Improved outcome	Piccionello et al. (2021)
Cranial cruciate ligament injury models				
Randomized placebo-controlled study ($n = 15$)/12 weeks	Allogeneic BM-MSCs (10×10^6), PRP (2 mL)	Gross and MRI	Improved outcome	Hexter et al. (2021b)
Randomized placebo-controlled comparative study ($n = 20$)/12 weeks	BM-MSCs (10×10^6), demineralized bone matrix (DBM, 4 mL), activated leukocyte-poor platelet-rich plasma	Tunnel wall sclerosis, trabecular bone analysis, μCT analysis	Improved outcome	Hexter et al. (2021a)

MSCs mesenchymal stem cells, *AF* amniotic fluid derived, *PB* peripheral blood derived, *BM* bone marrow derived, *MNCs* mononuclear cells, *n* total number of animals in all the groups of the study, *PRP* platelet rich plasma; Improved outcome: Indicates improved outcome with MSCs transplantation even if only one parameter improves; Study period: Indicates maximum time period for which study is conducted

(COMP) and low collagen type III (Crovace et al. 2008; Lacitignola et al. 2014). An interesting finding of the higher CD34+ cells in cell-treated tendons point towards MSCs potential chemotactic recruitment that can be seen up to sixth week (Lacitignola et al. 2014). These cells may be more effectively recruited upon mononuclear cell (MNC) transplantation as compared to the cultured BM-MSCs at eighth week (Crovace et al. 2008). CD34 tends to be the superficial marker being expressed by BM mononuclear cells that are lost while differentiating into the adult cell (Simmons and Torok-Storb 1991). Considering the critical weight bearing carried out by Achilles tendon, it is imperative to ensure healed tissue matrices and strength (that conveys collagen fibril formation and their orientation) are typical to the normal tendon. However, such extensive details are mostly lacking.

Deep digital flexor tendonitis (DDFT) models are created with enzyme (collagenase type I) (Martinello et al. 2013) or through surgery (Khan et al. 2018, 2020). MSCs from peripheral blood (PB) loaded in platelet-rich plasma (PRP) at day 7 (Martinello et al. 2013), BM (Khan et al. 2018) or synovium (Khan et al. 2020) at week 2 are transplanted. In the lateral branch DDFT lesion, transplantation of MSCs fails to show any gross and/or histological evidence of healing of the lesion up to 24 weeks. The cellular distribution occurs throughout the tendon synovial sheath but is restricted to it (Khan et al. 2018, 2020). BM-MSCs fail to reach to the lesion site with resultant lack of any therapeutic effect (Khan et al. 2018). The same group also fails to demonstrate observable benefits with the synovial MSCs (Sy-MSCs) transplantation and could be due to their heterogeneous nature exhibiting variabilities in proliferation, clonogenicity, multipotency and cell-surface marker expression. These cells mainly show chondrogenic differentiation, among trilineage and thus point their propensity towards chondrogenesis (Khan et al. 2020). Contrarily, PB-MSCs irrespective of the scaffold type (PRP) improves tendon repair (enhanced tissue remodelling and improved structural organization). As PRP shows no observable effect over the control, MSCs only remain to offer therapeutic effects (Martinello et al. 2013). Thus, model type (surgical vs. enzyme induced), actual MSCs (duly characterized) and their source may have a role to determine tendon healing.

Patellar tendon bone defect model mimics rotator cuff tears but lacks precision as the latter has associated degeneration. However, it allows assessment of healing enthesis as the extensor mechanism in sheep has no supporting structures. In such a model (patellar tendon distal 1 cm resected and sutured), BM-MNCs loaded on to the demineralized bone allograft improve mechanical and histological properties of the regenerated tendon bone tissue in comparison to the xenograft at 12th week. Further at the defect area, cell transplantation increases remodelling of bone matrix (demineralized) into tendon-like tissue and significantly higher fibrocartilage-based enthesis (Thangarajah et al. 2016). In the actual rotator cuff model (infraspinatus muscle), strength improves upon MSCs transplantation and surgical repair at 3 months. BM-MSCs appear superior (increased vascularity) over muscle MSCs. Tendon treated with BM-MSCs and muscle MSCs exhibits low fat to muscle ratio as compared to control under MRI (Coleman et al. 2011). Thus, MSCs and their source affect the repair of the rotator cuff muscle tendon.

In the common calcaneal tendonitis model (collagenase type I induced), stromal vascular fraction (SVF) transplanted at 2 weeks post injury improves outcome. These SVF transplanted tendons show improved fibre orientation and fibre oedema scores, besides the inflammatory and necrosis scores. Additionally, the healed tendon exhibits significantly higher Col I, factor VIII expression and lower collagen III expression. Overall, the transplantation of SVF improves tendon diameter and hardness comparable to the normal tendons (Piccionello et al. 2021). SVF contains MSCs and other cells in addition to the growth factors.

Tissue engineering might play an important role in diaphragmatic hernia repair. Congenital or acquired cattle/buffalo diaphragmatic hernia is common, and in extensive damages, tissue-engineered constructs are required. In order to ensure effective reconstruction, MSCs may be incorporated. In neonatal lamb diaphragmatic defect (ranging from 7.5×4.5 cm to 6×7 cm) models, amniotic fluid MSCs loaded on to the Teflon patch (Turner et al. 2011) or autologous acellular small intestinal submucosa (Fuchs et al. 2004) enhance tissue repair with significantly higher tensile strength, total collagen levels and blood vessels as compared to acellular bio-prosthetic grafts. The repair tissue stems from the surrounding native tissue (Fuchs et al. 2004; Turner et al. 2011). Thus, tissue-engineered constructs employing MSCs have a potential to repair diaphragmatic hernia defects.

Anterior cruciate ligament rupture is commonly treated surgically through tibial plateau levelling osteotomy (TPLO) or tibial tuberosity advancement (TTA). However, in both the cases, a major concern remains the bone healing. In two bone graft maturation studies post ACL resection, transplantation of BM-MSCs along with the PRP (Hexter et al. 2021b) or demineralized bone matrix/PRP (Hexter et al. 2021a) improves graft maturation and reduces bone tunnel widening in a 12-week period. However, reduction in bone tunnel widening with the cells is comparable with the one transplanted with the fibrin glue (Hexter et al. 2021a). Thus, the exact role of the MSCs may have to be evaluated. These studies have limitations of limited characterization of the cells and scaffolds like PRP and demineralized bone matrix. Further, continuous movement by the sheep during the study could impart differences in healing. Thus, further validation of mechanisms involved and in vivo therapeutic effects of well-characterized MSCs in different ailments is required.

Almost all these studies demonstrate MSCs (either alone or as tissue-engineered construct) ability to enhance tendon repair. These studies, however, are uncontrolled with variable cell sources, passage number, concentrations and dose repetitions. Further the treatment initiation too has remained variable starting either at earliest (acute model) or delayed (chronic model). The animal numbers and duration of studies have been limited to offer good statistical data and for a longer duration. Above all, the switch over to sheep clinical trials remains to be done to determine MSCs suitability, feasibility and finally the efficacy in tendon lesions. In relation to ligament, a single study shows improved healing; however, further extensive studies are desired for any logical conclusion.

3.5.3 Dog Studies

Dogs remain as one of the important animals in regenerative medicine that offers potential to provide proof-of-therapeutic principles both for veterinary and human medicine. Due to their importance in defence, as pets and above all their ethical treatment, it becomes imperative to develop advanced therapeutics for dog medicine as well. As such regenerative medicine employing MSCs is evaluated in them both under experimental tendon models and in clinical trials of tendon, ligament and muscle disorders.

3.5.3.1 Experimental Studies

A very limited literature is available on experimental studies involving MSCs as therapeutics in dog tendon and ligament regeneration (Table 3.2 and Fig. 3.1). Intrasynovial injuries are challenging for orthopaedician, and to evaluate novel therapeutics like MSCs in such conditions, experimental models on flexor digitorum profundus tendon resection and repair are conducted. AD-MSCs with or without connective tissue growth factor (cTGF) reduce the expression of inflammatory mediators (IL1B and IL6) and matrix degradation proteins (MMP3 and MMP13). cTGF may additively reduce the expression of the inflammatory (*IL1B* and *IL6* expression and inflammatory cell infiltration) and scar-associated (COL3A1) genes. An enhanced expression of IL4 (anti-inflammatory factor promoting M2 phenotype) and collagen occurs with the cell/cTGF transplantation as compared to the control. Furthermore, an increased tendon stem/progenitor cell (CD146+) at the tendon

Table 3.2 Dog in vivo MSCs preclinical experimental studies on various tendinopathies

Study design/ period	Biomaterial used & cell dose	Evaluation criteria	Overall result	References
Dog flexor tendon injury				
Randomized placebo-controlled study ($n = 10$)/ 14 days	AD-MSCs sheet cultured on collagen type I, 4-0 Supramid sutures loaded with connective tissue growth factor (cTGF) 14 days after injury and 16,000 cells/cm^2	Gene expression analysis, histology and immunohistochemistry	Improved outcome	Shen et al. (2018)
Unilateral infraspinatus tendon (Rotator cuff model)				
Randomized placebo-controlled study ($n = 42$)/ 6 weeks	BM-MSCs sheets (1×10^6 in 60 mm dish), tendon-fibrocartilage-bone composite (10×30 mm)	Histology, mechanical testing	Improved outcome	Liu et al. (2019)

MSCs mesenchymal stem cells, *AD* adipose tissue derived, *BM* bone marrow derived, *n* total number of animals in all the groups of the study; Improved outcome: Indicates improved outcome with MSCs transplantation even if only one parameter improves; Study period: Indicates maximum time period for which study is conducted surface and interior along the core suture tracks occurs with the cells but lacks in cTGF by day 14 (Shen et al. 2018). cTGF improves TSPC growth viability and differentiation and additionally regulates the expression of matrices like collagens and proteoglycans (Rui et al. 2019). cTGF potentially suppresses $iNOS^{+}$ M1 macrophages via enrichment of $CD146^{+}$ tendon stem/progenitor cells. However, in the above study, cTGF alone fails to recruit such cells. This might be due to the delayed evaluation at day 14 when its effect mostly occurs by the first week. Moreover, $CD146^{+}$ cells are available in very limited concentration ($<1\%$ of total cells in normal tendon), and as such cTGF alone may not be effective. $CD146^{+}$ cells are obvious in tendons treated with AD-MSCs/cTGF combination pointing combined effect either through recruitment or differentiation of the extrinsically loaded cells that remains to be seen (Lee et al. 2015). Thus, MSCs potentially reduce inflammation and rather enhance anti-inflammatory mediators and increase matrix secretion that may be enhanced with the cTGF. Considering the limited literature, further extensive studies are desired to reach to some logical conclusion.

Another challenging task is the prevention of rotator cuff repair failure occurring due to the inferior tendon-to-bone healing and is attributed to the tendon retraction, creating a gap that gets filled with the scar tissue (Liu et al. 2019). Thus, MSCs role is being evaluated for effective gap filling. Rotator cuff (infraspinatus tendon) transection and repair followed by immediate transplantation of BM-MSCs along with tendon-fibrocartilage-bone composite (10×30 mm) significantly enhances enthesis morphology and mechanical properties as compared to the scaffold only and control. The combined implantation favours the formation of histologically better tissue with higher fibrocartilage formation at tendon-bone interface in comparison to the control (Liu et al. 2019). An immediate implantation of the tissue-engineered construct prevents possible retraction that along with the extrinsic MSCs/scaffold might help in early growth and regeneration of the bone. However, for chronic rotator cuff tears that are more common, relevant studies are lacking. Besides, the single time point evaluation may not provide the details of progressive development post tissue-engineered construct implantation. MSCs together with the growth factors or engineered scaffolds have a great potential to enhance tendon repair; however, studies that are extensive in details regarding proper time of implantation, cell source and concentration, scaffolds, if any, are required.

3.5.3.2 Clinical Studies

3.5.3.2.1 Tendon Studies

As MSCs appear promising in experimental studies of tendon or ligament affections, their properties make way for their utilization in clinical cases (Table 3.3 and Fig. 3.1). Tendon injuries of gastrocnemius (Case et al. 2013) and supraspinatus muscle (Canapp Jr. et al. 2016a; McDougall et al. 2018) are treated with MSCs. A case of gastrocnemius tendon strain is treated intralesionally with BM-MSCs (>20 million) and evaluated for a period of 631 days. Clinically, lameness resolves with an increase in peak vertical force from 43% to 92% of the normal limb. Sonography reveals better echotexture; however, the fibre pattern of the tendon

Table 3.3 Dog in vivo MSCs clinical studies on various tendinopathies and ligament affections

Clinical case	Study design/period	Cell source, cell dose, route	Evaluation criteria	Overall result	References
Tendon injuries					
Gastrocnemius tendon injury	Case report ($n = 1$)/ 631 days	Autologous BM-MSC (cryopreserved $>20 \times 10^6$ implanted locally)	Serial orthopaedic examinations, ultrasonography and long-term force-plate gait analysis	Improved outcome	Case et al. (2013)
Supraspinatus tendinopathy	Retrospective study ($n = 55$)/90 days	AD-MSCs (5×10^6/mL) PRP (1-mL dose) Local injection	Clinical examination, ultrasonography	Improved outcome	Canapp Jr. et al. (2016a)
Supraspinatus tendinopathy	Retrospective case controlled study ($n = 41$)/90 days	BM-MNCs, PRP	Fibre pattern and echogenicity	Improved outcome	McDougall et al. (2018)
Patellar luxation and arthritis					
Grade IV patellar luxation, arthritis	Case report ($n = 1$)/ 5 years	Autologous AD-MSC (1×10^6), hyaluronic acid Local implantation	Lameness and radiography	Improved outcome	Yoon et al. (2012)
Cruciate ligament injury					
Partial cranial cruciate ligament tear treated with TPLO	Retrospective clinical comparative study ($n = 36$)/90 days	Bone marrow concentrate + PRP (2–4 mL), and AD-MSCs + PRP (1–2 mL), local joint implantation	Objective gait analysis, diagnostic arthroscopy, and validated functional questionnaire	Improved outcome	Canapp Jr. et al. 2016b)
Partial cranial cruciate ligament rupture	Case controlled clinical study ($n = 12$)/ 8 weeks	Autologous BM-MSCs, (2×10^6 intravenous and 5×10^6 (intra-articular)	Radiography, arthroscopy, serum and synovial cytokines and C-reactive protein	Improved outcome	Muir et al. (2016)
Cranial cruciate ligament rupture (TPLO)	Comparative study (blinded) ($n = 14$)/ 180 days	Allogeneic foetal adnexa MSCs (10×10^6), local implantation	Clinical score and gait and bone healing	Improved outcome	Taroni et al. (2017)
Cranial cruciate ligament rupture (TTA)	Case controlled double blinded study ($n = 9$)/ 120 days	AD-MSCs (15×10^6), local implantation	Radiography	Improved outcome	Rocha dos Santos et al. (2018)

(continued)

Table 3.3 (continued)

Clinical case	Study design/period	Cell source, cell dose, route	Evaluation criteria	Overall result	References
Cranial cruciate ligament rupture (TTA)	Pilot controlled study	Local implantation, joint injection	Clinical examination and quantification of radiographs	Improved outcome	Koprivec et al. (2021)

MSCs mesenchymal stem cells, *AD* adipose tissue derived, *BM* bone marrow derived, *MNCs* mononuclear cells, *PRP* platelet rich plasma, *TTA* tibial tuberosity advancement, *TPLO* tibial plateau levelling osteotomy, *n* total number of animals in all the groups of the study; Improved outcome: Indicates improved outcome with MSCs transplantation even if only one parameter improves; Study period: Indicates maximum time period for which study is conducted remains incomplete, well supported by incomplete improvement in tarsal extension angle and in persistent weight bearing (Case et al. 2013). In two retrospective studies on supraspinatus tendinopathy ($n = 55$), the effect of local transplantation of AD-MSCs (five million) in PRP (1 mL) (Canapp Jr. et al. 2016a) or bone marrow aspirate in PRP (McDougall et al. 2018) is evaluated up to 90 days through clinical examination and ultrasonography. The biomaterial treated cases show significant increase in total pressure index percentage (TPI %) at 90 days with almost 88% of cases showing TPI% comparable to the contralateral limb. Even the other 12% too show significant improvement. Overall, a significant reduction in tendon size cross-sectional area (CSA) occurs in comparison to the initial CSA of affected tendons with improved fibre pattern (Canapp Jr. et al. 2016a). Likewise, 90.6% of the cases in other studies show improvement in the fibre pattern and echogenicity by 90th day post treatment (McDougall et al. 2018). MSCs transplantation is in general correlated with the improvement in dog tendinopathies, and its standardization remains to be done. However, retrospective studies follow-up is not possible in all the cases and the evaluation does not remain blinded. There is a lack of literature that elucidates the actual therapeutic mechanisms of MSCs. Further, the lack of the control group and histopathological evaluation makes it compulsory to consider such lacunas as well before making any conclusion.

3.5.3.2.2 Ligament Studies

The effective homing of the cells to the ligaments and other joint structures needs a specific route of transplantation. Intra-articular injection appears effective as dog BM-MSCs engraft on to the cranial cruciate ligament (CrCL) (Linon et al. 2014). However, bone repair in CrCL injury cases treated surgically with the tibial tuberosity advancement may better be treated locally with MSCs as compared to the intra-articular implantation (Koprivec et al. 2021). In addition to bone healing, other main concern is the prevention of osteoarthritis.

In dog MSCs are clinically evaluated in the cranial cruciate ligament injuries (Canapp Jr. et al. 2016b; Muir et al. 2016; Taroni et al. 2017) and patellar luxation (Yoon et al. 2012). Surgical repair of tibial plateau levelling osteotomy (TPLO) (Linon et al. 2014; Canapp Jr. et al. 2016b; Muir et al. 2016; Taroni et al. 2017) or tibial tuberosity advancement (TTA) (Rocha dos Santos et al. 2018; Koprivec et al. 2021) is performed in the case of complete CrCL rupture. MSCs as adjunct is evaluated in partial (Canapp Jr. et al. 2016b; Muir et al. 2016) and complete (Taroni et al. 2017; Rocha dos Santos et al. 2018; Koprivec et al. 2021) CrCL tears. A long-term study (5 years) on a case report of a grade IV patellar luxation and associated arthritis locally treated with autologous AD-MSC (one million) along with hyaluronic acid demonstrates improved clinical condition with the time. Dog shows normal walking with reduced osteophyte and osteochondral lesions on radiographic examinations (Yoon et al. 2012). This promising study needs further details in relation to requirement of increased frequency of cell doses and their long term evaluation and validation.

In partial CrCL rupture, transplantation of BM-MSCs through intravenous and intra-articular routes leads to the reduction in serum C-reactive protein (CRP) (inflammatory indicator) at fourth week interval with an increased serum CXCL8 (promotes inflammation, MMPs) at eighth week. CRP and CXCL8 remain uncorrelated post MSCs transplantation. However, synovial CRP and IFNγ in the stifle joint decrease by eighth week. Thus, MSCs may show varied systemic responses including the immunomodulatory effects on macrophages and T lymphocytes. As higher CRP level in synovial joints at eighth week occurs, booster cell doses may be required (Muir et al. 2016). Considering the limited case number with no analysis on the synovial fluid immune cells, it becomes difficult to offer conclusive results, although reduction in inflammation may occur especially in early periods. Further, a retrospective comparative study on partial CrCL injury demonstrates that bone marrow concentrate in PRP offers comparable healing potential to that of the AD-MSCs in PRP at 90th day interval. Available arthroscopy of $n = 13$ out of $n = 36$ at 90 days interval reveals that $n = 9$ have fully healed CrCL. The healed tissue area shows marked neo-vascularization and a normal fibre pattern. Among others ($n = 4$], $n = 1$ shows improved healing and undergoes second dose of injection while rest $n = 3$ that fails to show response undergoes tibial plateau levelling osteotomy (Canapp Jr. et al. 2016b). Thus, MSCs with or without other factors show good prospects for ligament regeneration. However, in retrospective studies factors like variation in treatments and rehabilitation therapy, loss to follow-up, and incomplete documentation may not offer complete and uniform data, besides, may have variabilities in the individual observation. Furthermore, variability in the concentration of either cell type and in platelet count in PRP, besides, the severity of the cases demands furthermore uniform and detailed studies.

In complete CrCL tear, surgical procedures of TTA (Rocha dos Santos et al. 2018; Koprivec et al. 2021) and TPLO (Taroni et al. 2017) are followed by MSCs as adjunctive therapy. In comparison to the NSAIDs, foetal adnexa MSCs (ten million) offer better healing at early period (1 month) than gait and clinical scores to that of surgery/NSAIDs at later period (6 months) (Taroni et al. 2017). Another study also demonstrates that AD-MSCs (15 million) lead to higher ossification mean of 36.45% as compared to control at 1-month period. However, there is no reduction in overall morbidity period and animals show comparable signs at later period. An early bone healing recorded in cell-treated as compared to the control could occur through reduction in inflammation and enhanced matrix formation, although such evidential proof is not provided (Rocha dos Santos et al. 2018). Comparing various methods of implantation, animals that receive cells along with the silk fibroin fail to heal the osteotomy gap possibly due to the low absorption of silk fibroin. Cases that receive MSCs preconditioned with bone tissue and blood into the osteotomy gap through stifle joint recover clinically faster with better filling of the osteotomy defect than dogs who receive MSCs (Koprivec et al. 2021). Thus, it appears that MSCs do offer therapeutic effects in CrCL rupture treated surgically at an early period. However, in these studies, reduction of osteoarthritis by the MSCs remains to be evaluated. As one of the studies demonstrates insignificant differences in gait and clinical scores at later periods, focus on their effectiveness against the OA is desired. To effectuate early and long-lasting repair in reconstructive surgical procedures of CrCL, repeated doses of MSCs at higher concentrations may be transplanted. Additionally, various surgical procedures along with the cell transplantation may also be compared to confirm the better combination. Also it becomes difficult to restrict age for the clinical cases and as such variation that may occur remains to be evaluated (Rocha dos Santos et al. 2018).

3.5.3.2.3 Muscle Studies

There are no experimental in vivo studies on muscle repair with MSCs in dogs. Clinically muscle regenerative potential of MSCs is evaluated in semitendinosus myopathy (Brown et al. 2012; Gibson et al. 2017) and Duchenne's muscular dystrophy (Nitahara-Kasahara et al. 2012; Lorant et al. 2018) (Table 3.4 and Fig. 3.1). In the case of semitendinosus myopathy, local as well as intravenous transplantation of AD-MSCs effectuates significant healing of muscle in cases

Table 3.4 Dog in vivo MSCs clinical studies on various muscle affections

Study design/ period	Biomaterial type, cell dose	Evaluation criteria	Overall result	References
Traumatic semitendinosus muscle injuries				
Case report $n = 2$/ 19 weeks	Autologous AD-MSCs (4.7×10^6 locally and 4.7×10^6 IV) Autologous AD-MSCs (7.5×10^6 locally and 3.8×10^6 IV)	Ultrasonography, muscle biopsy, visual assessment of dog's	Improved outcome	Brown et al. (2012)
Uncontrolled retrospective comparative clinical trial ($n = 9$)/1 year	AD-MSCs	Ultrasonography, visual assessment score (VAS)	Improved outcome	Gibson et al. (2017)
Duchenne's muscular dystrophy				
Case report ($n = 8$)/ 12 weeks	Allogeneic MSCs (transfected with CD271+ (2×10^6 cells per intramuscular and intra-arterial injection)	Immunohistochemistry	Improved outcome	Nitahara-Kasahara et al. (2012)
Case report ($n = 16$)/ 9 months	Allogeneic muscle derived MSCs (6.3×10^7 to 1.2×10^8 per kg, intrafemoral injections), immunosuppression	Immunohistochemistry, histomorphometry, mononuclear cell proliferation, Western blotting	Improved outcome	Lorant et al. (2018)

MSCs mesenchymal stem cells, *AD* adipose tissue derived, *n* total number of animals in all the groups of the study; Improved outcome: Indicates improved outcome with MSCs transplantation even if only one parameter improves; Study period: Indicates maximum time period for which study is conducted

previously unresponsive to the traditional treatment. Lesion size significantly reduces with fibres organized and oriented in the plane of normal muscle (Brown et al. 2012). In a retrospective study ($n = 11$), AD-MSCs transplanted locally and intravenously lead to the normal gait in $n = 8$ dogs and an improved gait $n = 3$ cases. All $n = 8$ dogs return to active police work by the year. Being a retrospective study, specific cell concentration has not been maintained, evaluations have been variable and even the semitendinosus muscle rupture has been variable and either unilateral or bilateral (Gibson et al. 2017). MSCs although appear promising but their role (healing through differentiation or paracrine actions) remains to be confirmed. Besides, these studies are very limited with limited animal numbers that lack control group.

Observing the role of MSCs in Duchenne's muscular dystrophy, a single experimental model demonstrates human AD-MSCs reach to the affected muscles and express dystrophin, although muscle improvement remains to be recorded in non-immunosuppressed animals. The human dystrophin expression is visible up to 6 months post cell transplantation; however, by year end, it is lost (Vieira et al. 2012). In the case series, intra-arterial and intramuscular injected allogeneic BM-MSCs (CD271+) engraft to the affected muscle. Muscle-like cells develop at the area with an upregulated myosin heavy chain (dMyHC) and dystrophin genes at 8–12 weeks period in immunosuppressed animals (Nitahara-Kasahara et al. 2012). Enhanced dystrophin expression might occur due to the MSCs anti-inflammatory and immunosuppressive properties especially T-cell proliferation inhibition and ability to enhance normalization of physiological processes. CD271, a low-affinity nerve growth factor receptor/p75, plays a key role in satellite cell function (Deponti et al. 2009). Nerve growth factor through the receptor induces developmentally significant pathways required for myogenic differentiation and repair (Nitahara-Kasahara et al. 2012). Evaluating the need of immunosuppression, allogeneic muscle-derived MSCs along with immunosuppressant lead to the clinical improvement with muscle regeneration in muscular dystrophy while simple MSCs transplantation displays progressive impairment with the cellular response against these allogeneic cells (Lorant et al. 2018). Lack of immune response against the dystrophin in these studies might arise either due to the immunosuppression, its limited expression or dystrophin-expressing revertant fibres exerting tolerogenic effect. The argument to these reasons may fail when we consider pre-existing circulating anti-dystrophin T-cells available in these patients (Lorant et al. 2018). These preliminary studies on MSCs along with the immunosuppression appear promising in relation to the DMD; however, further extensive studies need to be undertaken to arrive at any conclusion.

3.5.4 Horse Studies

MSCs in horses are evaluated in experimental (Table 3.5 and Fig. 3.1) and clinical studies (Table 3.6 and Fig. 3.1). In experimental studies, MSCs therapeutic effects are mostly evaluated in induced superficial digital flexor tendonitis (Crovace et al.

Table 3.5 Horse in vivo MSCs preclinical experimental studies on various tendinopathies and ligament affections

Study design/ period	Biomaterial and cell dose	Evaluation criteria	Overall result	References
Superficial digital flexor tendonitis model				
Randomized placebo-controlled study ($n = 3$)/ 21 weeks	BM-MSCs (122×10^6), BM-MNCs (122×10^6), fibrinogen, 3 weeks post collagenase inj.	Ultrasonography, histological and immunohistochemistry	Improved outcome	Crovace et al. (2007)
Randomized controlled comparative study ($n = 12$)/ 8 weeks	IGF-I gene enhanced BM-MSCs (10×10^6, local implantation 5 days later collagenase injection)	Histological and immuno-histochemical scores, ultrasound	Improved outcome	Schnabel et al. (2009)
Randomized placebo-controlled study ($n = 6$)/ 18 weeks	Dog BM-MSCs (5.5×10^6) or BM-MNCs (1.2×10^8), intralesionally 3 weeks post collagenase inj.	Histological and immuno-histochemical scores	Improved outcome	Crovace et al. (2010)
Randomized placebo-controlled study ($n = 8$)/ 150 days	AD-MSCs (10×10^6 30 days later collagenase inj.)	Histological and immuno-histochemical scores, clinical and ultrasonographic features	Improved outcome	Carvalho et al. (2011)
Randomized placebo-controlled study ($n = 6$)/ 12 weeks	Autologous BM-MSCs (10×10^6 under ultrasound guidance 4 weeks post-surgical injury), serum	Mass average diameter (MAD), collagen fibril index (CFI), area dependant diameter (ADD)	Improved outcome	Caniglia et al. (2012)
Randomized placebo-controlled study ($n = 9$)/ 22 weeks	AD-MSCs (10×10^6 injected intralesionally after 2 weeks after surgical injury)	Ultrasonography and histology	Improved outcome	Conze et al. (2014)
Randomized placebo-controlled study ($n = 4$)/ 9 weeks	Autologous AD-MSCs (10×10^6 1 week and 3 week later surgical injury in separate cases)	MRI, histology	Cell engraftment after injection	Geburek et al. (2016)
Randomized placebo-controlled	AD-MSCs (10×10^6 injected intralesionally 14 days after	Ultrasonography, clinical, biochemical and biomechanical evaluation	Improved outcome	Geburek et al. (2017)

(continued)

Table 3.5 (continued)

Study design/ period	Biomaterial and cell dose	Evaluation criteria	Overall result	References
study ($n = 9$)/ 24 weeks	surgical injury), inactivated serum under ultrasound guided			
Randomized placebo-controlled comparative study ($n = 12$)/ 10 weeks	BM-MSCs (20×10^6), AD-MSCs (20×10^6), PRP (7 mL) 1 week later surgical injury	Ultrasonography, qPCR, histology	Improved outcome	Romero et al. (2017)
Randomized placebo-controlled study ($n = 8$)/ 112 days	Allogeneic tenogenic primed mesenchymal stem cells (tpMSCs) intralesional injection 1 week post-surgical injury	Clinical, ultrasound and ultrasound, macroscopically, histologically	Improved outcome	Depuydt et al. (2021)
Quadrilateral tendon lesions				
Randomized controlled comparative study ($n = 6$)/ 24 weeks	Allogeneic AD-MSCs (one million), autologous serum	Clinical, ultrasound, histology, MRI	Improved outcome	Ahrberg et al. (2018)
Suspensory ligament repair model				
Randomized placebo-controlled comparative study ($n = 6$)/ 6 days	BM MNCs, AD-MSCs	Histology and immunohistochemistry	Improved outcome with both the protocols	Garbin et al. (2019)

MSCs mesenchymal stem cells, *AD* adipose tissue derived, *BM* bone marrow derived, *MNCs* mononuclear cells, *PRP* platelet rich plasma, *n* total number of animals in all the groups of the study; Improved outcome: Indicates improved outcome with MSCs transplantation even if only one parameter improves; Study period: Indicates maximum time period for which study is conducted

2007, 2010; Schnabel et al. 2009; Carvalho et al. 2011; Caniglia et al. 2012; Conze et al. 2014; Geburek et al. 2016, 2017; Romero et al. 2017; Depuydt et al. 2021) with a single study each on quadriceps lateral tendonitis (Ahrberg et al. 2018) and the suspensory ligament (Garbin et al. 2019) (Table 3.5 and Fig. 3.1).

3.5.4.1 Experimental Studies

3.5.4.1.1 Tendon Studies

The experimental tendonitis studies are mostly conducted as randomized control studies induced with the collagenase or surgery. Surgical injury ensures maximal internal repair and minimizes the external repair (Caniglia et al. 2012). It compartmentalizes tendons usually seen in the degenerative strain-induced tendon disease (Schramme et al. 2010; Cadby et al. 2013).

MSCs derived from adipose tissue (Schnabel et al. 2009; Carvalho et al. 2011; Conze et al. 2014; Geburek et al. 2017; Ahrberg et al. 2018) and bone marrow (Crovace et al. 2007, 2010; Caniglia et al. 2012) are mostly utilized. Besides, BM-MNCs (Torricelli et al. 2011) or AD-SVF (Nixon et al. 2008) are also evaluated. Some of the studies compare BM-MSCs against BM-MNCs (Crovace et al. 2007, 2010) or AD-MSCs (Romero et al. 2017). BM-MSCs are injected intralesionally at variable concentration ranging from 5.5 million (Crovace et al. 2010) to 122 million (Crovace et al. 2007) while AD-MSCs are employed at ten million in most of the studies (Carvalho et al. 2011; Caniglia et al. 2012; Geburek et al. 2017). In all these studies, delayed cell transplantation is made to ensure tendonitis and to avoid cellular interaction with acute inflammatory reaction, in addition to the delayed availability of cells while culturing. It is considered that MSCs may better be transplanted in subacute phase (1–3 weeks post injury) when peak inflammation subsides and only a limited scar tissue exists (Koch et al. 2009). Variable animal numbers are made subject of the study with some employing bilateral tendonitis. Animal numbers range from $n = 3$ (Crovace et al. 2007) to $n = 12$ (Romero et al. 2017). Healing is evaluated through variable diagnostic modalities like histology and immunohistochemistry, sonography, MRI and clinical parameters for a period ranging from 9 weeks (Schnabel et al. 2009) to 24 weeks (Geburek et al. 2017; Ahrberg et al. 2018).

Autologous MSCs intralesional implantation ensures their 'homing' to the structures and the cells remain viable for varied periods. These cells tend to modulate the release of matrix proteins from the tendon surface (Garvican et al. 2014). Apart from autologous, allogeneic MSCs remain safe on direct tendon transplantation. A slight increase in the inflammation may occur upon MSCs transplantation but it subsides subsequently without any adverse effect on transplanted cells (Guest et al. 2008; Ahrberg et al. 2018; de Souza et al. 2018; Depuydt et al. 2021). The cells post local transplantation into the lesion remain available at the site for a period of 5 weeks (Guest et al. 2008), 9 weeks (Geburek et al. 2016) or 24 weeks (Burk et al. 2016) and even migrate to the normal tendon areas (Guest et al. 2008) or to the contralateral tendons (Geburek et al. 2016). Such a difference in survival period of the cells may be due to the differences in the local microenvironment.

Overall an improvement occurs in most of the tendinopathies like tendonitis or surgical lesions with the cell transplantation in comparison to the control. However, differences in individual studies in evaluations on healing patterns are reported. Non-significant differences with respect to the clinical and sonographic features occur, but histopathology and immunohistochemistry reveals a significant improvement in the organization of tendon fibre and Col I matrix of the cell-treated tendons in comparison to the control (Nixon et al. 2008; Carvalho et al. 2011). Further, inflammatory mediators are much reduced in cell-treated tendons as compared to the control. MSCs transplantation leads to the fibroplasia (tendon extracellular matrix with better fibre orientation and collagen I/III ratio) and marked neovascularization (Carvalho et al. 2011). The cell transplantation leads to the significantly higher tendon-specific collagen and Von Willebrand factor at 112 days post injection (Depuydt et al. 2021). Intralesional transplantation of AD-MSCs enhances the neovascularization at 2 weeks post-surgical lesion. Considering this phase as the active healing period, MSCs tend to play a significant role in early healing (Conze et al. 2014).

The suspension of these cells in inactivated autologous serum does not offer long-lasting effect on the signs of inflammation and is comparable to that acheived with the inactivated autologous serum. Transplantation of such biomaterials into the tendons lacks substantial improvement as per the ultrasonography, histology and/or biochemical and biomechanical features of the tendons up to 24 weeks when compared to the autologous serum. A significant increase in Glycosaminoglycans (GAGs) and hydroxylysylpyridinoline occurs in either biomaterial-treated tendons (Geburek et al. 2017). Considering the higher hydroxylysylpyridinoline content in healed lesions, it appears collagen crosslinking is better but an overall limited collagen formation in cell-treated tendon occurs. Comparable matrix in AD-MSCs and serum-treated tendons implies that the cells are not differentiated into the tenocytes. Further, higher GAG secretion implies incomplete tendon healing and as cells are transplanted at a delayed period (2 weeks later), their anti-inflammatory effects are obvious. Another study (that made minor mechanical injury and also injected collagenase) also confirms that cell or serum-transplanted tendons show comparable healing. An inadequate tissue formation in the mechanically induced lesions (surgical lesions) over a period of 24 weeks is evident with either biomaterial. However, contrary to the above study, early-phase (3–6 weeks) period evidences inflammatory reaction in cell-treated tendons (Ahrberg et al. 2018). Both inactivated (Geburek et al. 2017) and activated autologous serum (Ahrberg et al. 2018) used in these studies show comparable tendon-healing potential. Normal autologous serum kept in glass vials for 24 h contains IL-1ra and IL-10 that may potentially effectuate tendon repair (Hraha et al. 2011). However, such a response also occurs with the inactivated serum as well and might be due to the migration of the cells to the contralateral tendon (Burk et al. 2016). Lack of adequate healing with MSCs is also evident from other studies that show large fibrils of the bimodal fibres fail to regenerate at 4 months post cell transplantation (Caniglia et al. 2012).

In a similar way, BM-MSCs and AD-MSCs or PRP lead to better results as compared to the control. Again, an overall comparable improvement occurs with all the three biomaterials based on ultrasonography, histology and gene expression (collagen type I, decorin, tenascin and matrix metalloproteinases III) analysis for a period of 44 weeks. In early period (6 weeks), BM-MSCs-treated tendons show slightly better ultrasound findings as compared to the AD-MSCs and PRP wherein comparable findings occur at week 10. Further, in relation to the gene expression, BM-MSCs appear better (Romero et al. 2017). Thus, MSCs may offer beneficial effects through control of inflammation, decrease in cell density, connective tissue healing (neovascularization, perivascular cuffing and epitenon thickening), although typical tendon regeneration so far remains to be seen (Schnabel et al. 2009). Inflammation may be reduced through the interleukin 1 receptor antagonist and suppression of tumour necrosis factor α secretion by macrophage (Prockop 2007). Furthermore, the lack of large fibrils generation may be associated with the deficiency of the induction factors like TGF-β3 or suitable bio-scaffold (Liang et al. 2008; Kapacee et al. 2010).

In superficial digital flexor tendinopathy (collagenase induced), tendons transplanted with BM-MSCs and BM-MNCs express higher cartilage oligomeric matrix protein (COMP). Expression of type I collagen remains unaffected with low levels of type III collagen. The collagen fibre orientation patterns are normal in both the cell-treated tendons. Comparably, control- and fibrin-loaded tendons show low quantities of COMP and type I collagen but a large number of randomly oriented collagen type III fibres. Furthermore, the cell-loaded tendons additionally show the presence of CD34+ cells pointing the MSCs/MNCs engraft over the tendons and chemotactically recruit these CD34+ cells. A significant finding of the presence of CD34+ cells in the perivascular areas of the interfascicular zone of the BM-MNCs-treated tendon points toward their surviving clones and endotheliogenesis possibly through the endothelial cell differentiation. MSCs or MNCs have the potential to improve tendon healing but considering limited requirements with the MNCs and no time lag of culture, these cells may have opted in limited-resource clinics (Crovace et al. 2007, 2010).

Overall tendon healing with MSCs is better than the control but incomplete. Such a healing is comparable to that occurs with the autologous serum, PRP and MNCs in majority of the cases. Differences in the results in individual studies, especially in relation to the anti-inflammatory effects of MSCs might arise due to differences in the use of different burr or the enzymes for lesion creation, confinement of the horses or exercising them post-lesion induction. The studies need to be uniform with a focus on the mechanism of action of these biomaterials. A blinded study incorporating large animal numbers that too for a longer duration are desired in order to provide the evidence-based medicine.

3.5.4.1.2 Ligament Study

One of the main ligaments in horses that inures is the suspensory ligament injury. It remains untreated and most often is managed by pin firing. As such regenerative medicine may be evaluated. Only a single in vivo horse experimental ligament study utilizes MSCs. Suspensory ligament with fragment removed through biopsy in all the four limbs followed by local transplantation of AD-MSCs (one million) or the bone marrow cell suspension (0.8 mL after 2 days of injury) shows a comparable improvement but better than the control/placebo on the basis of histologic and immunohistochemical findings at day 6. The fibre alignment and neovascularization of the cell-treated tendons remain superior as compared to the control. Along with the collagen I, collagen III too is spotted under immunohistochemistry showing lesions under active healing process. However, effects on inflammation and longer duration effects remain to be evaluated (Garbin et al. 2019).

3.5.4.2 Clinical Studies

In horse, clinically tendonitis (Pacini et al. 2007; Del Bue et al. 2008; Kang et al. 2013; Renzi et al. 2013; Smith et al. 2013; Muttini et al. 2015; Rivera et al. 2020), desmitis (Spaas et al. 2011; Vandenberghe et al. 2015) or mixed tendonitis and desmitis (Renzi et al. 2013; Van Loon et al. 2014; Beerts et al. 2017) treated with MSCs are reported (Table 3.6 and Fig. 3.1).

Table 3.6 Horse in vivo MSCs clinical studies on various tendinopathies and ligament affections

Study design/ period	Biomaterial used & cell dose	Evaluation criteria	Overall result	References
Digital flexor tendinopathy				
Clinical case ($n = 1$)/ 6 weeks	Autologous BM-MSCs (6.4×10^5 cells in 1.5 mL plasma)	Ultrasound		
Clinical case series ($n = 26$)/ 12 months	Autologous BM-MSCs (9.5×10^6)	Ultrasound and ability to return to race	Improved outcome	Pacini et al. (2007)
Clinical case series ($n = 16$)/ 240 days	Allogeneic AD-MSCs + PRP (local intralesional injection)	Ultrasonography	Improved outcome	Del Bue et al. (2008)
Case series ($n = 141$)/ 2 years	Autologous BM-MSCs (5×10^6) intralesional injection	Clinical, ultrasonography, scintigraphy, histology	Improved outcome	Godwin et al. (2012)
Clinical case series ($n = 6$)/ 6 months	Autologous UCB-MSCs (2×10^7 intralesional injection, 3–4 weeks post diagnosis)	Ultrasonography	Improved outcome	Kang et al. (2013)
Clinical case series ($n = 19$)/ 24 months	AD-MSCs (2×10^6/ mL) PRP (2–6 mL) as per lesion size	Clinical and ultrasonography	Improved outcome	Renzi et al. (2013)
Clinical case series ($n = 12$)/ 6 months	Autologous BM-MSCs (1×10^7) suspended in 2 mL of marrow supernatant, growth factors, natural mechanical stimulus	Morphology, biomechanical, ultrasonography, histology	No observable effect	Smith et al. (2013)

(continued)

Table 3.6 (continued)

Study design/ period	Biomaterial used & cell dose	Evaluation criteria	Overall result	References
Clinical case series (acute, $n = 4$; chronic, $n = 2$)/ 6 months	Am-MSCs (7×10^6)	Ultrasonography, histology	Improved outcome	Muttini et al. (2015)
Clinical case series ($n = 10$)/ 4 months	Autologous AD-MSCs (0.6×10^6)	Ultrasonography	Improved outcome	Rivera et al. (2020)
Desmitis studies				
Case report ($n = 1$)/1 year	PB-MSCs (5×10^5)	Clinical, ultrasonography	Improved outcome	Spaas et al. (2011)
Case report ($n = 1$)/1 year	Allogeneic tenogenically differentiated PB-MSCs (2 doses of 0.5×10^6 12 weeks apart), PRP	American Association of Equine Practitioners (AAEP) scale and ultrasonography, radiography, CT	Improved outcome	Vandenberghe et al. (2015)
Case report ($n = 1$)/ 9 months	5-azacytidineand-resveratrol-treated MSCs derived microvesicles (MVsAZA/RES)	Clinical evaluation	Improved outcome	Kornicka-Garbowska et al. (2019)
Mixed lesion (tendonitis and desmitis)				
Comparative clinical study ($n = 33$)/ 15 months	Autologous BM-MSCs (2×10^6) in PRP per cm^2 of lesion	Clinical evaluation and ultrasonography	Improved outcome	Renzi et al. (2013)
Clinical case series ($n = 52$)/ 6 months	UCB-MSCs (2–10 × 10^6 single dose) & 20×10^6 (second dose in 5 horses at 4–36 weeks)	Clinical and ultrasonography	Improved outcome	Van Loon et al. (2014)
Clinical case series [SL ($n = 68$); SDFT ($n = 36$)]/ 24 months	Cryopreserved >50% viability allogeneic tenogenically induced peripheral blood MSCs (2–3 × 10^6) in 1 mL PRP	Clinical, lameness and ultrasonography	Improved outcome	Beerts et al. (2017)

MSCs mesenchymal stem cells, *AD* adipose tissue derived, *BM* bone marrow derived, *PB* peripheral blood derived, *UCB* umbilical cord blood derived, *MNCs* mononuclear cells, *PRP* platelet rich plasma, *SL* suspensory ligament, *n* total number of animals in all the groups of the study; Improved outcome: Indicates improved outcome with MSCs transplantation even if only one parameter improves; Study period: Indicates maximum time period for which study is conducted

3.5.4.2.1 Tendon Injuries

Clinically, superficial digital flexor tendinopathies are most commonly entertained for MSCs therapy. Autologous or allogeneic MSCs from bone marrow (Pacini et al. 2007; Smith et al. 2013) and adipose tissue (Del Bue et al. 2008; Renzi et al. 2013; Rivera et al. 2020) with dose ranging from 0.6 million (Rivera et al. 2020) to 10 million (Smith et al. 2013) are implanted. Single doses of xenogeneic ovine epithelial mesenchymal stem cells (seven million) are also made (Muttini et al. 2015). Some of the studies employ additional biomaterials like PRP (Renzi et al. 2013), and bone marrow and growth factors (Smith et al. 2013). Case-controlled studies are mostly conducted with variable animal numbers ranging from $n = 10$ (Rivera et al. 2020) to $n = 26$ (Pacini et al. 2007). One of the studies titrates dose as per the size of the lesion effective from 4 to 12 million cells (Renzi et al. 2013). The healing is evaluated through clinical observations, sonography and/or biomechanical testing for a period ranging from 4 months (Rivera et al. 2020) to 24 months (Renzi et al. 2013).

All these studies demonstrate enhanced healing with MSCs transplantation. Initially, in 2003, the use of autologous BM-MSCs loaded into the plasma was shown to obviate lameness without any adverse effect (Smith et al. 2003). Post that many equine tendinopathy studies employing MSCs either from autologous or allogeneic sources are conducted. Autologous BM-MSCs transplantation in incomplete thickness tendon defects helps in clinical recovery (sonographic features improved and animals are able to return to racing). $n = 9$ out of $n = 11$ shows clinical recover with normal fibre alignment and without any ectopic bone formation. $n = 1$ that fails to show clinical recovery is transplanted with insufficient autologous cells (less than one million) due to lack of recovery of the sufficient number of cells (Pacini et al. 2007). Similarly, another study demonstrates that $n = 14$ out of $n = 16$ traumatic tendon injuries post transplantation of allogeneic AD-MSCs loaded in PRP shows functional recovery and go back to the activity. $n = 1$ that fails to recover has chronic tendonitis and is being treated with various methods (infiltrations, blisters, etc.), and the other case shows delayed (8 weeks later) return to activity. However, some of the animals ($n = 6$) re-injure after sport activity (Del Bue et al. 2008) as also reported by others (Godwin et al. 2012). Allogeneic AD-MSCs are also supplemented with PRP and transplantation of such a combination recovers around 89.5% of the horses. These animals return to the normal activity and show quite low re-injury rate (10.5%) up to 2 years (Renzi et al. 2013). To determine exclusivity, cross-sectional surface area (CSA) grading/category of the clinical cases shows that transplantation of horse UC-MSCs into category 5 SDFTs leads to the improvement and the healed tendons show grade 1 after a month. These results when compared to Genovese et al. (1997) grouping show only 33% failure rate ($n = 2$ failure; $n = 4$ successful) as compared to the 90% normally seen in CSA grading (Kang et al. 2013). Clinical improvement and reduction in re-injury rate with MSCs or amniotic epithelial mesenchymal cell conform well to the histology and ultrasonography that improves in both acute and chronic tendinopathies with MSCs (Smith et al. 2009; Godwin et al. 2012; Muttini et al. 2015). Thus, a minimal standard cell population if transplanted can effectively improve tendinopathy, although variations within the cases may occur especially in relation to the chronicity and type of treatment offered prior to the cell transplantation.

Apart from the fibre arrangement, autologous BM-MSCs loaded in BM concentrate are correlated with the improvement in the tendon stiffness with elasticity comparable to that of the normal contralateral tendon. In addition to the mechanical factors, histological and biochemical composition (GAGs and MMP-13) too improve, although complete sonographic anatomy of a normal tendon is currently lacking. Reduction in GAGs and MMP-13 points limited proliferation of the tissue and its matrix degradation, respectively, and is correlated with the better improvement of the tendon (Smith et al. 2013). The mechanism involved in the normalization of GAGs, however, remains to be seen. Normally, proteoglycan degradation enzymes like ADAMTS4 and ADAMTS5 are involved in GAG reduction. Since MSCs enhance GAG accumulation and therefore simultaneously might prevent its degradation through inactivation of these enzymes (Corps et al. 2012). A recent study demonstrates the effect of the microenvironment on the secretion of matrix metalloproteinases and tissue inhibitor matrix metalloproteinases (Doll et al. 2021). Further, it is immaterial to evaluate hydroxyproline among the cell treated and control tendons as this does not distinguish between the collagen types that develop in either case. Thus, transplantation of MSCs with or without other biomaterials helps in gradual reduction in the scar length of the injured tendon (Rivera et al. 2020). However, the role specification of these biomaterials remains to be made.

3.5.4.2.2 Ligament Injuries

Either tenogenic-induced MSCs (Vandenberghe et al. 2015) or resveratrol-treated MSCs microvesicles (Kornicka-Garbowska et al. 2019) are used for the suspensory ligament injuries. Tenogenically induced PB-MSCs are used along with the PRP in chronic-injured suspensory ligament (SL) (new bone formation at the proximal region of insertion of the suspensory ligament) (Vandenberghe et al. 2015). Healing evaluation is conducted through various parameters for a period of 9 months (Kornicka-Garbowska et al. 2019) or 1 year (Vandenberghe et al. 2015). MSCs intralesional transplantation improves desmitis-associated lameness. The lameness reduces to grade 1 (out of 4 American Association of Equine Practitioners, AAEP grade) as animal trots sound by fourth week after second dose, i.e. 16th week post first injection of tenogenically induced PB-MSCs along with the PRP. The echogenicity of the lesion site and normal tendon remains undifferentiated by this time. At week 32, the animal goes back to the previous performance level (Vandenberghe et al. 2015). Delayed result in this study could be due to the established new bone compromising the results of the first injection. The second dose might have reactivated the healing process and led to the recovery. Besides, the tenogenically induced cells may not have effective anti-inflammatory effects to enhance early healing. However, the narrative behind using tenogenically induced cells remains to prevent any new bone formation as chronic lesions have associated calcified lesions. Furthermore, to avoid such possible adverse effects of the cells, microvesicles derived from 5-azacytidine and resveratrol-treated AD-MSCs (from equine metabolic syndrome patients) also enhance ligament healing resulting in increased lesion filling and improvement of angiogenesis and elasticity in injured tissue of a horse at 12 months (Kornicka-Garbowska et al. 2019). In equine metabolic syndrome, MSCs are senescent with high oxidative stress and have limited differentiation. Treatment with 5-azacytidine and resveratrol restores normal physiological functions and rejuvenates these cells. Overall MSCs appear quite promising in desmitis of the suspensory ligament.

3.5.4.2.3 Mixed Tendon and Ligament Injury Studies

In studies reporting tendonitis and desmitis, MSCs either alone (Spaas et al. 2011; Van Loon et al. 2014) or along with PRP (Renzi et al. 2013; Beerts et al. 2017) are transplanted. One of the studies employs double dose of cells in some cases (Van Loon et al. 2014), while other transplants tenogenically differentiated PB-MSCs (Beerts et al. 2017). All these studies demonstrate a positive outcome with the transplantation of these biomaterials. In a single case, post PB-MSCs transplantation, tendon and ligament swelling subsides and lameness disappears at 5 weeks. The lesions get filled to some extent but with some heterogenic lesions at the edges. At 9 weeks period (exercise allowed), sonography shows smooth echogenicity and later at 18 weeks period, mare starts gradual trotting and cantering. By 6 months, the mare reaches the original level of competition in show jumping. For a follow-up of 1 year, non-recurrence of the symptoms occurs (Spaas et al. 2011). Umbilical cord blood-derived MSCs lead to an improvement in 77% (40/52) animals with the potential to return to normal activity, irrespective of animal age. Among, 20/23 SDFT cases and 15/22 cases of desmitis have a positive outcome at sixth month. A very interesting finding but contrary to the one above is the limited improvement in aged patients, although statistically the finding is non-significant. Besides, some cases show improvement after second dose, and such differences arise due to the differences in severity (Van Loon et al. 2014). A further extensive study demonstrates that 83.8% of SDFT and 79.2% of SL return to the previous performance levels in 12th month period. Such figures at 24th month are 82.4% and 85.7% for SL and SDFT cases, respectively. Thus, cell therapy significantly reduces re-injury rate (Beerts et al. 2017). In another extensive study, $n = 97$ horses have 101 injuries, out of which 61 injuries are of soft tissue cases and 40 of orthopaedic nature. Among these cases, 52 cases of soft tissue injuries and 29 cases of orthopaedic cases return back to work after cell transplantation. Besides, non-significant correlation with respect to the age/sex/breed and severity of lesion with the outcome is seen (Ferris et al. 2009). Comparing BM-MSCs treatment to pin firing, gradual but effective healing observed under sonography occurs, and by 24th week post cell implantation, the tendons are anatomically restored. Out of $n = 18$ cell treated, $n = 13$ return to race competitions while $n = 0$ among $n = 13$ pin fired horses (Renzi et al. 2013). Thus, MSCs significantly improve clinical cases of tendonitis and desmitis that may have to be reported depending upon the case severity.

In experimental or clinical studies, there is a variability in results with the cell sources, dose and route of implantation. Incorporation of other biomaterials along with the cells also makes studies diverse. The cellular concentration along with other factors tends to have an effect on the healing potential as one million up to ten million (Smith 2008; Caniglia et al. 2012) cell dose fails to enhance tendon healing while still higher doses of 10 million (Smith et al. 2009) or 20 million (Van Loon et al. 2014) improve matrix synthesis that is comparable to the normal tendon. It is worth mentioning here that other basic managemental aspects need to be followed for better and effective outcomes.

3.6 Conclusions

In tendon/ligament or muscle, specific stem cells from their tissue may be more amenable for respective repair or healing effects. However, the limited sources, limitations in standardization of their harvesting sites, culture methods and characterization criterions remain a challenge. As MSCs express relevant markers and have diverse sources with ability to differentiate and secrete pro-healing factors, these cells remain vanguard among all the stem cell-based therapies. MSCs appear fruitful for tendinopathies, desmitis or overall in joint affections. Varied kinds of the tendon and ligament models have been experimentally created in sheep, dog and horse. All these experimental studies demonstrate MSCs potential healing effect although well short of the complete improvement. MSCs have also been transplanted along with other biomaterials in which case therapeutic mechanisms of each biological agent remain to be delineated. Apart from experimental studies, clinical trials in dog and horse favour MSCs clinical utility in tendon, muscle and ligament injuries. These cells enhance repair in muscle affections and tendinopathies and desmitis and significantly reduce re-injury rates. However, to standardize the stem cell therapeutics, it is necessary to have uniform case-controlled extensive studies that employ well-characterized MSCs from specific cell sources, culture methods, passage numbers and at specific concentrations. This must be followed by specific administration protocols for a specific lesion both in size and nature and for specific periods. Evaluations have to be more advanced in order to confirm the cell therapeutic mechanisms, their fate and above all the healed tissue properties. Further, comparative studies of MSCs from various sources and against the currently available therapeutic options are desired. The clinical studies may be performed with a placebo control and the animal numbers and duration of studies must be sufficient enough to compare and understand the results for logical conclusions.

References

Ahrberg AB, Horstmeier C, Berner D, Brehm W, Gittel C, Hillmann A, Josten C, Rossi G, Schubert S, Winter K, Burk J (2018) Effects of mesenchymal stromal cells versus serum on tendon healing in a controlled experimental trial in an equine model. BMC Musculoskelet Disord 19:230

Alsook SMK, Gabriel A, Piret J, Waroux O, Tonus C, Connan D, Baise E, Antoine N (2015) Tissues from equine cadaver ligaments up to 72 hours of post-mortem: a promising reservoir of stem cells. Stem Cell Res Ther 6(1):253

Beerts C, Suls M, Broeckx SY, Seys B, Vandenberghe A, Declercq J, Duchateau L, Vidal MA, Spaas JH (2017) Tenogenically induced allogeneic peripheral blood mesenchymal stem cells in allogeneic platelet-rich plasma: 2-year follow-up after tendon or ligament treatment in horses. Front Vet Sci 4:158

Bi Y, Ehirchiou D, Kilts TM et al (2007) Identification of tendon stem/progenitor cells and the role of the extracellular matrix in their niche. Nat Med 13(10):1219–1227

Brandão JS, Alvarenga ML, Pfeifer JPH, Dos Santos VH, FonsecaAlves CE, Rodrigues M, Laufer-Amorim R, Castillo JAL, Alves ALG (2018) Allogeneic mesenchymal stem cell transplantation in healthy equine superficial digital flexor tendon: a study of the local inflammatory response. Res Vet Sci 118:423–430

Brown SG, Harman RJ, Black LL (2012) Adipose-derived stem cell therapy for severe muscle tears in working German shepherds: two case reports. Stem Cell Discov 2(2):41

Burk J, Berner D, Brehm W, Hillmann A, Horstmeier C, Josten C, Paebst F, Rossi G, Schubert S, Ahrberg AB (2016) Long-term cell tracking following local injection of mesenchymal stromal cells in the equine model of induced tendon disease. Cell Transplant 25(12):2199–2211

Cadby JA, David F, van de Lest C, Bosch G, van Weeren PR, Snedeker JG, van Schie HT (2013) Further characterisation of an experimental model of tendinopathy in the horse. Equine Vet J 45(5):642–648

Cadby JA, Buehler E, Godbout C, van Weeren PR, Snedeker JG (2014) Differences between the cell populations from the peritenon and the tendon core with regard to their potential implication in tendon repair. PLoS One 9(3):e92474

Čamernik K, Marc J, Zupan J (2019) Human skeletal muscle-derived mesenchymal stem/stromal cell isolation and growth kinetics analysis. Methods Mol Biol 2045:119–129

Canapp SO Jr, Canapp DA, Ibrahim V, Carr BJ, Cox C, Barrett JG (2016a) The use of adipose-derived progenitor cells and platelet-rich plasma combination for the treatment of supraspinatus tendinopathy in 55 dogs: a retrospective study. Front Vet Sci 3:61

Canapp SO Jr, Leasure CS, Cox C, Ibrahim V, Carr BJ (2016b) Partial cranial cruciate ligament tears treated with stem cell and platelet-rich plasma combination therapy in 36 dogs: a retrospective study. Front Vet Sci 3:112

Caniglia CJ, Schramme MC, Smith RK (2012) The effect of intralesional injection of bone marrow derived mesenchymal stem cells and bone marrow supernatant on collagen fibril size in a surgical model of equine superficial digital flexor tendonitis. Equine Vet J 44:587–593

Carlo B, Alberto C, Ilaria F, Assunta P, Pietro R (2019) Muscle stem cells: what's new in orthopedics? Acta Biomed 90(Suppl 1):8–13

Carvalho AM, Alves ALG, Gomes de Oliveira PG, Álvarez LEC, Renée CAH, Elenice D (2011) Use of adipose tissue-derived mesenchymal stem cells for experimental tendinitis therapy in equines. J Equine Vet Sci 31:26–34

Case JB, Palmer R, Valdes-Martinez A, Egger EL, Haussler KK (2013) Gastrocnemius tendon strain in a dog treated with autologous mesenchymal stem cells and a custom orthosis. Vet Surg 42:355–360

Chen J, Xu J, Wang A, Zheng M (2009) Scaffolds for tendon and ligament repair: review of the efficacy of commercial products. Expert Rev Med Devices 6:61–73

Choi K-H, Yoon JW, Kim M, Lee HJ, Jeong J, Ryu M, Jo C, Lee C-K (2021) Muscle stem cell isolation and *in vitro* culture for meat production: a methodological review. Compr Rev Food Sci Food Saf 20(1):429–457

Coleman SH, Ehtishami JR, Kisiday JD, Altchek DW, Warren RF, Turner AS (2011) The effects of mesenchymal stem cells on rotator cuff muscle in a chronic injury model in sheep. Paper No. 167 • 55th Annual Meeting of the Orthopaedic Research Society

Conze P, van Schie HTM, van Weeren R, Staszyk C, Conrad S, Skutella T, Hopster K, Rohn K, Stadler P, Geburek F (2014) Effect of autologous adipose tissue-derived mesenchymal stem cells on neovascularization of artificial equine tendon lesions. Regen Med 9(6):743–757

Corps AN, Robinson AH, Harrall RL, Avery NC, Curry VA, Hazleman BL, Riley GP (2012) Changes in matrix protein biochemistry and the expression of mRNA encoding matrix proteins and metalloproteinases in posterior tibialis tendinopathy. Ann Rheum Dis 71(5):746–752

Costa-Almeida R, Calejo I, Gomes ME (2019) Mesenchymal stem cells empowering tendon regenerative therapies. Int J Mol Sci 20:3002

Crovace A, Lacitignola L, De Siena R, Rossi G, Francioso E (2007) Cell therapy for tendon repair in horses: an experimental study. Vet Res Commun 31(Suppl 1):281–283

Crovace A, Lacitignola L, Francioso E, Rossi G (2008) Histology and immunohistochemistry study of sheep tendon grafted with cBMSCs and BMMNCs after collagenase-induced tendinitis. Vet Comp Orthop Traumatol 21(4):329–336

Crovace A, Lacitignola L, Rossi G, Francioso E (2010) Histological and immunohistochemical evaluation of autologous cultured bone marrow mesenchymal stem cells and bone marrow mononucleated cells in collagenase-induced tendinitis of equine superficial digital flexor tendon. Vet Med Int 2010:1

de Souza JB, Alvarenga ML, Pfeifer JPH, dos Santos VH, Fonseca-Alves CE, Rodrigues M, Laufer-Amorim R, Lucas Castillo JA, Garcia Alves AL (2018) Allogeneic mesenchymal stem cell transplantation in healthy equine superficial digital flexor tendon: a study of the local inflammatory response. Res Vet Sci 118:423–430

Del Bue M, Ricco S, Ramoni R, Conti V, Gnudi G, Grolli S (2008) Equine adipose-tissue derived mesenchymal stem cells and platelet concentrates: their association in vitro and in vivo. Vet Res Commun 32:S51–S55

Deponti D, Buono R, Catanzaro G, De Palma C, Longhi R, Meneveri R, Bresolin N, Bassi MT, Cossu G, Clementi E, Brunelli S (2009) The low-affinity receptor for neurotrophins p75NTR plays a key role for satellite cell function in muscle repair acting via RhoA. Mol Biol Cell 20(16):3620–3627

Depuydt E, Broeckx SY, Van Hecke L, Chiers K, Van Brantegem L, van Schie H, Beerts C, Spaas JH, Pille F, Martens A (2021) The evaluation of equine allogeneic tenogenic primed mesenchymal stem cells in a surgically induced superficial digital flexor tendon lesion model. Front Vet Sci 8:641441

Doll CU, Niebert S, Burk J (2021) Mesenchymal stromal cells adapt to chronic tendon disease environment with an initial reduction in matrix remodeling. Int J Mol Sci 22(23):12798

Dudhia J, Becerra P, Valdés MA, Neves F, Hartman NG, Smith RK (2015) In vivo imaging and tracking of technetium-99m labeled bone marrow mesenchymal stem cells in equine tendinopathy. J Vis Exp 106:e52748

Durgam SS, Stewart AA, Pondenis HC, Gutierrez-Nibeyro SM, Evans RB, Stewart MC (2012) Comparison of equine tendon-and bone marrow-derived cells cultured on tendon matrix with or without insulin-like growth factor-I supplementation. Am J Vet Res 73:153–161

El-Habta R, Kingham PJ, Backman LJ (2018) Adipose stem cells enhance myoblast proliferation via acetylcholine and extracellular signal-regulated kinase 1/2 signaling. Muscle Nerve 57:305–311

Ferris D, Frisbie DD, Kisiday J et al (2009) Clinical evaluation of bone marrow-derived mesenchymal stem cells in naturally occurring joint disease. Regen Med 4:S16

Fuchs JR, Kaviani A, Oh JT, LaVan D, Udagawa T, Jennings RW, Wilson JM, Fauza DO (2004) Diaphragmatic reconstruction with autologous tendon engineered from mesenchymal amniocytes. J Pediatr Surg 39(6):834–838

Galloway MT, Lalley AL, Shearn JT (2013) The role of mechanical loading in tendon development, maintenance, injury, and repair. J Bone Joint Surg 95(17):1620–1628

Garbin LC, Soares AS, Mendes HMF, Cassali GD, Vasconcelos AC, Faleiros RR (2019) Histological and immunohistochemical evaluation of two cell therapy protocols in equine suspensory ligament repair. Arq Bras Med Vet Zootec 71:1469–1476

Garvican ER, Dudhia J, Alves A-L, Clements LE, Du Plessis F, Smith RKW (2014) Mesenchymal stem cells modulate release of matrix proteins from tendon surfaces in vitro: a potential beneficial therapeutic effect. Regen Med 9(3):295–308

Geburek F, Mundle K, Conrad S, Hellige M, Walliser U, van Schie HTM, van Weeren R, Skutella T, Stadler PM (2016) Tracking of autologous adipose tissue derived mesenchymal stromal cells with in vivo magnetic resonance imaging and histology after intralesional treatment of artificial equine tendon lesions—a pilot study. Stem Cell Res Ther 7:21

Geburek F, Roggel F, van Schie HTM, Beineke A, Estrada R, Weber K, Hellige M, Rohn K, Jagodzinski M, Welke B, Hurschler C, Conrad S, Skutella T, van de Lest C, van Weeren R, Stadler PM (2017) Effect of single intralesional treatment of surgically induced equine superficial digital flexor tendon core lesions with adipose-derived mesenchymal stromal cells: a controlled experimental trial. Stem Cell Res Ther 8:129

Gelberman RH, Linderman SW, Jayaram R, Dikina AD, Sakiyama-Elbert H, Alsberg E, Thomopoulos S, Shen H (2017) Combined administration of ASCs and BMP-12 promotes an M2 macrophage phenotype and enhances tendon healing. Clin Orthopaed Relat Res 475(9): 2318–2331

Genovese R, Longo K, Brethold B, Jorgenson J (1997) Quantitative sonographic assessment in the clinical management of superficial digital flexor injuries in thoroughbred racehorses. Proc Am Assoc Equine Pract 43:285–290

Gibson MA, Brown SG, Brown NO (2017) Semitendinosus myopathy and treatment with adipose-derived stem cells in working German shepherd police dogs. Can Vet J 58:241–246

Godwin EE, Young NJ, Dudhia J, Beamish IC, Smith RKW (2012) Implantation of bone marrow-derived mesenchymal stem cells demonstrates improved outcome in horses with overstrain injury of the superficial digital flexor tendon. Equine Vet J 44(1):25–32

Gomiero C, Bertolutti G, Martinello T, Van Bruaene N, Broeckx SY, Patruno M et al (2016) Tenogenic induction of equine mesenchymal stem cells by means of growth factors and low-level laser technology. Vet Res Commun 40:39–48

Guest DJ, Smith MR, Allen WR (2008) Monitoring the fate of autologous and allogeneic mesenchymal progenitor cells injected into the superficial digital flexor tendon of horses: preliminary study. Equine Vet J 40:178–181

Gugjoo MB, Amarpal, Chandra V, Wani MY, Dhama K, Sharma GT (2018) Mesenchymal stem cell research in veterinary medicine. Curr Stem Cell Res Ther 13(8):645–657

Gugjoo MB, Amarpal, Makhdoomi DM, Sharma GT (2019) Equine mesenchymal stem cells: properties, sources, characterization and potential therapeutic applications. J Equine Vet Sci 72:16–27

Gugjoo MB, Amarpal AAI, Aithal HP, Kinjavdekar P, Saikumar GS, Sharma GT (2020) Allogeneic mesenchymal stem cells and growth factors in gel scaffold repair osteochondral defect in rabbit. Regen Med 15(2):1261–1275. https://doi.org/10.2217/rme-2018-0138

Gulati BR, Kumar R, Mohanty N, Kumar P, Somasundaram RK, Yadav PS (2013) Bone morphogenetic protein-12 induces tenogenic differentiation of mesenchymal stem cells derived from equine amniotic fluid. Cells Tissues Organs 198:377–389

Heinemeier KM, Kjaer M (2011) In vivo investigation of tendon responses to mechanical loading. J Musculoskelet Neuronal Interact 11(2):115–123

Heinemeier KM, Skovgaard D, Bayer ML et al (2012) Uphill running improves rat Achilles tendon tissue mechanical properties and alters gene expression without inducing pathological changes. J Appl Physiol 113(5):827–836

Hexter AT, Karali A, Kao A, Tozzi G, Heidari N, Petrie A, Boyd A, Kalaskar DM, Pendegrass C, Rodeo S, Haddad F, Blunn G (2021a) Effect of demineralized bone matrix, bone marrow mesenchymal stromal cells, and platelet-rich plasma on bone tunnel healing after anterior cruciate ligament reconstruction: a comparative micro-computed tomography study in a tendon allograft sheep model. Orthop J Sports Med 9(9):23259671211034166

Hexter AT, Sanghani-Kerai A, Heidari N, Kalaskar DM, Boyd A, Pendegrass C, Rodeo SA, Haddad FS, Blunn GW (2021b) Mesenchymal stromal cells and platelet-rich plasma promote tendon allograft healing in ovine anterior cruciate ligament reconstruction. Knee Surg Sports Traumatol Arthrosc 29(11):3678–3688

Hraha TH, Doremus KM, McIlwraith CW, Frisbie DD (2011) Autologous conditioned serum: the comparative cytokine profiles of two commercial methods (IRAP and IRAP II) using equine blood. Equine Vet J 43:516–521

Im G-I, Kim T-K (2020) Stem cells for the regeneration of tendon and ligament: a perspective. Int J Stem Cells 13(3):335–341

Javanshir S, Soltani FY, Dowlati G, Parham A, Naderi-Meshkin H (2020) Induction of tenogenic differentiation of equine adipose-derived mesenchymal stem cells by platelet-derived growth factor-BB and growth differentiation factor-6. Mol Biol Rep 47(9):6855–6862

Johnson JA, Austin C, Breur GJ (1994) Incidence of canine appendicular musculo-skeletal disorders in 16 veterinary teaching hospitals from 1980 through 1989. Vet Comp Orthop Traumatol 7:56–69

Kajikawa Y, Morihara T, Watanabe N, Sakamoto H, Matsuda K, Kobayashi M, Oshima Y, Yoshida A, Kawata M, Kubo T (2007) GFP chimeric models exhibited a biphasic pattern of mesenchymal cell invasion in tendon healing. J Cell Physiol 210(3):684–691

Kang J-G, Park S-B, Seo M-S, Kim H-S, Chae J-S, Kang K-S (2013) Characterization and clinical application of mesenchymal stem cells from equine umbilical cord blood. J Vet Sci 14(3): 367–371

Kapacee Z, Yeung CYC, Lu Y, Crabtree D, Holmes DF, Kadler KE (2010) Synthesis of embryonic tendon-like tissue by human marrow stromal/mesenchymal stem cells requires a three-dimensional environment and transforming growth factor β3. Matrix Biol 29:668–677

Kasashima Y, Takahashi T, Smith RKW, Goodship AE, Kuwano A, Ueno T, Hirano S (2004) Prevalence of superficial digital flexor tendonitis and suspensory desmitis in Japanese Thoroughbred flat racehorses in 1999. Equine Vet J 36(4):346–350

Kaux J-F, Crielaard J-M (2013) Platelet-rich plasma application in the management of chronic tendinopathies. Acta Orthop Belg 79:10–15

Kawai M, Aida H, Hiraga A, Miyata H (2013) Muscle satellite cells are activated after exercise to exhaustion in Thoroughbred horses. Equine Vet J 45(4):512–517

Khan MR, Smith RKW, DeGodoy R, Dakin S, David F, Goodship AE, Dudhia J (2013) Superparamagnetic iron oxide labeling of mesenchymal stem cells for tracking in a sheep model of intra-thecal tendon injury. Eur Cells Mater 26(7):64

Khan MR, Dudhia J, David FH, De Godoy R, Mehra V, Hughes G, Dakin SG, Carr AJ, Goodship AE, Smith RKW (2018) Bone marrow mesenchymal stem cells do not enhance intra-synovial tendon healing despite engraftment and homing to niches within the synovium. Stem Cell Res Ther 9:169

Khan MR, Smith RK, David F, Lam R, Hughes G, DeGodoy R, Carr AJ, Goodship AE, Dudhia J (2020) Evaluation of the effects of synovial multipotent cells on deep digital flexor tendon repair in a large animal model of intra-synovial tendinopathy. J Orthop Res 38(1):128–138

Koch TG, Berg LC, Betts DH (2009) Current and future regenerative medicine—principles, concepts, and therapeutic use of stem cell therapy and tissue engineering in equine medicine. Can Vet J 50:155–165

Kono S, Kazama T, Kano K, Harada K, Uechi M, Matsumoto T (2014) Phenotypic and functional properties of feline dedifferentiated fat cells and adipose-derived stem cells. Vet J 199:88–96

Koprivec S, Novak M, Bernik S, Voga M, Mohorič L, Majdič G (2021) Treatment of cranial cruciate ligament injuries in dogs using a combination of tibial tuberosity advancement procedure and autologous mesenchymal stem cells/multipotent mesenchymal stromal cells—a pilot study. Acta Vet Hung 68(4):405–412

Korchunjit W, Laikul A, Taylor J, Watchararat K, Ritruechai P, Supokawej A, Wongtawan T (2019) Characterisation and allogeneic transplantation of equine bone marrow derived multipotent mesenchymal stromal cells collected from cadavers. J Equine Vet Sci 73:15–23

Kornicka-Garbowska K, Pędziwiatr R, Woźniak P, Kucharczyk K, Marycz K (2019) Microvesicles isolated from 5-azacytidineand-resveratrol-treated mesenchymal stem cells for the treatment of suspensory ligament injury in horse—a case report. Stem Cell Res Ther 10:394

Kulesza A, Burdzinska A, Szczepanska I, Zarychta-Wisniewska W, Pajak B, Bojarczuk K, Dybowski B, Paczek L (2016) The mutual interactions between mesenchymal stem cells andmyoblasts in an autologous co-culture model. PLoS One 11(8):e0161693

Lacitignola L, Staffieri F, Rossi G, Francioso E, Crovace A (2014) Survival of bone marrow mesenchymal stem cells labelled with red fluorescent protein in an sheep model of collagenase induced tendinitis. Vet Comp Orthop Traumatol 27(3):204–209

Lavagnino M, Wall ME, Little D, Banes AJ, Guilak F, Arnoczky SP (2015) Tendon mechanobiology: current knowledge and future research opportunities. J Orthop Res 33(6): 813–822

Lee JC, Mitchell AWM, Healy JC (2012) Imaging of muscle injuries. Br J Radiol 85(1016): 1173–1185

Lee CH, Lee FY, Tarafder S, Kao K, Jun Y, Yang G, Mao JJ (2015) Harnessing endogenous stem/ progenitor cells for tendon regeneration. J Clin Invest 125(7):2690–2701

Lee E-M, Kim A-Y, Lee E-J, Park J-K, Park S-I, Cho S-G, Kim HK, Kim S-Y, Jeong K-S (2016) Generation of equine-induced pluripotent stem cells and analysis of their therapeutic potential for muscle injuries. Cell Transplant 25(11):2003–2016

Lee KJ, Clegg PD, Comerford EJ, Canty-Laird EG (2017) Ligament-derived stem cells: identification, characterisation, and therapeutic application. Stem Cells International 2017:1919845

Letouzey V, Tan KS, Deane JA, Ulrich D, Gurung S, Ong YR, Gargett CE (2015) Isolation and characterisation of mesenchymal stem/stromal cells in the sheep endometrium. PLoS One 10(5): e0127531

Liang R, Woo SL-Y, Nguyen TD, Ping-Cheng L, Almarza A (2008) Effects of a bioscaffold on collagen fibrillogenesis in healing medial collateral ligament in rabbits. J Orthop Res 8:1098–1104

Linon E, Spreng D, Rytz U, Forterre S (2014) Engraftment of autologous bone marrow cells into the injured cranial cruciate ligament in dogs. Vet J 202(3):448–454

Liu Q, Yu Y, Reisdorf RL, Qi J, Lu C-K, Berglund LJ, Amadio PC, Moran SL, Steinmann SP, An K-N, Gingery A, Zhao C (2019) Engineered tendon-fibrocartilage-bone composite and bone marrow-derived mesenchymal stem cell sheet augmentation promotes rotator cuff healing in a non-weight-bearing canine model. Biomaterials 192:189–198

Lombana KG, Goodrich LR, Phillips JN, Kisiday JD, RupleCzerniak A, McIlwraith CW (2015) An investigation of equine mesenchymal stem cell characteristics from different harvest sites: more similar than not. Front Vet Sci 2:67

Longo UG, Lamberti A, Maffulli N, Denaro V (2010) Tendon augmentation grafts: a systematic review. Br Med Bull 94:165–188

Lorant J, Larcher T, Jaulin N, Hedan B, Lardenois A et al (2018) Vascular delivery of allogeneic MuStem cells in dystrophic dogs requires only short-term immunosuppression to avoid host immunity and generate clinical/tissue benefits. Cell Transplant 27(7):1096–1110

Lovati AB, Corradetti B, Cremonesi F, Bizzaro D, Consiglio AL (2012) Tenogenic differentiation of equine mesenchymal progenitor cells under indirect co-culture. Int J Artif Organs 35(11): 996–1005

Martinello T, Bronzini I, Perazzi A, Testoni S, De Benedictis GM, Negro A, Caporale G, Mascarello F, Iacopetti I, Patruno M (2013) Effects of in vivo applications of peripheral blood-derived mesenchymal stromal cells (PB-MSCs) and platelet-rich plasma (PRP) on experimentally injured deep digital flexor tendons of sheep. J Orthop Res 31(2):306–314

Mathewson MA, Kwan A, Eng CM, Lieber RL, Ward SR (2014) Comparison of rotator cuff 12 muscle architecture between humans and other selected vertebrate species. J Exp Biol 13: 261–273

McDougall RA, Canapp SO, Canapp DA (2018) Ultrasonographic findings in 41 dogs treated with bone marrow aspirate concentrate and platelet rich plasma for a supraspinatus tendinopathy: a retrospective study. Front Vet Sci 5:98

Melzer M, Schubert S, Müller SF, Geyer J, Hagen A, Niebert S, Burk J (2021) Rho/ROCK inhibition promotes TGF-β3-induced tenogenic differentiation in mesenchymal stromal cells. Stem Cells Int 2021:8284690

Mienaltowski MJ, Adams SM, Birk DE (2013) Regional differences in stem cell/progenitor cell populations from the mouse Achilles tendon. Tissue Eng A 19(1–2):199–210

Milner DJ, Bionaz M, Monaco E, Cameron JA, Wheeler MB (2018) Myogenic potential of mesenchymal stem cells isolated from porcine adipose tissue. Cell Tissue Res 372:507–522

Mishra A, Tummala P, King A et al (2009) Buffered platelet-rich plasma enhances mesenchymal stem cell proliferation and chondrogenic differentiation. Tissue Eng 15(3):431–435

Mohanty N, Gulati BR, Kumar R, Gera S, Kumar P, Somasundaram RK et al (2014) Immunophenotypic characterization and tenogenic differentiation of mesenchymal stromal cells isolated from equine umbilical cord blood. In Vitro Cell Dev Biol Anim 50:538–548

Muir P, Hans EC, Racette M, Volstad N, Sample SJ, Heaton C, Holzman G, Schaefer SL, Bloom DD, Bleedorn JA et al (2016) Autologous bone marrow-derived mesenchymal stem cells modulate molecular markers of inflammation in dogs with cruciate ligament rupture. PLoS One 11(8):e0159095

Muttini A, Russo V, Rossi E, Mattioli M, Barboni B, Tosi U, Maffulli N, Valbonetti L, Abate M (2015) Pilot experimental study on amniotic epithelial mesenchymal cell transplantation in natural occurring tendinopathy in horses. Ultrasonographic and histological comparison. Muscles Ligaments Tendons J 5(1):5–11

Nakamura A, Takeda S (2011) Mammalian models of Duchenne muscular dystrophy: pathological characteristics and therapeutic applications. J Biomed Biotechnol 2011:184393

Nitahara-Kasahara Y, Hayashita-Kinoh H, Ohshima-Hosoyama S, Okada H, Wada-Maeda M, Nakamura A, Okada T, Takeda S (2012) Long-term engraftment of multipotent mesenchymal stromal cells that differentiate to form myogenic cells in dogs with Duchenne muscular dystrophy. Mol Ther 20(1):168–177

Nixon A, Dahlgren L, Haupt J, Yeager A, Ward D (2008) Effect of adipose-derived nucleated cell fractions on tendon repair in horses with collagenase-induced tendinitis. Am J Vet Res 69:928–937

Nourissat G, Berenbaum F, Duprez D (2015) Tendon injury: from biology to tendon repair. Nat Rev Rheumatol 11:223–233

Oh HJ, Park JE, Kim MJ, Hong SG, Ra JC, Jo JY, Kang SK, Jang G, Lee BC (2011) Recloned dogs derived from adipose stem cells of a transgenic cloned beagle. Theriogenology 75:1221–1231

Okamura LH, Cordero P, Palomino J, Parraguez VH, Torres CG, Peralta OA (2017) Myogenic differentiation potential of mesenchymal stem cells derived from fetal bovine bone marrow. Anim Biotechnol 29:1–11. https://doi.org/10.1080/10495398.2016.1276926

Oliva R, Núñez I, Segunda MN, Peralta OA (2021) Tenogenic potential of equine fetal mesenchymal stem cells under the in vitro effect of bone morphogenetic protein-12 (BMP-12). J Equine Vet Sci 104:103681

Omae H, Zhao C, Sun YL (2009) Multilayer tendon slices seeded with bone marrow stromal cells: a novel composite for tendon engineering. J Orthop Res 27:937–942

Pacini S, Spinabella S, Trombi L, Fazzi R, Galimberti S, Dini F, Carlucci F, Petrini M (2007) Suspension of bone marrow-derived undifferentiated mesenchymal stromal cells for repair of superficial digital flexor tendon in race horses. Tissue Eng 13(12):2949–2955

Patruno M, Gomiero C, Sacchetto R, Topel O, Negro A, Martinello T (2017) Tat-MyoD fused proteins, together with C2c12 conditioned medium, are able to induce equine adult mesenchymal stem cells towards the myogenic fate. Vet Res Commun 41(3):211–217

Piccionello AP, Riccio V, Senesi L, Volta A, Pennasilico L, Botto R, Rossi G, Tambella AM, Galosi L, Marini C, Vullo C, Gigante A, Zavan B, De Francesco F, Riccio M (2021) Adipose micro-grafts enhance tendinopathy healing in ovine model: an in vivo experimental perspective study. Stem Cells Transl Med 10(11):1544–1560

Prockop DJ (2007) 'Stemness' does not explain the repair of many tissues by mesenchymal stem/multipotent stromal cells (MSCs). Clin Pharm Ther 82:241–243

Quam VG, Altmann NN, Brokken MT, Durgam SS (2021) Zonal characterization and differential trilineage potentials of equine intrasynovial deep digital flexor tendon-derived cells. BMC Vet Res 17:138

Raabe O, Shell K, Fietz D, Freitag C, Ohrndorf A, Christ HJ et al (2013) Tenogenic differentiation of equine adipose-tissue-derived stem cells under the influence of tensile strain, growth differentiation factors and various oxygen tensions. Cell Tissue Res 352:509–521

Rajpar I, Barrett JG (2020) Multi-differentiation potential is necessary for optimal tenogenesis of tendon stem cells. Stem Cell Res Ther 11(1):152

Ramirez-Espinosa JJ, Gonzalez-Davalos L, Shimada A, Pina E, Varela-Echavarria A, Mora O (2015) Bovine *(Bos taurus)* bone marrow mesenchymal cell differentiation to adipogenic and myogenic lineages. Cells Tissues Organs 201:51–64

Renzi S, Ricco C, Dotti S, Sesso L, Grolli S, Cornali M, Carlin S, Patruno M, Cinotti S, Ferrari M (2013) Autologous bone marrow mesenchymal stromal cells for regeneration of injured equine ligaments and tendons: a clinical report. Res Vet Sci 95(1):272–277

Rink BE, Amilon KR, Esteves CL, French HM, Watson E, Aurich C, Donadeu FX (2017) Isolation and characterization of equine endometrial mesenchymal stromal cells. Stem Cell Res Ther 8: 166

Rivera C, Tuemmers C, Bañados R, Vidal-Seguel N, Montiel-Eulefi E (2020) Reduction of recurrent tendinitis scar using autologous mesenchymal stem cells derived from adipose tissue from the base of the tail in Holsteiner horses (Equus ferus caballus). Int J Morphol 38(1): 186–192

Rocha dos Santos C, Filgueiras RR, Malard PF, da Cunha Barreto-Vianna AR, Nogueira K, da Silva Leite C, Mendes de Lima EM (2018) Mesenchymal stem cells in osteotomy repair after tibial tuberosity advancement in dogs with cranial cruciate ligament injury. J Exp Orthop 5:17

Romero A, Barrachina L, Ranera B, Remacha AR, Moreno B, de Blas I, Sanz A, Vázquez FJ, Vitoria A, Junquera C, Zaragoza P, Rodellar C (2017) Comparison of autologous bone marrow and adipose tissue derived mesenchymal stem cells, and platelet rich plasma, for treating surgically induced lesions of the equine superficial digital flexor tendon. Vet J 224:76–84

Roth SP, Glauche SM, Plenge A, Erbe I, Heller S, Burk J (2017) Automated freeze-thaw cycles for decellularization of tendon tissue—a pilot study. B M C Biotechnol 17:13

Roth SP, Schubert S, Scheibe P, Groß C, Brehm W, Burk J (2018) Growth factor-mediated tenogenic induction of multipotent mesenchymal stromal cells is altered by the microenvironment of tendon matrix. Cell Transplant 27(10):1434–1450

Roth SP, Brehm W, Groß C, Scheibe P, Schubert S, Burk J (2019) Transforming growth factor beta 3-loaded decellularized equine tendon matrix for orthopedic tissue engineering. Int J Mol Sci 20(21):5474

Rui Y-F, Chen M-H, Li Y-J, Xiao L-F, Geng P, Wang P, Xu Z-Y, Zhang X-P, Dai G-C (2019) CTGF attenuates tendon-derived stem/progenitor cell aging. Stem Cells Int 2019:6257537

Russo V, El Khatib M, di Marcantonio L, Ancora M, Wyrwa R, Mauro A, Walter T, Weisser J, Citeroni MR, Lazzaro F, Di Federico M, Berardinelli P, Cammà C, Schnabelrauch M, Barboni B (2020) Tendon biomimetic electrospun PLGA fleeces induce an early epithelial-mesenchymal transition and tenogenic differentiation on amniotic epithelial stem cells. Cells 9:303

Sassoon AA, Ozasa Y, Chikenji T, Sun Y-L, Larson DR, Maas ML, Zhao C, Jen J, Amadio PC (2012) Skeletal muscle and bone marrow derived stromal cells: a comparison of tenocyte differentiation capabilities. J Orthop Res 30(11):1710–1718

Schnabel L, Lynch M, Van der Meulen M, Yeager A, Kornatowski M, Nixon A (2009) Mesenchymal stem cells and insulin-like growth factor-I gene-enhanced mesenchymal stem cells improve structural aspects of healing in equine flexor digitorum superficialis tendons. J Orthop Res 27: 1392–1398

Schneider PR, Buhrmann C, Mobasheri A, Matis U, Shakibaei M (2011) Three-dimensional high-density co-culture with primary tenocytes induces tenogenic differentiation in mesenchymal stem cells. J Orthop Res 29:1351–1360

Schneider M, Angele P, Järvinen TAH, Docheva D (2017) Rescue plan for Achilles: therapeutics steering the fate and functions of stem cells in tendon wound healing. Adv Drug Deliv Rev

Schramme M, Hunter S, Campbell N, Blikslager A, Smith R (2010) A surgical tendonitis model in horses: techinque, clinical, ultrasonographic and histological characterisation. Vet Comp Orthop 23(4):231–239

Schulze-Tanzil G (2019) Intraarticular ligament degeneration is interrelated with cartilage and bone destruction in osteoarthritis. Cell 8(9):990

Sharma P, Maffulli N (2005) Tendon injury and tendinopathy: healing and repair. J Bone Joint Surg Am 87(1):187–202

Shen H, Jayaram R, Yoneda S, Linderman SW, Sakiyama-Elbert SE, Xia Y, Gelberman RH, Thomopoulos S (2018) The effect of adipose-derived stem cell sheets and CTGF on early flexor tendon healing in a canine model. Sci Rep 8:11078

Shojaee A, Parham A (2019) Strategies of tenogenic differentiation of equine stem cells for tendon repair: current status and challenges. Stem Cell Res Ther 10:181

Simmons PJ, Torok-Storb B (1991) CD34 expression by stromal precursors in normal human adult bone marrow. Blood 78:2848–2853

Smith RKW (2008) Mesenchymal stem cell therapy for equine tendinopathy. Disabil Rehabil 30: 1752–1758

Smith RKW, Korda M, Blunn GW, Goodship AE (2003) Isolation and implantation of autologous equine mesenchymal stem cells from bone marrow into the superficial digital flexor tendon as a potential novel treatment. Equine Vet J 35:99–102

Smith R, Young N, Dudhia J, Kasashima Y, Clegg PD, Goodship A (2009) Effectiveness of bone marrow-derived mesenchymal progenitor cells for naturally occurring tendinopathy in the horse. Regen Med 4(Suppl 6):25–26

Smith MM, Ravi V, Smith SM, Hunyor S, Dart AJ, Sonnabend DH, Little CB (2011) Heterologous labeled mesenchymal stem cells injected into surgically-induced tendon lesions preferentially migrate towards muscle in a sheep model of shoulder tendinopathy. Poster No. 1706 • ORS 2011 Annual Meeting

Smith RKW, Werling NJ, Dakin SG, Alam R, Goodship AE, Dudhia J (2013) Beneficial effects of autologous bone marrow-derived mesenchymal stem cells in naturally occurring tendinopathy. PLoS One 8(9):e75697

Snedeker JG, Foolen J (2017) Tendon injury and repair—a perspective on the basic mechanisms of tendon disease and future clinical therapy. Acta Biomater 63:18–36

Spaas JH, Gambacurta A, Polettini M, Broeckx S, Van Hoeck F, De Schauwer C, Van De Walle GR, Van Soom A (2011) Purification and expansion of stem cells from equine peripheral blood, with clinical applications. Vlaams Diergeneeskundig Tijdschrift 80:129–135

Tan Q, Lui PPY, Lee YW (2013) In vivo identity of tendon stem cells and the roles of stem cells in tendon healing. Stem Cells Dev 22(23):3128–3140

Taroni M, Cabon Q, Fèbre M, Cachon T, Saulnier N, Carozzo C, Maddens S, Labadie F, Robert C, Viguier E (2017) Evaluation of the effect of a single intra-articular injection of allogeneic neonatal mesenchymal stromal cells compared to Oral non-steroidal anti-inflammatory treatment on the postoperative musculoskeletal status and gait of dogs over a 6-month period after tibial plateau leveling osteotomy: a pilot study. Front Vet Sci 4:83

Tempfer H, Wagner A, Gehwolf R et al (2009) Perivascular cells of the supraspinatus tendon express both tendon- and stem cell-related markers. Histochem Cell Biol 131(6):733–741

Thangarajah T, Shahbazi S, Pendegrass CJ, Lambert S, Alexander S, Blunn GW (2016) Tendon reattachment to bone in an sheep tendon defect model of retraction using allogenic and xenogenic demineralised bone matrix incorporated with mesenchymal stem cells. PLoS One 11(9):e0161473

Torricelli P, Fini M, Filardo G, Tschon M, Pischedda M, Pacorini A, Kon E, Giardino R (2011) Regenerative medicine for the treatment of musculoskeletal overuse injuries in competition horses. Int Orthop 35:1569–1576

Tripathi AK, Ramani UV, Ahir VB, Rank DN, Joshi CG (2010) A modified enrichment protocol for adult caprine skeletal muscle stem cell. Cytotechnology 62:483–488

Turner CG, Klein JD, Steigman SA, Ahmed A, Zurakowski D, Eriksson E, Fauza DO (2011) Preclinical regulatory validation of an engineered diaphragmatic tendon made with amniotic mesenchymal stem cells. J Pediatr Surg 46(1):57–61

Tyrnenopoulou P, Karayannopoulou M, Angelopoulou S, Pyrros A, Mparous E, Koliakos G, Diakakis N (2016) Successful management of an equine carpal chip fracture by intra-articularly injected adipose-derived stromal vascular fraction after arthroscopic removal. Iran J Vet Res 17(1):59–61

Van Loon VJF, Scheffer CJW, Genn HJ, Hoogendoorn AC, Greve JW (2014) Clinical follow-up of horses treated with allogeneic equine mesenchymal stem cells derived from umbilical cord blood for different tendon and ligament disorders. Vet Q 34:92–97

Vandenberghe A, Broeckx SY, Beerts C, Seys B, Zimmerman M, Verweire I, Suls M, Spaas JH (2015) Tenogenically induced allogeneic mesenchymal stem cells for the treatment of proximal suspensory ligament desmitis in a horse. Front Vet Sci 2:49

Vieira NM, Brandalise V, Zucconi E, Secco M, Strauss BE, Zatz M (2010) Isolation, characterization, and differentiation potential of canine adipose-derived stem cells. Cell Transplant 19:279–289

Vieira NM, Valadares M, Zucconi E, Secco M, Bueno CR, Brandalise V, Assoni A, Gomes J, Landini V, Andrade T, Caetano HV, Vainzof M, Zatz M (2012) Human adipose-derived mesenchymal stromal cells injected systemically into GRMD dogs without immunosuppression are able to reach the host muscle and express human dystrophin. Cell Transplant 21(7): 1407–1417

Violini S, Ramelli P, Pisani LF, Gorni C, Mariani P (2009) Horse bone marrow mesenchymal stem cells express embryo stem cell markers and show the ability for tenogenic differentiation by in vitro exposure to BMP-12. BMC Cell Biol 10:29

Wang D, Pun CCM, Huang S, Tang TCM, Ho KKW, Rothrauff BB, Yung PSH, Blocki AM, Ker EDF, Tuan RS (2020) Tendon-derived extracellular matrix induces mesenchymal stem cell tenogenesis via an integrin/transforming growth factor-β crosstalk-mediated mechanism. FASEB J 34(6):8172–8186

Wijekoon HMS, Toyota K, Kim S, Fang J, Bwalya EC, Hosoya K, Okumura M (2017) Differentiation potential of synoviocytes derived from joints with cranial cruciate ligament rupture and medial patella luxation in dogs. Res Vet Sci 114:370–377

Williamson KA, Lee KJ, Humphreys WJ, Comerford EJ, Clegg PD, Canty-Laird EG (2015) Restricted differentiation potential of progenitor cell populations obtained from the equine superficial digital flexor tendon (SDFT). J Orthop Res 33(6):849–858

Wu X, Wang S, Chen B, An X (2010) Muscle-derived stem cells: isolation, characterization, differentiation, and application in cell and gene therapy. Cell Tissue Res 340:549–567

Yamasaki A, Omura T, Murata D, Kobayashi M, Sunaga T, Kusano K, Ueno Y, Kuramoto T, Hobo S, Misumi K (2018) A pilot study of regenerative therapy by implanting synovium-derived mesenchymal stromal cells in equine osteochondral defect models. J Equine Sci 29(4): 117–122

Yang J, Zhao Q, Wang K, Liu H, Ma C, Huang H, Liu Y (2016) Isolation and biological characterization of tendon-derived stem cells from fetal bovine. Vitro Cell Dev Biol Anim 52(8):846–856

Yoon HY, Lee JH, Jeong SW (2012) Long-term follow-up after implantation of autologous adipose tissue derived mesenchymal stem cells to treat a dog with stifle joint osteoarthrosis. J Vet Clin 29:82–86

Young HE, Steele TA, Bray RA, Detmer K, Blake LW, Lucas PW, Black AC Jr (1999) Human pluripotent and progenitor cells display cell surface cluster differentiation markers CD10, CD13, CD56, and MHC class-I. Proc Soc Exp Biol Med 221:63–71

Zelzer E, Blitz E, Killian ML, Thomopoulos S (2014) Tendon-to-bone attachment: from development to maturity. Birth Defects Res C Embryo Today 102(1):101–112
Zhang J, Wang JHC (2010a) Characterization of differential properties of rabbit tendon stem cells and tenocytes. BMC Musculoskelet Disord 11(1):10
Zhang J, Wang JH (2010b) Platelet-rich plasma releasate promotes differentiation of tendon stem cells into active tenocytes. Am J Sports Med 38(12):2477–2486
Zhou Z, Akinbiyi T, Xu L et al (2010) Tendon-derived stem/progenitor cell aging: defective self-renewal and altered fate. Aging Cell 9(5):911–915

Mesenchymal Stem Cells Therapeutic Applications in Peripheral Nervous System Disorders

4

Abstract

Peripheral nerve affections are not uncommon in veterinary species. Numerous nerves are affected with predilection for some specific nerves. Currently, no specific and effective treatment is available. To address the challenge, regenerative medicine incorporating mesenchymal stem cells (MSCs) and scaffolds are being evaluated. Peripheral nerve injury is studied in experimental and clinical cases. Tissue-engineered constructs employing MSCs are mainly studied for sciatic nerve injuries. The limited available literature on applications of MSCs in nerve affections in veterinary species shows promising role, however, consensus lacks on their definitive applications. The current chapter discusses peripheral nerve injury models and critically evaluates MSCs therapeutic effects.

Keywords

Experimental model · Large animal · Mesenchymal stem cell · Peripheral nerve injury · Sciatic nerve

4.1 Introduction

Peripheral nervous system comprises three cell types including the neuronal, glial and stromal. Through peripheral nerves (motor, sensory and autonomic), signals are conveyed between the higher centre (brain/spinal cord) and the body parts (Menorca et al. 2013). Peripheral nerves due to their widespread body distribution are prone to injuries. Peripheral nerve injury (PNI) is a commonly encountered clinical condition that tends to produce debilitating effects, and in about 50% cases permanent disability sets in (Rodrigues et al. 2012). Affected peripheral nerves may lead to impaired motor, sensory and/or autonomic functionality depending upon the severity of the injury (Siemionow and Brzezicki 2009; Isaacs 2013; Rosberg et al. 2013). In

M. B. Gugjoo, *Therapeutic Applications of Mesenchymal Stem Cells in Veterinary Medicine*, https://doi.org/10.1007/978-981-19-3277-9_4

humans, nerve affections vary from 1.64 to 5% based upon the involved aetiology (Taylor et al. 2008; Thorsen et al. 2012). Peripheral neuropathy occurs either through injury or systemic diseases (Choulam 2017). Road traffic accidents, dislocations, fractures, faulty drug injection, peripheral nerve anaesthesia techniques, patient position (recumbency) and even gunshot wounds commonly produce PNI (Jacobson and Schrader 1987; Hewson et al. 2018; Al-Massri et al. 2020). Apart from the injuries, peripheral neuropathy in veterinary species can also arise from systemic diseases like metabolic diseases, toxins, genetic disorders, infections, auto-immune diseases or tumours (Pilkington et al. 2001; Kim et al. 2003; Schwarz et al. 2008; James et al. 2010; Barigye et al. 2016; Choulam 2017). In dogs, fracture or iatrogenic-induced nerve injury mainly affects peripheral nerves. Injury to peripheral nerve means the loss or diminished function of the particular area to which it innervates (Moore 2016). Some of the peripheral nerves may be more commonly affected like suprascapular and radial nerve among horse forelimb nerve affections (Emond et al. 2016) and sciatic and obturator nerves in cattle hindlimb nerve affections (Poulton et al. 2019). Neuropathy affects the quality of life and poses the burden of treatment costs and associated morbidities (Thorsen et al. 2012). In veterinary species, especially large animals, it tends to make animals recumbent, and thus, animal culling remains the only option.

4.2 Peripheral Nerve Injury

Peripheral nerve injury (PNI) encompasses injuries to all the nerves of brachial- and lumbo-sacral plexus, and other nerves like spinal accessory nerve (11th cranial nerve). The injury is classified/graded mainly through Seddon's (1943) or Sunderland's (1951) recommendations. Seddon's classification divides nerve injury into three classes involving class I (neurapraxia, mild form of injury), class II (axonotmesis, neuroma-in-continuity) and class III (neurotmesis, axonal and connective tissue transection). Sunderland classifies nerve injury into a total of five grades among which first and second grades are comparable to that of Seddon's classes I and II. In grade III, axon and its myelin sheath are transected, but perineurium remains intact. Severe forms of nerve injury are grade IV and V; in the former grade, only epineurium remains intact while the same is lost along with other structures in the latter.

Neurapraxia does not exhibit axonal degeneration, although myelin sheath may be locally degenerated without undergoing Wallerian degeneration. It may lead to partial sensory-motor impairments in distal sites of damaged nerves. The condition, however, resolves spontaneously within a few days to few months (Chhabra et al. 2014). Axonotmesis exhibits axonal degeneration; nerve continuity is maintained through supporting structures like Schwann tubes, perineurium and endoneurium. Contrary to that of neurapraxia, Wallerian degeneration in distal areas occurs in axonotmesis. It leads to autonomic function deficit and impairment of sensory-motor activity. The recovery may be seen after months of injury and may be mediated by reactivation of endogenous Schwann cells and endoneurial tubes (Sunderland 1951).

In grade III, axonal loss, demyelination and endoneurial disruption occur, but epineurium and perineurium remain intact. In grade IV, axonal and myelin damage, and perineurial and endoneurial disruption are seen but epineurium is preserved. The neurotmesis of Seddon's classification is comparable to grade V of Sunderland. In grade V, axonal and myelin damage, epineurium, perineurium and endoneurium are disrupted. Although the classification of Sunderland is extensive, the differential diagnosis between IV and V becomes difficult. There is a possibility of complete recovery in classes/ grades I and II and only partial in grade III. In grades IV and V, surgical intervention is desired (Sunderland 1951). Apart from these two gradings, one more grade (grade VI) has been added by Mackinnon and Dellon, wherein a mixed type of nerve injury occurs and may vary from grades I to V (Alvites et al. 2018).

4.3 Nerve Homeostasis Potential

A sequential and substantial physiological change occurs post-PNI. With the compromise in blood–nerve barrier an increase in the permeability occurs. This initiates an inflammatory response leading to the Schwann cell proliferation and local macrophage activation (Woodhoo et al. 2009; Forostyak et al. 2013). Additionally, Schwann cells de-differentiate that along with the potential outgrowth of neurons distal to the injury prompt an axonal re-growth/repair (Fex Svennigsen and Dahlin 2013). The plastic de-differentiation of Schwann cells is exhibited by their ability to show regeneration and myelination, and myelinophagy (Rodrigues et al. 2012). The Schwann cell (SC) regeneration and/or myelination is activated through cross-talk between the Schwann cells and the surrounding extracellular matrix and nerve cells (Grochmal and Midha 2014; Sullivan et al. 2016). In the first instance, the cell body undergoes changes like chromatin fragmentation, nuclear material diffusion and Nissl body dissolution due to the retrograde signals of damaged nerve stumps. At the same stage, immature SC markers like glial fibrillary acidic protein (GFAP), neural cell adhesion molecule (NCAM) and p75 neurotrophin receptor (p75NTR) expression increase. SCs de-differentiate into myelin synthesis SCs and non-myelin synthesis SCs. The former cell type re-differentiates towards myelinating phenotype leading to myelination. Newly secreted myelin remains thinner and shorter than native myelin of the healthy nerve (Kolar and Kingham 2014; Gomez-Sanchez et al. 2017). The latter cell type facilitates axonal regeneration and myelinophagy. However, myelin synthesis (myelin basic protein and myelin-associated glycoprotein-related genes, Egr2 and cholesterol synthesis) mostly remain reduced at this stage. This stage is followed by the reparative phase, wherein various growth-promoting factors [neurotrophin-3 (NT3), brain-derived neurotrophic factor (BDNF), vascular endothelial growth factor (VEGF) and glial cell line-derived neurotrophic factor (GDNF)] released promote axonal re-growth and neural survival. In addition to the axonal growth, SCs release chemo-attractants and inflammatory cytokines [monocyte chemotactic protein1 (MCP1), tumour necrosis factor α (TNFα) and interleukin-1] to promote macrophage recruitment. Macrophages clear the myelin debris and thereby axonal recuperation (Jessen and Mirsky 2016). Among various transcription factors of the peripheral nerve repair process, currently c-Jun stands tall in signalling pathways (Jessen and Mirsky 2016). There is a limited insight in understanding of the healing power of peripheral nerves. This includes the gene expression of nerve cells and their correlation with respect to the limited healing power and almost nil power in CNS. In case of extensive damage, such processes fall short to repair the injured nerve. The available microenvironment may also hamper such repair processes.

In order to satisfy their demand, extrinsic SCs too enhance repair (Lundborg 2004; Li et al. 2006). To use cultured SCs in sufficient concentrations, an extended period is required with the current protocols, and thus, the nerve repair may be delayed (Keilhoff et al. 2000; Calderon-Martinez et al. 2002). Delay in nerve repair would be deleterious clinically as neuronal cell death might further increase (Hart et al. 2002). Unlike stem cells, allogeneic SCs are not immuno-evasive and elicit an immune response. As SC harvest essentially requires the autologous donor, problems of sensory loss, scarring and in some cases neuroma arise (Keilhoff et al. 2006).

4.4 Surgical Procedure and Limitations Thereof

For neural tissue loss, primary end-to-end repair (epineural) is the gold standard while in the case of critical size neural tissue loss (>1 cm) grafting (autologous or allogeneic), nerve transfers and artificial nerve conduits are utilized depending upon the situation. Autologous nerve grafting is the gold standard for such large nerve defects (Bassilios Habre et al. 2018; Zhang et al. 2021). Although surgical techniques are currently widely used, significant associated limitations are limited autologous graft availability, immune rejection in allogeneic grafts, pain, and delayed and/or incomplete return of functional activity. As such regenerative medicines involving supplementary application of exogenous neurotrophic factors, bioengineered scaffolds, electrical stimulation, and cell-based therapies like mesenchymal stem cells are being undertaken (Lien et al. 2020).

4.5 Why MSCs in PNI?

MSCs are considered to provide therapeutic effects through differentiation and/or secretion of microvesicles (exosomes) and various trophic factors that along with the anti-inflammatory/immuno-modulatory effects promote anabolic actions (Gugjoo et al. 2019, 2020a, b, 2022). MSCs differentiation into neural-like cells is detailed and discussed in the next chapter (Chapter No. 5). These cells may provide analgesia after homing to the dorsal spinal cord or dorsal root ganglions of the injured area. This analgesic effect may be independent of the MSCs source, timing and route of implantation (Wang et al. 2020). MSCs immuno-evasive nature and immunomodulation make it possible to use them from allogeneic or even xenogeneic sources (Gugjoo et al. 2020b). MSCs paracrine factors including the soluble factors and exosomes may vary with the tissue source. The soluble factors include numerous growth and neurogenic factors that can play an important role in nerve protection and repair. These secreted soluble neurotrophic factors include ciliary neurotrophic factor (CNTF), hepatocyte growth factor (HGF), brain-derived neurotrophic factor (BDNF), glial cell-derived neurotrophic factor (GDNF), platelet-derived growth factor (PDGF), nerve growth factor (NGF), neurotrophin-3, 4/5 (NT-3, 4/5), insulin-like growth factor 1 (IGF1), basic fibroblast growth factor (FGF2), epidermal growth factor (EGF), Pigment epithelium-derived factor (PEDF) and erythropoietin (EPO) (Mead et al. 2015; Fontanilla et al. 2016; Noverina et al. 2019; Usategui-Martín et al. 2020). All these factors might promote axonal regeneration or axonal outgrowth, myelin formation and inhibition of neural cell apoptosis that ultimately recovers denervated muscles (Konala et al. 2016; Sowa et al. 2016; Salehi et al. 2017). Exosomes contain lipids and proteins, besides the nucleic acid materials like mRNA, microRNA (miRNA) that can potentially restore essential information required in maintaining the physiology and in preventing the pathology. MSCs may provide these exosomes to the neighbouring cells via cell–cell contact through gap junctions (Kourembanas 2015; Zhang et al. 2021). These exosomes may protect neuronal damage that may arise with the glutamate through PI3-Akt signalling pathway (Wei et al. 2016). Scaffolds loaded with the MSCs may ensure adjacent tissue interference-free healing of the nerve, and thus, regulation of the local microenvironment with the resultant lack of any neuroma (Ma et al. 2013; Zhang et al. 2021). Despite such potential features, wide variability in results in individual studies and as per the species exists.

4.6 In Vivo Experimental Studies

To provide an insight and an understanding in the nerve regeneration process, in vivo experimental studies are being conducted. Regenerative medicine that employs stem cells is being extended to understand the pathophysiology of nerve affections and hasten its healing. Small-lab animal studies involving rat, mouse and rabbit are extensively undertaken. These animal studies provide proof-of-therapeutics for human and veterinary medicine. There is a paucity of the literature on in vivo MSCs peripheral nerve regeneration in veterinary species (cattle, sheep, goat, cat, dog and horse). Among various species, dog studies to some extent are undertaken with a single study each on horse (Villagrán et al. 2016) and sheep (Casañas et al. 2014) (Table 4.1). Sheep nerves are comparable in size and regeneration behaviour is also comparable to that of humans (Glasby et al. 1993; Hems et al. 1994; Fullarton et al. 2001; Starritt et al. 2011). Sheep median nerve size is comparable to ulnar and median nerves in humans (Jeans et al. 2007). The femoral nerve in sheep has the same length as the ulnar nerve in humans (Glasby et al. 1990). Among various nerves, sciatic nerve is the most commonly studied in rodents while in sheep median nerve has mostly been made as a subject (Diogo et al. 2017), being easily accessible and large in size (Varejão et al. 2004; Nichols et al. 2005; Savastano et al. 2014).

The experimental PNI models are conducted either as nerve compression (axonotmesis) or transection (Neurotmesis). The compression lesion evaluates the nerve changes, proximal and distal to the lesion. It also evaluates changes that may ensue in the neuronal cell bodies and innervated anatomical structures like muscles. The PNI model evaluates the effect of drugs or cells like stem cells on healing/ regeneration effects. The compression can be obtained either by crush or ligation of the nerve. The compression interrupts the axonal continuity (axonotmesis or even grade III PNI) but continuity of the nerve is maintained through peripheral connective tissue scaffolds like epineurium or perineurium. Such an injury does not hamper axonal regrowth as the pathway towards distal injured nerve end remains intact (Geuna et al. 2009). The crushing causes limited anatomical damage to the nerve. It does not require extensive microsurgical skills, has limited inter-individual variability and has higher reproducibility (Chen et al. 2008; Toth et al. 2008; Lou et al. 2012; Long et al. 2005; Sta et al. 2014; Wright et al. 2014). Due to these benefits, it is one of the commonly studied experimental models to investigate regenerative changes over the course of time (Bridge et al. 1994; Savastano et al. 2014). It also provides the basis of proof-of-therapeutic principle for a therapeutic agent in veterinary and/or human medicine (Fleming et al. 2007; Amado et al. 2008; Baptista et al. 2008; Dadon-Nachum et al. 2011; Kilic et al. 2013; Wang et al. 2014).

The experimental nerve crush injury model offers advantages in terms of feasibility and reproducibility, and it has limited translational potential for two main reasons. Most surgical lesions in veterinary species at least show partial nerve transection/laceration and often have a different clinical background. Laboratory animal axonal regeneration happens to be spontaneous but fails to occur in humans and large animals and may be due to the frequent extensive fibrosis at the lesion site. The crush lesions therefore require surgery to remove the damaged tissue and maintain continuity through a conduit (Tos et al. 2012). Axonal regeneration in nerve transection tends to occur in lab animals, while in large animals/human axons unless surgically repaired may not regenerate and can undergo chronic denervation. To mimic the experimental model to the large mammal regeneration, severed nerves may be sutured to the muscle to prevent axonal regeneration. These types of models are mostly used to study chronic denervation with a little focus on therapeutics (Karsidag et al. 2012; Moimas et al. 2013; Blom et al. 2014). To translate experimental results into the clinical settings, a uniform and resembling nerve repair and regeneration model need to be devised and evaluated (Geuna 2015).

4.6.1 Peripheral Nerve Injury Models

Nerve regenerative potential of MSCs is studied in animal models of sciatic (mixed), recurrent laryngeal (mixed), facial (mixed), femoral (mixed), ulnar (mixed), hypogastric (autonomic) and ramus communicans nerves (connecting a sympathetic ganglion with a spinal nerve). The cells from various sources like bone marrow (BM), adipose tissue (AD) and umbilical cord (UC) are utilized, although BM-MSCs have mostly been evaluated. These cells are employed with or without nerve conduits/scaffolds and other biomaterials (Shalaby et al. 2017; Sun et al. 2018). MSCs harvested from xenogeneic, allogeneic and/or autologous sources are used. The cellular concentration remains variable. In the case of nerve resection models (neurotmesis), the conduit/scaffold is incorporated to fill the gap of the resected nerve while compression injury does not necessarily incorporate scaffolds. The cells are implanted locally into the epineurium, perineurium and/or the conduit or parenterally into muscle, intravenous or intraperitoneal routes. Even the duration of studies too remains variable with studies conducted for as low as 4 weeks to a maximum of 6–12 months.

There is a lack of consensus on the optimal MSCs posology to regenerate and relieve nerve sufferings in lab animals. In controlling neurological pain, MSCs may not be affected by type, time, dose and route of delivery (Wang et al. 2020). However, nerve regeneration and complete recovery from neuropathy remains to be taken care of.

4.6.2 Large Animal Peripheral Nerve Resection Models

Among various veterinary species, dog, horse and sheep nerve resection models are evaluated through the prism of regenerative medicine employing MSCs along with

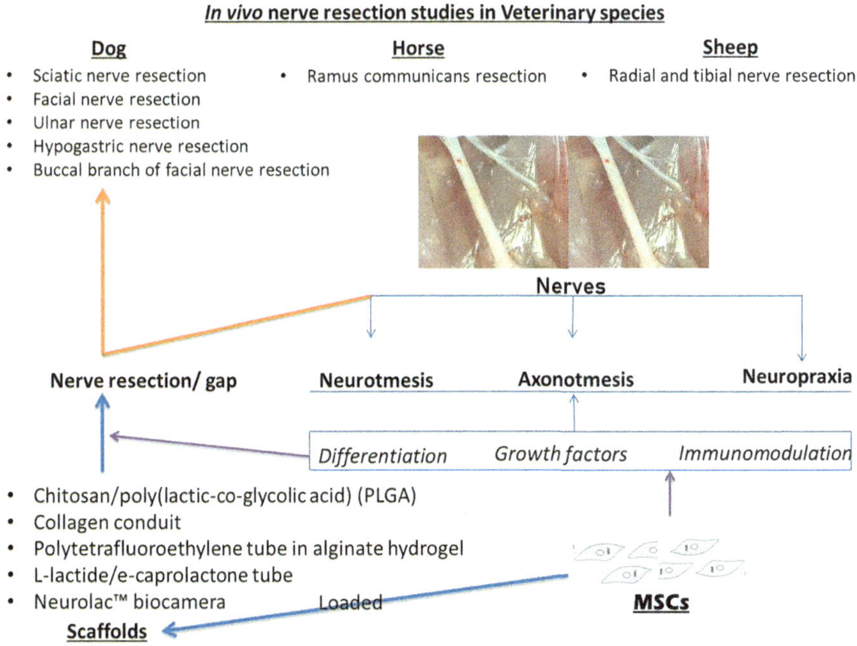

Fig. 4.1 Tissue engineering employing mesenchymal stem cells and scaffolds for nerve regeneration in peripheral nerves resection models in dog, horse and sheep

the scaffolds (Fig. 4.1). No MSCs based study on compression injury could be located in large animal veterinary species. Below are discussed the nerve resection studies conducted in veterinary species.

4.6.2.1 Sciatic Nerve Resection Model

Sciatic nerve resection model forms the major portion of the dog literature (Table 4.1) with defects size ranging from 3 cm (Chen et al. 2019) to 6 cm (Xue et al. 2017). Mostly BM-MSCs are utilized together with the scaffolds like chitosan/ poly(lactic-co-glycolic acid) (PLGA) (Ding et al. 2010; Xue et al. 2012), acellular matrix (Xue et al. 2017), longitudinally oriented collagen conduit (LOCC) (Cui et al. 2018) and acellular nerve allograft (ANA) along with testosterone propionate (TP) (Chen et al. 2019) to form a conduit. The healing potential is evaluated with variable diagnostic parameters including behavioural assessment, electrophysiology, retrograde fluorogold tracing, histological examination and transmission electron microscopy for a period ranging from as low as 5 months (Chen et al. 2019) to as high as 1 year (Xue et al. 2012, 2017). Implantations of tissue-engineered constructs (MSCs and scaffolds) survive for some time in the area. Most of the studies correlate these biomaterials with the improvement in the healing. The neural tube/scaffolds may show a positive role but beyond a certain gap there is need of support cells (de Boer et al. 2012). Improvement in large animals occurs at earliest by 3 months. In relation to the clinical recovery, motor function (symmetric or dynamic parameters) in cell-scaffold-treated animals approach that of the autografts (Xue et al. 2012, 2017). An improvement in nerve regeneration evidenced by the recovery of sensory and motor function, compound muscle action potential (CMAP) amplitude and histomorphometry occurs (Ding et al. 2010; Xue et al. 2017). However, electrophysiological recovery may be comparable (Xue et al. 2017) or remains below (Cui et al. 2018) to that occurs with the autograft. CMAP amplitude is proportionate to that of the number of nerve fibres innervating the muscle. It also correlates with the myelinated nerve fibre diameter and myelin sheath thickness. Improvement in healing through augmented blood perfusion and increased intraneural vascularity together with the higher density of myelinated nerve fibres and thicker myelin sheath is also shown (Ding et al. 2010; Cui et al. 2018). The myelin sheath formation in the middle part of the conduits like LOCC may also occur (Cui et al. 2018). It is worth mentioning that most of the studies fail to show significant improvement comparable to the autograft in all the evaluation criteria. Some of the studies show poor histology (Xue et al. 2012) while other shows histological features (including the presence of SCs) with the use of BM-MSCs/acellular nerve allograft/testosterone comparable to that of the autograft (Chen et al. 2019). Androgens like testosterone are considered to play a neuroprotective role as they facilitate neuron and axon regeneration through the regulation of morphology and function of astrocytes (Li et al. 2008). The androgen signals in relation to the BM-MSCs differentiation too occur (Huang et al. 2014). Although there appears a promising role of tissue engineering in these limited sciatic nerve repair studies, the current status demands further extensive research to achieve complete regeneration.

Table 4.1 MSCs preclinical experimental studies in various peripheral nerve models in dog horse and sheep

Study design/period	Biomaterial used/ route	Evaluation criteria	Overall result	References
Dog sciatic nerve resection model				
Randomized placebo-controlled comparative study ($n = 24$)/6 months	Autologous BM-MSCs (5×10^7), chitosan/poly(lactic-co-glycolic acid) (PLGA)/local implantation	Electrophysiological assessment, Fluoro-gold retrograde tracing and histological investigation	Improved outcome	Ding et al. (2010)
Randomized placebo-controlled study ($n = 20$)/ 12 months	Autologous BM-MSCs (1×10^8/mL) chitosan conduit inserted with poly(lactic-co-glycolic acid) (PLGA) fibres/local implantation	Behavioural analysis, electrophysiology, retrograde fluorogold tracing and histological examination, transmission electron microscopy	Improved outcome	Xue et al. (2012)
Randomized placebo-controlled study ($n = 24$)/ 12 months	BM-MSCs (1×10^6/mL), acellular matrix (ACM)/local implantation	Behavioural, functional and histology	Improved outcome	Xue et al. (2017)
Randomized placebo-controlled comparative study ($n = 20$)/9 months	Xenogeneic UC-MSCs (UC-MSC, 1×10^6/mL), longitudinally oriented collagen conduit (LOCC)/local implantation	Electrophysiological, electron microscopy, histological analysis	Improved outcome	Cui et al. (2018)
Randomized placebo-controlled study ($n = 28$)/ 5 months	Allogeneic BM-MSCs (1×10^7/mL implanted locally), acellular nerve allograft (ANA), testosterone propionate (TP, injected into the lateral crus)	Electrophysiological assessment, histological evaluation	Improved outcome	Chen et al. (2019)
Dog facial nerve resection model				
Randomized placebo-controlled study ($n = 7$)/ 12 weeks	Autologous AD-MSCs (2×10^6 cells/mL), polytetrafluoroethylene tube encapsulated in alginate hydrogel/ local implantation	Electrophysiological assessment, histomorphometric analysis	Improved outcome	Ghoreishian et al. (2013)
Dog ulnar nerve defect model				
Randomized placebo-controlled study ($n = 8$)/ 24 weeks	Autologous BM-MSCs (3×10^7), L-lactide/e-caprolactone tube/local implantation	Immunohistochemistry, electrophysiological examination, morphometric measurements, wet weight of hypothenar muscles	Improved outcome	Kaizawa et al. (2016)

(continued)

Table 4.1 (continued)

Study design/period	Biomaterial used/ route	Evaluation criteria	Overall result	References
Dog hypogastric nerve resection model				
Randomized controlled comparative study ($n = 9$)/ 12 weeks	BM-MSCs, lactic and glycolic acids (PLGA)	Histopathology	Improved outcome	Li et al. (2009)
Randomized placebo-controlled study ($n = 15$)/ 12 weeks	BM-MSCs copolymer of lactic, glycolic acids (PLGA) tube, collagen protein sponge/local implantation	Morphology, HE staining, neurofilament immunohistochemistry staining, electron microscope scanning and measurement quantity	Improved outcome	Li et al. (2017)
Dog buccal branch of facial nerve resection model				
Randomized placebo-controlled study ($n = 9$)/ 10 weeks	BM-MSCs, silicone tubes embedded with collagen type I sponge/local implantation	Histopathology examination	Improved outcome	Abdel Azeem et al. (2014)
Horse ramus communicans resection model				
Randomized placebo-controlled study ($n = 3$)/ 45 days	Allogeneic BM-MSCs (10×10^6)/fascia surrounding nerve	Histology	Non-obvious improved outcome	Villagrán et al. (2016)
Sheep radial and tibial nerve resection model				
Randomized experimental study ($n = 7$)/6 months	BM-MSCs ($30–50 \times 10^6$), repairment with Neurolac™ biocamera scaffold, PRP/local implantation	Histological, immunohistochemistry and morphometry	Improved outcome	Casañas et al. (2014)

MSCs mesenchymal stem cells, *BM* bone marrow derived, *AD* adipose tissue derived, *UC* umbilical cord derived, *n* total number of animals in all the groups of the study; Improved outcome: Indicates improved outcome with MSCs transplantation even if only one parameter improves; Study period: Indicates maximum time period for which study is conducted

4.6.2.2 Miscellaneous Nerve Resection Models

A single study of nerve resection each on dog facial nerve (Ghoreishian et al. 2013), ulnar nerve (Kaizawa et al. 2016) and buccal branch of the facial nerve (Abdel Azeem et al. 2014) and two studies of hypogastric (Li et al. 2009, 2017) nerves utilize autologous or allogeneic AD-MSCs and BM-MSCs. These studies evaluate nerve repair through the electrophysiological assessment, histomorphometric analysis, for 10–24 weeks. In sheep and horse, a single study on ramus communicans (Villagrán et al. 2016) and radial and ulnar nerve resection (Casañas et al. 2014), respectively, exists that utilize MSCs. In horse ramus communicans resection model, fascia suturing is followed by BM-MSCs transplantation along with the scaffolds (Table 4.1).

In dog ulnar nerve resection, some of the l-lactide/e-caprolactone loaded BM-MSCs (35%)-treated cases express glial markers at eighth week. Functional recovery occurs in the form of improved nerve amplitude and conduction velocity, but it is significantly lower than the autograft-treated animals at 12 weeks. However, at 24 weeks period, non-significant differences in relation to the nerve amplitude and conduction velocity, myelinated axon number and diameter and muscle weight occur (Kaizawa et al. 2016). Similar types of results are also observed in facial nerve repair models. Although axonal diameter is lower by 29% in cell-loaded scaffold nerves, maximal amplitude of action potential and nerve conduction velocity are better as compared to the control at 12 weeks period. Furthermore, a larger number of NF-positive fibres (statistically non-significant) with a limited pattern of axonal fibres are also observed in cell-loaded scaffolds than the control (Ghoreishian et al. 2013). In the buccal branch of the right side of the facial nerve resection model, up to 6 weeks, no observable improvement is recognized. By eighth week, a majority of the dogs show facial movement of the enervated area and the salivary drooling reduces moderately. By tenth week, left over dogs also show such an improvement. The clinical observations correlate well with the histological observations. At sixth week, MSCs proliferate and at eighth week the multiple blood vessels and nerve-like tissue appear. The nerve-like tissue is well organized with well-dilated blood vessels by tenth week (Abdel Azeem et al. 2014). Similarly, sheep radial and tibial resection models show improved results in histomorphometry and histology (myelinated nervous fibres and conduction increasing) with cell-loaded scaffolds. These results are more significant at 6 months than at 3 months (Casañas et al. 2014).

The neural tube/scaffold might potentially be acting as a barrier against mesodermal external environment. This helps in preventing the fibrosis during nerve regeneration, and additionally provides mechanical support and retains endoneurial fluids. Incorporation of the MSCs might help in cellular differentiation and elongation (like SCs) and in promoting the neural microenvironment by preventing inflammation and immune response. These cells may also help in growth of the blood vessels. Although the results appear promising, varied factors restrict their clinical usage. One of the factors is the safety of the particular scaffold type and its combination with the cells. All the above studies show improvement with the time, and thus, further extended studies are desired. However, one of the large animal (horse) studies (Villagrán et al. 2016) fail to show any improvement as compared to the control both in early and long duration.

There are limitations in relation to evaluation of nerve repair in these studies. These studies are very limited in details including the MSCs engraftment, their differentiation and/or paracrine action. No specific role of MSCs is attributed as simultaneous utilization of variable types of scaffolds provide raw data. No detailed information in relation to the anti-inflammatory actions and/or the gene and protein expressions typical to nerves is available. The animal numbers are small enough to offer statistically correlated data and the duration of studies is too limited to confirm long-lasting effects. Factors like surgical technique, nerve anatomical region (horse distal aspect of the limb has a limited available vasculature), method of cellular delivery and labelling and tracking of cells may also play a role. Thus, MSCs optimal posology with respect to the specific suitable nerve injury model remains to be seen.

4.7 Conclusion(s)

Peripheral nerve regeneration poses great challenge in animals and humans. For effective regeneration and pain relief, regenerative medicine employing MSCs is evaluated. These regenerative medicine studies restrict to the experimental conditions only. In veterinary species, the literature is quite limited with no clinical trials. Most resection models are evaluated especially in the sciatic nerve of the dog. All these studies are preliminary, and in order to arrive at any conclusion, it is imperative to conduct the studies that focus on the particular nerve at a time and conduct uniform studies. The uniform studies incorporate MSCs from particular source, passage, concentration, route and frequency. MSCs typical role remains to be evaluated. The scaffolds and other biomaterials incorporated should also be uniform and evaluated in a comparative manner. Evaluation criterion and period of study should be extensive to arrive at any conclusion. After attaining data, the studies may be conducted to compare safety, efficacy and the final outcome of the therapy for an evidence-based medicine.

References

Abdel Azeem A, Abel-Rahman MA, Mohammed IA, Aboul-Ezz EHA, Abdel-Hameed MA (2014) Evaluation of motor nerve regeneration using guided tubes seeded with stem cells "an experimental study". Egypt Dent J 60:146–156

Al-Massri KF, Ahmed LA, El-Abhar HS (2020) Mesenchymal stem cells in chemotherapy-induced peripheral neuropathy: a new challenging approach that requires further investigations. J Tissue Eng Regen Med 14:108–122

Alvites R, Caseiro AR, Pedrosa SS, Branquinho MV, Ronchi G, Geuna S, Varejão ASP, Maurício AC (2018) Peripheral nerve injury and axonotmesis: state of the art and recent advances. Cogent Med 5(1):1466404

Amado S, Simões MJ, Armada da Silva PAS, Luís AL, Shirosaki Y, Lopes MA, Santos JD, Fregnan F, Gambarotta G, Raimondo S, Fornaro M, Veloso AP, Varejão ASP, Maurício AC,

Geuna S (2008) Use of hybrid chitosan membranes and N1E-115 cells for promoting nerve regeneration in an axonotmesis rat model. Biomaterials 29:4409–4419

Baptista AF, Gomes JR, Oliveira JT, Santos SM, Vannier-Santos MA, Martinez AM (2008) High- and low-frequency transcutaneous electrical nerve stimulation delay sciatic nerve regeneration after crush lesion in the mouse. J Peripher Nerv Syst 13:71–80

Barigye R, Davis S, Hunt R, Hunt N, Walsh S, Elliott N, Burnup C, Aumann S, Day C, Dyrting K, Weir R, Melville LF (2016) Viral neurotropism, peripheral neuropathy and other morphological abnormalities in bovine ephemeral fever virus-infected downer cattle. Aust Vet J 94(10): 362–370

Bassilios Habre S, Bond G, Jing XL, Kostopoulos E, Wallace RD, Konofaos P (2018) The surgical management of nerve gaps: present and future. Ann Plast Surg 80:252–261

Blom CL, Mårtensson LB, Dahlin LB (2014) Nerve injury-induced c-Jun activation in Schwann cells is JNK independent. Biomed Res Int 2014:392971

Bridge PM, Ball DJ, Mackinnon SE, Nakao Y, Brandt K, Hunter DA, Hertl C (1994) Nerve crush injuries—a model for axonotmesis. Exp Neurol 127:284–290

Calderon-Martinez D, Garavito Z, Spinel C, Hurtado H (2002) Schwann cell-enriched cultures from adult human peripheral nerve: a technique combining short enzymatic dissociation and treatment with cytosine arabinoside (AraC). J Neurosci Methods 114:1

Casañas J, de la Torre J, Solerc F, Garcíab F, Rodellard C, Pumarolac M, Climente J, Solerf R, Orozco L (2014) Peripheral nerve regeneration after experimental section in sheep radial and tibial nerves using synthetic nerve grafts, including expanded bone marrow mesenchymal cells: morphological and neurophysiological results. Injury 45(S4):S2–S6

Chen L, Qin J, Cheng C, Niu S, Liu Y, Shi S, Liu H, Shen A (2008) Spatiotemporal expression of SSeCKS in injured rat sciatic nerve. Anat Rec 291:527–537

Chen C, Tian Y, Wang J, Zhang X, Nan L, Dai P, Gao Y, Zheng S, Liu W, Zhang Y (2019) Testosterone propionate can promote effects of acellular nerve allograft-seeded bone marrow mesenchymal stem cells on repairing canine sciatic nerve. J Tissue Eng Regen Med 1–17

Chhabra A et al (2014) Peripheral nerve injury grading simplified on MR neurography: as referenced to Seddon and Sunderland classifications. Indian J Radiol Imaging 24(3):217–224

Choulam A (2017) Peripheral neuropathy & animal models of peripheral neuropathy: a review. Asian J Pharm Technol Innov 03(10):42–60

Cui Y, Yao Y, Zhao Y, Xiao Z, Cao Z, Han S, Dai J (2018) Functional collagen conduits combined with human mesenchymal stem cells promote regeneration after sciatic nerve transection in dogs. J Tissue Eng Regen Med 12(5):1285–1296

Dadon-Nachum M, Melamed E, Offen D (2011) Stem cells treatment for sciatic nerve injury. Expert Opin Biol Ther 11:1591–1597

de Boer R, Borntraeger A, Knight AM et al (2012) Short- and long-term peripheral nerve regeneration using a poly-lactic-co-glycolic-acid scaffold containing nerve growth factor and glial cell line-derived neurotrophic factor releasing microspheres. J Biomed Mater Res A 100(8): 2139–2146

Ding F, Wu J, Yang Y, Hu W, Zhu Q, Tang X, Liu J, Gu X (2010) Use of tissue-engineered nerve grafts consisting of a chitosan/poly(lactic-co-glycolic acid)-based scaffold included with bone marrow mesenchymal cells for bridging 50-mm dog sciatic nerve gaps. Tissue Eng Part A 16(12):3779–3790

Diogo CC, Camassa JA, Pereira JE, da Costa LM, Filipe V, Couto PA, Geuna S, Maurício AC, Varejão AS (2017) The use of sheep as a model for studying peripheral nerve regeneration following nerve injury: review of the literature. Neurol Res 39(10):926–939

Emond A-L, Bertoni L, Seignour M, Coudry V, Denoix J-M (2016) Peripheral neuropathy of a forelimb in horses: 27 cases (2000–2013). J Am Vet Med Assoc 249(10):1187–1195

Fex Svennigsen A, Dahlin LB (2013) Repair of the peripheral Nerve—Remyelination that Works. Brain Sci 3:1182–1197

Fleming CE, Saraiva MJ, Sousa MM (2007) Transthyretin enhances nerve regeneration. J Neurochem 103:831–839

Fontanilla CV, Gu HY, Liu QP, Zhu TZ, Zhou CW, Johnstone BH, March KL, Pascuzzi RM, Farlow MR, Du YS (2016) Adipose-derived stem cell conditioned media extends survival time of a mouse model of amyotrophic lateral sclerosis. Sci Rep 5:16953

Forostyak S, Jendelova P, Sykova E (2013) The role of mesenchymal stromal cells in spinal cord injury, regenerative medicine and possible clinical applications. Biochimie 95(12):2257–2270

Fullarton AC, Myles LM, Lenihan DV et al (2001) Obstetric brachial plexus palsy: a comparison of the degree of recovery after repair of a C6 ventral root avulsion in newborn and adult sheep. Br J Plast Surg 54:697–704

Geuna S (2015) The sciatic nerve injury model in pre-clinical research. J Neurosci Methods 243: 39–46

Geuna S, Raimondo S, Ronchi G, Di Scipio F, Tos P, Czaja K, Fornaro M (2009) Histology of the peripheral nerve and changes occurring during nerve regeneration. Int Rev Neurobiol 87:27–46

Ghoreishian M, Rezaei M, Beni BH (2013) Facial nerve repair with Gore-Tex tube and adipose-derived stem cells: an animal study in dogs. J Oral Maxillofac Surg 71:577–587

Glasby MA, Gilmour JA, Gschmeissner SE et al (1990) The repair of large peripheral nerves using skeletal muscle autografts: a comparison with cable grafts in the sheep femoral nerve. Br J Plast Surg 43:169–178

Glasby MA, Mountain RE, Murray JA (1993) Repair of the facial nerve using freeze-thawed muscle autografts: a surgical model in the sheep. Arch Otolaryngol Head Neck Surg 119:461–465

Gomez-Sanchez JA et al (2017) After nerve injury, lineage tracing shows that myelin and Remak Schwann cells elongate extensively and branch to form repair Schwann cells, which shorten radically on remyelination. J Neurosci 37(37):9086–9099

Grochmal J, Midha R (2014) Recent advances in stem cell-mediated peripheral nerve repair. Cells Tissues Organs 200(1):13–22

Gugjoo MB, Amarpal, Makhdoomi DM, Sharma GT (2019) Equine mesenchymal stem cells: properties, sources, characterization and potential therapeutic applications. J Equine Vet Sci 72:16–27

Gugjoo MB, Hussain S, Amarpal, Shah RA, Dhama K (2020a) Mesenchymal stem cell mediated Immuno-modulatory and anti-inflammatory mechanisms in immune and allergic disorders. Recent Patents Inflamm Allergy Drug Discov 14:1

Gugjoo MB, Pal A, Sharma GT (2020b) Mesenchymal stem cell and its properties. In: Gugjoo MB, Pal A (eds) Mesenchymal stem cell in veterinary sciences. Springer, Singapore, pp 13–26

Gugjoo MB, Farooq F, Sakeena Q, Dar ER, Khan S, Amarpal A, Parrah JD, Makhdoomi DM, Dhama K, Sharma GT (2022) Stem Cell Therapeutics in veterinary medicine in India. Indian J Anim Sci 92(5):533–544

Hart AMK, Brannstrom T, Wiberg M, Terenghi G (2002) Primary sensory neurons and satellite cells after peripheral axotomy in the adult rat. Timecourse of cell death and elimination. Exp Brain Res 142:308

Hems TE, Clutton RE, Glasby MA (1994) Repair of avulsed cervical nerve roots. An experimental study in sheep. J Bone Joint Surg Br 76:818–823

Hewson DW, Bedworth NM, Hardman JG (2018) Peripheral nerve injury arising in anaesthesia practice. Anaesthesia 73(Suppl 1):51–60

Huang CK, Luo J, Lee SO, Chang C (2014) Concise review: androgen receptor differential roles in stem/progenitor cells including prostate, embryonic, stromal, and hematopoietic lineages. Stem Cells 32:2299–2308

Isaacs J (2013) Major peripheral nerve injuries. Hand Clin 29:371–382

Jacobson A, Schrader SC (1987) Peripheral nerve injury associated with fracture or fracture-dislocation of the pelvis in dogs and cats: 34 cases (1978-1982). J Am Vet Med Assoc 190(5):569–572

James FM, Engiles JB, Beech J (2010) Meningitis, cranial neuritis, and radiculoneuritis associated with *Borrelia burgdorferi* infection in a horse. J Am Vet Med Assoc 237:1180–1185

Jeans LA, Gilchrist T, Healy D (2007) Peripheral nerve repair by means of a flexible biodegradable glass fibre wrap: a comparison with microsurgical epineurial repair. J Plast Reconstr Aesthet Surg 60:1302–1308

Jessen KR, Mirsky R (2016) The repair Schwann cell and its function in regenerating nerves. J Physiol 594(13):3521–3531

Kaizawa Y, Kakinoki R, Ikeguchi R, Ohta S, Noguchi T, Oda H, Matsuda S (2016) Bridging a 30 mm defect in the canine ulnar nerve using vessel-containing conduits with implantation of bone marrow stromal cells. Microsurgery 36(4):316–324

Karsidag S, Akcal A, Sahin S, Karsidag S, Kabukcuoglu F, Ugurlu K (2012) Neurophysiological and morphological responses to treatment with acetyl-L-carnitine in a sciatic nerve injury model: preliminary data. J Hand Surg Eur 37:529–536

Keilhoff G, Fansa H, Smalla KH, Schneider W, Wolf G (2000) Neuroma: a donor-age independent source of human Schwann cells for tissue engineered nerve grafts. Neuroreport 11:3805

Keilhoff G, Stang F, Goihl A, Wolf G, Fansa H (2006) Transdifferentiated mesenchymal stem cells as alternative therapy in supporting nerve regeneration and myelination. Cell Mol Neurobiol 26:1233

Kilic A, Ojo B, Rajfer RA, Konopka G, Hagg D, Jang E, Akelina Y, Mao JJ, Rosenwasser MP, Tang P (2013) Effect of white adipose tissue flap and insulin-like growth factor-1 on nerve regeneration in rats. Microsurgery 33:367–375

Kim D-Y, Cho D-Y, Kim DY, Lee J, Taylor HW (2003) Malignant peripheral nerve sheath tumor with divergent mesenchymal differentiations in a dog. J Vet Diagn Investig 15:174–178

Kolar M, Kingham PJ (2014) Peripheral nerve tissue engineering, pp 468–497

Konala VBR, Mamidi MK, Bhonde R, Das AK, Pochampally R, Pal R (2016) The current landscape of the mesenchymal stromal cell secretome: a new paradigm for cell-free regeneration. Cytotherapy 18(1):13–24

Kourembanas S (2015) Exosomes: vehicles of intercellular signaling, biomarkers, and vectors of cell therapy. Annu Rev Physiol 77:13–27

Li Q, Ping P, Jiang H, Liu K (2006) Nerve conduit filled with GDNF gene-modified Schwann cells enhances regeneration of the peripheral nerve. Microsurgery 26:116

Li ZK, Ke H, Ni LM, Li QH (2008) Effects of androgen on phosphacan and NG2 proteoglycan expression and neurite regeneration in neonatal rats with hypoxic-ischemic brain damage. Zhongguo dang dai er ke za zhi Chin J Contemp Pediatr 10:357–361

Li H, Yan F, Lei L, Li Y, Xiao Y (2009) Application of autologous cryopreserved bone marrow mesenchymal stem cells for periodontal regeneration in dogs. Cells Tissues Organs 190:94–101

Li Y, Xu W, Cheng L-Y (2017) Adipose-derived mesenchymal stem cells accelerate nerve regeneration and functional recovery in a rat model of recurrent laryngeal nerve injury. Neural Regen Res 12(9):1544–1550

Lien BV, Brown NJ, Ransom SC, Lehrich BM, Shahrestani S, Tafreshi AR, Ransom RC, Sahyouni R (2020) Enhancing peripheral nerve regeneration with neurotrophic factors and bioengineered scaffolds: a basic science and clinical perspective. J Peripher Nerv Syst 25(4):320–334

Long X, Olszewski M, Huang W, Kletzel M (2005) Neural cell differentiation in vitro from adult human bone marrow mesenchymal stem cells. Stem Cells Dev 14:65–69

Lou D, Sun B, Wei H, Deng X, Chen H, Xu D, Li G, Xu H, Wang Y (2012) Spatiotemporal expression of testicular protein kinase 1 after rat sciatic nerve injury. J Mol Neurosci 47:180–191

Lundborg G (2004) Alternatives to autologous nerve grafts. Handchir Mikrochir Plast Chir 36:1

Ma T, Gong K, Ao Q, Yan Y, Song B, Huang H, Zhang X, Gong Y (2013) Intracerebral transplantation of adipose-derived mesenchymal stem cells alternatively activates microglia and ameliorates neuropathological deficits in Alzheimer's disease mice. Cell Transplant 22 (Suppl 1):S113–S126

Mead B, Berry M, Logan A, Scott RA, Leadbeater W, Scheven BA (2015) Stem cell treatment of degenerative eye disease. Stem Cell Res 14:243–257

Menorca RMG, Fussell TS, Elfar JC (2013) Peripheral nerve trauma: mechanisms of injury and recovery. Hand Clin 29(3):317–330

Moimas S, Novati F, Ronchi G, Zacchigna S, Fregnan F, Zentilin L, Papa G, Giacca M, Geuna S, Perroteau I, Arnež ZM, Raimondo S (2013) Effect of vascular endothelial growth factor gene therapy on post-traumatic peripheral nerve regeneration and denervation-related muscle atrophy. Gene Ther 20:1014–1021

Moore SA (2016) Managing neuropathic pain in dogs. Front Vet Sci 3:12

Nichols CM, Myckatyn TM, Rickman SR, Fox IK, Hadlock T, Mackinnon SE (2005) Choosing the correct functional assay: a comprehensive assessment of functional tests in the rat. Behav Brain Res 163:143–158

Noverina R, Widowati W, Ayuningtyas W, Kurniawan D, Afifah E, Laksmitawati DR, Rinendyaputri R, Rilianawati R, Faried A, Bachtiar I, Wirakusumah FF (2019) Growth factors profile in conditioned medium human adipose tissue-derived mesenchymal stem cells (CM-hATMSCs). Clin Nutr Exp 2019(24):34–44

Pilkington A, Buchanan D, Jamal GA, Gillham R, Hansen S, Kidd M, Hurley JF, Soutar CA (2001) An epidemiological study of the relations between exposure to organophosphate pesticides and indices of chronic peripheral neuropathy and neuropsychological abnormalities in sheep farmers and dippers. Occup Environ Med 58:702–710

Poulton PJ, Fisher AD, Mansell PD, Pyman MF (2019) Clinical findings from 104 cases of calving paralysis in dairy cows from Gippsland, Australia. N Z Vet J 67(4):214–218

Rodrigues MC, Rodrigues AA Jr, Glover LE, Voltarelli J, Borlongan CV (2012) Peripheral nerve repair with cultured Schwann cells: getting closer to the clinics. ScientificWorldJournal 2012: 413091

Rosberg HE, Carlsson KS, Cederlund RI, Ramel E, Dahlin LB (2013) Costs and outcome for serious hand and arm injuries during the first year after trauma—a prospective study. BMC Public Health 13:501

Salehi H, Amirpour N, Razavi S, Esfandiari E, Zavar R (2017) Overview of retinal differentiation potential of mesenchymal stem cells: a promising approach for retinal cell therapy. Ann Anat 210:52–63

Savastano LE, Laurito SR, Fitt MR, Rasmussen JA, Gonzalez Polo V, Patterson SI (2014) Sciatic nerve injury: a simple and subtle model for investigating many aspects of nervous system damage and recovery. J Neurosci Methods 227:166–180

Schwarz BC, Sallmutter T, Nell B (2008) Keratoconjunctivitis sicca attributable to parasympathetic facial nerve dysfunction associated with hypothyroidism in a horse. J Am Vet Med Assoc 233: 1761–1766

Seddon HJ (1943) Three types of nerve injuries. Brain 66:237

Shalaby SM, El-Shal AS, Ahmed FE, Shaban SF, Wahdan RA, Kandel WA, Senger MS (2017) Combined Wharton's jelly derived mesenchymal stem cells and nerve guidance conduit: a potential promising therapy for peripheral nerve injuries. Int J Biochem Cell Biol 86:67–76

Siemionow M, Brzezicki G (2009) Current techniques and concepts in peripheral nerve repair. Int Rev Neurobiol 87:141–172

Sowa Y, Kishida T, Imura T, Numajiri T, Nishino K, Tabata Y, Mazda O (2016) Adipose-derived stem cells promote peripheral nerve regeneration in vivo without differentiation into Schwann-like lineage. Plast Reconstr Surg 137:318e–330e

Sta M, Cappaert NL, Ramekers D, Baas F, Wadman WJ (2014) The functional and morphological characteristics of sciatic nerve degeneration and regeneration after crush injury in rats. J Neurosci Methods 222:189–198

Starritt NE, Kettle SA, Glasby MA (2011) Sutureless repair of the facial nerve using biodegradable glass fabric. Laryngoscope 121:1614–1619

Sullivan R et al (2016) Peripheral nerve injury: stem cell therapy and peripheral nerve transfer. Int J Mol Sci 17(12):2101

Sun Y, Zhang R, Mao X, Zhang M (2018) Research of acellular xenogeneic nerve combined with adipose-derived stem cells and platelet rich plasma in repair of rabbit facial nerve injury. Chin J Reparative Reconstr Surg 32(6):736–744

Sunderland S (1951) A classification of peripheral nerve injuries producing loss of function. Brain 74(4):491–516

Taylor CA et al (2008) The incidence of peripheral nerve injury in extremity trauma. Am J Phys Med Rehabil 87(5):381–385

Thorsen F et al (2012) Digital nerve injuries: epidemiology, results, costs, and impact on daily life. J Plast Surg Hand Surg 46(3–4):184–190

Tos P, Battiston B, Ciclamini D, Geuna S, Artiaco S (2012) Primary repair of crush nerve injuries by means of biological tubulization with muscle-vein-combined grafts. Microsurgery 32:358–363

Toth C, Martinez JA, Liu WQ, Diggle J, Guo GF, Ramji N, Mi R, Hoke A, Zochodne DW (2008) Local erythropoietin signaling enhances regeneration in peripheral axons. Neuroscience 154: 767–783

Usategui-Martín R, Puertas-Neyra K, García-Gutiérrez MT, Fuentes M, Pastor JC, Fernandez-Bueno I (2020) Human mesenchymal stem cell secretome exhibits a neuroprotective effect over in vitro retinal photoreceptor degeneration. Mol Ther Methods Clin Dev 2020(17):1155–1166

Varejão ASP, Cabrita AM, Meek MF, Bulas-Cruz J, Melo-Pinto P, Raimondo S, Geuna S, Giacobini-Robecchi MG (2004) Functional and morphological assessment of a standardized rat sciatic nerve crush injury with a non-serrated clamp. J Neurotrauma 21:1652–1670

Villagrán C, Schumacher J, Donnell R, Dhar MS (2016) A novel model for acute peripheral nerve injury in the horse and evaluation of the effect of mesenchymal stromal cells applied in situ on nerve regeneration: a preliminary study. Front Vet Sci 3:80

Wang CZ, Chen YJ, Wang YH, Yeh ML, Huang MH, Ho ML, Liang JI, Chen CH (2014) Low-level laser irradiation improves functional recovery and nerve regeneration in sciatic nerve crush rat injury model. PLoS One 9(8):e103348

Wang Q, He H, Xie S, Wei Q, He C (2020) Mesenchymal stem cells transplantation for neuropathic pain induced by peripheral nerve injury in animal models: a systematic review. Stem Cells Dev 29(22):1420–1428

Wei JJ, Chen YF, Xue CL, Ma BT, Shen YM, Guan J, Bao XJ, Wu H, Han Q, Wang RZ, Zhao CH (2016) Protection of nerve injury with exosome extracted from mesenchymal stem cell. Zhongguo Yi Xue Ke Xue Yuan Xue Bao 38:33–36

Woodhoo A, Alonso MBD, Droggiti A, Turmaine M, D'Antonio M, Parkinson DB, Wilton DK, Al-Shawi R, Simons P, Shen J, Guillemot F, Radtke F, Meijer D, Feltri ML, Wrabetz L, Mirsky R, Jessen KR (2009) Notch controls embryonic Schwann cell differentiation, postnatal myelination and adult plasticity. Nat Neurosci 12(7):839–847

Wright MC, Mi R, Connor E, Reed N, Vyas A, Alspalter M, Coppola G, Geschwind DH, Brushart TM, Höke A (2014) Novel roles for osteopontin and clusterin in peripheral motor and sensory axon regeneration. J Neurosci 34:1689–1700

Xue C, Hu N, Gu Y, Yang Y, Liu Y, Liu J, Ding F, Gu X (2012) Joint use of a chitosan/PLGA scaffold and MSCs to bridge an extra large gap in dog sciatic nerve. Neurorehabil Neural Repair 26(1):96–106

Xue C, Ren H, Zhu H, Gu X, Guo Q, Zhou Y, Huang J, Wang S, Zha G, Gu J, Yang Y, Gu Y, Gu X (2017) Bone marrow mesenchymal stem cells-derived acellular matrix-coated chitosan/silk scaffold for neural tissue regeneration. J Mater Chem B 5(6):1246–1257

Zhang R-C, Du W-Q, Zhang J-Y, Yu S-X, Lu F-Z, Ding H-M, Cheng Y-B, Ren C, Geng D-Q (2021) Mesenchymal stem cell treatment for peripheral nerve injury: a narrative review. Neural Regen Res 16(11):2170–2176

Mesenchymal Stem Cells Therapeutic Applications in Central Nervous System Disorders

5

Abstract

Mammalian nervous system is the most complex and little understood system of the body. An ailment to any structure of the central nervous system often leads to the mal- or non-function of the particular body organ or system controlled centrally. Currently, no therapeutic modality addresses these ailments in an effective way. As such regenerative medicine employing stem cells are being evaluated. Central nervous system is not devoid of stem cells; however, the limited concentration and their specific localization together with the environmental effect make their actions little effective. As such mesenchymal stem cells (MSCs) are being evaluated due to their characteristic properties like multiplication, differentiation, homing, immunomodulation/anti-inflammatory, antioxidant and reduction of apoptosis. In humans, more extensive studies are conducted as compared to the animals. There are numerous studies demonstrating MSCs in vitro neural-like cell differentiation, although some questions remain to be answered. In vivo applications are extended to various affections involving the meningoencephalomyelitis of unknown origin (MUO), canine distemper, global hypoxia injury, myelomeningocele (MMC) Parkinson's disease (PD), spinal cord injury (SCI), intervertebral disc disease and Wobbler syndrome in animals. In all these studies, MSCs potential therapeutic role occurs through immunomodulation/anti-inflammatory actions (inhibiting astrogliosis and microgliosis), angiogenesis and growth and proliferation of local neural cells. Such a potential is further enhanced with the gene transfection and addition of scaffolds. However, the literature available on MSCs is limited enough to provide evidence-based medicine in case of different neural affections. The current chapter focuses on MSCs properties in relation to the central nervous system and their potential therapeutic applications in different clinical ailments.

M. B. Gugjoo, *Therapeutic Applications of Mesenchymal Stem Cells in Veterinary Medicine*, https://doi.org/10.1007/978-981-19-3277-9_5

Keywords

CNS · Mesenchymal stem cells · Wobbler syndrome · Parkinson's disease · Spinal cord injury · Intervertebral disc disease

5.1 Introduction

The mammalian nervous system is the most complex and least-understood body system and is mainly divided into the central and peripheral nervous systems. Central nervous system (CNS) constitutes the brain and spinal cord, while the peripheral nervous system (PNS) comprises all the nerves functioning peripherally. Nervous tissue, especially CNS is generally considered to lack regenerative potential, and once affected/damaged, permanent dysfunction and disability set in. Currently, a very limited understanding on the aetiopathological basis of the neural tissue ailments exists. Its extremely delicate and sensitive nature against any insult limits the therapeutic aspect. There are numerous conditions considered as potential causes of neural affections and some commonly well-established causes are vascular, infectious, chronically progressive, hereditary and traumatic pathologies. There are many other unknown aetiologies involved. Above all the physiological and molecular basis of the neural tissue affections remain to be understood completely.

5.2 Regeneration Potential of PNS vs. CNS

Peripheral nerves due to their widespread distribution are more prone to injuries as compared to the CNS but simultaneously their regeneration power is comparably better. A peripheral nerve is a bunch of axons that are encased in various sheaths for protection and effective conduction. The main nerve cell body is closer to the central nervous system so that axons compared to the cell body are more amenable to insults. Part of the axon below trauma is actually affected and the cell body effectuating the regeneration through transcriptional changes remains intact. Retrograde injury signals initiated in the peripheral nerves are considered to activate the nerve cell body (Abe and Cavalli 2008). Upregulation of immediate early genes like c-Jun and preceded by the expression of regeneration-associated genes like growth-associated protein 43 (GAP-43) are considered to be the key components for peripheral nerve regeneration (Denny 2006; Liu et al. 2011). Additionally, the Schwann cells (SCs) available in the peripheral nervous system (PNS) dedifferentiate, secrete trophic factors and rapidly degrade myelin debris (Kim et al. 2014). Schwann cells along with the inflammatory cells like macrophages phagocytize debris of myelin and degenerating axon to prevent obstruction of new axonal growth (Rotshenker 2011). Overall, peripheral nerve healing progresses very slowly and the main concern is to establish the right connections. The repaired/healed peripheral nerve threshold required for electrical conduction remains impaired. This may be due to the establishment of less number of axons, less efficient network, cross-wired connections, damage to the sheath-forming cells and the scar tissue that increases electrical resistance. Peripheral nerve regenerative potential varies with the type of nerve and its size, defect length and time to start of treatment (Sulaiman and Gordon 2013; Griffin et al. 2014; Sullivan et al. 2016).

In CNS injury, the actual nerve cell bodies present are more likely to get damaged along with the oligodendrocytes. The loss of the nerve cell bodies means loss of regenerative potential (Abe and Cavalli 2008). Further, oligodendrocytes unlike Schwann cells may not regenerate spontaneously; therefore, myelin sheath may not sustain leading to loss of normal conduction properties. Glial response (astrocytosis/gliosis) results in scar formation that blocks the possibility of axonal regrowth (Fawcett and Asher 1999; Hannila and Filbin 2008). Inflammation and immune response to the damaged neural tissue make the environment hostile which further limits the regeneration activities (Mata et al. 2021). Contrary to the previous notion, recent studies demonstrate the intrinsic system of homeostasis in the CNS. This may be achieved through native neural stem cells (NSCs) and oligodendroglia precursor cells (OPCs). NSCs give rise to both the neurons and glial cells, while OPCs generate glial cells (Temple and Alvarez-Buylla 1999). The neurogenic regions harbouring multipotent stem cells in the adult mammalian brain are subventricular zone (SVZ) facing the lateral ventricles and hippocampal dentate gyrus (Morshead et al. 1994; Palmer et al. 1997; Temple and Alvarez-Buylla 1999; Okano 2006) and non-neurogenic zones like in spinal cord (Weiss et al. 1996; Shihabuddin et al. 1997). OPCs are located mostly in the area of the cerebellum and spinal cord (Levine et al. 1993; Engel and Wolswijk 1996). In CNS diseases, especially the ones involving myelin degeneration like canine distemper and/or multiple sclerosis, myelin repair activities may occur through these cells. However, considering the percent availability of these cells, myelin regeneration capacity remains limited and might be due to the NSCs restriction to discrete locations, fewer numbers and low efficiency or is restricted in space by a microenvironment that does not support neuronal differentiation. Shorter life of newly generated neurons is possibly due to the absence of the formation of any functional synapses (Gonzalez et al. 2016). The stem cell plasticity might be an aide to retrain repaired misconnections for a less efficient but to some extent workable condition (Mendes-Pinheiro et al. 2020). This plasticity along with the late axonal regeneration may be one of the possible reasons for recovery after spinal cord injuries or the arousal from the coma. Although not well established, the retrograde peripheral injury signals may also arise from CNS neurons like retina (Yin et al. 2003). The field is therefore open for further insights into the mechanisms that could stimulate resident neural stem cells to differentiate into a specific cell lineage, besides secreting factors that could make the environment conducive and pro-to-regeneration (Temple and Alvarez-Buylla 1999).

Nervous tissue physiology is little understood, and consequently, the therapeutics for its disorders are limited. Understanding nervous tissue physiology and the basis of its derangement may aid in developing therapeutics. Currently, no effective treatment is available to address nerve affections and as such regenerative medicine employing stem cells is being evaluated. Stem cells are characterized by their self-renewal, multiplication and differentiation properties. Due to their characteristic features, an all-in-one solution for ailments of varied pathologies seems possible. Among various stem cell types, mesenchymal stem cells (MSCs) are preferred, being widely available in almost all the body tissues and foetal membranes, are immuno-evasive and able to self-renew (although less extensive), secrete pro-healing factors and differentiate into other cell types. Apart from these properties, MSCs are mostly barred from ethical issues and have limited tendency to develop teratoma unlike that of the pluripotent stem cells (embryonic stem cells and induced pluripotent stem cells). MSCs have the ability to migrate and home into the ailing tissues. The therapeutic mechanism is mostly attributed to their ability to secrete plethora of factors either soluble or microvesicles like angiogenic, antioxidant, anti-apoptosis and immunomodulatory biomolecules or less through transdifferentiation into Schwann-like cell phenotype (Nakamura et al. 2015; Sharma et al. 2017; Kim et al. 2019; Al Delfi et al. 2020). MSC-conditioned medium may also improve the growth and differentiation of the neural stem cells under in vitro study (Nakamura et al. 2020). Various preliminary lab animal studies favour and demonstrate MSCs potential to heal nervous tissues and restore its functions. The current chapter is aimed to discuss MSCs in vitro neural differentiation and in vivo applications in CNS diseases of domestic and pet animals.

5.2.1 In Vitro MSCs Neurogenic Studies

MSCs under specific culture environments undergo neural-like cell differentiation although with some reservations to the extent of even being questioned. MSCs may differentiate into various lineages including the terminally differentiated neural-like cells upon exposure to various biological and pharmacological factors, mechanical cues and/or electrical forces (Titushkin et al. 2010; Choi et al. 2014, 2017; Javanmard et al. 2016; Mou et al. 2017). Co-culture with the neural tissues (olfactory ensheathing cells or Schwann cells) or fluids (CSF) may predispose MSCs towards neural lineages (Furno et al. 2013; Ge et al. 2015). The microenvironment cues activate MSCs to transdifferentiate into particular neural phenotypes viz., glial-like cells including oligodendrocytes, microglia, astrocytes and Schwann cells and neural-like cells like dopaminergic, and voltage- and glutamate-responsive cells (Fig. 5.1). MSCs neuronal differentiation is claimed on the basis of change in their morphology, mRNA expression or protein translation of neural specific markers, neurotrophic factors and ion channel markers. Even ectonucleotidase activity also changes with neural induction of MSCs (Woodbury et al. 2000; Rivers et al. 2006; Bahat-Stroomza et al. 2009; Radtke et al. 2009; Jiang et al. 2010; Kim et al. 2014; Xiong et al. 2014; Roszek et al. 2016; Blecker et al. 2017; Singh et al. 2017; Edamura et al. 2020).

An extensive literature exists on the lab animal MSCs transdifferentiation towards the neural-like cells. However, for common veterinary species, limited literature is available, especially in relation to the extensiveness of the details of neural

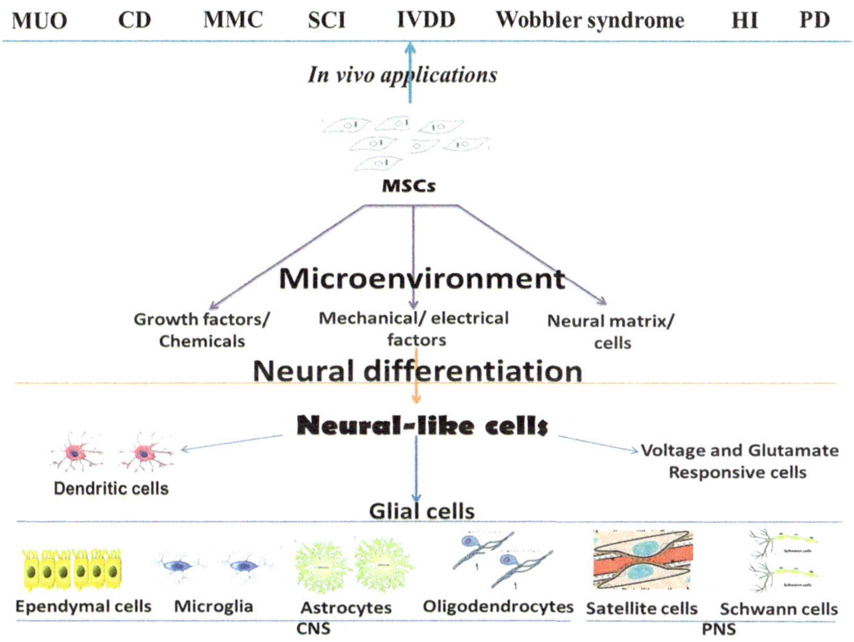

Fig. 5.1 MSCs transdifferentiation into various neural-like cells and their applications in various neurological affections conducted in veterinary species. *MUO* meningoencephalitis of unknown origin, *CD* Canine distemper, *MMC* Myelomeningocele, *SCI* spinal cord injury, *IVDD* intervertebral disc disease, *HI* hypoxia ischaemia, *PD* Parkinson's disease

differentiation. Various commercially available neural differentiation cocktails are claimed to effectuate MSCs to express neural-specific factors (Cardoso et al. 2012; Lyahyai et al. 2012; Mediano et al. 2015; Edamura et al. 2020). The chemical or growth factor cocktail composition, however, mostly remains concealed in such media. Notable growth factors utilized for MSCs neural-like cell differentiation include fibroblast growth factor (FGF) and epidermal growth factor (EGF) (Kim et al. 2014; Edamura et al. 2020). Chemical ingredients like dimethyl sulfoxide (DMSO), butylated hydroxyanisole (BHA), hydrocortisone, insulin, 3-Isobutyl-1-Metylxanthine (IBMX), dibutyryl cyclic AMP (dbcAMP) and adenosine 3′5′-cyclic monophosphate sodium salt monohydrate, and beta-mercaptoethanol and transretinoic acid, apart from the normal culture ingredients like low glucose Dulbecco's modified Eagle's medium (L-DMEM), foetal bovine serum (FBS) and basal eagle medium (BME) also effectuate MSCs neural-like cell differentiation (Ryu et al. 2009, 2012; Oh et al. 2011; Uranio et al. 2011; Lange-Consiglio et al. 2012; Oda et al. 2013; Gao et al. 2014; Sun et al. 2014; Xiong et al. 2014; Blecker et al. 2017; Hillmann et al. 2016). Additionally, introduction of specific inhibitors like BIX-01294, a specific inhibitor of methyltransferase G9a responsible for methylation of histone H3 lysine 9 (H3K9) or Glycogen synthase kinase 3 inhibitor (kenpaullone) ensures goat AD-MSCs (Wang et al. 2018) and buffalo Am-MSCs

(Deng et al. 2018) transdifferentiation into neural-like cells, respectively. Methyltransferase G9a inhibition-dependent MSCs transdifferentiation occurs through Nanog regulatory pathway (Wang et al. 2018; Mihevc et al. 2020). Overall, the combination of growth factors like FGF and EGF and the chemicals like IBMX and dbcAMP may better induce MSCs neural-like cell differentiation, although it remains to be confirmed in all the veterinary species. Combination of growth factors and chemicals induce neural-like cell differentiation of dog BM-MSCs with appreciable morphology, long-term viability, significant availability and ability to subculture. These dog neural-like cells, in contrast to rats, are both immature as well as mature. This potentially might be the reason behind the lack of production of integral neurons. Furthermore, the functional assessment of these neural-like cells remains to be conducted (Oda et al. 2013).

5.2.2 Neural-Like Cell Morphology, Expression Analysis and Electrical Activity

Neural-induced MSCs show increasingly elongated morphology with bipolar elongations harbouring oval nuclei that are visible as early as 3 h of culture and increase with the increase in culture time. The differentiated cells appear morphologically as neuron-like cells containing Nissl bodies or glial-like with arterial tree-like processes. Furthermore, these cells establish a synapse-like connection with the adjacent cells (Tharasanit et al. 2011; Edamura et al. 2012; Oda et al. 2013; Villagrán et al. 2014; Blecker et al. 2017; Gomez-Sanchez et al. 2017; Mihevc et al. 2020). The differentiated cells express neuronal/glial markers like Nestin (NES), vimentin, beta-III tubulin, neurofilament H (NF-H), glial fibrillary acidic protein (GFAP), neurofilament 200 (N200), neuronal microtubule-associated protein 2 (MAP2) and NF160, Tau, glial fibrillary acidic protein (GFAP), neurotrophin 3 (NT3), neurofilament light peptide (NEFL*)* and medium chain (NEFM*)*. These cells also express neurotrophic factors like brain-derived neurotrophic factor (BDNF), glial cell-line-derived neurotrophic factor (GDNF) and nerve growth factor (NGF), although in variable capacity (Seo et al. 2009; Cardoso et al. 2012; Uranio et al. 2014; Villagrán et al. 2014; Blecker et al. 2017). The NES expression varies but is mostly considered to express in an early period (up to 3 h) (Woodbury et al. 2000; Blecker et al. 2017), although other studies on human, murine, sheep and cat demonstrate its expression later in the culture as well (Safford et al. 2002; Long et al. 2005; Lyahyai et al. 2012; Munoz et al. 2012). This might be due to the differences in the neural induction media, tissue source and species involved.

In MSCs not only the above-mentioned neural markers but also others and several ion channel markers also increase in neural induction media. The protein expression levels of NES and neurofilament-L (NF-L) increase significantly in dog BM-MSCs. However, the neuron-like cells derived from dog BM-MSCs lack Ca_2^+ influx characteristic of spiking neurons (Nakano et al. 2013). Basic fibroblast growth factor (bFGF) induces an increase in mRNA expression of neuronal marker mRNAs (MAP2, NF-L and NSE) and proteins (NF-L and NSE). Such a factor contributes to the differentiation of dog BM-MSCs into voltage- and glutamate-responsive neuron-like cells. The pathways involved in bFGF-induced neuronal differentiation of dog BM-MSCs include fibroblast growth factor receptors (FGFR) and the PI3K/Akt signalling (Nakano et al. 2015; Edamura et al. 2020). However, to confirm complete MSCs neuronal differentiation, detailed electrophysiological function (like patch clamp technique) and functions of the released neurotransmitter are desirable. Apart from expression analysis, substantial change in ectonucleotidase (ecto-NTPDases) activity of dog AD-MSCs too occurs upon neural induction (Roszek et al. 2016). Some of the chemicals like lead (Pb) may inhibit such a differentiation potential of MSCs (Kermani et al. 2008).

5.2.3 MSCs Source-Based Differences and Activity in Disease

A human MSCs neural induction study demonstrates higher NES expression at day 5 as compared to day 18. The same study demonstrates that AD-MSCs show higher NES expression as compared to other cell types (BM-MSC, SD-MSC and UC-MSCs). By day 18, neural induction leads to the higher expression of neurofilament genes *NEFL* and *NEFM* in AD-MSCs and Skin-MSCs, respectively, as compared to that of BM-MSCs and UC-MSCs. Moreover, dopaminergic neuron marker NURR1, astrocyte marker S100B and neurotrophic factor NT-3 significantly increase in AD-MSCs (Urrutia et al. 2019). The NES location in neural-like cells may change with passaging. Some of the induced equine cells may display a perinuclear location and others show a more diffuse cytoplasmic expression. The NES expression decreases with passaging, and as it is expressed by highly proliferating cells, it may signify lower proliferation of the equine MSCs with extended passaging (Villagrán et al. 2014). Among various neural markers, MAP2 and beta-III tubulin are detected both in undifferentiated and differentiated dog BM- and AD-MSCs, and equine BM-MSCs (Kamishina et al. 2006; Villagrán et al. 2014; Blecker et al. 2017). Their expression, however, increases in the neural induction media as compared to basal media detected by gene expression analysis and immunohistochemical analysis. The expression again decreases after 72 h. mRNA expression of GFAP in dog AD-MSCs may not be observed but is localized through immunohistochemistry and Western blotting. This may occur due to the differences in GFAP isoforms and thus lack of a specific primer sequence (Kamphuis et al. 2012; Blecker et al. 2017). Some of the dog UC-MSCs and Wj-MSCs fail to express MAP2 and beta-III tubulin in basal medium but show GFAP, Tuj-1 and NF160 expressions in basal and differentiation medium (Seo et al. 2009, 2012). Dog BM-MSCs may show higher expression of beta III tubulin as compared to the AD-MSCs (Chung et al. 2013). Similarly, dog dental pulp MSCs express higher expression of NT-3, NGF and BDNF as compared to BM-MSCs and fibroblasts among which NT-3 expression occurs significantly higher (Naito et al. 2015). Further, dog UC-MSCs show significantly higher percentage of cells stained with GFAP and NeuN as compared to dog Wj-MSCs and dog AD-MSCs, respectively (Ryu et al. 2012). The expression of neurotrophin-like BDNF may increase in a time-dependent manner, while the expression of GDNF increases only for a shorter period (up to 72 h). The expression of NGF remains minimal and comparable to the control cells (Blecker et al. 2017).

As there is a variation in induced MSCs gene expression with respect to the time period, its implication needs to be studied for better understanding of neural-like cell differentiation. Among notable ones is the expression of NES and Map-2 at the earlier period of differentiation (24-h culture period), while at the later differentiation period (144 h), tropomyosin receptor kinase A (TRKA) and prion protein cellular (PrP^c) expressions are seen (Lyahyai et al. 2012; Dueñas et al. 2014). PrP^c is considered important for neuritogenesis. Further, some of the cells show basal expression of neural markers while others lack it. The variations may occur on the basis of species and tissue sources and differentiation state as well as in correlation with the induction media and detection methods (Blecker et al. 2017). The protocol variation too affects duration of MSCs neural-like cell features and maintenance of such features requires continued culturing in maintenance medium (Rismanchi et al. 2003; Krabbe et al. 2005).

In a scrapie-infected sheep, PB-MSCs show altered morphology with significant reduction in neural marker expression (NES and beta tubulin III) in neural induction media. Such a downregulation may also occur in healthy BM-MSCs. In basal media, the expression of NES and beta tubulin III reduce in scrapie-infected ovine BM-MSCs and PB-MSCs, respectively (Mediano et al. 2015). The scrapie-infected sheep PB-MSCs tend to reduce proliferation to prevent entry of the prion, while at the same time, the PB-MSCs differentiation into neural-like cells have full access to the prion proteins for maintenance and proliferation (García-Mendívil et al. 2021).

5.2.4 Conflicting Reports

MSCs actual neuronal differentiation from different sources is being questioned. The common notion remains that the morphological changes and alterations are merely artefacts and are stress -induced. Further, MSCs express neural-specific genes even under normal culture medium (Tondreau et al. 2004; Cardoso et al. 2012, 2017a, b; Corradetti et al. 2013; Gao et al. 2014; Sun et al. 2014; Hu et al. 2015; Deng et al. 2018). Some claim that these cells undergo morphological changes through shrinkage rather than neuronal outgrowth formation (Lu et al. 2004). These cells may be stressed with ingredients like BME, DMSO and BHA and undergo morphological changes as early as 2 h. This rapidity of change towards neuron-like morphology but retaining the ability to return back questions the cellular changes including gene and protein expression related to differentiation commitment (Lu et al. 2004; Neuhuber et al. 2004). Even the concept of cell–cell fusion and exhibition of neural markers by MSCs under in vitro and in vivo conditions is also reported in murine studies (Terada et al. 2002; Ying et al. 2002; Vassilopoulos et al. 2003; Weimann et al. 2003). Furthermore, MSCs might be contaminated with the neural progenitor cells that actually lead to the neural cell growth and marker expression (Song and Sanchez-Ramos 2003; Krabbe et al. 2005). To define their actual neuronal differentiation, some standard criteria need to be set. The cells should bear typical neural-like cell morphology, post-mitotic features with a single axon and numerous dendrites and a clear polarity. These cells should generate action potential and communicate through synapses via appropriate neurotransmitters. The gene expression and protein translation specific to neurons must be ensured. Ensuring the induction into mature neurons is also a challenge for functional activity. Furthermore, MSCs from a particular source have to be evaluated in depth to get better understanding and to evaluate the effect of source. Above all these cells have to be evaluated under an in vivo system to confirm their safety, feasibility and efficacy.

5.3 In Vivo Studies

Central nervous tissue undergoes diverse ailments arising out of trauma, infection, ageing or through other ailments of known or unknown aetiology. All these affections are termed as neurological disorders. In all these conditions, the main pathophysiological changes that develop are inflammation, demyelination and/or terminally the neural cell loss. CNS disorders are often accompanied by sequelae that pose a big therapeutic challenge (Cole and Bentz 2014). Currently, no effective treatment is available to address the nervous tissue challenges. Regenerative medicine employing stem cells is being evaluated for potential therapeutic applications. Among various stem cell types, the potential applications of mesenchymal stem cells (MSCs) with or without scaffolds and growth factors in the last two decades have intensively been focussed. These cells have primarily been correlated through the trophic and immunomodulatory effects rather than neuronal replacement in animal models or clinical trials (Andrzejewska et al. 2021). The cells may be transfected with the genes of interest or supplemented with scaffolds to effectuate neural tissue formation, although with the limited scope in brain diseases. Scaffolds especially 3D ones may appear better as compared to the 2D by mimicking the actual in vivo environment. Recently, 3D bioprinting has made it possible to develop such scaffolds more aptly that mimic actual tissue dimension and design. These 3D tissue scaffolds may be utilized as study models and for the clinical applications (Bedira et al. 2020). However, development of 3D bio-printed neural tissue that could cater the need remains a difficult challenge. The current topic mainly focuses on the MSCs therapeutic applications with or without other biomaterials. Lab animal studies focussed for human medicine are very diverse but studies on veterinary species are very limited and restricted to only some of the specific neural affections. MSCs applications are mostly evaluated under experimental models and only few clinical studies are conducted. The current chapter discusses MSCs potential applications in brain and spinal cord affections in veterinary species (Tables 5.1, 5.2, 5.3, 5.4 and 5.5).

Table 5.1 Dog MSCs clinical studies on MUO and canine distemper

Study design/period	Biomaterial	Route of implantation	Diagnostic methods	Results	Reference
Meningoencephalomyelitis of unknown origin (MUO) in dog					
Case report ($n = 8$)/ 24 months	Autologous BM-MSC (2×10^6, 4×10^6 [(3 dogs), IT and IC] and 2×10^6, 0.5×10^6 [(4 dogs), IT and IV]	Intrathecal (IT in cisterna magna)/right carotid artery (IC) and intrathecal and intravenous	MRI, neurological examination and histopathology	Improved outcome	Zeira et al. (2015)
Canine distemper					
Case series (Double-blind randomized trial, $n = 8$)/ 15 days	Foetal olfactory epithelium stem cells (1.0×10^6/kg b. wt.)	Intravenous	RT-PCR and laboratory investigations	No clinical improvement	Pinheiro et al. (2016)
Case report ($n = 4$)/1 year	Autologous AD-MSCs (1.0×10^7, 3 doses at monthly interval)	Intravenous	Neurological examination	Delayed improvement	Pinheiro et al. (2019)
Case report ($n = 7$)/ 180 days	Allogeneic AD-MSCs (5.0×10^6 cells/3.0 mL, three doses on day 1, 15 and 30)	Intravenous	Walking characteristics, urinary or faecal incontinence, state of consciousness, myoclonus and occurrence of seizures	Improved outcome	Gonçalves et al. (2018)

MSCs mesenchymal stem cells, *BM* bone marrow derived, *AD* adipose tissue derived, *IT* intrathecal, *IV* intravenous, *IC* intracarotid, *n* total number of animals in all the groups of the study. Positive outcome: indicates improved outcome with MSCs transplantation even if only one parameter improves. Study period: indicates maximum time period for which study is conducted

Table 5.2 MSCs preclinical experimental studies on myelomeningocele and global hypoxia injury in sheep

Study design/defect area	Biomaterial used/cell dose	Evaluation criteria	Overall result	References
Sheep model of myelomeningocele				
Randomized placebo controlled (n=4)/6 lumbar vertebrae removed	AM-MSCs (5×10^5 (early gestation MSCs; 1×10^6 term gestation MSCs) in collagen gel, foetal membrane (FM) patch (25 days after defect)	Gross and histology assessment	Improved outcome	Brown et al. (2016)
Randomized placebo controlled (n=12)/6 lumbar vertebrae	Human placental MSCs (5×10^5, 25 days after defect creation), collagen	Motor functional, histological and immunohistochemical analysis	Improved outcome	Wang et al. (2015)
Randomized placebo controlled (n=21)	Placental MSCs, porcine small intestine-derived extracellular matrix	Sheep locomotor rating (SLR) scale and histology of spinal cord	Improved outcome	Kabagambe et al. (2017a, b)
Randomized placebo controlled (n=28)/6 lumbar vertebrae removed	Xenogeneic (human) placental MSCs 3×10^5 cells/cm^2 of scaffold, Biodesign® Dural graft	Postnatal motor function, magnetic resonance imaging and histologic examination of the spinal cord	Improved outcome	Galganski et al. (2020)
Randomized placebo controlled (n=28)	Allogeneic UC-MSCs (2.76×10^6, 15 days after defect creation), fibrin patch	Locomotor rating score, histopathology	Improved outcome	Guilbaud et al. (2022)
Sheep global hypoxia injury				
Randomized placebo controlled (n=24)	Xenogeneic (Human) BM-MSCs derived extracellular vesicles (BM-MSCs-EV) (4×10^7 (two doses of 2×10^7 intrauterine inj.; first dose 1 h after UCO and other 4 days later)	Electrophysiological parameters and histology of brain	Improved outcome	Ophelders et al. (2016)

MSCs mesenchymal stem cells, *AM* amniotic membrane derived, *n* total number of animals in all the groups of the study, Positive outcome: indicates improved outcome with MSCs transplantation even if only one parameter improves. Study period: Indicates maximum time period for which study is conducted

Table 5.3 MSCs preclinical experimental studies on intervertebral disc disease (IVDD) in veterinary species

Study design/ Defect area/ Period	Biomaterial	Evaluation criteria	Overall result	References
Sheep IVDD model				
Randomized placebo-controlled study (*n*=6)/ microdiscectomy at three levels: L2/L3, L3/L4, L4/L5/6 months	Mesenchymal progenitor cells (5×10^6), pentosane polysulphate, gelatin/fibrin scaffold (immediate transplantation)	Radiography, MRI, biochemical and histology	Improved outcome	Oehme et al. (2014)
Randomized placebo-controlled comparative study (*n*=6)/annular incision at three levels: L2/L3, L3/L4, L4/L5/6 months	Mesenchymal progenitor cells (5×10^6), pentosane polysulphate, Gelatin/fibrin scaffold (12 week later transplantation)	Radiography, disc morphology assessment, biochemical and histology	Improved outcome	Oehme et al. (2016)
Randomized placebo-controlled study (*n*=8)/ Posterolateral annulotomy at 3 levels: L3/L4, L4/L5, L5/L6/12 months	Allogeneic BM-MSCs (1×10^6, 6 months after defect creation (6 months later transplantation)	Radiography, MRI, biochemical and histology	Improved outcome	Freeman et al. (2016)
Randomized placebo controlled (*n*=45)/L1–L2, L3–L4 and L5–L6/6 months	Allogeneic BM-MSCs (1×10^7)	Histology and gene expression	Improved outcome	Dowdell et al. (2017)
Randomized placebo controlled (*n*=18)/L2–3 and L3–4/6 months	Pentosan polysulphate-primed mesenchymal progenitor cells (pMSCs) and MSCs (5×10^5)	3T MRI, standard radiography, morphologically, histologically, biochemically for their proteoglycans (PGs), collagen and DNA content	Improved outcome	Daly et al. (2018)
Randomized placebo-controlled comparative study (*n*=36)/22 weeks	Allogeneic BM-MSCs (1×10^6)	Disc height, GAGs content, qPCR	Improved outcome	Shu et al. (2018)

(continued)

Table 5.3 (continued)

Study design/ Defect area/ Period	Biomaterial	Evaluation criteria	Overall result	References
Randomized placebo controlled (n=15)/6 weeks/ L2/L3, L3/L4, L4/L5	Allogeneic BM-MSCs (10^6 MSCs/mL), high-density collagen (HDC) gel (immediate transplantation)	Histopathology, MRI, disc height measurement	Improved outcome	Hussain et al. (2019)
Dog IVDD				
Randomized placebo controlled (n=18)/L3/4, L4/5 and L5/6/ 12 weeks	Autologous BM-MSCs (1×10^6, 4 weeks after nucleotomy)	Radiology, MRI, macroscopy, histology, immunohistochemistry, RT-PCR (FaSL)	Improved outcome	Hiyama et al. (2008)
Goat IVDD model				
Randomized placebo controlled comparative study (n=6)/T1-L5/ 12 weeks	BM-MSCs, platelet rich plasma, gelatine sponge	HE and mason trichrome staining, Alcian Blue periodic acid Schiff and collagen II staining	Improved outcome	Xu et al. (2017a, b)
Randomized placebo controlled (n=10)/12 weeks/ Lumbar (L1–L5 chondroitinase ABC)	BM-MSCs, hydrogel (comprising dextran, chitosan and teleostean)	Histopathology, microcomputed tomography, tumour necrosis factor-α expression	Improved outcome	Zhang et al. (2021)

MSCs mesenchymal stem cells, *BM* bone marrow derived, *n* total number of animals in all the groups of the study; Positive outcome: indicates improved outcome with MSC transplantation even if only one parameter improves. Study period: Indicates maximum time period for which study is conducted

5.3.1 Neurological Affections and MSCs Mechanism of Actions

Neurological ailments may be either genetic like Huntington disease or acquired that arise through hypoxia as in stroke and traumatic injury. In some diseases like Parkinson's disease and amyotrophic lateral sclerosis, aetiological factors are currently unknown. There are some diseases that are immune mediated like multiple sclerosis. Stroke leads to immediate compromise in blood circulation and cascade of brain damage, while traumatic brain injury is associated with the immediate compromise in the blood–brain barrier, brain swelling and cranial haemorrhage and in late stage may show cell death, axonal damage and diminished neurogenesis through ionic imbalance, excitatory amino acids release, overload of calcium and

Table 5.4 Dog MSCs preclinical experimental studies on spinal cord injuries (SCI)

Study design/period	Defect site/route	Biomaterial used/cell dose	Evaluation criteria	Overall result	References
Dog spinal cord injury models					
Randomized placebo controlled (n=25)/8 weeks	L4/cells injected at three locations (centre of the injury, 1.0 mm proximal and 1.0 mm distal to the injury, 3.0 mm in depth). Subcutaneous injection of rmhGCSF for 7 days	Allogeneic UCB-MSCs (1×10^6 after a week of SCI), recombinant methionyl human granulocyte colony-stimulating factor (rmhGCSF @ 5 μg/animal/day)	Behavioural assessment (Olby score), MRI, somatosensory evoked potentials and histopathological examinations	Improved outcome	Lim et al. (2007)
Randomized placebo controlled comparative study (n=30)/5 weeks	T13-L2/intrathecally transplanted through the lumbar spinal cord into the injured lesion	Autologous and allogeneic BM-MSCs (1×10^7 cells 7 days after SCI)	Behavioural analysis (Olby score), MRI, histopathological and immunohistochemical analysis, RT-PCR	Improved outcome	Jung et al. (2009)
Randomized placebo controlled (n=10)/55 weeks	L2–3/percutaneous transplantation of MSCs under fluoroscopic guidance at injured area (some animals) and at 3 locations [cranially (L1–2), and caudally (L3–4) to the injured segment, and directly to the injured segment]	Xenogeneic (human) UCB–MSCs (1×10^6 7 days after SCI)	Behavioural testing (modified Tarlov scale), histological and immunohistochemical analysis	Improved outcome	Lee et al. (2009)
Randomized placebo controlled (n=11)/9 weeks	L1/transplantation at three locations using a 30 gauge needle (middle of the injury site, proximal and distal margins; 3 mm of depth)	Allogenic AD-MSCs (1×10^6 1 week after SCI)	Behavioural (Olby score) assessments, somatosensory evoked potential (SEP), MRI, histopathology and immunohistochemistry	Improved outcome	Ryu et al. (2009)

Randomized placebo controlled (n=10)/1 year	L2 and L3/transplantation at three separate spinal cord parenchyma: cranially (L1–L2), caudally (L3–L4) to the injured segment, and directly to the injured segment (L2–L3)	Xenogeneic (human) UC-MSCs (1×10^6 after a week and 2×10^6 after 8 weeks of SCI in 2 dogs)	Behavioural assessment [Basso, Beattie and Bresnahan (BBB) Scale], MRI, histology and immunohistochemistry	Clinically unimproved outcome	Lee et al. (2011)
Randomized placebo controlled comparative study n=21/8 weeks	L1/transplantation at injured area	Allogeneic UC-MSCs (1×10^6 at 12 h, 1 week and 2 week implantation post SCI)	Behavioural assessment (Olby and Tarlov scores), histology, immunohistochemistry, Western blotting	Improved outcome	Park et al. (2011)
Randomized placebo controlled (n=12)/8 weeks	L1/transplantation at the injured area of L1 spinal cord parenchyma with intact dura	Neural differentiated AD-MSCs (NMSCs) and Matrigel (1×10^7 1 week after SCI)	Behavioural assessment (Tarlov scale), Histopathologic and immunofluorescence and Western blot	Improved outcome	Park et al. (2012)
Randomized placebo-controlled comparative study (n=20)/8 weeks	L1/transplantation at cranial, epicentre and caudal lesions	*BM-MSCs, AD-MSCs, WJ-MSCs, UC-MSCs, Matrigel* (1×10^5 cells/μL suspends in 60 μL of Matrigel 1 week after injury)	Behavioural assessment (Olby modified Tarlov scores), Histopathological, immunofluorescence, qPCR	Improved outcome	Ryu et al. (2012)
Randomized placebo controlled (n=16)/7 days	L1/intravenous (3 doses in 3 days)	Methylprednisolone sodium succinate (MPSS), allogeneic AD-MSCs (1×10^7 6 h post injury)	Revised Tarlov scale, oxidative and inflammatory markers	Improved outcome	Kim et al. (2015)
Randomized placebo-controlled comparative study (n=16)/8 weeks	L1/three region implantation at middle, proximal and distal margins of the injury site)	Allogeneic AD-MSCs (1×10^7), chondroitinase ABC (CABC) after 3 weeks	Behavioural, histopathological and immunofluorescence and WB	Improved outcome (more with cell and CABC combination)	Lee et al. (2015)

(continued)

Table 5.4 (continued)

Study design/period	Defect site/route	Biomaterial used/cell dose	Evaluation criteria	Overall result	References
Randomized placebo controlled (n=18)/36 weeks	T8/transplantation at spinal cord resection area	Xenogeneic (human) placental MSCs (1×10^7), linear ordered collagen scaffold (LOCS)	Behavioural assessment (Olby scores), immunocytochemistry, quantitative analyses	Improved outcome	Han et al. (2017)
Randomized placebo controlled (n=8)/8 weeks	L1/injected at the SCI site in three locations (middle of the injury site and at proximal and distal margins) to depths of 3 mm by using a 30-gauge needle	AD-MSCs (haemoxygenase-1 overexpression) or GFP AD-MSCs (10×10^6 1 week after SCI)	Behavioural assessment (BBB), histopathology, immunofluorescence and Western blotting	Improved outcome	Lee et al. (2017)
Randomized placebo-controlled comparative study (n=18)/38 weeks	T8/transfected and treated with biomaterials	Xenogeneic UC-MSCs (1×10^7), NeuroRegen scaffold (collagen scaffold) after 2 months	Behavioural assessment (BBB), histopathology and TEM	Improved outcome (higher with cell laden scaffold than scaffold only)	Li et al. (2017)
Randomized placebo-controlled comparative study (n=16)/8 weeks	L1/3 injections of 50 μL each were made: two at the epicentre and one in the centre of the injured spinal cord segment	BDNF-AD-MSCs, HO-1-AD-MSCs and GFPAD-MSCs (1×10^7) In group implanted with combination of cells, half of each cell population was injected	Behavioural Assessment [Basso, Beattie, and Bresnahan, (cBBB)], Western blot analysis, immunohistology and histopathological assessments	Improved outcome	Khan et al. (2018)
Randomized placebo-controlled comparative study (n=15)/6.5 months	T10/transfected and treated with biomaterials	Allogeneic Adv-TrkC transfected BM-MSCs (5×10^5 per scaffold), three-dimensional gelatin sponge scaffold	Behavioural assessment, MRI, DTI detection, anterograde labelling of corticospinal tract, motor evoked potential (MEP) detection, immunofluorescence	Improved outcome	Wu et al. (2018)

Randomized placebo-controlled comparative study (n=12)/4 weeks	L4/intrathecally	BM-MSCs (2×10^6 all injected intrathecally), neural differentiated BM-MSCs (2×10^6 all injected intrathecally), Cell free exosomes in a dose of 100 μg/kg body	Clinical, histology,	Improved outcome	Gabr et al. (2020)
Randomized placebo Pilot study (n=12)/6 months	At transfected site (4–5 mm): T8–T11	Xenogeneic UC-MSCs (1×10^7), collagen/heparin sulphate scaffold after haemostasis	Locomotion, motor evoked potential, MRI, diffusion tensor imaging and urodynamic parameters	Improved outcome	Deng et al. (2021)
Randomized placebo comparative study (n=12)/ 6 months		Cryopreserved MSCs, HO-1 MSCs, heat shock protein	Locomotion, inflammatory markers, histopathology	Improved outcome	Kim et al. (2022)

MSCs mesenchymal stem cells, *AD* adipose tissue derived, *AM* amniotic membrane derived, *BM* bone marrow derived, *BDNF* brain derived neurotrophic factor, *HO-1* haemoxygenase, *UC* umbilical cord, *UCB* umbilical cord blood, *n* total number of animals in all the groups of the study; Positive outcome: indicates improved outcome with MSCs transplantation even if only one parameter improves. Study period: Indicates maximum time period for which study is conducted

Table 5.5 Dog MSCs clinical studies on spinal cord injuries (SCI) and disc disease in dog and cat

Study design/period	Defect site/biomaterial	Route	Evaluation criteria	Overall result	References
Chronic spinal cord injury in dog					
Non-blinded controlled clinical study (n=23)/ 35 months	T10-L3, autologous BM-MSCs (1.4×10^6 and 5.6×10^6 1 and 3 months after surgery)	Subarachnoid space implantation between L5 and L6	TSCIS system (involves evaluation of each pelvic limb separately and has 3 components: gait, proprioceptive positioning and nociception)	Improved outcome	Nishida et al. (2011)
Case report (n=7)/ 90 days	T10-L3/allogeneic BM-MSCs (1×10^6)	Intralesional and caudal to lesion implantation	MRI and Olby scale	Improved outcome	Sarmento et al. (2014)
Case report (Pilot clinical study, n=6)/ 16 week	Allogeneic AD-MSCs (1×10^7)	Intralesional implantation	Functional Evaluation of Locomotor Activity	Improved outcome in 1 case	Escalhão et al. (2017)
Case report (n=3)/ 60 days	Xenogeneic (human) dental pulp stem cells (1×10^6 post hemilaminectomy)	DP-MSCs injected in three different sites: directly at spinal cord injury, 2 mm cranial and 2 mm caudal to the lesion	Limb function and MRI	Improved outcome	Feitosa et al. (2017)
Case report (n=4)/ 6 months	(T11-L1, T13-L2, T11-T12, T13-L1)/allogeneic BM-MSCs conditioned medium (concentration 1×10^6 BMMSC/1 mL conditioned medium)	Intravenously (25–35 mL/h)	Clinical evaluation (Olby score), goniometry measurements	Improved outcome	Vikartovska et al. (2020)
Randomized placebo-controlled study (n=8)/ 3 months	Thoracolumbar IVDD (chronic 1 year at least)/allogeneic Am-MSCs (2×10^6)	Local implantation 1st dose, 2nd and 3rd doses epidural injection	Functional and neurological evaluation, MRI	Improved outcome	Orlandin et al. (2021)

Spinal cord injury in dog					
Case report ($n=1$)/ 2 years	Autologous bone marrow mononuclear cells (BM-MNCs) (20×10^6 and 4.16×10^6 BM-MNCs), thermos-reversible gelatin polymer	First dose local post at T12 left hemilaminectomy and 2nd dose (intravenous implantation after 19 days)	Neurological examination (Olby scores)	Improved outcome	William et al. (2011)
Case report ($n=44$)/ 45 da	Allogeneic BM-MSCs	Local implantation at lesion	Clinical evaluation (Olby score)	Improved outcome	Bhat et al. (2019)
Case report ($n=27$)/ 60 days	AD-MSCs (1×10^7 two doses in a week's interval), acupuncture	Lumbar puncture in the space (L7–S1)	Electromyography	Improved outcome	Krueger et al. (2019)
Controlled, randomized double-blinded trial ($n=85$)/ 6 months	T10 and L4/Olfactory ensheathing cells ($3.20 \pm 0.49 \times 10^6$, $2.82 \pm 0.49 \times 10^6$), fibroblasts ($0.21 \pm 0.06 \times 10^6$)	Intralesional	Gait analysis, somatosensory-evoked potential, transcranial magnetic motor-evoked potentials, urodynamics	Improved outcome	Granger et al. (2012)
Intervertebral disc disease with loss of pain perception in dog					
Case report/ 10 weeks	L2–L3/xenogeneic UCB-MSCs (1×10^6)	Local transplantation (L1–L2, L2–L3 and L3–L4)	Modified Tarlov scale, MRI	Improved outcome	Chung et al. (2013)
Case reports ($n=4$)/ 18 months	T12-L5/Autologous BM-MSCs (1×10^6), collagen Gelfoam scaffold	Local lesion implantation post-second hemilaminectomy in the same location as the initial decompression surgery	Clinically, radiographically, nuclear MRI	Improved outcome	Penha et al. (2014)
Case report ($n=13$)/ 12 months	Autologous neurogenically induced BM-MSCs (5.0×10^6, 2 doses after 21 days of interval post hemilaminectomy/ laminectomy)	Percutaneous three different point transplantation	Texas Spinal Cord Injury Scale (TSCIS), somatosensory evoked potentials (SEP) and motor evoked potentials (MEP), MRI	Improved outcome	Besalti et al. (2015)

(continued)

Table 5.5 (continued)

Study design/period	Defect site/biomaterial	Route	Evaluation criteria	Overall result	References
Case report (*n*=9)/ 6 months	T11/12, T12/13, T13/L1, L1/2, L2/3 and L3/4/allogeneic AD-MSCs (1×10^7)	Local implantation at three locations (the middle of the injury site and the proximal and distal margins)	Clinical evaluation	Improved outcome	Kim et al. (2016)
Case report (*n*=1)/ 90 days	Hansen type I IVDD (T12–T13)/allogeneic BM-MSCs (1×10^6)	Local implantation at injury site (four doses fortnightly)	Clinical evaluation	Improved outcome	Sharun et al. (2021)
Case report (*n*=1)/ 4.5 years	AD-MSCs (91.8×10^6; paravertebrally 30.6×10^6 cells and 15.3×10^6 cells in knee)	Paravertebral injection at L7–S1 and knee	Clinical evaluation	Improved outcome	Mrkovački et al. (2021)
Cat lumbar compression fracture					
Case report (*n*=1)/ 3 months	L1–L5/autologous BM-MSCs (7×10^8), collagen I gel and physiotherapy	Local implantation	Clinical evaluation	Improved outcome	Penha et al. (2012)

MSCs mesenchymal stem cells, *AD* adipose tissue derived, *BM* bone marrow derived, *UCB* umbilical cord blood, *n* total number of animals in all the groups of the study; Positive outcome: indicates improved outcome with MSCs transplantation even if only one parameter improves. Study period: Indicates maximum time period for which study is conducted dysfunction of the mitochondria. Brain ischemia affects all the cells of the brain and its associated structures like vascular endothelial cells. In certain specific chronic neurodegenerative diseases like Alzheimer's disease, there may be loss of cholinergic neurons due to the accumulation of chemicals like amyloid β (Aβ) plaques and its induced neuroinflammation. Multiple sclerosis, an inflammatory and demyelinating autoimmune disease of CNS, is characterized by the formation of autoreactive lymphocytes and antigen-presenting cells (APCs) in the body that recruit to the CNS leading to the astrogliosis and microgliosis. The neuroinflammation destroys the myelin sheaths and axons leading to multifocal degenerative changes (Andrzejewska et al. 2021).

MSCs, in comparison to the other stem cells, have diverse and wider sources with no associated ethical or teratogenic effects. These cells may differentiate into diverse cell types of the nervous system including the neuron-like or the glial cells as discussed above. However, available in vivo studies so far fail to establish such findings. Their therapeutic effects are mainly being achieved through the paracrine actions. These cells upregulate neurotrophic factors [brain-derived neurotrophic factor (BDNF), nerve growth factor (NGF), insulin-like growth factor (IGF)-1 and vascular endothelial growth factor (VEGF)] to protect neurons and their integrity (Ma et al. 2013). MSCs through secretion of vascular factors like VEGF may promote angiogenesis and/or vasculogenesis (Liao et al. 2009; Bao et al. 2013). Besides these cells may engraft to the injured area and protect cerebral microvasculature against the ischemic–reperfusion injury. This may be ensured through mitochondrial transfer to the damaged endothelial cells (Liu et al. 2019). In case of traumatic brain injury, MSCs may stabilize blood–brain barrier (BBB) through interaction with pericytes, astrocytes and neurons (Sheikh et al. 2019). An important MSCs function is diminishing the inflammation by reducing astrogliosis and microgliosis, and pro-inflammatory cytokines (IL-1α, IL-1β, IL-6 and TNF-α) and simultaneously enhancing anti-inflammatory cytokines (IL-4, IL-10, INF-β) (Xin et al. 2013; Boido et al. 2014; Karlupia et al. 2014; Dabrowska et al. 2019; de Munter et al. 2020; Salehi et al. 2020). MSCs might reprogram pro-inflammatory microglia (M1) to the anti-inflammatory phenotype M2 (Zhang et al. 2013; Zanier et al. 2014; Xu et al. 2017a, b). These cells may additionally prevent synaptic protein loss by restoring the levels of synaptotagmin-1 (SYT1), synaptophysin (SYP) and glutamic acid decarboxylase 65 synaptic markers (Zappa Villar et al. 2020). MSC may also improve remyelination and axonal recovery (Jaramillo-Merchán et al. 2013; Nessler et al. 2013). These cells may enhance ependymal cells (neural progenitor cells) to proliferate and undergo neuronal differentiation (Li et al. 2017). These studies are mostly conducted in lab animals and are mostly lacking in veterinary species. Below are discussed the potential clinical ailments wherein MSCs may offer therapeutic benefits.

5.3.2 Meningoencephalomyelitis of Unknown Origin (MUO)

MUO is an inflammatory affection of the central nervous system (CNS) without the involvement of known infectious aetiology (Granger et al. 2010). Dog MUO is classified as granulomatous meningoencephalitis, necrotizing meningoencephalitis or eosinophilic meningoencephalitis, though the condition is rare in cats wherein infectious cause of encephalitis is mainly reported (Nessler et al. 2020). MSCs ability to secrete various anti-inflammatory factors (IDO, PGE2, IL-6 and VEGF) that increases upon stimulation with the inflammatory mediators (IFNγ and TNFα) or macrophages makes them potential candidates for therapeutics in inflammatory disorders of brain or spinal cord (Amorim et al. 2020).

To evaluate the effect of MSCs in this condition (Table 5.1), a single available clinical study undertakes $n=8$ relapsed cases of canine idiopathic autoimmune inflammatory disorders of the CNS. The condition in these cases prior to stem cell therapy would relapse after prednisone stoppage and respond poorly to a second full dose. The owners asking for euthanasia after apprising of the stem cell therapy took it as the last option. Intracarotid and intra-arterial transplantation of autologous BM-MSCs appears safe as none of the cases shows any adverse reaction. One dog shows hyperthermia but resolves on its own. Transplantation of these cells together with antiepileptic drugs reduces the seizure frequency in 66% cases (2/3) and seizure intensity in 33% (1/3) cases. Apart from seizures, improvement in clinical symptoms like paraspinal hyperesthesia of the cervical region, behavioural changes, depression, proprioception and postural reactions, ataxia, head tilt, and circling also occurs. At 6 months, $n=4$ dogs appear completely normal, $n=1$ retains mild ataxia, $n=2$ needs to continue anti-epileptic treatment but otherwise appear normal, and $n=1$ dies for reasons not related to its neurologic disease. Histological examination of CNS of a dead dog shows lesions localized to white matter of the telencephalon, caudal brain stem, cerebellum and cervical spinal cord. These lesions are characterized by dense aggregates of inflammatory cells arranged in perivascular cuffs. The results with intrathecal along with the intra-arterial transplantation to that of the intrathecal and intravenous transplantation are comparable, although the former route shows early response. The former route might allow faster and better circulation of the cells with the arterial blood flow and cerebrospinal fluid (Zeira et al. 2015). This study demonstrates potential therapeutic effects of MSCs in dog MUO; however, long term extensive studies incorporating large numbers of animals are desired. Individual differences with the variable degree of severity may have an important bearing on the outcome. Further, use of anti-epileptic drugs may have a role to play in subsiding the symptoms as no control group exists in the study. MSCs though appear promising but considering paucity of the literature any inference drawn will be exaggeration.

5.3.3 Canine Distemper (CD)

Canine distemper (CD) is caused by the single-strand RNA canine distemper virus (Morbillivirus). It has various forms including the nervous one. Neuro-invasion of the virus mainly occurs via infected mononuclear cells through the blood–brain barrier that infects the resident epithelial and endothelial cells. It may also directly affect endothelial cells prior to trafficking via leukocytes. Additionally, the virus may spread from pia matter to the distal grey matter (Lempp et al. 2014). The neurologic stage of CD exhibits CNS inflammation, demyelination and axonal injury (Lempp et al. 2014; Ulrich et al. 2014; Pinheiro et al. 2019). In CD, neuropathological changes bear resemblance to that of the human multiple sclerosis and thus can potentially offer an effective translational model to study pathogenesis of myelin loss and its possible prevention through regenerative medicine employing MSCs (Beineke et al. 2009; Ulrich et al. 2014; Pinheiro et al. 2019).

One of the early studies on acute form of CD fails to show therapeutic benefits of intravenously transplanted foetal olfactory epithelial stem cells (one million per kg body weight). A comparable systemic or neurological signs and mortality rate in case of cell-treated and untreated animals is reported on day 15 of the study (Pinheiro et al. 2016). However, recent studies demonstrate improvement in neurological signs with the application of autologous AD-MSCs (Pinheiro et al. 2019) or allogeneic AD-MSCs (Gonçalves et al. 2018) (Table 5.1). The study that transplants three doses of AD-MSCs (ten million) at monthly interval reports no observable improvement up to 1-year period; however, Grade V myoclonus intensity in $n=3/4$ animals is reduced to a Grade IV (moderate level) and animals regain grade I functional ambulation. All the four dogs start to move independently (Grades I and II) at 1 year after cell implantation. In 50% of animals, the myoclonic severity becomes mild (Grade III) (Pinheiro et al. 2019). Another similar study that utilizes three doses of allogeneic AD-MSCs (five million) fortnightly demonstrates attenuation of the clinical signs as early as 15 days with noticeable improvement at day 30. This improvement persists throughout 180 days of observation based on the devised evaluation parameters. Among various parameters, statistically significant differences are seen in walking, urinary and faecal incontinence. Further decrease in seizures also occurs (Gonçalves et al. 2018). Some unpublished studies (available reports of thesis) too demonstrate some improvement in neurological symptoms with BM-MSCs and AD-MSCs (Brito 2015; Monteiro 2017). Amongst, one study evaluates allogeneic AD-MSCs in case of chronic phase CD and reports $n=13$ out of $n=30$ animals show reduced neurological scale score, though most animals remain tetra paretic (Monteiro 2017). Seizures are the consequences of the demyelization, microglial-free radical formation (destroy synaptic transmission and neurons) and excessive glutamate release and its receptor binding. As above studies demonstrate reduction in seizures, MSCs paracrine actions could reduce inflammation and improve neural cell fusion, neuronal and glial cell recruitment and axonal extension. Besides, MSCs may also regulate glutamate uptake and angiogenesis (D'intino et al. 2006; Colpo et al. 2015). Most of the above studies demonstrate improvement in CD neurological signs with MSCs though to the limited extent only. In case of the acute phase of CD, cells fail to show any therapeutic effect in an immediate period but higher and multiple doses of cells along with different permutations and combinations may prove fruitful. Besides effective cellular routes may be evaluated for better outcomes. Evaluations through advanced diagnostic techniques like MRI, electro-encephalography among others may be employed for a clear clinical picture of the cases. Variability in stem cells types especially the olfactory epithelial stem cells and MSCs, their route and doses besides the clinical condition (acute or chronic) may determine the outcome. Limited animal size in these studies additionally make the relevance statistically poor. MSCs may alleviate severity of the disabling lesions of demyelinating leukoencephalitis but do not actually regenerate the lost nerve tissue.

5.3.4 Myelomeningocele (MMC)

MMC or spina bifida, a congenital defect due to incomplete development of the caudal part of the neural tube, occurs due to genetic and environmental factors. Protrusion of the meninges is termed a meningocele and if accompanied by the spinal cord it is termed myelomeningocele (MMC). The incidence of the condition has reduced with the intake of folic acid in humans but it still exists in animals (Kaufman 2004; von Koch et al. 2005; Testoni et al. 2010; Ployart et al. 2013). The typical signs of MMC in humans are hydrocephalus, intellectual disability and paralysis and associated conditions like bladder incontinence and Arnold Chiari Malformation II (Osaka et al. 1978; Heffez et al. 1990; Lomax-Bream et al. 2007). Although unclear, the pathogenesis involved is considered to be the defective neural tube development and the subsequent exposure of the open spinal cord to the amniotic fluid (AF) during development (Danzer et al. 2008). Palliative surgical treatment of MMC offered after birth usually fails to prevent clinical symptoms (Bowman et al. 2009; Roach et al. 2011) but intrauterine repair of defects may improve neurologic outcome and reduce disability. The intrauterine treatment reduces CSF loss and prevents direct exposure of neural tissues to the amniotic fluid (AF) (Farmer et al. 2003; Adzick et al. 2011). Among various animals, sheep are utilized as model animals for human MMC to evaluate stem cell efficacy (Table 5.2).

Amniotic fluid MSCs (AF-MSCs), the stem cell types during gestation, activate healing of induced foetal defects as genes related to musculoskeletal tissues (PAX3 and CD34, HOXD4 and HOXC8) are upregulated in these cells. Following the surgically induced MMC (16th week), the AF has an abrupt increase of PAX3-positive progenitor cells with characteristics similar to the activated satellite cells. HOXD4 and HOXC8 are highly expressed in AF-MSCs both after the MMC damage and its repair with biomatrix, strongly suggests their involvement in the repair (Ceccarelli et al. 2015). Considering such an increase in their proliferation and differentiation potential, MSCs may potentially be used as therapeutic agents in repairing and/or preventing the effects of MMC.

To evaluate MSCs effect, various sheep experimental studies are conducted with almost all favouring their therapeutic role. The models studied are created at mid-gestation in sheep (75 days) and are transplanted with the cells at 15–25 days post defect creation. The transplanted cells are mostly xenogeneic (human) or allogeneic placental MSCs along with grafts (Table 5.2). The healing at birth is evaluated through various diagnostic tests involving motor function, histology, immunohistochemistry, MRI and electrophysiology. The motor functional scores improve, and reduction in seizures with MSCs application is demonstrated (Wang et al. 2015; Brown et al. 2016; Guilbaud et al. 2022). MSCs transplantation is correlated with the development of the large neurons (Wang et al. 2015; Brown et al. 2016; Kabagambe et al. 2017a, b; Galganski et al. 2020; Guilbaud et al. 2022) and reduced fibrosis (Guilbaud et al. 2022). Placental MSCs, mostly used in these studies, are better able to secrete brain-derived neurotrophic factor (BDNF) and hepatocyte growth factor (HGF) as compared to the adult BM-MSCs (Martinez et al. 2014). BDNF as name signifies is a neurotrophic factor that widely affects nervous system in its synaptic plasticity, nerve fibre regrowth and inflammation (Weishaupt et al. 2012; Lu et al. 2014). HGF is considered a potential angiogenic factor and tends to work in nervous system development in synergy with the neurotrophic factors (Bussolino et al. 1992; Maina and Klein 1999; Bai et al. 2012). Tissue inhibitor matrix metalloproteinase-1 (TIMP-1) is highly secreted in placental MSCs than BM-MSCs, and thus, the former cells may be more fruitful for MMC (Tejima et al. 2009; Souvenir et al. 2011).

As with each developmental phase the potential of stem cells to differentiate decreases, placental MSCs at different gestation periods also show variable potential. Early-gestational AM-MSCs show higher efficacy than the complete-term AM-MSCs. A lamb spinal cord (six lumbar vertebrae removed) repaired with foetal amniotic membranes (FM) loaded with early gestation AM-MSCs develop three times higher cross-sectional area than the other lamb at epicentre. The lamb repaired with FM only and FM plus term-gestation AM-MSCS both have some grey matter preservation at the epicentre, but the lambs repaired with early-gestation AM-MSCs show a fourfold increase in the area of grey matter. Skin closure only lamb shows no visible large neurons, while the lambs repaired with FM only and FM plus term AM-MSCs have increased preservation of neurons. The lambs repaired with early-gestation AM-MSCs have a dramatic increase in the number of large neurons. The same study also shows that early-gestation AM-MSCs have higher expression of transcription factor Sox2, while term-gestation AM-MSCs fail to express it (Brown et al. 2016). As this factor is associated with a more primordial state including those in the neural progenitor cells, its expression in early-gestation AM-MSCs might point towards better pluripotency potential. However, there are studies that have demonstrated Sox2 expression in late gestation as well (Somal et al. 2016). In this study, only a single case was used in all the groups and therefore results cannot be extrapolated.

In all these studies, although overall improvement in clinical signs is reported, the complete clinical improvement remains to occur. There has to be further extensive studies that focus on mechanism of action of MSCs and incorporate these cells from uniform sources, and are of specific concentration, passage, route involving multiple doses and compared with various relevant parameters to draw any conclusion. Scaffold incorporation may also have a bearing on the outcome of these studies. Further, animal numbers must be adequate to present results statistically and the duration of studies must be longer for evaluating the healing changes that may occur.

5.3.5 Global Hypoxia Injury

Hypoxia–ischemia (HI) in the developing brain tends to affect viability and neurological morbidity in infants (Jellema et al. 2015). In animals, much emphasis might not be provided but human medicine faces its crisis as currently no medicine is available to cure. In a sheep umbilical cord occlusion model, BM-MSC-derived extracellular vesicles transplanted 1 h and later at 4 days after occlusion reduce duration and frequency of seizures while seizures continue in control. As the study demonstrates prevention of the damage to the baroreceptor reflex sensitivity, it may be attributed to the protective action against HI-induced white matter injury. Besides, as the neuroinflammation remains unaffected, MSCs anti-inflammatory action is not considered to have played the main role rather their anti-apoptotic and angiogenic properties (Ophelders et al. 2016) (Table 5.2). Thus, MSCs may provide some respite although variably and therefore further studies are desired.

5.3.6 Parkinson's Disease (PD)

Parkinson's disease (PD), a chronic, progressive, neurodegenerative disease with a multifactorial aetiology, is characterized by bradykinesia, rigidity, tremor and postural instability (Salem 2019). Neurodegeneration is characterized by the loss of nigrostriatal dopaminergic innervation but is not restricted to the area and may involve widespread neural tissue (Kouli et al. 2018). PD seen in dogs could be correlated to that of the human condition.

Apart from the lab animal studies, human clinical trials are undertaken with results demonstrating amelioration of some of the clinical symptoms (Zhang et al. 2008; Moloney et al. 2010; Venkataramana et al. 2010; Boika et al. 2020). A significant reduction in the severity of motor and non-motor PD symptoms is demonstrated after MSCs intravenous and combined intravenous and intranasal transplantation (Boika et al. 2020). Intranasal-transplanted MSCs tend to migrate to the brain within a day and tend to remain localized for about 3 weeks at the lesion (Shanko et al. 2018; Zafranskaya et al. 2019). Number of MSCs trials in PD are undertaken citing their in vitro tendency to transdifferentiate to neural cell types including the dopaminergic cells and others. These cells are also able to secrete various neurotrophins (NGF, BDNF, NGF, IGF-1 and VEGF) and structurally organize the individual brain cells and the neuronal network (Koniusz et al. 2016). Even studies employing the genetically transfected MSCs and dopamine-differentiated MSCs are also undertaken (Barzilay et al. 2009; Shetty et al. 2009).

However, complete recovery remains to be seen. Such MSCs studies may be undertaken in animals to evaluate amelioration of the symptoms, if any.

5.3.7 Intervertebral Disc Disease (IVDD)

Disc disease like herniation is increasingly being reported (Yao 2009). In dogs, clinically IVDD (either burst or bulge of discs) mostly occurs at or near the thoracolumbar junction and produces chronic complete paraplegia (Olby 1999). The surgical procedure used to rectify lumbar disc herniation involves removal of the nerve nucleus and treating lumbar disc herniation to reduce the suffering (Quah et al. 2014). Nerve nucleus removal destroys surrounding annulus tissue causing recurrence with resultant projection of the nucleus from the damaged annulus that presses the nerve (Pan et al. 2014; Xiao 2014). Annulus suturing may help to address the issue but the lack of the sufficient tissue makes it difficult (Ahlgren et al. 2000). Therefore, the annulus fibrosus may be repaired using tissue engineering methods to reduce the recurrence rate of surgery. Goat is considered to be the suitable IVDD model animal for humans due to their comparable size of the lumbar discs (~5 mm disc height) and similarities in their disc mechanical properties and biochemical composition (Beckstein et al. 2008). However, only two studies evaluate MSCs role in goat IVDD model (Xu et al. 2017a, b; Zhang et al. 2021). Among various animal species, MSCs role in IVDD is evaluated in sheep, goat and dog (Table 5.3). In sheep IVDD, multifidus muscle also gets affected, considered as one of the causes of lower back pain and thus provides resemblance to the human IVDD.

In goat, disc degeneration of lumbar vertebrae is made through creation of surgical defect (lamina decompression) in annulus fibrosus (1×1 cm) (Xu et al. 2017a, b) or through injection of the chondroitinase ABC in nucleus pulposus (L1-L5) (Zhang et al. 2021). Sheep IVDD model of the lumbar vertebrae is made through microdiscectomy (Oehme et al. 2014; Daly et al. 2018) or annular incisions (Freeman et al. 2016; Oehme et al. 2016; Shu et al. 2018; Hussain et al. 2019). One of the ovine models is cervical discectomy (Goldschlager et al. 2010). In dogs, IVDD is induced with the disc aspiration (Hiyama et al. 2008).

5.3.7.1 MSCs Dose and Survival

Transplanted MSCs engraft the area and are visible up to 6 months; however, the functional differentiation remains to be evaluated in sheep model of annular incisions (Freeman et al. 2016). In dog surgically induced IVDD, MSCs may survive for 3 weeks (Hang et al. 2017). Considering the optimal dose, a single dog study shows that one million cells appear superior (in disc height, annular structure) over 1 lakh or ten million cells (Serigano et al. 2010).

5.3.7.2 Goat Studies

In goat, transplantation of BM-MSCs along with the hydrogel scaffold (composed of dextran, chitosan and teleostean) (Zhang et al. 2021) or hydrogel scaffold (gelatin sponge) plus biomaterials like platelet-rich plasma (PRP) (Xu et al. 2017a, b) in two different models show improvement. In surgical annulus fibrosis model, the cell density and cellular gross morphology improve from 3 weeks and are closer to those of the sham group at 12 weeks. A gradual increase in fibroblast population and tissue occurs with chondrocytes taking part in the annulus repair (Xu et al. 2017a, b). The disc height index (DHI) increases by 10% as earliest as 2 weeks (Zhang et al. 2021). Further in these small subset of discs, large endplate disruptions are also observed that remains to be understood whether they are model or species specific (Ashinsky et al. 2020). Disc height is a good parameter evaluated clinically for it can easily be assessed on MRI/radiography and its height reduction is associated with the nerve impingement and foraminal stenosis developing clinical signs. However, pain parameters may also be evaluated for clinical efficacy. Besides the model does not focus on the extent of damage reaching to the adjacent areas and the mechanism of the action of the biomaterials implanted. Above all, the shorter study period (15 days or 12 weeks) does not evaluate any long-term benefits of the biomaterials.

5.3.7.3 Sheep Studies

In sheep lumbar disc models, the cellular transplantation with or without scaffolds or other biomaterials is mostly delayed in the case of annular incisions while in case of microdiscectomy the cells are transplanted immediately after defect creation (Table 5.3). In these models, transplanted cells improve DHI and disc height or MRI Pfirrmann grade (Freeman et al. 2016; Oehme et al. 2016; Hussain et al. 2019; Shu et al. 2018). Histological improvement and an increase in proteoglycan contents in the cell-treated defects also occur (Oehme et al. 2014, 2016; Freeman et al. 2016; Daly et al. 2018; Hussain et al. 2019; Shu et al. 2018). Comparing the low dose (0.1 million cells) to the higher dose (0.5 million cells) along with the pentosan polysulphate (PPS), no observable higher potency in relation to the low dose occurs (Oehme et al. 2016). This shows that the maintenance of the higher cell population at the local site is mandatory for their efficacy. Considering the cervical motion preservation in the sheep anterior cervical discectomy models, transplantation of MSCs along with the PPS is able to produce cartilaginous tissue at 3-month period that could replace the intervertebral disc (Goldschlager et al. 2010).

MRI Pfirrmann scores (objective assessment) and disc height index remain inconsistent with the histological/biochemical evaluations (subjective assessment) (Daly et al. 2018; Zhang et al. 2021). This could be due to the limited gain well below the detectable limit of semi-quantitative Pfirrmann grading or that the hydrogel MR properties could have confounded MRI signal intensity in the nucleus pulposus.

It is worth mentioning that repair with surgical procedure is necessary to prevent further compressive ischaemic damage. As decompression surgery is must and if followed by cell implantation, a significant outcome can be expected. Considering MSCs differentiation potential towards different lineages on the basis of microenvironment, chondrocyte concentration might have increased in the area. Besides the PRP being a stimulating factor for BM-MSCs, it might help in cellular proliferation and therefore the improved results (Kim et al. 2014). Scaffolds like gelatin sponges or hydrogels or pentosane polysulphate (PPS) may offer histocompatibility, have osteoporosity and other organizational structures that might help BM-MSCs to grow and proliferate (Buckley et al. 2018). PPS (Goldschlager et al. 2010; Oehme et al. 2016) and chitosan (Zhang et al. 2021) may selectively direct specific differentiation of MSCs to chondrocytes or limit inflammatory action in the area, respectively. MSCs tend to secrete a plethora of growth factors, antioxidants, anti-apoptotic factors, besides anti-inflammatory factors and tissue-inhibitor matrix metalloproteinases (TIMPs) might possibly improve healing. However, all these studies incorporate limited number of animals and employ variable models. The transplanted cells are from variable sources and injected at different concentrations and above all the studies are conducted for shorter duration. Thus, further extensive studies are desired to evaluate the outcome on a long-term basis.

5.3.8 Spinal Cord Injury (SCI)

Spinal cord injury (SCI) is a common clinical condition in animals and humans and carry a huge impact on an individual's life. SCI resembles in humans and dogs (Fluehmann et al. 2006; Sarmento et al. 2014) and can occur in road traffic accidents, fall from height or from intervertebral disc diseases (IVDD). In dogs, IVD extrusion (IVDE) is by far the most common cause for SCI (Olby 2010) and represents about 2% of all the diseases (Webb et al. 2010). SCI tends to produce ischemia, oxidative damage (free-radical formation) and finally cytotoxicity. Since CNS carries limited regenerative potential, SCI progresses to neuronal death leading to the permanent state of sensory and motor loss (Hulsebosch 2002). It is therefore imperative to develop advanced therapeutic options that can optimize and promote recovery after injury. The aim remains to reduce/prevent further progression of injury through modulation of the local microenvironment/niche and enhance remyelination and neuronal growth through regeneration (Myckatyn et al. 2004). Among various options, cell transplantation, especially the stem cell, is considered a potential mode to treat CNS affections (Gugjoo et al. 2020). This is attributed to their specialized properties of self-renewal, differentiation, anti-inflammatory and/or immunomodulatory properties (Roszek et al. 2016). Among various animal species, dog has preferably been used as an experimental model of SCI to evaluate MSCs therapeutic potential. Clinical trials are also conducted in the same species (Table 5.4).

5.3.8.1 Experimental SCI Studies

SCI models are created mostly by compression (ischemic injury), while few studies also involve spinal cord resection (Table 5.4). Ischaemic SCI is created with balloon catheter compression (mimicking compression or contusion caused by intervertebral disc herniation) mostly at lumbar area, although few studies involve thoracic part as well. These studies are conducted as acute or subacute and chronic based upon the duration of implantation of the biomaterials after injury. In acute SCI models, immediate or within a few hours to a week, biological materials are transplanted, while in case of subacute after a week, and in chronic models, transplantation is done after 3 weeks. Clinically, dog SCI does not restrict to lumbar but thoracic vertebrae are also affected.

These above studies employ autologous, allogeneic or xenogeneic MSCs derived from varied tissues including placental (UC, UCB, WJ) and adult tissues (AD, BM, dental pulp and olfactory mucosa). Placental-derived MSCs are mostly utilized in experimental studies while mostly BM- and/or AD-MSCs in clinical cases. Apart from the primitive undifferentiated MSCs, neural-inducted MSCs (Park et al. 2012; Gabr et al. 2020), MSC-derived neural network tissue (Wu et al. 2018), HO-1 and BDNF-transfected MSCs (Lee et al. 2017; Khan et al. 2018), heat-shock proteins and/or HO-1-transfected MSCs (Kim et al. 2022), MSCs along with chondroitinase ABC (Lee et al. 2015) and MSCs exosomes (Gabr et al. 2020) are evaluated in experimental SCI models as detailed in Table 5.4. Varied MSCs concentrations ranging from one million to ten million cells are implanted locally, mostly after a gap of 1 week post SCI. The local cell transplantation is made at three different locations viz., at lesion, and cranial and caudal to the lesion. Some studies make intravenous transplantations either sole route or as adjunct (Kim et al. 2015; Lee et al. 2016; Khan et al. 2019; Kim et al. 2022). Considering alternative intravenous route, the cells engraft to the local lesions of the spinal cord and it may be beneficial to transplant cells immediately without getting any effect of the local hyper inflammatory effect (Khan et al. 2019). However, cell dose may have to be on the higher side. Additionally, MSCs implantation in these studies is also supported with other drugs/chemicals like methylprednisolone, methionyl human granulocyte colony-stimulating factor (rmhGCSF), chondroitinase ABC, in addition to the scaffolds like Matrigel and collagen, NeuroRegen scaffold (bovine aponeurosis) and collagen/heparin sulphate.

These experimental studies evaluate repair for variable periods ranging from as short as a few days to as long as 1 year. Evaluation is done through behavioural assessment including Olby, Tarlov and the Basso, Beattie, Bresnahan (BBB) scores, Texas Spinal Cord Injury Scale (TSCIS), histology and/or immunohistochemistry, Western blotting, polymerase chain reaction, magnetic resonance imaging (MRI), electromyography, somatosensory evoked potentials (SEP), motor-evoked potentials (MEP) and transcranial magnetic motor-evoked potentials.

MSCs transplantation correlates with the improvement in symptoms and repair in both acute and chronic compression models of SCI, although complete regeneration and recovery remains to be seen. In compressive SCI model, the behavioural scores and nerve conduction velocity improve as early as 2 weeks after MSCs transplantation (Lim et al. 2007; Jung et al. 2009; Lee et al. 2009, 2011; Ryu et al. 2009, 2012; Park et al. 2011; Kim et al. 2015; Khan et al. 2018). Histopathological findings show reduction in further inflammatory reaction and neuronal formation along with axonal growth in the injured areas of the spinal cord (Lim et al. 2007; Jung et al. 2009; Lee et al. 2009, 2011; Ryu et al. 2009, 2012; Park et al. 2011, 2012; Lee et al. 2017; Khan et al. 2018; Gabr et al. 2020). MSCs transplanted lesions show enhanced expression of neural factors like NeuN, glial fibrillary acidic protein and von Willebrand factor, in addition to the neurotrophic factors (Jung et al. 2009; Lee et al. 2009, 2011; Ryu et al. 2009; Park et al. 2011; Khan et al. 2018). Furthermore, reduction in inflammatory factors and upregulation of anti-inflammatory factors also occur with the cell transplantation (Khan et al. 2018, 2019). MSCs transplantation upregulate cytokines like SDF-1, VEGF and TGF-β (Jung et al. 2009). MRI too demonstrates improved regenerative features on T1-weighted images, although T2-weighted images show mixed signals (Lim et al. 2007; Jung et al. 2009; Ryu et al. 2009; Lee et al. 2011). Improvement in the locomotor function may be less associated with the actual neurogenesis but mainly to the maturation of newly generated neurons into the specific 5-HT-positive neurons, their possible re-myelination and functional synapses (increased synaptophysin) (Li et al. 2017).

In the case of the spinal cord resection model, dogs treated with xenogeneic placental MSCs laden in collagen-based scaffolds show improved locomotor support and endogenous recovery. Animals transplanted with biomaterials support their weight using the hind limbs and show improvement in locomotor function (Han et al. 2017; Deng et al. 2021). Reduction in cystic cavities and a large amount of Tuj-1-labelled immature neurons in cell-treated animals also occur (Han et al. 2017; Deng et al. 2021). At around 1 year, the newly generated neurons may mature into 5-HT-positive neurons and many regenerated axon fibres might exhibit remyelination and synapse formation (Li et al. 2017; Wu et al. 2018).

5.3.8.2 Comparative Studies

MSCs transplantation at 1 week offers better recovery than at 12 h or 2 weeks post SCI. It points out that the cell transplantation should be avoided in peak inflammatory periods and their effect may be less pronounced in chronic cases (Park et al. 2011). Comparing autologous MSCs to that of the allogeneic cells, non-significant differences in the improvement in spinal cord occur although autologous cells may have an edge over allogeneic in longer survivability in the defect area (Jung et al. 2009). MSCs based on the source although show in vitro variabilities, but in vivo studies may not show significant difference in spinal cord healing potential. Transplantation of MSCs derived from adipose tissue (AD), bone marrow (BM), Wharton's jelly (WJ) and umbilical cord blood (UCB) show comparable ability to improve locomotor functions; however, UCB-MSCs may be more amenable towards induction of nerve regeneration and anti-inflammatory activities. The poor correlation between neuroprotection and functional recovery could be due to the poor rearrangement of the axons and nerve cells to carry normal nerve conduction. Also, the normal neural cell numbers developed could fall short of minimal neurons to offer enhanced functional recovery rates (Ryu et al. 2012).

Neural-differentiated MSCs and BM mononuclear cells (BM-MNCs) too improve SCI healing (William et al. 2011; Besalti et al. 2015). Among BM-MSCs, neural-inducted BM-MSCs and BM-MSCs exosomes, the exosomes induce marked improvement in the grey (large, regenerated neurons with multiple processes) and white matter structures (apparent normal white matter with many nerve fibres) as compared to the actual cells. However, all the three types of treatments lead to remyelination and augmented neural regeneration, resulting in improved neurological functions in a dog clipping contusion spinal cord model (Gabr et al. 2020). Implantation of chondroitinase ABC along with the allogeneic AD-MSCs in chronic

SCI shows better outcome especially in relation to the increased expression of digested chondroitin sulphate proteoglycans (2B6), β3 tubulin and neurofilament (NF)-M as compared to the simple MSC-transplanted animals (Lee et al. 2015). Chondroitinase ABC (ChABC) degrades chondroitin sulphate proteoglycans (CSPGs) and is associated with the improvement in functional recovery as it may promote axonal regeneration and reduce the inhibitory effects on neurite growth (Lee et al. 2010). Injection of recombinant methionyl human granulocyte colony-stimulating factor (rmhGCSF) along with the transplantation of allogenic UCB-MSCs has no edge over UCB-MSCs transplantation in relieving SCI. rmhGCSF is generally considered to activate bone marrow stem cells and effectuate their migration but such an activity could not be achieved through the intravenous route. Further, its transplantation for 7 days might have additionally trigger MSCs downregulation (Lim et al. 2007).

5.3.8.3 MSCs Transfection and SCI Repair

To explore transfected MSCs and compare their potential to that of the normal MSCs, BDNF (Lee et al. 2016) and haemoxygenase-1 (HO-1) (Lee et al. 2017; Khan et al. 2019; Kim et al. 2022), relevant to the SCI repair, or their combined transfection of MSCs (Khan et al. 2018) has been conducted. Transfected MSCs potentiate SCI repair both in chronic and subacute models. Such a transfection of AD-MSCs results in their better ability to offer functional improvement and reduce fibrotic changes as compared to AD-MSCs. Transplantation of the MSCs transfected with both the BDNF and HO-1 improves functional ability in terms of a higher cBBB score as compared to the transplantation of the individual factor-transfected MSCs. Further, the neuroregeneration is evidenced by the increase in factors like Tuj-1, GAP-43 and NF-M (Khan et al. 2018). HO-1 transfection of MSCs downregulates the expression of the pro-inflammatory cytokines like IL6, TNF-α and IL-1β and simultaneously upregulates anti-inflammatory markers like IL-10 and HO-1 (Khan et al. 2019). BDNF-transfected MSCs show enhanced BDNF expression, while HO-1 upregulates the BDNF expression through activation of ERK and PI3K-AKT pathways (Hung et al. 2010) which activates the neuroregeneration and survival pathways like TrkB, PI3K/Akt signalling pathways (Qi et al. 2014). Additionally, haemoxygenase-1 is known for its strong antioxidative and anti-inflammatory effects, and also modulates immune response and tends to maintain microcirculation (Zhang et al. 2014). Contrarily, BDNF does not have anti-inflammatory effects rather their pro-neural regeneration potential may prevent inflammation. Although BDNF-transfected MSCs may work in an inflammatory environment, significant effects may occur in the absence of the inflammation (Lee et al. 2016). Transfection of MSCs with combination of BDNF and HO-1 potentiates the expression of galactocerebrosidase precursor (GALC), a marker for oligodendrocytes and remyelination. This marker in contrast is downregulated in chronic and subacute SCI model animals transplanted with the BDNF (Lee et al. 2016) and HO-1 (Lee et al. 2017) transfected MSCs, respectively. This shows BDNF-transfected MSCs behave differently in chronic SCI. Overall, transfected factors like HO-1 and BDNF potentiate MSCs therapeutic action through creation of conducive microenvironment.

Reduction in secondary injuries post cell transplantation may occur through variable mechanisms. One of the MSCs potential therapeutic effects is their ability to inhibit lipid peroxidation and protein oxidation (Kim et al. 2015). These cells also decrease the levels of the inflammatory mediators like COX-2, IL-6 and TNFα and this might activate astrocytes, usually considered to prevent further injury. Such an activation may, however, favour astrogliosis leading to glial scar formation, a barrier for axonal regeneration. Some of the large animal studies fail to show that glial scarring hampers axonal regeneration (Kim et al. 2015; Li et al. 2017; Khan et al. 2019). One of these studies that utilizes bovine aponeurosis or Matrigel scaffold along with MSCs attributes scaffold property in inhibition of the glial scarring (Park et al. 2012; Li et al. 2017) and in promoting the neurotrophic factors such as NGF and NT-3 (Park et al. 2012). Local transplantation of BDNF-transfected MSCs along with chondroitinase ABC and also MSCs intravenous transplantation leads to the significant decrease in the expression levels of TNF-α, IL-6, COX-2, GFAP and GalC. It also increases the expression levels of BDNF, β3-tubulin neurofilament medium and nestin (Lee et al. 2016). Thus, MSCs may restrict astrogliosis and macrophage movement to the lesion and simultaneously show neuronal improvement. The anti-inflammatory effect of MSCs in the SCI area may additionally play a significant role in proliferation, migration and differentiation of endogenous spinal cord-derived neural progenitor cells (Ryu et al. 2012). Transplantation of MSCs along with the scaffold-like collagen, collagen/heparin sulphate and Matrigel also seem to offer better results over simple MSCs, especially in axonal growth (Ryu et al. 2012; Han et al. 2017; Li et al. 2017; Deng et al. 2021). Scaffolds might ensure cell protection and enhance neural growth with limited degradation due to the inflammatory environment (Ryu et al. 2012; Han et al. 2017). These may guide axonal regeneration and decrease glial scar deposition after SCI (Deng et al. 2021).

All these studies show positive outcomes upon MSCs transplantation in the form of neuronal protection rather than neuronal regeneration. Such a potential further improves with the gene transfection of these cells or along with the scaffolds. However, complete recovery remains a challenge. A big challenge remains to ensure normal fibre connections in the spinal cord which usually are very limited. In relation to MSCs engraftment, differentiation and mechanisms involved, limited literature exists. The studies lack sufficient animal numbers to draw statistical analysis and the studies are not of sufficiently long duration as are required in neural regeneration. Further, the coherence in studies for a particular outcome remains to be seen.

5.3.8.4 Clinical Intervertebral Disc or Spinal Cord Injury Studies

MSCs therapeutic effects are evaluated in the dog natural cases of acute or chronic cases of IVDD and SCI. Many chronic SCI studies evaluate MSCs (Nishida et al. 2011; Sarmento et al. 2014; Escalhão et al. 2017; Feitosa et al. 2017; Vikartovska et al. 2020; Orlandin et al. 2021); however, in some studies, spinal cord injury is not classified (Bhat et al. 2019; Krueger et al. 2019). Some of the studies restrict word to disc degeneration with loss of deep pain or broader term of SCI is used (Penha et al.

2014; Besalti et al. 2015; Kim et al. 2016; Steffen et al. 2018; Sharun et al. 2021) that could be either acute or chronic condition. Furthermore, one of the studies employs fibrocartilaginous embolic myelopathy (FCEM) for MSCs transplantation (Chung et al. 2013). These studies are mostly uncontrolled case reports and have been conducted for a variable duration (Table 5.5).

In SCI or IVDD, decompression surgery is being conducted to relieve the compression if any on the spinal cord and restore the lost reflux. However, some of the cases do not respond, especially treated later than 72 h of loss of proprioception. In order to improve their life standard, regenerative medicine employing MSCs is evaluated. Majority of the studies perform decompression surgery; however, some of the studies directly utilize MSCs (Bhat et al. 2019; Krueger et al. 2019; Sharun et al. 2021). Mostly autologous/allogeneic BM-MSCs and in some studies AD-MSCs, and a single study each on Am-MSCs, UCB-MSCs, xenogeneic (human) dental pulp stem cells, olfactory ensheathing cells has also been reported. One of the studies transplants neural-inducted MSCs (Besalti et al. 2015). These cells are transplanted mostly at local injury sites either once or multiple times. BM-MSC-conditioned medium (CM) has also been injected through intravenous route (Vikartovska et al. 2020) and one of the studies uses BM mononuclear cells through intravenous route, in addition to the local transplantation (William et al. 2011). Some of the studies inject multiple doses: two doses of allogeneic BM-MSCs (Bhat et al. 2019), three doses of allogeneic Am-MSCs (Orlandin et al. 2021) or four doses of allogeneic BM-MSCs at the site of injury (Sharun et al. 2021). In one of the IVDD studies, Am-MSCs are transplanted at a local site initially followed by epidural doses at 15th and 45th day after surgery (Orlandin et al. 2021). One of the studies makes combined use of cells and acupuncture (Krueger et al. 2019). In contrast to the experimental model, recovery from clinical SCI mostly relies on clinical evaluation and diagnostic imaging like MRI and electromyography. Evaluation is made for variable duration ranging from 45 days (Bhat et al. 2019) to a maximum of 2 years (Nishida et al. 2011; William et al. 2011).

In chronic SCI, transplantation of MSCs or their conditioned medium is correlated with the significant neurological improvement but with variable nociceptive response. A variable locomotor response like an improved Olby scores with partial improvement in the range of joint motion observed through goniometry is demonstrated in some of the clinical cases (Nishida et al. 2011; Sarmento et al. 2014; Escalhão et al. 2017; Feitosa et al. 2017; Vikartovska et al. 2020; Orlandin et al. 2021). In some cases, bladder function (Vikartovska et al. 2020), defecation power and tail movement (Sarmento et al. 2014) also improve. These findings do not comply with those of the diagnostic imaging techniques like MRI (Feitosa et al. 2017; Orlandin et al. 2021). Ability to walk and coordinate thoracic and pelvic limbs suggests development of the axonal connections within the injured area of the spinal cord. Although preservation of as low as 5–10% of the descending axons can adequately drive the local circuits for basic locomotion (Basso et al. 1996). Thus, locomotor response may not be a good indicator for actual and extensive nerve regeneration.

In case of spinal cord injury, improved clinical outcome is seen as early as 10 days after surgico-cell treatment. The conscious reflexes recover simultaneous to that of the moderate improvement in intestine and urinary bladder functions (Penha et al. 2014). Acupuncture adjunctively improves such an outcome as evidenced by electromyography (Krueger et al. 2019). Urinary disorders after SCI occur due to the disruption of the supraspinal and afferent inputs, and intraspinal circuitry reorganization (Herrera et al. 2010). As such improvement in such disruptions might occur. However, as positive MRI findings are again negative post cell transplantation (Penha et al. 2014; Krueger et al. 2019), only a partial recovery in urinary bladder control can be expected.

In IVDD, considerable improvement is seen after cell transplantation (Kim et al. 2016; Steffen et al. 2018; Sharun et al. 2021; Mrkovački et al. 2021). However, one of the studies (lumbosacral intervertebral discs) shows such an improvement in surgical treatments as well. This significant improvement might occur due to the decompression. In the same study, improved Pfirrmann grade on MRI is also reported. A significant unwanted finding of the localization of Schmorl's nodes (SN) that are protrusions of disc material into the vertebral body also occurs. Such a disc material may contact the subchondral bone marrow and incite an inflammation. Thus, disc extrusion prevention may not necessarily promote disc regeneration (Steffen et al. 2018). Contrary to this study, thoracolumbar disc disease mostly improves in surgico-cell-treated animals as compared to the surgical-only treated ones (Kim et al. 2016). However, in a study that utilizes neural-inducted MSCs for intervertebral disk disease fails to get better results except for the gait improvement (Besalti et al. 2015). This might be due to the lack of a favourable environment for the differentiated cells.

In case of fibrocartilaginous embolic myelopathy, transplantation of human umbilical cord MSCs improves locomotor functions. This condition is generally considered to recover within two weeks although if the improvement fails, prognosis remains poor. As in this study improvement occurs even after the normalization period, MSCs might play the therapeutic role (Chung et al. 2013).

A cat clinical study of lumbar compression fracture is surgically corrected by hemilaminectomy that is followed by MSCs (700 million in collagen gel) transplantation. Surgico-cell-treated case shows better clinical signs up to 75 days post treatment. Thereafter, stagnation in improvement occurs and quite possibly due to the remaining compressive lesion (Penha et al. 2012).

Overall, application of MSCs appears promising in SCI and IVDD, although the prior duration of injury (Chronicity) might affect the outcome. All the above studies have mostly shown that MSCs application improves clinical signs to some extent, although histological and MRI findings are not up to the mark to call MSCs as a definitive therapy. A wide variation in results as per the case/model type occurs, besides, only slight improvement in locomotion asks for further advances in the field. Gene transfection of the MSCs and incorporation of scaffolds although improves MSCs therapeutic efficacy but still falls short of the clinical standards. MSCs remain to be evaluated for their engraftment, mechanisms like differentiation, anti-inflammatory actions and other paracrine actions. Thus, there is a need for further uniform studies that evaluate the effect of model, cell source and concentration, combination of routes, scaffolds and time of implantation. A limitation in animal number and study duration makes results less illustrative. Further, studies that are case-controlled and tend to correlate findings from various diagnostic methods are desired for making MSCs therapy in SCI as evidence-based medicine.

5.3.9 Wobbler Syndrome in Horse

Wobbler syndrome (WS) is a problem that is commonly reported in horses and dogs among veterinary species. It deems them unsuitable for any work. It is characterized by the cervical vertebral canal narrowing leading to the spinal cord compression. In dogs, WS may be disc-associated or bone-associated. The former condition is a multifactorial involving primary developmental abnormalities and secondary degenerative changes that lead to vertebral canal stenosis and spinal cord compression. The latter condition is a congenital or developmental abnormality of the cervical vertebrae or arises as sequelae of the degenerative joint disease leading to the progressive dorsal and/or lateral spinal cord compression (da Costa 2010; De Decker et al. 2012; Knell et al. 2021). In horse, two forms are seen viz., static narrowing of the cervical vertebral canal and dynamic compression of the spinal cord by the cervical vertebrae (Pezzanite and Easley 2019).

There is no effective treatment and the condition is mostly managed by palliative treatment with steroidal and nonsteroidal anti-inflammatory drugs together with the restricted diet and exercise (Cequier et al. 2021). In rare cases, surgery opted fails to offer complete recovery and mostly remains eventful (Pezzanite and Easley 2019). Thus, there is a need to develop alternative effective therapy that could help in complete recovery of the animals. MSCs appear promising in diverse musculoskeletal conditions and as such may be evaluated in the WS as well. Currently, only two horse studies evaluate MSCs safety in Wobbler syndrome. Intrathecal transplantation of autologous BM-MSCs appears safe (Maia et al. 2015). However, atlanto-occipital or lumbosacral transplantation of AD-MSCs (@ 100 million) fails to alter neurological status in healthy or cervical vertebral stenotic myelopathy (CVSM)-affected horses. Atlanto-occipital injection may be an effective route for cell distribution towards the lesions (Barberini et al. 2018). As currently no insight in such a condition is available, MSCs studies are desired to be conducted in dog and horse WS.

Apart from these above-discussed conditions, there are numerous other clinical nervous disorders that may be studied in light of the MSCs therapeutics.

5.4 Conclusion(s)

In vitro studies demonstrate MSCs ability to undergo morphological changes and gene upregulation comparable to neural-like cells, although such studies do not surely confirm their neural differentiation and are very limited in veterinary species.

MSCs have been evaluated in various neurological affections experimentally or in clinical conditions. MSCs therapeutic effects in veterinary species have been evaluated in MUO, CD, IVDD and SCI and are proven to be useful to some extent, while therapeutic effects of MSCs on some of the clinical conditions like Wobbler syndrome have failed to demonstrate any therapeutic effect. In relation to the efficacy of MSCs in clinical conditions, only SCI and CD have been studied. A positive clinical impact occurs with MSCs and such a potential slightly improves with the gene transfection and use of scaffolds, although not to the extent of complete clinical recovery. Overall, these studies are too limited and random to draw any conclusive report. There are wide variations in cell source and passage number, their concentration, dose and route, besides the frequency of implantation. Further, models vary and no comparative study in relation to the effectiveness of the model is available. Scaffolds used have to be understood in relation to its characteristics. Furthermore, the effect of other adjunctive therapeutics including the physiotherapy and routine neurogenic drugs may also be evaluated alongside the stem cell treatment.

References

Abe N, Cavalli V (2008) Nerve injury signaling. Curr Opin Neurobiol 18(3):276–283

Adzick NS, Thom EA, Spong CY et al (2011) A randomized trial of prenatal versus postnatal repair of myelomeningocele. N Engl J Med 364:993–1004

Ahlgren BD, Lui W, Herkowitz HN, Panjabi MM, Guiboux JP (2000) Effect of annular repair on the healing strength of the intervertebral disc: a sheep model. Spine 25:2165–2170

Al Delfi IRT, Wood CR, Johnson LDV, Snow MD, Innes JF, Myint P, Johnson WEB (2020) An in vitro comparison of the neurotrophic and angiogenic activity of human and canine adipose-derived mesenchymal stem cells (MSCs): translating MSC-Based therapies for spinal cord injury. Biomolecules 10:1301

Amorim RM, Clark K, Walker NJ, Kumar P, Herout K, Borjesson DL, Wang A (2020) Placenta-derived multipotent mesenchymal stromal cells: a promising potential cell based therapy for canine inflammatory brain disease. Stem Cell Res Ther 11:304

Andrzejewska A, Dabrowska S, Lukomska B, Janowski M (2021) Mesenchymal stem cells for neurological disorders. Adv Sci 8(7):2002944

Ashinsky BG, Bonnevie ED, Mandalapu SA, Pickup S, Wang C, Han L, Mauck RL, Smith HE, Gullbrand SE (2020) Intervertebral disc degeneration is associated with aberrant endplate remodeling and reduced small molecule transport. J Bone Miner Res 35(8):1572–1581

Bahat-Stroomza B, Barhum Y, Levy YS, Karpov O, Bulvik S, Melamed E et al (2009) Induction of adult human bone marrow mesenchymal stromal cells into functional astrocyte-like cells: potential for restorative treatment in Parkinson's disease. J Mol Neurosci 39:199–210

Bai L, Lennon DP, Caplan AI et al (2012) Hepatocyte growth factor mediates mesenchymal stem cell–induced recovery in multiple sclerosis models. Nat Neurosci 15:862–870

Bao XJ, Liu FY, Lu S, Han Q, Feng M, Wei JJ, Li GL, Zhao RC, Wang RZ (2013) Transplantation of Flk-1+ human bone marrow-derived mesenchymal stem cells promotes behavioral recovery and anti-inflammatory and angiogenesis effects in an intracerebral hemorrhage rat model. Int J Mol Med 31(5):1087–1096

Barberini DJ, Aleman M, Aristizabal F, Spriet M, Clark KC, Walker NJ, Galuppo LD, Amorim RM, Woolard KD, Borjesson DL (2018) Safety and tracking of intrathecal allogeneic mesenchymal stem cell transplantation in healthy and diseased horses. Stem Cell Res Ther 9:96

Barzilay R, Ben-Zur T, Bulvik S, Melamed E, Offen D (2009) Lentiviral delivery of LMX1a enhances dopaminergic phenotype in differentiated human bone marrow mesenchymal stem cell. Stem Cells Dev 18:591–601

Basso DM, Beattie MS, Bresnahan JC (1996) Graded histological and locomotor outcomes after spinal cord contusion using the NYU weight-drop device versus transection. Exp Neurol 139: 244–256

Beckstein JC, Sen S, Schaer TP, Vresilovic EJ, Elliott DM (2008) Comparison of animal discs used in disc research to human lumbar disc: axial compression mechanics and glycosaminoglycan content. Spine 33:E166–E173

Bedira T, Ulaga S, Ustundagb CB, Gunduz O (2020) 3D bioprinting applications in neural tissue engineering for spinal cord injury repair. Mater Sci Eng C 110:110741

Beineke A, Puff C, Seehusen F, Baumgärtner W (2009) Pathogenesis and immunopathology of systemic and nervous canine distemper. Vet Immunol Immunopathol 127:1–18

Besalti O, Can P, Akpinar E, Aktas Z, Elcin AE, Elcin YM (2015) Intraspinal transplantation of autologous neurogenically-induced bone marrow-derived mesenchymal stem cells in the treatment of paraplegic dogs without deep pain perception secondary to intervertebral disk disease. Turk Neurosurg 25:625–632

Bhat IA, Sivanarayanan TB, Somal A, Pandey S, Bharti MK, Panda BSK, Indu B, Verma M, Anand J, Sonwane A et al (2019) An allogenic therapeutic strategy for canine spinal cord injury using mesenchymal stem cells. J Cell Physiol 234(3):2705–2718

Blecker D, Elashry MI, Heimann M, Wenisch S, Arnhold S (2017) New insights into the neural differentiation potential of canine adipose tissue-derived mesenchymal stem cells. Anat Histol Embryol 46(3):304–315

Boido M, Piras A, Valsecchi V, Spigolon G, Mareschi K, Ferrero I, Vizzini A, Temi S, Mazzini L, Fagioli F (2014) Human mesenchymal stromal cell transplantation modulates neuroinflammatory milieu in a mouse model of amyotrophic lateral sclerosis. Cytotherapy 16: 1059–1072

Boika A, Aleinikava N, Chyzhyk V, Zafranskaya M, Nizheharodava D, Ponomarev V (2020) Mesenchymal stem cells in Parkinson's disease: motor and nonmotor symptoms in the early posttransplant period. Surg Neurol Int 11:380

Bowman RM, Mohan A, Ito J et al (2009) Tethered cord release: a long-term study in 114 patients. J Neurosurg Pediatr 3:181–187

Brito HFV (2015) Use of bone marrow mononuclear cells for the treatment of neurological sequelae of canine distemper, Thesis (Doctorate). Federal University of Parana, Curitiba, Parana, Brazil. p 79. https://acervodigital.ufpr.br/handle/1884/41363

Brown EG, Keller BA, Lankford L, Pivetti CD, Hirose S, Farmer DL, Wang A (2016) Age does matter: a pilot comparison of placenta-derived stromal cells for in utero repair of myelomeningocele using a lamb model. Fetal Diagn Ther 39(3):179–185

Buckley CT, Hoyland JA, Fujii K, Pandit A, Iatridis JC, Grad S (2018) Critical aspects and challenges for intervertebral disc repair and regeneration-harnessing advances in tissue engineering. JOR Spine 1(3):e1029

Bussolino F, Di Renzo MF, Ziche M et al (1992) Hepatocyte growth factor is a potent angiogenic factor which stimulates endothelial cell motility and growth. J Cell Biol 119:629–641

Cardoso TC, Ferrari HF, Garcia AF, Novais JB, Silva-Frade C, Ferrarezi MC, Andrade AL, Gamiero R (2012) Isolation and characterization of Wharton's jelly-derived multipotent mesenchymal stromal cells obtained from bovine umbilical cord and maintained in defined serum free three-dimensional system. BMC Biotechnol 12:18

Cardoso MT, Pinheiro AO, Vidane AS, Casals JB, de Oliveira VC, Gonçalves NJN, Martins DS, Ambrósio CE (2017a) Characterization of teratogenic potential and gene expression in canine and feline amniotic membrane-derived stem cells. Reprod Domest Anim 52(s2):58–64

Cardoso TC, Okamura LH, Baptistella JC, Gameiro R, Ferreira HL, Marinho M, Flores EF (2017b) Isolation, characterization and immunomodulatory-associated gene transcription of Wharton's jelly-derived multipotent mesenchymal stromal cells at different trimesters of cow pregnancy. Cell Tissue Res 367(2):243–256

Ceccarelli G, Pozzo E, Scorletti F, Benedetti L, Cusella G, Ronzoni FL, Sahakyan V, Zambaiti E, Mimmi MC, Calcaterra V, Deprest J, Sampaolesi M, Pelizzo G (2015) Molecular signature of amniotic fluid derived stem cells in the fetal sheep model of myelomeningocele. J Pediatr Surg 50(9):1521–1527

Cequier A, Sanz C, Rodellar C, Barrachina L (2021) The usefulness of mesenchymal stem cells beyond the musculoskeletal system in horses. Animals (Basel) 11(4):931

Choi Y-K, Lee DH, Seo Y-K, Jung H, Park J-K, Cho H (2014) Stimulation of neural differentiation in human bone marrow mesenchymal stem cells by extremely low-frequency electromagnetic fields incorporated with MNPs. Appl Biochem Biotechnol 174(4):1233–1245

Choi Y-K, Urnukhsaikhan E, Yoon H-H, Seo Y-K, Cho H, Jeong J-S, Kim S-C, Park J-K (2017) Combined effect of pulsed electromagnetic field and sound wave on in vitro and in vivo neural differentiation of human mesenchymal stem cells. Biotechnol Prog 33(1):201–211

Chung W-H, Park S-A, Lee J-H, Chung D-J, Yang W-J, Kang E-H, Choi C-B, Chang H-S, Kim D-H, Hwang S-H, Han H, Kim H-Y (2013) Percutaneous transplantation of human umbilical cord-derived mesenchymal stem cells in a dog suspected to have fibrocartilaginous embolic myelopathy. J Vet Sci 14(4):495–497

Cole C, Bentz B (2014) Treatment of equine nervous system disorders. In: Maxwell L (ed) Equine pharmacology. Wiley, Hoboken, NJ, pp 192–217

Colpo GD, Ascoli BM, Wollenhaupt-Aguiar B, Pfaffenseller B et al (2015) Mesenchymal stem cells for the treatment of neurodegenerative and psychiatric disorders. An Acad Bras Cienc 87: 1435–1449

Corradetti B, Meucci A, Bizzaro D, Cremonesi F, Lange Consiglio A (2013) Mesenchymal stem cells from amnion and amniotic fluid in the bovine. Reproduction 145:391–400

D'intino G, Vaccari F, Sivilia S, Ecagliarini A et al (2006) A molecular study of hippocampus in dogs with convulsion during canine distemper virus encephalitis. Brain Res 1098:186–195

da Costa RC (2010) Cervical spondylomyelopathy (wobbler syndrome) in dogs. Vet Clin North Am Small Anim Pract 40(5):881–913

Dabrowska S, Andrzejewska A, Strzemecki D, Muraca M, Janowski M, Lukomska B (2019) Human bone marrow mesenchymal stem cell-derived extracellular vesicles attenuate neuroinflammation evoked by focal brain injury in rats. J Neuroinflammation 16(1):216

Daly CD, Ghosh P, Zannettino ACW, Badal T, Shimmon R, Jenkin G, Oehme D, Jain K, Sher I, Vais A, Cohen C, Chandra RV, Goldschlager T (2018) Mesenchymal progenitor cells primed with pentosan polysulfate promote lumbar intervertebral disc regeneration in an sheep model of microdiscectomy. Spine J 18(3):491–506

Danzer E, Finkel RS, Rintoul NE et al (2008) Reversal of hindbrain herniation after maternal-fetal surgery for myelomeningocele subsequently impacts on brain stem function. Neuropediatrics 39:359–362

De Decker S, da Costa RC, Volk HA, Van Ham LML (2012) Current insights and controversies in the pathogenesis and diagnosis of disc-associated cervical spondylomyelopathy in dogs. Vet Rec 171(21):531–537

de Munter JPJM, Shafarevich I, Liundup A, Pavlov D, Wolters EC, Gorlova A, Veniaminova E, Umriukhin A, Kalueff A, Svistunov A, Kramer BW, Lesch KP, Strekalova T (2020) Neuro-cells therapy improves motor outcomes and suppresses inflammation during experimental syndrome of amyotrophic lateral sclerosis in mice. CNS Neurosci Ther 26:504–517

Deng Y, Huang G, Zou L, Nong T, Yang X, Cui J, Wei Y, Yang S, Shi D (2018) Isolation and characterization of buffalo (*bubalus bubalis*) amniotic mesenchymal stem cells derived from amnion from the first trimester pregnancy. J Vet Med Sci 80(4):710–719

Deng W-S, Yang K, Liang B, Liu Y-F, Chen X-Y, Zhang S (2021) Collagen/heparin sulfate scaffold combined with mesenchymal stem cells treatment for canines with spinal cord injury: a pilot feasibility study. J Orthop Surg 29(2):1–11

Denny JB (2006) Molecular mechanisms, biological actions, and neuropharmacology of the growth associated protein GAP-43. Curr Neuropharmacol 4:293–304

Dowdell J, Erwin M, Choma T, Vaccaro A, Iatridis J, Cho SK (2017) Intervertebral disk degeneration and repair. Neurosurgery 80(3S):S46–S54

Dueñas F, Becerra V, Cortes Y, Vidal S, Sáenz L, Palomino J, De Los RM, Peralta OA (2014) Hepatogenic and neurogenic differentiation of bone marrow mesenchymal stem cells from abattoir-derived bovine fetuses. BMC Vet Res 10:154

Edamura K, Kuriyama K, Kato K, Nakano R, Teshima K, Asano K, Sato T, Tanaka S (2012) Proliferation capacity, neuronal differentiation potency and microstructures after the differentiation of canine bone marrow stromal cells into neurons. J Vet Med Sci 74:923–927

Edamura K, Takahashi Y, Fujii A, Masuhiro Y, Narita T, Seki M, Asano K (2020) Recombinant canine basic fibroblast growth factor-induced differentiation of canine bone marrow mesenchymal stem cells into voltage- and glutamate-responsive neuron-like cells. Regener Ther 15:121–128

Engel U, Wolswijk G (1996) Oligodendrocyte-type-2 astrocyte (O-2A) progenitor cells derived from adult rat spinal cord: in vitro characteristics and response to PDGF, bFGF and NT-3. Glia 16:16–26

Escalhão CCM, Ramos IP, Hochman-Mendez C, Brunswick THK, Lopes Souza SA, Gutfilen B, dos Santos Goldenberg RC, Coelho-Sampaio T (2017) Safety of allogeneic canine adipose tissue-derived mesenchymal stem cell intraspinal transplantation in dogs with chronic spinal cord injury. Stem Cells Int 2017:3053759

Farmer DL, von Koch CS, Peacock WJ et al (2003) In utero repair of myelomeningocele: experimental pathophysiology, initial clinical experience, and outcomes. Arch Surg 138:872–878

Fawcett JW, Asher RA (1999) The glial scar and CNS repair. Brain Res Bull 49:377

Feitosa MLT, Sarmento CAP, BocabelloII RZ, Beltrão-Braga PCB, Pignatari GC, Giglio RF, Angelica Miglino M, Orlandin JR, Ambrósio CE (2017) Transplantation of human immature dental pulp stem cell in dogs with chronic spinal cord injury. Acta Cir Bras 32(7):540–549

Fluehmann G, Doherr MG, Jaggy A (2006) Canine neurological diseases in a referral hospital population between 1989 and 2000 in Switzerland. J Small Anim Pract 47:582–587

Freeman BJ, Kuliwaba JS, Jones CF, Shu CC, Colloca CJ, Zarrinkalam MR, Mulaibrahimovic A, Gronthos S, Zannettino AC, Howell S (2016) Allogeneic mesenchymal precursor cells promote healing in postero-lateral annular lesions and improve indices of lumbar intervertebral disc degeneration in an sheep model. Spine 41(17):1331–1339

Furno DL, Pellitteri R, Graziano ACE, Giuffrida R, Vancheri C, Gili E, Cardile V (2013) Differentiation of human adipose stem cells into neural phenotype by neuroblastoma- or olfactory ensheathing cells-conditioned medium. J Cell Physiol 228(11):2109–2118

Gabr H, Elkheir WA, Fares AE, Farghali HAM, Mahmoud BE, Madbouly MA, Gamal N, Hamaad AM, Elkheir AA, Hassan RM (2020) Free mesenchymal stem cell-associated exosomes induce better neuroregeneration than mesenchymal stem cells and neural differentiated mesenchymal stem cells in canine model of spinal cord injury. J Stem Cell Res Ther 10(6):463

Galganski LA, Priyadarsini K, Vanover MA, Pivetti CD, Anderson JE, Lankford L, Paxton ZJ, Chung K, Lee C, Hegazi MS, Yamashiro KJ, Wang A, Farmer DL (2020) In utero treatment of myelomeningocele with placental mesenchymal stromal cells—selection of an optimal cell line in preparation for clinical trials. J Pediatr Surg 55(9):1941–1946

Gao Y, Zhu Z, Zhao Y, Hua J, Ma Y, Guan W (2014) Multi-lineage potential research of bovine amniotic fluid mesenchymal stem cells. Int J Mol Sci 15:3698–3710

García-Mendívil L, Mediano DR, Hernaiz A, Sanz-Rubio D, Vázquez FJ, Marín B, López-Pérez Ó, Otero A, Badiola JJ, Zaragoza P et al (2021) Effect of scrapie prion infection in ovine bone marrow-derived mesenchymal stem cells and ovine mesenchymal stem cell-derived neurons. Animals 11:1137

Ge W, Ren C, Duan X, Geng D, Zhang C, Liu X, Chen H, Wan M, Geng R (2015) Differentiation of mesenchymal stem cells into neural stem cells using cerebrospinal fluid. Cell Biochem Biophys 71(1):449–455

Goldschlager T, Ghosh P, Zannettino A, Gronthos S, Rosenfeld JV, Itescu S, Jenkin G (2010) Cervical motion preservation using mesenchymal progenitor cells and pentosan polysulfate, a novel chondrogenic agent: preliminary study in an ovine model. Neurosurg Focus 28(6):E4

Gomez-Sanchez JA et al (2017) After nerve injury, lineage tracing shows that myelin and Remak Schwann cells elongate extensively and branch to form repair Schwann cells, which shorten radically on remyelination. J Neurosci 37(37):9086–9099

Gonçalves DSV, Gomes MVS, Guterra VLP, Lucchi-Rodrigues AF, Mathias CHT, Maestri LFP, Argôlo-Neto NM, Monteiro BS (2018) Mesenchymal stem cell infusion for the treatment of neurological sequelae of canine distemper virus: a clinical study. Genet Mol Res 17:GMR18088

Gonzalez R, Hamblin MH, Lee J-P (2016) Neural stem cell transplantation and CNS diseases. CNS Neurol Disord Drug Targets 15(8):881–886

Granger N, Smith PM, Jeffery ND (2010) Clinical findings and treatment of non-infectious meningoencephalomyelitis in dogs: a systematic review of 457 published cases from 1962 to 2008. Vet J 184:290–297

Granger N, Blamires H, Franklin RJM, Jeffery ND (2012) Autologous olfactory mucosal cell transplants in clinical spinal cord injury: a randomized double-blinded trial in a canine translational model. Brain 135:3227–3237

Griffin MF et al (2014) Peripheral nerve injury: principles for repair and regeneration. Open Orthop J 8:199–203

Gugjoo MB, Amarpal FMR, Shah RA, Mir MS, Sharma GT (2020) Goat mesenchymal stem cell basic research and potential applications. Small Rum Res 183:106045

Guilbaud L, Dugas A, Weber M, Deflers C, Lallemant P, Lilin T, Adam C, Cras A, Mebarki M, Zérah M, Faivre L, Larghero J, Jouannic J-M (2022) In utero treatment of myelomeningocele with allogenic umbilical cord-derived mesenchymal stromal cells in an ovine model. Curr Res Transl Med 70:103314

Han S, Xiao Z, Li X, Zhao H, Wang B, Qiu Z, Li Z, Mei X, Xu B, Fan C, Chen B, Han J, Gu Y, Yang H, Shi Q, Dai J (2017) Human placenta-derived mesenchymal stem cells loaded on linear ordered collagen scaffold improves functional recovery after completely transected spinal cord injury in canine. Sci China Life Sci 61:2

Hannila SS, Filbin MT (2008) The role of cyclic AMP signaling in promoting axonal regeneration after spinal cord injury. Exp Neurol 209(2):321–332

Hang D, Li F, Che W, Wu X, Wan Y, Wang J, Zheng Y (2017) One-stage positron emission tomography and magnetic resonance imaging to assess mesenchymal stem cell survival in a canine model of intervertebral disc degeneration. Stem Cells Dev 26(18):1334–1343

Heffez DS, Aryanpur J, Hutchins GM et al (1990) The paralysis associated with myelomeningocele: clinical and experimental data implicating a preventable spinal cord injury. Neurosurgery 26:987–992

Herrera JJ, Haywood-Watson RJ 2nd, Grill RJ (2010) Acute and chronic deficits in the urinary bladder after spinal contusion injury in the adult rat. J Neurotrauma 27(2):423–431

Hillmann A, Ahrberg AB, Brehm W, Heller S, Josten C, Paebst F, Burk J (2016) Comparative characterization of human and equine mesenchymal stromal cells: a basis for translational studies in the equine model. Cell Transplant 25:109–124

Hiyama A, Mochida J, Iwashina T, Omi H, Watanabe T, Serigano K, Tamura F, Sakai D (2008) Transplantation of mesenchymal stem cells in a canine disc degeneration model. J Orthop Res 26(5):589–600

Hu P, Pu Y, Li X, Zhu Z, Zhao Y, Guan W, Ma Y (2015) Isolation, *in vitro* culture and identification of a new type of mesenchymal stem cell derived from fetal bovine lung tissues. Mol Med Rep 12:3331–3338

Hulsebosch CE (2002) Recent advances in pathophysiology and treatment of spinal cord injury. Am J Physiol 26:238–255

Hung SY, Liou HC, Fu WM (2010) The mechanism of heme oxygenase-1 action involved in the enhancement of neurotrophic factor expression. Neuropharmacology 58(2):321–329

Hussain I, Sloan SR Jr, Wipplinger C, Navarro-Ramirez R, Zubkov M, Kim E, Kirnaz S, Bonassar LJ, Härtl R (2019) Mesenchymal stem cell-seeded high-density collagen gel for annular repair: 6-week results from in vivo sheep models. Neurosurgery 85(2):E350–E359

Jaramillo-Merchán J, Jones J, Ivorra J, Pastor D, Viso-León M, Armengól JA et al (2013) Mesenchymal stromal-cell transplants induce oligodendrocyte progenitor migration and remyelination in a chronic demyelination model. Cell Death Dis 4(8):e779

Javanmard F, Azadbakht M, Pourmoradi M (2016) EXPERIMENTAL STUDY. The effect of hydrostatic pressure on staurosporine-induced neural differentiation in mouse bone marrow-derived mesenchymal stem cells. Bratisl Med J 117(5):283–289

Jellema RK, Ophelders DRMG, Zwanenburg A, Nikiforou M, Delhaas T, Andriessen P, Mays RW, Deans R, Germeraad WTV, Wolfs TGAM, Kramer BW (2015) Multipotent adult progenitor cells for hypoxic-ischemic injury in the preterm brain. J Neuroinflammation 12:241

Jiang J, Lv Z, Gu Y, Li J, Xu L, Xu W et al (2010) Adult rat mesenchymal stem cells differentiate into neuronal-like phenotype and express a variety of neuro-regulatory molecules in vitro. Neurosci Res 66:46–52

Jung D-I, Ha J, Kang B-T, Kim J-W, Quan F-S, Lee J-H, Woo E-J, Park H-M (2009) A comparison of autologous and allogenic bone marrow-derived mesenchymal stem cell transplantation in canine spinal cord injury. J Neurol Sci 285:67–77

Kabagambe SK, Chen YJ, Farmer DL (2017a) Fetal surgery for myelomeningocele: a review of current clinical practice and translational research. Minerva Pediatr 69:59–65

Kabagambe SK, Jensen GW, Chen YJ, Vanover MA, Farmer DL (2017b) Fetal surgery for myelomeningocele: a systematic review and meta-analysis of outcomes in fetoscopic versus open repair. Fetal Diagn Ther 43:161–174. https://doi.org/10.1159/000479505

Kamishina H, Deng J, Oji T, Cheeseman JA, Clemmons RM (2006) Expression of neural markers on bone-marrow–derived canine mesenchymal stem cells. Am J Vet Res 67:1921–1928

Kamphuis W, Mamber C, Moeton M, Kooijman L, Sluijs JA, Jansen AHP, Verveer M, de Groot LR, Smith VD, Rangarajan S, Rodriguez JJ, Orre M, Hol EM (2012) GFAP isoforms in adult mouse brain with a focus on neurogenic astrocytes and reactive astrogliosis in mouse models of Alzheimer disease. PLoS One 7:e42823

Karlupia N, Manley NC, Prasad K, Schäfer R, Steinberg GK (2014) Intraarterial transplantation of human umbilical cord blood mononuclear cells is more efficacious and safer compared with umbilical cord mesenchymal stromal cells in a rodent stroke model. Stem Cell Res Ther 5(2):45

Kaufman BA (2004) Neural tube defects. Pediatr Clin North Am 51:389–419

Kermani S, Karbalaie K, Madani SH, Jahangirnejad AA, Eslaminejad MB, Nasr-Esfahani MH, Baharvand H (2008) Effect of lead on proliferation and neural differentiation of mouse bone marrow-mesenchymal stem cells. Toxicol In Vitro 22(4):995–1001

Khan IU, Yoon Y, Kim A, Jo KR, Choi KU, Jung T, Kim N, Son YS, Kim WH, Kweon O-K (2018) Improved healing after the co-transplantation of HO-1 and BDNF overexpressed mesenchymal stem cells in the subacute spinal cord injury of dogs. Cell Transplant 27:1140–1153

Khan IU, Yoon Y, Choi KU, Jo KR, Kim N, Lee E, Kim WH, Kweon O-K (2019) Therapeutic effects of intravenous injection of fresh and frozen thawed HO-1-overexpressed Ad-MSCs in dogs with acute spinal cord injury. Stem Cells Int 2019:8537541

Kim JK, Lee HJ, Park HT (2014) Two faces of Schwann cell dedifferentiation in peripheral neurodegenerative diseases: pro-demyelinating and axon-preservative functions. Neural Regen Res 9:1952–1954

Kim Y, Jo S-H, Kim WH, Kweon O-K (2015) Antioxidant and anti-inflammatory effects of intravenously injected adipose derived mesenchymal stem cells in dogs with acute spinal cord injury. Stem Cell Res Ther 6:229

Kim Y, Lee SH, Kim WH, Kweon O-K (2016) Transplantation of adipose derived mesenchymal stem cells for acute thoracolumbar disc disease with no deep pain perception in dogs. J Vet Sci 17(1):123–126

Kim S-M, Li Q, An J-H, Chae H-K, Yang J-I, Ryu M-O, Nam A, Song W-J, Youn H-Y (2019) Enhanced angiogenic activity of dimethyloxalylglycine-treated canine adipose tissue-derived mesenchymal stem cells. J Vet Med Sci 81(11):1663–1670

Kim WK, Kim WH, Kweon O-K, Kang B-J (2022) Heat-shock proteins can potentiate the therapeutic ability of cryopreserved mesenchymal stem cells for the treatment of acute spinal cord injury in dogs. Stem Cell Rev Rep 18:1461–1477. https://doi.org/10.1007/s12015-021-10316-6

Knell SC, Smolders LA, Pozzi A (2021) *Ex vivo* evaluation of the dynamic morphometry of the caudal cervical intervertebral disc spaces of small dogs and cats. Front Vet Sci 8:706452

Koniusz S, Andrzejewska A, Muraca M, Srivastava AK, Janowski M, Lukomska B et al (2016) Extracellular vesicles in physiology, pathology, and therapy of the immune and central nervous system, with focus on extracellular vesicles derived from mesenchymal stem cells as therapeutic tools. Front Cell Neurosci 10:109

Kouli A, Torsney KM, Kuan W-L, Stoker TB, Greenland JC (2018) Parkinson's disease: etiology, neuropathology, and pathogenesis. National Library of Medicine, Bethesda, MD

Krabbe C, Zimmer J, Meyer M (2005) Neural trans-differentiation of mesenchymal stem cell: a review. APMIS 113:831–844

Krueger E, Magrib LMS, Botelhoc AS, Bachd FS, Rebellatod CLK, Fracarod L, Fragosod FYI, Villanova JRJA, Brofmand PRS, Popović-Maneski L (2019) Effects of low-intensity electrical stimulation and adipose derived stem cells transplantation on the time-domain analysis-based electromyographic signals in dogs with SCI. Neurosci Lett 696:38–45

Lange-Consiglio A, Corradetti B, BizzaroD MM, Ressel L, Tassan S, Parolini O, Cremones F (2012) Characterization and potential applications of progenitor-like cells isolated from horse amniotic membrane. Tissue Eng Reg Med 8(6):622–663

Lee J-H, Chang H-S, Kang E-H, Chung D-J, Choi C-B, Lee J-H, Hwang S-H, Han H, Kim H-Y (2009) Percutaneous transplantation of human umbilical cord blood–derived multipotent stem cells in a canine model of spinal cord injury Laboratory investigation. J Neurosurg Spine 11: 749–757

Lee H, McKeon RJ, Bellamkonda RV (2010) Sustained delivery of thermostabilized chABC enhances axonal sprouting and functional recovery after spinal cord injury. Proc Natl Acad Sci U S A 107:3340–3345

Lee WS, Suzuki Y, Graves SS, Iwata M, Venkataraman GM, Mielcarek M, Peterson LJ, Ikehara S, Torok-Storb B, Storb R (2011) Canine bone-marrow–derived mesenchymal stromal cells suppress alloreactive lymphocyte proliferation in vitro but fail to enhance engraftment in canine bone marrow transplantation. Biol Blood Marrow Transplant 17:465–475

Lee SH, Kim Y, Rhew D, Kuk M, Kim M, Kim WH, Kweon O-K (2015) Effect of the combination of mesenchymal stromal cells and chondroitinase ABC on chronic spinal cord injury. Cytotherapy 17:1374–1383

Lee SH, Kim Y, Rhew D, Kim A, Jo KR, Yoon Y, Choi KU, Jung T, Kim WH, Kweon OK (2016) Impact of local injection of brain-derived neurotrophic factor-expressing mesenchymal stromal cells (MSCs) combined with intravenous MSC delivery in a canine model of chronic spinal cord injury. Cytotherapy S1465-3249(16):30540

Lee SH, Kim Y, Rhew D, Kim A, Jo KR, Yoon Y, Choi KU, Jung T, Kim WH, Kweon O-K (2017) Effect of canine mesenchymal stromal cells overexpressing heme oxygenase-1 in spinal cord injury. J Vet Sci 18(3):377–386

Lempp C, Spitzbarth I, Puff C, Cana A, Kegler K, Techangamsuwan S, Baumgärtner W, Seehusen F (2014) New aspects of the pathogenesis of canine distemper leukoencephalitis. Viruses 6(7): 2571–2601

Levine JM, Stincone F, Lee Y-S (1993) Development and differentiation of glial precursor cells in the rat cerebellum. Glia 7:307–321

Li X, Tan J, Xiao Z, Zhao Y, Han S, Liu D, Yin W, Li J, Li J, WanggouS CB, Ren C, Jiang X, Dai J (2017) Transplantation of hUC-MSCs seeded collagen scaffolds reduces scar formation and promotes functional recovery in canines with chronic spinal cord injury. Sci Rep 7:43559

Liao W, Zhong J, Yu J, Xie J, Liu Y, Du L, Yang S, Liu P, Xu J, Wang J, Han Z, Han ZC (2009) Therapeutic benefit of human umbilical cord derived mesenchymal stromal cells in intracerebral hemorrhage rat: implications of anti-inflammation and angiogenesis. Cell Physiol Biochem 24(3–4):307–316

Lim JH, Byeon YE, Ryu HH, Jeong YH, Lee YW, Kim WH, Kang K-S, Kweon O-K (2007) Transplantation of canine umbilical cord blood-derived mesenchymal stem cells in experimentally induced spinal cord injured dogs. J Vet Sci 8:275–282

Liu Z, Huang D, Zhang M, Chen Z, Jin J, Huang S, Zhang Z, Wang Z, Chen L, Chen L, Xu Y (2011) Cocaine- and amphetamine-regulated transcript promotes the differentiation of mouse bone marrow-derived mesenchymal stem cells into neural cells. BMC Neurosci 12:67

Liu K, Guo L, Zhou Z, Pan M, Yan C (2019) Mesenchymal stem cells transfer mitochondria into cerebral microvasculature and promote recovery from ischemic stroke. Microvasc Res 123:74–80

Lomax-Bream LE, Barnes M, Copeland K et al (2007) The impact of spina bifida on development across the first 3 years. Dev Neuropsychol 31:1–20

Long X, Olszewski M, Huang W, Kletzel M (2005) Neural cell differentiation in vitro from adult human bone marrow mesenchymal stem cells. Stem Cells Dev 14:65–69

Lu P, Blesch A, Tuszynski MH (2004) Induction of bone marrow stromal cells to neurons: differentiation, transdifferentiation, or artifact? J Neurosci Res 77:174–191

Lu B, Nagappan G, Lu Y (2014) BDNF and synaptic plasticity, cognitive function, and dysfunction. Handb Exp Pharmacol 220:223–250

Lyahyai J, Mediano DR, Ranera B, Sanz A, Remacha AR, Bolea R, Zaragoza P, Rodellar C, Martín-Burriel I (2012) Isolation and characterization of ovine mesenchymal stem cells derived from peripheral blood. BMC Vet Res 8:169

Ma T, Gong K, Ao Q, Yan Y, Song B, Huang H, Zhang X, Gong Y (2013) Intracerebral transplantation of adipose-derived mesenchymal stem cells alternatively activates microglia and ameliorates neuropathological deficits in Alzheimer's disease mice. Cell Transplant 22 (Suppl 1):S113–S126. https://doi.org/10.3727/096368913X672181

Maia L, da Cruz Landim-Alvarenga F, Taffarel MO, de Moraes CN, Machado GF, Melo GD, Amorim RM (2015) Feasibility and safety of intrathecal transplantation of autologous bone marrow mesenchymal stem cells in horses. BMC Vet Res 11:63

Maina F, Klein R (1999) Hepatocyte growth factor, a versatile signal for developing neurons. *Nat Neurosci* 2:213–217

Martinez AM, Goulart CD, Ramalho BD et al (2014) Neurotrauma and mesenchymal stem cells treatment: From experimental studies to clinical trials. World J Stem Cells 6:179–194

Mata R, Yao Y, Cao W, Ding J, Zhou T, Zhai Z, Gao C (2021) The dynamic inflammatory tissue microenvironment: signality and disease therapy by biomaterials. Research 2021:418951

Mediano DR, Sanz-Rubio D, Bolea R, Marín B, Vázquez FJ, Remacha AR, Pez-Pérez O, Fernández-Borges N, Castilla J, Zaragoza P, Badiola JJ, Rodellar C, Martín-Burriel I (2015) Characterization of mesenchymal stem cells in sheep naturally infected with scrapie. J Gen Virol 96:3715–3726

Mendes-Pinheiro B, Marote A, Marques CR, Teixeira FG, Ribeiro JC, Salgado AJ (2020) Applications of the stem cell secretome in regenerative medicine. In: El-Hashash AHK (ed) Mesenchymal stem cells in human health and diseases. Elsevier, Amsterdam, pp 79–114

Mihevc SP, Grgich VK, Kopitar AN, Mohoric N, Majdic G (2020) Neural differentiation of canine mesenchymal stem cells/multipotent mesenchymal stromal cells. BMC Vet Res 16:282

Moloney T, Rooney G, Barry F, Howard L, Dowd E (2010) Potential of rat bone marrow-derived mesenchymal stem cells as vehicles for delivery of neurotrophins to the Parkinsonian rat brain. Brain Res 1359:33–43

Monteiro BA (2017) Effects of Therapy with Mesenchymal Stem Cells on Disorders of the Nervous System of dogs, Thesis (Doctorate). Faculty of Veterinary Medicine and Zootechnics, Paulista State University, Botucatu Campus, Sao Paulo, Brazil. 68p. https://repositorio.unesp.br/handle/11449/151543

Morshead CM, Reynolds BA, Craig CG, McBurney MW, Staines WA, Morassutti D, Weiss S, Van der Kooy D (1994) Neural stem cells in the adult mammalian forebrain: a relatively quiescent subpopulation of subependymal cells. Neuron 13:1071–1082

Mou X, Wang S, Liu X, Guo W, Li J, Qiu J, Yu X, Wang ZL, Liu X, Geng Z, Liu H (2017) Static pressure-induced neural differentiation of mesenchymal stem cells. Nanoscale 9(28): 10031–10037

Mrkovački J, Dražilov SS, Spasovski V, Fazlagić A, Pavlović S, Nikčević G (2021) Case report: successful therapy of spontaneously occurring canine degenerative lumbosacral stenosis using autologous adipose tissue-derived mesenchymal stem cells. Front Vet Sci 8:732073

Munoz JL, Greco SG, Patel SA, Sherman LS, Bhatt S, Bhatt RS, Shrensel JA, Guan Y-Z, Xie G, Ye J-H, Rameshwar P, Siegel A (2012) Feline bone marrow-derived mesenchymal stromal cells (MSCs) show similar phenotype and functions with regards to neuronal differentiation as human MSCs. Differentiation 84(2):214–222

Myckatyn TM, Mackinnon SE, McDonald JW (2004) Stem cell transplantation and other novel techniques for promoting recovery from spinal cord injury. Transpl Immunol 12:343–358

Naito E, Kudo D, Sekine S-I, Watanabe K, Kobatake Y, Tamaoki N, Inden M, Iida K, Ito Y, Hozumi I, Shibata T, Maeda S, Kamishina H (2015) Characterization of canine dental pulp cells and their neuroregenerative potential. In Vitro Cell Dev Biol Animal 51:1012–1022

Nakamura Y, Miyaki S, Ishitobi H, Matsuyama S, Nakasa T, Kamei N, Akimoto T, Higashi Y, Ochi M (2015) Mesenchymal-stem-cell-derived exosomes accelerate skeletal muscle regeneration. FEBS Lett 589(11):1257–1265

Nakamura M, Nishida H, Yoshizaki K, Akiyoshi H, Hatoya S, Sugiura K, Inaba T (2020) Canine mesenchymal stromal cell-conditioned medium promotes survival and neurite outgrowth of neural stem cells. J Vet Med Sci 82(5):668–672

Nakano R, Edamura K, Sugiya H, Narita T, Okabayashi K, Moritomo T, Teshima K, Asano K, Nakayama T (2013) Evaluation of mRNA expression levels and electrophysiological function of neuron-like cells derived from canine bone marrow stromal cells. Am J Vet Res 74:1311–1320

Nakano R, Edamura K, Nakayama T, Teshima K, Asano K, Narita T, Okabayashi K, Sugiya H (2015) Differentiation of canine bone marrow stromal cells into voltage- and glutamate responsive neuron-like cells by basic fibroblast growth factor. J Vet Med Sci 77(1):27–35

Nessler J, Bénardais K, Gudi V, Hoffmann A, Salinas Tejedor L, Janßen S, Prajeeth CK, Baumgärtner W, Kavelaars A, Heijnen CJ, van Velthoven C, Hansmann F, Skripuletz T, Stangel M (2013) Effects of murine and human bone marrow-derived mesenchymal stem cells on cuprizone induced demyelination. PLoS One 8:e69795

Nessler J, Wohlsein P, Junginger J, Hansmann F, Erath J, Söbbeler F, Dziallas P, Tipold A (2020) Meningoencephalomyelitis of unknown origin in cats: a case series describing clinical and pathological findings. Front Vet Sci 7:291. https://doi.org/10.3389/fvets.2020.00291

Neuhuber B, Gallo G, Howard L, Kostur L, MacKay A, Fischer I (2004) Reevaluation of in vitro differentiation protocols for bone marrow stromal cells: disruption of the actin cytoskeleton induces rapid morphological changes and mimics neuronal phenotype. J Neurosci Res 7:192–204

Nishida H, Nakayama M, Tanaka H, Kitamura M, Hatoya S, Sugiura K, Suzuki Y, Ide C, Inaba T (2011) Evaluation of transplantation of autologous bone marrow stromal cells into the cerebrospinal fluid for treatment of chronic spinal cord injury in dogs. Am J Vet Res 72:1118–1123

Oda Y, Tani K, Kanei T, Haraguchi T, Itamoto K, Nakazawa H, Taura Y (2013) Characterization of neuronlike cells derived from canine bone marrow stromal cells. Vet Res Commun 37:133–138

Oehme D, Ghosh P, Shimmon S, Wu J, McDonald C, Troupis JM, Goldschlager T, Rosenfeld JV, Jenkin G (2014) Mesenchymal progenitor cells combined with pentosan polysulfate mediating disc regeneration at the time of microdiscectomy: a preliminary study in an ovine model. J Neurosurg Spine 20(6):657–669

Oehme D, Ghosh P, Goldschlager T, Itescu S, Shimon S, Wu J, McDonald C, Troupis JM, Rosenfeld JV, Jenkin G (2016) Reconstitution of degenerated ovine lumbar discs by STRO-

3-positive allogeneic mesenchymal precursor cells combined with pentosan polysulfate. J Neurosurg Spine 24(5):715–726

Oh HJ, Park JE, Kim MJ, Hong SG, Ra JC, Jo JY, Kang SK, Jang G, Lee BC (2011) Recloned dogs derived from adipose stem cells of a transgenic cloned beagle. Theriogenology 75:1221–1231

Okano H (2006) Adult neural stem cells and central nervous system repair. Ernst Schering Res Found Workshop 60:215–228

Olby N (1999) Current concepts in the management of acute spinal cord injury. J Vet Intern Med 13(5):399–407

Olby N (2010) The pathogenesis and treatment of acute spinal cord injuries in dogs. Vet Clin North Am Small AnimPract 40:791–807

Ophelders DRMG, Wolfs TGAM, Jellema RK, Zwanenburg A, Andriessen P, Delhaas T, Ludwig A-K, Radtke S, Peters V, Janssen L, Giebel B, Kramer BW (2016) Mesenchymal stromal cell-derived extracellular vesicles protect the fetal brain after hypoxia-ischemia. Stem Cells Transl Med 5:754–763

Orlandin JR, da Silva GI, de FátimaSallum LS, Cagnim AF, Casals JB, Carregaro AB, Freitas SH, Machado LC, Reis Castiglioni MC, Garcia Alves AL, de Vasconcelos Machado VM, Ambrósio CE (2021) Treatment of chronic spinal cord injury in dogs using amniotic membrane-derived stem cells: preliminary results. Stem Cells Clon 14:39–49

Osaka K, Tanimura T, Hirayama A et al (1978) Myelomeningocele before birth. J Neurosurg 49: 711–724

Palmer TD, Takahashi J, Gage FH (1997) The adult rat hippocampus contains primordial neural stem cells. Mol Cell Neurosci 8:389–404

Pan B, Fu CD, Ge YL, Li ZY, Zhang ZJ, Liu HM et al (2014) Unilateral pedicle screw fixation and transforaminal lumbar interbody fusion through paraspinal muscle approach for recurrent lumbar disc herniation combined with lumbar instability. Zhongguo Gu Shang 27:712–716

Park S-S, Byeon Y-E, Ryu H-H, Kang B-J, Kim YS, Kim W-H, Kang K-S, Han H-J, Kweon O-K (2011) Comparison of canine umbilical cord blood-derived mesenchymal stem cell transplantation times: involvement of astrogliosis, inflammation, intracellular actin cytoskeleton pathways, and neurotrophin-3. Cell Transplant 20:1867–1880

Park SS, Lee YJ, Lee SH, Lee D, Choi K, Kim WH, Kweon OK, Han HJ (2012) Functional recovery after spinal cord injury in dogs treated with a combination of matrigel and neural-induced adipose-derived mesenchymal stem cells. Cytotherapy 14:584–597

Penha EM, Aguiar PHP, Barrouin-Melo SM, de Lima RS, da Silveira ACC, Otelo ARS, Pinheiro CMB, Ribeiro-dos-Santos R, Soares MBP (2012) Clinical neurofunctional rehabilitation of a cat with spinal cord injury after hemilaminectomy and autologous stem cell transplantation. Int J Stem Cells 5(2):146–150

Penha EM, Meira CS, Guimarães ET, Mendonça MVP, Gravely FA, Pinheiro CMB, Pinheiro TMB, Barrouin-Melo SM, Ribeiro-dos-Santos R, Soares MBP (2014) Use of autologous mesenchymal stem cells derived from bone marrow for the treatment of naturally injured spinal cord in dogs. Stem Cells Int 2014:437521

Pezzanite L, Easley J (2019) Update on surgical treatment of wobblers. Vet Clin North Am Equine Pract 35(2):299–309

Pinheiro AO, Cardoso MT, Vidane AS, Casals JB, Passarelli D, Alencar ALF, Sousa RLM, Fantinato-Neto P, Oliveira VC, Lara VM et al (2016) Controversial results of therapy with mesenchymal stem cells in the acute phase of canine distemper disease. Genet Mol Res 15(2). https://doi.org/10.4238/gmr.15028310.

Pinheiro LL, de Lima AR, Martins DM, de Oliveira EHC, Souza MPC, de Carvalho Miranda CMF, Beltrão-Braga PCB, Russo FB, Conceição Pignatari G, da Silva FE, Branco E (2019) Mesenchymal stem cells in dogs with demyelinating leukoencephalitis as an experimental model of multiple sclerosis. Heliyon 5(6):e01857

Ployart S, Doran I, Bomassi E, Bille C, Libermann S (2013) Myelomeningocele and a dermoid sinus-like lesion in a French bulldog. Can Vet J 54:1133–1136

Qi D, Ouyang C, Wang Y, Zhang S, Ma X, Song Y, Yu H, Tang J, Fu W, Sheng L (2014) Ho-1 attenuates hippocampal neurons injury via the activation of BDNF–TRKB–PI3K/AKT signaling pathway in stroke. Brain Res 1577:69–76

Quah C, Syme G, Swamy GN, Nanjayan S, Fowler A, Calthorpe D (2014) Obesity and recurrent intervertebral disc prolapse after lumbar microdiscectomy. Ann R Coll Surg Engl 96:140–143. https://doi.org/10.1308/003588414X13814021676873

Radtke C, Schmitz B, Spies M, Kocsis JD, Vogt PM (2009) Peripheral glial cell differentiation from neurospheres derived from adipose mesenchymal stem cells. Int J Dev Neurosci 27(8):817–823

Rismanchi N, Floyd CL, Berman RF, Lyeth BG (2003) Cell death and long-term maintenance of neuron-like state after differentiation of rat bone marrow stromal cells: a comparison of protocols. Brain Res 991:46–55

Rivers FJ, Couillard-Despres S, Pedre X, Ploetz S, Caioni M, Lois C et al (2006) Mesenchymal stem cells instruct oligodendrogenic fate decision on adult neural stem cells. Stem Cells 24: 2209–2219

Roach JW, Short BF, Saltzman HM (2011) Adult consequences of spina bifida: a cohort study. Clin Orthop Relat Res 469:1246–1252

Roszek K, Makowska N, Czarnecka J, Porowinska D, Dabrowski M, Danielewska J, Nowak W (2016) Canine adipose-derived stem cells: purinergic characterization and neurogenic potential for therapeutic applications. J Cell Biochem 9999:1–8

Rotshenker S (2011) Wallerian degeneration: the innate-immune response to traumatic nerve injury. J Neuroinflammation 8:109

Ryu H-H, Lim J-H, Byeon Y-E, Park J-R, Seo M-S, Lee Y-W, Kim WH, Kang K-S, Kweon O-K (2009) Functional recovery and neural differentiation after transplantation of allogenic adipose-derived stem cells in a canine model of acute spinal cord injury. J Vet Sci 10(4):273–284

Ryu H-H, Kang B-J, Park S-S, Kim Y, Sung G-J, Woo H-M, Kim WH, Kweon O-K (2012) Comparison of Mesenchymal stem cells derived from fat, bone marrow, Wharton's jelly, and umbilical cord blood for treating spinal cord injuries in dogs. J Vet Med Sci 74(12):1617–1630

Safford KM, Hicok KC, Safford SD, Halvorsen Y-DC, Wilkison WO, Gimble JM, Rice HE (2002) Neurogenic differentiation of murine and human adipose-derived stromal cells. Biochem Biophys Res Commun 294:371–379

Salehi MS, Pandamooz S, Safari A, Jurek B, Tamadon A, Namavar MR, Dianatpour M, Dargahi L, Azarpira N, Fattahi S, Shid Moosavi SM, Keshavarz S, Khodabandeh Z, Zare S, Nazari S, Heidari M, Izadi S, Poursadeghfard M, Borhani-Haghighi A (2020) Epidermal neural crest stem cell transplantation as a promising therapeutic strategy for ischemic stroke. CNS Neurosci Ther 26(7):670–681

Salem NA (2019) Mesenchymal stem cell based therapy for Parkinson's disease. Int J Stem Cell Res Ther 6:062

Sarmento CAP, Rodrigues MN, Bocabello RZ, Mess AM, Miglino MA (2014) Pilot study: bone marrow stem cells as a treatment for dogs with chronic spinal cord injury. Regener Med Res 2:9

Seo MS, Jeong YH, Park JR, Park SB, Rho KH, Kim HS, Yu KR, Lee SH, Jung JW, Lee YS, Kang KS (2009) Isolation and characterization of canine umbilical cord blood-derived mesenchymal stem cells. J Vet Sci 10:181–187

Seo MS, Park SB, Kang KS (2012) Isolation and characterization of canine Wharton's jelly-derived Mesenchymal stem cells. Cell Transplant 21:1493–1502

Serigano K, Sakai D, Hiyama A, Tamura F, Tanaka M, Mochida J (2010) Effect of cell number on mesenchymal stem cell transplantation in a canine disc degeneration model. J Orthop Res 28: 1267–1275

Shanko Y, Navitskaya V, Zamaro A, Zafranskaya M, Krivenko S, Koulchitsky S et al (2018) Somatotopic principle of perineural implantation of stem cells in patients with brain injuries. J Neurol Stroke 8:259–261

Sharma N, SWK DLH, Ghosh M, Sodhi SS, Singh AK, Kim NE et al (2017) A PiggyBac mediated approach for lactoferricin gene transfer in bovine mammary epithelial stem cells for management of bovine mastitis. Oncotarget 8(61):104272

Sharun K, Kumar R, Chandra V, Saxena AC, Pawde AM, Kinjavdekar P, Dhama K, Amarpal I, Sharma GT (2021) Percutaneous transplantation of allogenic bone marrow-derived mesenchymal stem cells for the management of paraplegia secondary to Hansen type I intervertebral disc herniation in a Beagle dog. Iran J Vet Res 22:161–166. https://doi.org/10.22099/IJVR.2021.38613.5620

Sheikh AM, Yano S, Mitaki S, Haque MA, Yamaguchi S, Nagai A (2019) A mesenchymal stem cell line (B10) increases angiogenesis in a rat MCAO model. Exp Neurol 311:182–193

Shetty P, Ravindran G, Sarang S, Thakur A, Rao H et al (2009) Clinical grade mesenchymal stem cells transdifferentiated under xenofree conditions alleviates motor deficiencies in a rat model of Parkinson's disease. Cell Biol Int 33:830–838

Shihabuddin LS, Ray J, Gage FH (1997) FGF-2 is sufficient to isolate progenitors found in the adult mammalian spinal cord. Exp Neurol 148:577–586

Shu CC, Dart A, Bell R, Dart C, Clarke E, Smith MM, Little CB, Melrose J (2018) Efficacy of administered mesenchymal stem cells in the initiation and co-ordination of repair processes by resident disc cells in an ovine (Ovis aries) large destabilizing lesion model of experimental disc degeneration. JOR Spine 1(4):e1037

Singh M, Kakkar A, Sharma R, Kharbanda OP, Monga N, Kumar M, Chowdhary S, Airan B, Mohanty S (2017) Synergistic effect of BDNF and FGF2 in efficient generation of functional dopaminergic neurons from human mesenchymal stem cells. Sci Rep 7:10378

Somal A, Bhat IA, Baiju I, Pandey S, Panda BSK, Thakur N, Sarkar M, Chandra V, Saikumar G, Sharma GT (2016) A comparative study of growth kinetics, in vitro differentiation potential and molecular characterization of fetal adnexa derived caprine mesenchymal stem cells. PLoS One 11(6):e0156821

Song S, Sanchez-Ramos J (2003) Brain as the sea of marrow. Exp Neurol 184:54–60

Souvenir R, Fathali N, Ostrowski RP et al (2011) Tissue inhibitor of matrix metalloproteinase-1 mediates erythropoietin-induced neuroprotection in hypoxia ischemia. Neurobiol Dis 44:28–37

Steffen F, Bertolo A, Affentranger R, Ferguson SJ, Stoyanov J (2018) Treatment of naturally degenerated canine lumbosacral intervertebral discs with autologous mesenchymal stromal cells and collagen microcarriers: a prospective clinical study. Cell Transplant 28:201–211. https://doi.org/10.1177/0963689718815459

Sulaiman W, Gordon T (2013) Neurobiology of peripheral nerve injury, regeneration, and functional recovery: from bench top research to bedside application. Ochsner J 13(1):100–108

Sullivan R et al (2016) Peripheral nerve injury: stem cell therapy and peripheral nerve transfer. Int J Mol Sci 17(12):2101

Sun T, Yu C, Gao Y, Zhao C, Hua J, Cai L, Guan W, Ma Y (2014) Establishment and biological characterization of a dermal mesenchymal stem cells line from bovine. Biosci Rep 34(2):139–146

Tejima E, Guo S, Murata Y et al (2009) Neuroprotective effects of overexpressing tissue inhibitor of metalloproteinase TIMP-1. J Neurotrauma 26:1935–1941

Temple S, Alvarez-Buylla A (1999) Stem cells in the adult mammalian central nervous system. Curr Opin Neurobiol 9:135–141

Terada N, Hamazaki T, Ok M, Hoki M, Mastalerz DM, Nakano Y et al (2002) Bone marrow cells adopt the phenotype of other cells by spontaneous cell fusion. Nature 416:542–545

Testoni S, Dall'apria A, Gallo D, Gentile A (2010) Spina bifida in calves: description of seven cases. Large Anim Rev 16(2):63–66

Tharasanit T, Phutikanit N, Wangdee C, Soontornvipart K, Tantrajak S, Kaewamatawong T, Suwimonteerabutr J, Supaphol P, Techakumphu M (2011) Differentiation potentials of canine bone marrow mesenchymal stem cells. Thailand J Vet Med 41(1):79–86

Titushkin I, Sun S, Shin J, Cho M (2010) Physicochemical control of adult stem cell differentiation: shedding light on potential molecular mechanisms. J Biomed Biotechnol 2010:743476. https://doi.org/10.1155/2010/743476

Tondreau T, Lagneaux L, Dejeneffe M, Massy M, Mortier C, Delforge A, Bron D (2004) Bone marrow-derived mesenchymal stem cells already express specific neural proteins before any differentiation. Differentiation 72(7):319–326

Ulrich R, Puff C, Wewetzer K (2014) Transcriptional changes in canine distemper virus-induced demyelinating leukoencephalitis favor a biphasic mode of demyelination. PLoS One 9:e95917

Uranio MF, Valentini L, Lange-Consiglio A, Caira M, Guaricci AC, L'abbate A, Catacchio CR, Ventura M, Cremonesi F, Dell'aquila ME (2011) Isolation, proliferation, cytogenetic, and molecular characterization and in vitro differentiation potency of canine stem cells from foetal adnexa: a comparative study of amniotic fluid, amnion, and umbilical cord matrix. Mol Reprod Dev 78:361–373

Uranio MF, Dell'aquila ME, Caira M, Guaricci AC, Ventura M, Catacchio C, Martino NA, Valentini L (2014) Characterization and in vitro differentiation potency of early-passage canine amnion- and umbilical cord-derived mesenchymal stem cells as related to gestational age. Mol Reprod Dev 81:539–551

Urrutia DN, Caviedes P, Mardones R, Minguell JJ, Vega-Letter AM, Jofre CM (2019) Comparative study of the neural differentiation capacity of mesenchymal stromal cells from different tissue sources: an approach for their use in neural regeneration therapies. PLoS One 14(3):e0213032

Vassilopoulos G, Wang PR, Russell DW (2003) Transplanted bone marrow regenerates liver by cell fusion. Nature 422:901–904

Venkataramana N, Kumar S, Balaraju S, Radhakrishnan R, Bansal A et al (2010) Open labeled study of unilateral autologous bone-marrow-derived mesenchymal stem cell transplantation in Parkinson's disease. Transl Res 155:62–70

Vikartovska Z, Kuricova M, Farbakova J, Liptak T, Mudronova D, Humenik F, Madari A, Maloveska M, Sykova E, Cizkova D (2020) Stem cell conditioned medium treatment for canine spinal cord injury: pilot feasibility study. Int J Mol Sci 21:5129

Villagrán CC, Amelse L, Neilsen N, Dunlap J, Dhar M (2014) Differentiation of equine mesenchymal stromal cells into cells of neural lineage: potential for clinical applications. Stem Cells Int 2014:891518

von Koch CS, Compagnone N, Hirose S, Yoder S, Harrison MR, Farmer DL (2005) Myelomeningocele: characterization of a surgically induced sheep model and its central nervous system similarities and differences to the human disease. Am J Obstet Gynecol 193:1456–1462

Wang A, Brown EG, Lankford L, Keller BA, Pivetti CD, Sitkin NA, Beattie MS, Bresnahan JC, Farmer DL (2015) Placental mesenchymal stromal cells rescue ambulation in sheep myelomeningocele. Stem Cells Transl Med 4:659–669

Wang Q, Wang X, Lai D, Deng J, Hou Z, Liang H, Liu D (2018) BIX-01294 promotes the differentiation of adipose mesenchymal stem cells into adipocytes and neural cells in Arbas cashmere goats. Res Vet Sci 119:9–18

Webb AA, Ngan S, Fowler JD (2010) Spinal cord injury I: a synopsis of the basic science. Can Vet J 51:485–492

Weimann JM, Charlton CA, Brazelton TR, Hackman RC, Blau HM (2003) Contribution of transplanted bone marrow cells to Purkinje neurons in adult brains. Proc Natl Acad Sci U S A 100:2088–2093

Weishaupt N, Blesch A, Fouad K (2012) BDNF: the career of a multifaceted neurotrophin in spinal cord injury. Exp Neurol 238:254–264

Weiss S, Dunne C, Hewson J, Wohl C, Wheatley M, Peterson AC, Reynolds BA (1996) Multipotent CNS stem cells are present in the adult mammalian spinal cord and ventricular neuroaxis. J Neurosci 16:7599–7609

William JB, Prabakaran R, Ayyappan S, Puskhinraj H, Rao D, Manjunath SR, Thamaraikannan P, Dedeepiya VD, Kuroda S, Yoshioka H et al (2011) Functional recovery of spinal cord injury following application of Intralesional bone marrow mononuclear cells embedded in polymer scaffold–two year follow-up in a canine. J Stem Cell Res Ther 1:110

Woodbury D, Schwarz EJ, Prockop DJ, Black IB (2000) Adult rat and human bone marrow stromal cells differentiate into neurons. J Neurosci Res 61:364–370

Wu G-H, Shi H-J, Che M-T, Huang M-Y, Wei Q-S, Feng B, Ma Y-H, Wang L-J, Jiang B, Wang Y-Q, Hang I, Ling E-A, Zeng X, Zeng Y-S (2018) Recovery of paralyzed limb motor function in canine with complete spinal cord injury following implantation of MSC-derived neural network tissue. Biomaterials 181:15–34

Xiao C (2014) The research process of recurrent lumbar disc herniation and its surgical treatment. J Clin Med Lit 2:132–138

Xin H, Chopp M, Shen LH et al (2013) Multipotent mesenchymal stromal cells decrease transforming growth factor β1 expression in microglia/macrophages and down-regulate plasminogen activator inhibitor 1 expression in astrocytes after stroke. Neurosci Lett 542:81–86

Xiong H, Bai C, Wu S, Gao Y, Lu T, Hu Q, Guan W, Ma Y (2014) Biological characterization of mesenchymal stem cells from bovine umbilical cord. Anim Cells Syst 1(1):59–67

Xu C, Fu F, Li X, Zhang S (2017a) Mesenchymal stem cells maintain the microenvironment of central nervous system by regulating the polarization of macrophages/microglia after traumatic brain injury. Int J Neurosci 127:1124–1135

Xu X, Hu J, Lu H (2017b) Histological observation of a gelatin sponge transplant loaded with bone marrow-derived mesenchymal stem cells combined with platelet-rich plasma in repairing an annulus defect. PLoS One 12(2):e0171500

Yao G (2009) Therapeutic choice of lumbar intervertebral disc herniation. China J Orthop Trauma 22:247–249

Yin Y, Cui Q, Li Y, Irwin N, Fischer D, Harvey AR, Benowitz LI (2003) Macrophage-derived factors stimulate optic nerve regeneration. J Neurosci 23(6):2284–2293

Ying Q, Nichols J, Evans EP, Smith AG (2002) Changing potency by spontaneous fusion. Nature 416:545–547

Zafranskaya MM, Nizhegorodova DB, Aleynikova NE, Kuznetsova TE, Vanslav MI, Ignatovich TV et al (2019) The migration of multipotent mesenchymal stromal cells after systemic and local administration in an experimental model of Parkinson's disease. Ann Clin Exp Neurol 13: 32–40

Zanier ER, Pischiutta F, Riganti L, Marchesi F, Turola E, Fumagalli S, Perego C, Parotto E, Vinci P, Veglianese P, D'Amico G, Verderio C, De Simoni MG (2014) Bone marrow mesenchymal stromal cells drive protective M2 microglia polarization after brain trauma. Neurotherapeutics 11(3):679–695

Zappa Villar MF, López Hanotte J, Pardo J, Morel GR, Mazzolini G, García MG, Reggiani PC (2020) Mesenchymal stem cells therapy improved the Streptozotocin-induced behavioral and hippocampal impairment in rats. Mol Neurobiol 57:600–615

Zeira O, Asiag N, Aralla M, Ghezzi E, Pettinari L, Martinelli L, Zahirpour D, Dumas MP, Lupi D, Scaccia S et al (2015) Adult autologous mesenchymal stem cells for the treatment of suspected non-infectious inflammatory diseases of the canine central nervous system: safety, feasibility and preliminary clinical findings. J Neuroinflammation 12:181

Zhang S, Zou Z, Jiang X, Xu R, Zhang W et al (2008) The therapeutic effects of tyrosine hydroxylase gene transfected hematopoietic stem cells in a rat model of Parkinson's disease. Cell Mol Neurobiol 28:529–543

Zhang R, Liu Y, Yan K, Chen L, Chen XR, Li P, Chen FF, Jiang XD (2013) Anti-inflammatory and immunomodulatory mechanisms of mesenchymal stem cell transplantation in experimental traumatic brain injury. J Neuroinflammation 23(10):106. https://doi.org/10.1186/1742-2094-

Zhang LF, Qi J, Zuo G, Jia P, Shen X, Shao J, Kang H, Yang H, Deng L (2014) Osteoblast-secreted factors promote proliferation and osteogenic differentiation of bone marrow stromal cells via VEGF/heme-oxygenase-1 pathway. PLoS One 9(6):e99946

Zhang C, Gullbrand SE, Schaer TP, Boorman S, Elliott DM, Chen W, Dodge GR, Mauck RL, Malhotra NR, Smith LJ (2021) Combined hydrogel and mesenchymal stem cell therapy for moderate-severity disc degeneration in goats. Tissue Eng Part A 27(1-2):117–128

Mesenchymal Stem Cells Therapeutic Applications in Cardiovascular Disorders

6

Abstract

Diseases of the heart are often serious considering the physiological role played by it. There are many heart diseases (HDs) that lack effective treatment and most of the diseases need lifelong palliative treatment. To ensure actual cardiac tissue regeneration and/or prevent any underlying ailment, regenerative medicine employing stem cells especially mesenchymal stem cells (MSCs) is being evaluated. Under in vitro environments, MSCs potentially differentiate into cardiomyocyte-like cells and secrete various paracrine pro-healing factors. In numerous in vivo experimental studies, MSCs therapeutic benefits too are reported. In veterinary cardiology, a limited literature on therapeutic efficacy of MSCs is available. Small animals like sheep and dog are utilized as translational model animals for humans especially for the ischaemic cardiomyopathies (myocardial infarction), valvular affections and biological pacemaker. However, clinical therapeutics based on MSCs remain to be standardized. Clinically, MSCs are studied for dog cardiomyopathy or myxomatous mitral valve disease (MMVD) without any significant improvement. In cat chronic degenerative valvular heart disease, an improved outcome is reported. The current chapter discusses the literature on MSCs potential therapeutic role for various cardiovascular ailments.

Keywords

Cardiovascular system · Mesenchymal stem cell · Ischaemic cardiomyopathy · Non-ischaemic cardiomyopathy · Valvular disease · Vascular patch

M. B. Gugjoo, *Therapeutic Applications of Mesenchymal Stem Cells in Veterinary Medicine*, https://doi.org/10.1007/978-981-19-3277-9_6

6.1 Introduction

Heart disease (HD) is an affection of any sort to the heart. It may range from a simple physiological condition with a good prognosis like sinus arrhythmia to that of the condition with guarded prognosis such as heart failure (HF). The aetiology behind the condition actually determines the overall prognosis. HD may be congenital or acquired and can arise as the structural (dilatation cardiomyopathy, hypertrophic cardiomyopathy, septal defects, valvular defects), functional (AV block, mitral and aortic valve insufficiency) and/or electrical disorders (atrial flutter, fibrillation, ventricular tachycardia). In animals, HD reporting has increased with the advancement in diagnostic techniques (Gugjoo et al. 2013, 2014) and in small animals like dogs and cats, heart ailments are now reported higher than previously perceived (Darke 1985; Gugjoo et al. 2014). In horses, heart ailments are also not rare but many ailments are mostly mild and well-tolerated. However, in moderate-to-severe heart conditions, life expectancy together with quality of life and safety decreases significantly (Bonagura 2019). Common myocardial diseases in horses are dilatation cardiomyopathy (DCM) and myocarditis (Nout et al. 2003). In cattle, prognosis of HDs remains guarded to poor even in the absence of the signs of heart failure (Baisan et al. 2021).

HDs distribution as per the prevalence varies among human and veterinary species. Ischaemic cardiomyopathies (myocardial infarction) are one of the most common HDs seen in humans contrary to animals where non-ischaemic cardiomyopathies are more commonly reported clinically (Driehuys et al. 1998). Ischaemic cardiomyopathies have a limited rescue period and often lead to sudden collapse. This along with the limited post mortem analysis in animals may fail to show the actual prevalence of ischaemic cardiomyopathies in animals. Non-ischaemic cardiomyopathies may be non-infectious, e.g. DCM or infection like *Parvovirus* and *Trypanosoma*, and *Aphthovirus*-induced myocarditis in dog and cattle, respectively (Aktas et al. 2015; Ford et al. 2017).

Among various HDs, mitral valve disease (MVD) followed in order by DCM is one of the most common ailments reported in dogs. In cats, hypertrophic cardiomyopathy is by far the most common HD reported. However, other HDs are also reported in small animals and around 94.05% diseases are acquired and only 5.95% are congenital (Baisan et al. 2021). In horses, heart failure arises with affections like valvular lesions (Miller and Holmes 1985; Reef et al. 1998), pericarditis (Worth and Reef 1998), neoplasia (Dill et al. 1986), congenital defects (Reef 1995), aorto-cardiac fistulas (Marr et al. 1998), myocarditis (Reef 1999), infiltrative cardiomyopathy secondary to systemic amyloidosis (Nout et al. 2003) and endocarditis (Whitlock et al. 1978). However, congestive heart failure, a rare condition in horses compared to small animals, mostly occurs with mitral regurgitation (Miller and Holmes 1985; Davis et al. 2002). In cattle, limited specification of HDs exists with few reports on particular diseases like traumatic reticulopericarditis (Gugjoo et al. 2021), bacterial endocarditis (Healy 1996), DCM (Nart et al. 2004), ventricular septal defect (Buczinski et al. 2006) and congenital heart disease (Buczinski et al.

2005). Among various species, small animals followed by horse and cattle and at least small ruminants are considered clinically relevant towards the HDs. This might be due to the economic considerations for management of HDs as these small and large ruminants are mostly reared for reproductive and productive traits.

Currently, HDs are mostly managed through palliative treatment that reduces symptoms and prevents disease progression. In order to develop the therapeutics that cures the HDs and actually heals the damaged heart, regenerative medicine involving mesenchymal stem cells (MSCs) is being evaluated. Under regenerative medicine, animals like dogs and sheep have been utilized as model animals for human translational cardiovascular stem cell studies. MSCs evaluation has mainly been focussed on the myocardial infarction in animal models, a condition most commonly seen in humans. A very limited literature on therapeutic application of MSCs in veterinary medicine is available. The current chapter discusses the MSCs characteristics and potentialities against the cardiovascular system affections.

6.2 Why MSCs?

Among various stem cells, MSCs availability from numerous sources, their characteristics and little ethical and teratogenic issues makes them an attractive option for cardiovascular regenerative medicine. Compared to other cell-based therapies, MSCs are immuno-evasive (lack MHC-II and T cell co-stimulatory molecules) and are able to 'home in' to the inflammatory areas (Gugjoo et al. 2019, 2020). MSCs are able to provide therapeutic effects through secretion of paracrine soluble secretory factors/microvesicles and/or through the direct differentiation, although the latter mechanism is currently questionable (Gugjoo et al. 2018, 2019). It is worth mentioning here that MSCs from diverse sources may show variable properties. These cells may show therapeutic properties variably through growth factors, secretion of matrices, angiogenic properties and immunomodulatory activities (Lee et al. 2011; Nakanishi et al. 2011; Carrade et al. 2012; Amable et al. 2014; Cassano et al. 2018; Donders et al. 2018; Gugjoo et al. 2020) (Fig. 6.1).

6.3 MSCs Cardiogenic Studies

6.3.1 Differentiation

MSCs are demonstrated to undergo cardiomyocyte-like differentiation in the presence of various chemicals, growth factors or the scaffolds under ex vivo studies. However, such a cardiogenic differentiation under an in vivo environment remains to be established. MSCs undergo changes in morphology and are able to secrete cardiomyocyte-specific factors upon exposure to specific chemicals/matrices. Fibroblast growth factor (FGF), insulin-like growth factor-1 (IGF-1) and bone morphogenetic protein-2 (BMP-2) under in vitro culture may facilitate dog BM-MSCs to express cardiogenic factors (desmin, antimyocyte enhancer factor-2 and Nkx2.5)

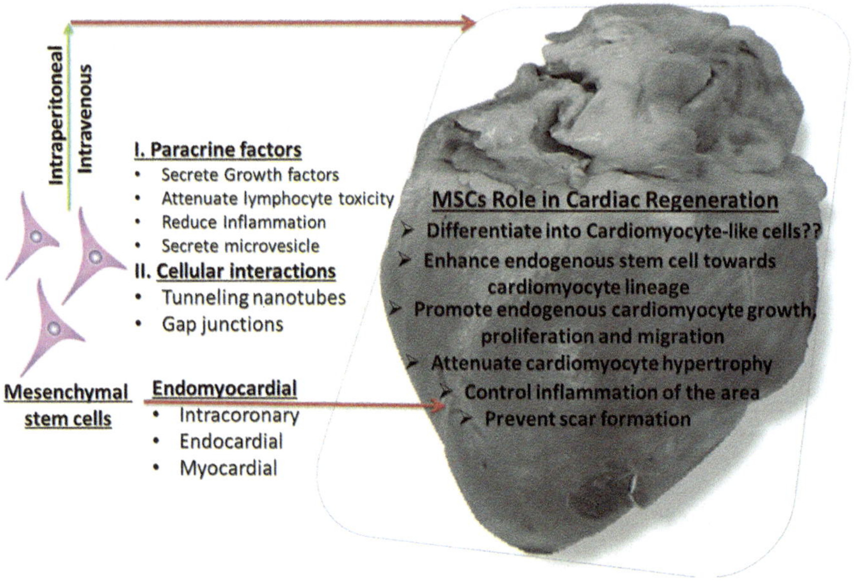

Fig. 6.1 Mesenchymal stem cell (MSCs) route of implantation and their possible therapeutic effects

(Bartunek et al. 2007). MSCs exposure to chemicals like 5-azacytidine (inhibits DNA methyltransferase) stimulates their morphological change from fibroblast shape to rod-like colonies having spindle-like cells. The presence of microfilaments and atrial granules around the nucleus is shown under ultramicroscopy, and specific proteins including the sarcomeric α-actinin, cardiac myosin heavy head and troponin I are localized under immunohistochemistry (Hao et al. 2015).

Chemicals like reversine (enhance cell reprogramming and induce dedifferentiation of the multiple terminally differentiated cells) and 5-azacytidine induce sheep BM-MSCs towards cardiomyocyte-like cell lineage. These differentiated cells express cardiac-related genes like atrial natriuretic peptide (ANP), myosin heavy chain 6 (MYH6) and troponin I, besides the fibroblast growth factor (FGF) signalling genes (Soltani et al. 2016). 5-azacytidine (5-AC) induces cardiogenic differentiation of MSCs through MEK/ERK signalling pathway (Qian et al. 2012). However, cardiomyocyte-like cell differentiation with 5-AC remains to be validated in all the studies and with all the cardiomyocyte features as one of the studies fails to demonstrate spontaneous beating of the cells and changes in cardiogenic expression (Liu et al. 2003). Indirect co-culture of MSCs with the 5-AC and reversine significantly increases the expression of the FGF pathway, MAPK1 and FGFR1. An indirect co-culture of reversine addition followed by 5-AC results in BM-MSCs plasticity and finally to an increased cardiac differentiation. Additionally, connexin 43 protein, an important constituent to develop tunnelling for mitochondrial transfer, is also expressed in the combination of 5-AC and reversine in these sheep cells (Soltani et al. 2016). 5-AC may induce epigenetic changes that activate several silent genes related to cardiac differentiation through inhibition of DNA methylation (Chiu and Blau 1985) or through induction of changes in signalling pathways such as FGF that are involved in cardiac differentiation (Qian et al. 2012).

In order to evaluate the efficiency of various differentiation protocols (5-aza group, myocardial lysates and myocardial induced fluid), Nkx2.5 (inhibits myocyte differentiation and myotube formation) is expressed by MSCs in all these groups. Myocardial lysates may, however, have an edge over other differentiation protocols and a comparable expression may be seen with 5-aza and myocardial-induced fluid. These differentiated cells also express myocardial structural proteins (α-actin, desmin, b-MHC and troponin-I) that undifferentiated MSCs fail to express (Ma et al. 2013). Apart from myocardiocyte-like differentiation, these cells are transfected with some specific genes to function like a pacemaker. Dog BM-MSCs transfected with mouse hyperpolarization-activated cyclic nucleotide-gated channel 4 (mHCN4) gene are able to produce high levels of Cs^+-sensitive current activated in the diastolic potential range, confirming the sinoatrial (SA) node current known as funny current (I_f) (Jun et al. 2012). Apart from the HCN transfection, dog BM-MSCs transfected with Tbx18 (role in embryonic sinoatrial node development) show twelve- and four-fold HCN4 protein and connexin 45 (gap junction protein) expression, respectively. These cells upon co-culture with atrial cells are able to produce I_f that has a similar trend to that of SA node but are limited in density (Xiao et al. 2018). Lack of the exact I_f density could be due to the requirement of additional modulatory elements including the protein–protein and protein–phospholipid interactions, and cytoplasmic factors modulation. Furthermore, the I_f channel is only an immature channel in the early differentiation stage. Therefore, MSCs offer variable physiological actions depending upon the available microenvironment. However, typical features remain to be established, and contrasting reports on actual cardiomyocyte-like differentiation make it less understandable.

6.3.2 Paracrine Actions

MSCs secrete diverse soluble factors or extracellular vesicles that mediate cardioprotective and regenerative effects. MSCs may promote endogenous cardiomyocyte survival, proliferation and migration through Akt-mediated pathway and SDF-1 secretion (Mirotsou et al. 2007; Hatzistergos et al. 2010, 2016; Beigi et al. 2013). SDF-1 may additionally facilitate endogenous c-kit+ cardiac stem and/or progenitor recruitment towards the cardiomyocyte lineage (Loffredo et al. 2011; Suzuki et al. 2011). However, continued SDF-1 secretion may not be a suitable option as antagonism of SDF-1 too favours cardiomyocyte differentiation of the cardiac precursor cells (Chung et al. 2015).

An important criterion in myocardial regeneration is the complete lack of fibrous tissue. MSCs modulate cardiac fibroblast phenotype and their extracellular matrix degradability, thereby reducing scar formation (Mias et al. 2009). Secretion of pro-healing factors like HGF ensures cardioprotection (Shabbir et al. 2010). MSCs exposure to the fibrous tissue (collagen I) dysregulates myo-proliferation and thereby inhibits the expression of inflammatory and growth factor genes and subsequent myoblast formation (Alexakis et al. 2007; De Lisio et al. 2014). This may occur through secretion of Nkx2.5 that inhibits myocyte differentiation and myotube formation (Ma et al. 2013). MSCs might also attenuate cardiomyocyte hypertrophy through normalization of proteins involved in hypertrophy-related signalling. These include phosphatidylinositol 3-kinase-α (PI3K-α), PI3K-γ, extracellular signal-related kinase (ERK) and phosphorylated ERK. Furthermore, MSCs may prevent imbalance of calcium-handling proteins, normalize upregulation of pro-apoptotic protein BCL-2 and reduce apoptosis of cardiomyocytes (Zhao et al. 2012). MSCs attenuate both innate as well as the acquired immunity of the body. These cells attenuate lymphocyte toxicity and reduce the expression of tumour necrosis factor alpha (TNF-α) and interleukin (IL)-6 protein. These cells additionally inhibit nuclear factor kB (NF-kB) while simultaneously stimulating IL-10 expression in the peri-infarct myocardium (Uccelli et al. 2007; Amable et al. 2014; Golpanian et al. 2016). Inflammatory mediators like IL-6 additionally enhance MSCs engraftment to infarcted myocardium and their potential to reduce infarct (Bortolotti et al. 2017). Such processes might ensure regenerated cardiac tissue free of fibrous tissue, a potential cause for arrhythmia.

In addition to the paracrine actions, MSCs intercellular connections through tunnelling nanotubes or gap junctions may exchange mitochondria, microvesicles, macromolecular complexes and exosomes to the resident cells, differentiated and/or stem cells (Bian et al. 2014; Cervio et al. 2015; Sinclair et al. 2016; Zhang et al. 2016a, b). Mitochondrial transfer from MSCs to other cells may occur via intrinsic expression of the mitochondrial Rho GTPase1 (MIRO1). The efficiency of such a transfer increases with tunnelling nanotube formation via TNT-a/NF-kB-signalling pathway activation (Ahmad et al. 2014). MSC-derived microvesicles (containing functional proteins, miRNAs, mRNAs) may produce neoangiogenesis and functional improvement in infarcted hearts (having hypoxia) (Bian et al. 2014; Cervio et al. 2015). However, the detailed account of these results is lacking in veterinary species and are mostly reported from rodent studies.

6.4 In Vivo Studies

To ensure safe and effective use of MSCs in ischaemic HDs like myocardial infarction, indirect routes like intravenous and intraperitoneal or direct endomyocardial and intracoronary transplantations are being evaluated (Fig. 6.1). Such transplantations are evaluated in adults or intrauterine models. Intravenously transplanted xenogeneic AD-MSCs engraft to the lesions developed through the radiofrequency catheter ablation (RFCA). These cells differentiate into cardiomyocyte-like cells expressing anti-α-actinin, anti-cardiac troponin-I, anti-connexin 43 and anti-VEGFR-2 (Kim et al. 2011). Achievement of effective cell concentration with such transplantation at the lesion site remains to be determined. Intracoronary (IC) infusion of MSCs is contentious due to their physical properties that pose threat of occlusion of the coronary microvasculature. IC infusion though appears safe but has a potential to develop ischaemia. IC transplantation of autologous MSCs or mixture of MSCs/mononuclear cells (MNCs) ensures their engraftment up to 2 weeks with minimal adverse effect on the coronary blood perfusion of dogs (Jin et al. 2005; Minguell et al. 2010). Only minor ST-T abnormalities are recorded without any evidence of injury or infarction up to 1 week. Even cardiac tissue mostly remains histologically normal except a single dog that has discrete inflammatory changes, with mild mononuclear infiltration at 1 month (Minguell et al. 2010). Further, IC transplantation of allogeneic BM-MSCs is demonstrated as safe, feasible, and markedly effective in sheep-induced acute myocardial infarction (Zhao et al. 2012; Houtgraaf et al. 2013). Contrary to these reports, intracoronary arterial injection of autologous MSCs into dogs leads to the acute myocardial ischaemia and subacute myocardial microinfarction (Vulliet et al. 2004). This complication has been attributed to the larger size of the early passage dog MSCs, although such a feature does not seem to be the sole reason.

In ischaemic injury models, trans-endocardial cell delivery appears superior to that of the IC infusion. The dog cells are well retained with an overall improved functional outcome (Perin et al. 2008). For direct endomyocardial transplantation to the viable or non-viable tissue border zone, left ventricular assisted device (LVAD) under electromechanical mapping or 3D echo may be followed (Cheng et al. 2013; Zheng et al. 2013). To further improve the delivery without any adverse effect, retrograde coronary venous infusion may alternatively be utilized (Wang et al. 2014, 2015). Retrograde transfusion ensures cell transplantation through non-diseased coronary veins, which parallel the arterial system, for selective delivery to ischemic myocardium (Jain et al. 2006). For HDs other than ischaemic conditions, MSCs are mostly transplanted through IC (Sousa et al. 2011; Pogue et al. 2013) or intravenous (Petchdee and Sompeewong 2016; Yang et al. 2021) routes under clinical settings.

MSCs safe and result-oriented transplantation in healthy and diseased hearts may not be affected solely by the route; other factors including the cellular concentration and rate of their transplantation may have equal role to play. Single-dose IC transplantation of higher cell concentration (75 million) develops myocardial infarction than the lower cell concentration (25 million) in acute MI models (Dixon et al. 2009; Hamamoto et al. 2009). Further, MSCs total dose and infusion rate may have bearing on the heart functionality and anatomy. Healthy heart may safely be infused with 1.25 million MSCs per min, while a dose of 2.5 million per min may prove catastrophic. It is worth mentioning that higher doses may be infused at lower rates as compared to the lower doses. Up to 25 million total doses, one million cells per minute appears safe and for a total dose of 40 million, 0.5 million per min appears safe, beyond which the blood flow to heart might become sluggish (Houtgraaf et al. 2013). Furthermore, intracoronary transplantation protocol and the cell preparation may also have bearing on the outcome (Grieve et al. 2010). In acute MI, it may not be advisable to use extremely higher doses intramyocardially as the cells show ceiling effect and might actually injure the heart muscle (Dixon et al. 2009; Hamamoto et al. 2009). In contrast, chronic MI may demand higher cell concentration although rate and route have to be appropriate (Cheng et al. 2013).

Considering the appropriate route of MSCs transplantation for congenital or developmental heart diseases, currently no effective route seems to be appropriate. AD-MSCs or BM-MSCs transplanted intramyocardially survive at the myocardial infarction site, while the cells transplanted intraperitoneally may (Airey et al. 2004) or may not reach to the site (Emmert et al. 2013). Intraperitoneal transplantation of MSCs at 55–62 days of gestation in sheep may reach to the heart and even differentiate into the Purkinje fibres (Airey et al. 2004). To enhance cellular distribution properly to the foetal heart, an aortic flow-stop procedure shows favourable distribution, although the foetal viability rate post cell transplantation remains remarkably low (Margaryan et al. 2020). Thus, the in utero standard route of MSCs transplantation remains to be established and as such further studies and advancements in this area are desired.

6.4.1 Cardiomyopathy

Cardiomyopathies are ischaemic or non-ischaemic depending upon the cause. In the former, blood flow is compromised, while in the latter, causes other than compromise in blood flow exist.

6.4.1.1 Ischaemic Cardiomyopathy

Transplantation of MSCs in myocardial infarction (MI) models of sheep, dogs and goats, irrespective of the source (allogeneic/xenogeneic/autologous; bone marrow or blood derived), is associated with improvement. An increase in vascularity in and around the infarct area occurs. MSCs transplantation prevents compensatory cardiomyocyte hypertrophy, inhibit fibrosis through remodelling and contractile strain alteration. Such actions thereby increases the perfusion and reduces infarct size and finally improves the cardiac functional parameters, both global as well as the regional. However, none of the studies could actually confirm MSCs typical cardiomyocyte-like differentiation and myocardial regeneration within the infarct zone (Fig. 6.1 and Table 6.1). One of the studies shows cardiomyocyte-like cells of variable developmental stages infiltrate in the border zone of MSC-transplanted myocardium (Rabbani et al. 2017). Besides, imbalance of the calcium-handling proteins such as sarcoplasmic reticulum Ca^{2+} adenosine triphosphatase (SERCA2a), phospholamban (PLB) and sodium/calcium exchanger type 1 (NCX-1) induced by MI may also be prevented by MSCs (Zhao et al. 2012). Compensatory cardiomyocyte hypertrophy (increase cardiac failure) attenuation may occur through the normalization of the various hypertrophy-related signalling proteins like phosphatidylinositol 3-kinase α (PI3Kα), PI3Kγ, ERK and phosphorylated ERK in the adjacent zone of MI (Zhao et al. 2012). Thus, a ray of hope for cardiac regeneration with MSCs appears.

6.4.1.1.1 Sheep Studies

Sheep is often used as a translational model animal for human cardiovascular surgery (DiVincenti et al. 2014). As such, numerous sheep MSCs studies on cardiovascular regeneration are conducted. These mainly include ischaemic models like myocardial

Table 6.1 MSCs preclinical experimental studies on myocardial infarction in veterinary species

Study type/number of animals/study period	Biomaterial used	Evaluation criteria	Overall result	References
Sheep acute myocardial infarction (Ischaemic model)				
Randomized placebo-controlled comparative study/$n = 46/8$ weeks	S'tro-3+ BM-MSCs (25×10^6; 75×10^6; 225×10^6; 450×10^6)	Echocardiography, histology	Improved outcome (ceiling effect of cell concentration). Higher cell dose may lead to MI	Dixon et al. (2009)
Randomized placebo-controlled comparative study/$n = 46/8$ weeks	Stro-3+ BM-MSCs (25×10^6; 75×10^6; 225×10^6; 450×10^6)	Echocardiography, immune-histochemistry	Improved outcome (ceiling effect of cell concentration). Higher cell dose may lead to MI	Hamamoto et al. (2009)
Randomized placebo-controlled study/$n = 10/12$ weeks	Allogeneic BM-MSCs [2×10^8 (6 injections each containing 3.33×10^7)]	Histology, Western blot	Improved outcome	Zhao et al. (2012)
Randomized placebo-controlled comparative study/$n = 68/12$ weeks	BM-MSCs (12.5×10^6; 25×10^6; 37.5×10^6)	Invasive hemodynamics, echocardiography	Improved outcome	Houtgraaf et al. (2013)
Randomized placebo-controlled study/$n = 10/4$ weeks	Human MSCs (1×10^6 per kg b. wt.)	Echocardiography and perfusion scintigraphy	Improved outcome	Dayan et al. (2016)
Randomized placebo-controlled comparative study/$n = 18/8$ weeks	Human HIF-1α transfected allogeneic BM-MSCs, simple BM-MSCs (cells @ 2×10^7)	Cardiac magnetic resonance, echocardiography, histology	Improved outcome (transfected cells better)	Hnatiuk et al. (2016)
Randomized placebo-controlled comparative study/$n = 18/8$ weeks	BM-MSCs and endothelial cells	Echocardiography, electron microscopy and immune-histochemistry	Improved outcome	Rabbani et al. (2017)
Sheep subacute myocardial infarction (Ischaemic model)				
Randomized placebo-controlled comparative study/$n = 32/4$ weeks	VEGF transfected BM-MSCs, BM-MSCs, VEGF (cells @ 2×10^7)	Magnetic resonance, echocardiography	Improved outcome (VEGF transfected cells better)	Locatelli et al. (2015)

(continued)

Table 6.1 (continued)

Study type/number of animals/study period	Biomaterial used	Evaluation criteria	Overall result	References
Sheep chronic myocardial infarction (Ischaemic model)				
Randomized placebo-controlled comparative study/$n = 39$/8 weeks	Allogeneic MSCs (25×10^6; 75×10^6; 225×10^6)	Histological, echocardiography	Improved outcome better with higher cell concentration)	Cheng et al. (2013)
Sheep non-ischaemic cardiomyopathy				
Randomized placebo-controlled study/$n = 15$/8 weeks	Allogeneic BM-MSCs [$0.5 \pm 0.2 \times 10^6$ cells; 20 injections (109×10^6 total dose)]	Cardiac magnetic resonance, echocardiography, histology	Improved outcome	Psaltis et al. (2010)
Cat non-ischaemic cardiomyopathy				
Randomized placebo-controlled comparative study/$n = 39$/66 days	Autologous cardiac stem cells (CSCs), MSCs (cells @ 1×10^6)	Echocardiography, invasive hemodynamics, immunohistochemistry	Improved outcome (slightly better with MSCs)	Taghavi et al. (2015)
Dog acute myocardial infarction (Ischaemic model)				
Randomized placebo-controlled comparative study/$n = 24$/28 days	BM-MSCs (2×10^7) and ultrasound-guided microbubble destruction (UTMD)	Echocardiography, histology	Improved outcome (better with combination of MSCs and UTMD)	Chang et al. (2015)
Randomized placebo-controlled study/$n = 12$/10 weeks	BM-MSCs (1×10^7)	SPECT, histopathology, immunohistochemistry	Improved outcome	Hao et al. (2015)
Randomized placebo-controlled comparative study/$n = 28$/40 days	MSCs (1×10^7), bFGF (40 mL of 200 mg/L)	Speckle-tracking echocardiography, HP	Improved outcome (better with combination of MSCs and bFGF)	Sun et al. (2016)
Dog subacute myocardial infarction (Ischaemic model)				
Randomized placebo-controlled study/$n = 16$/21 days	MSCs (100×10^6)	Histology, immunohistochemistry	Improved outcome	Vela et al. (2009)

Randomized placebo-controlled comparative study/n = 18/4 weeks	BM-MSCs [(1×10^8 cells), bFGF (200 ng/mL)]	ECHO, histopathology, immunohistochemistry	Improved outcome (better with combination of MSCs and bFGF)	Wang et al. (2015)
Dog chronic myocardial infarction (Ischaemic model)				
Randomized placebo-controlled comparative study/n = 12/30 days	Allogeneic BM-MScs (100×10^6/10 mL saline)	Echo, histopathology, immunohistochemistry	Improved outcome	Silva et al. (2005)
Randomized comparative study/n = 30/12 weeks	Autologous BM-MSCs ($168 \pm 24 \times 10^6$), cardiac differentiated BM-MSCs ($147 \pm 96 \times 10^6$)	Immuno-histochemistry, echocardiography	Improved outcome (better with differentiated cells)	Bartunek et al. (2007)
Randomized placebo-controlled comparative study/n = 24/17 weeks	BM-MSCs ($227 \pm 32 \times 10^6$) and BMNCs ($232 \pm 40 \times 10^6$)	Echocardiography, MRI, immunohistochemistry, macromorphometry, PCR	Improved outcome with BM-MNCs than MSCs	Mathieu et al. (2009)

MSCs mesenchymal stem cells, *AD* adipose tissue derived, *BM* bone marrow derived, *HIF-1 α* hypoxia inducible factor 1-α, *VEGF* vascular endothelial growth factor, *bFGF* basic fibroblast growth factor, *n* total number of animals in all the groups of the study; Positive outcome: indicates improved outcome with MSC transplantation even if only one parameter improves. Study period: Indicates maximum time period for which study is conducted infarction (MI), although non-ischaemic cardiomyopathies like DCM have also been made (Table 6.1 and Fig. 6.2).

Sheep MI models are mostly created through ligation of coronary arteries. Based on the cell transplantation time, these models are categorized as acute, subacute and chronic. In acute models, cell transplantation is made immediately or within a few hours of the infarct creation, while in subacute and chronic models, cell transplantation is made after a week and around 4 weeks post infarction, respectively. In all the models, either allogeneic or xenogeneic (human) cells are transplanted. The cells are implanted either directly to the defect and surrounding area or through the intravenous route. The animal number ranges from $n = 10$ (Zhao et al. 2012) to $n = 68$ (Houtgraaf et al. 2013), although total animal number may not be important, rather number of animals in the particular study group. MSCs therapeutic effects are evaluated at least for 1 month (Dayan et al. 2016) to the maximum of 3 months (Zhao et al. 2012). Evaluation of cardiac repair has variably been made through cardiac magnetic resonance, echocardiography, histology, immunohistochemistry, invasive haemodynamic and western blotting (WB).

As mentioned above, MSCs tend to show beneficial effects but fall short of the typical cardiac tissue regeneration. In order to evaluate the effect of diverse factors that govern MSCs effects, various comparative and modified studies are conducted. In acute MI models, cell doses (@ 25 million, 75 million, 225 million and 450 million) are transplanted intramyocardially (borderzone 1 h later). The lowest cell dose (25 million) appears more effective in improving the myocardial anatomy and function at 8 weeks (Dixon et al. 2009; Hamamoto et al. 2009). Low-dose MSCs are more effective in altering the collagen dynamics through matrix metalloproteinases (MMPs) and tissue inhibitor MMPs (Dixon et al. 2009) and in enhancing the vascularity in the border zone (Hamamoto et al. 2009). Reduced efficacy upon borderzone myocardial transplantation of the larger doses may arise due to their competition for limited nutrients or the cells may inflict an inflammatory environment that reduces their activity (Hamamoto et al. 2009). As MSCs are depleted by the eighth week post transplantation, MSCs engraftment remains irrelevant for their concentration, and therapeutic benefits might occur mostly through the paracrine actions. Further, higher cell transplanted MIs show increased myocyte cross-sectional area (indicating hypertrophy), while the degree of border zone in lower cell transplanted MIs is unaffected (Dixon et al. 2009). In chronic Mis, higher dose (225 million) transplantations increase arteriole density and wall thickening in both the infarct as well as the border zones. Such increase in vascularity is correlated with an incremental increase in ejection fraction as compared to the low doses (25 and 75 million) and control (Cheng et al. 2013). In chronic MI, higher cell doses are required to activate cellular pathways and enhance angiogenesis.

To improve MSCs cardiac regeneration potential, transfection of MSCs with suitable genes is also done. One of the acute MI studies demonstrates that transplantation of allogeneic human hypoxia inducible factor 1-α (HIF1-α) transfected BM-MSCs (@ 20 million) better reduces the infarct volume and improves angio/arteriogenesis as compared to the similar dose of normal BM-MSCs. The duration of engraftment of these transfected cells also increases in comparison to the

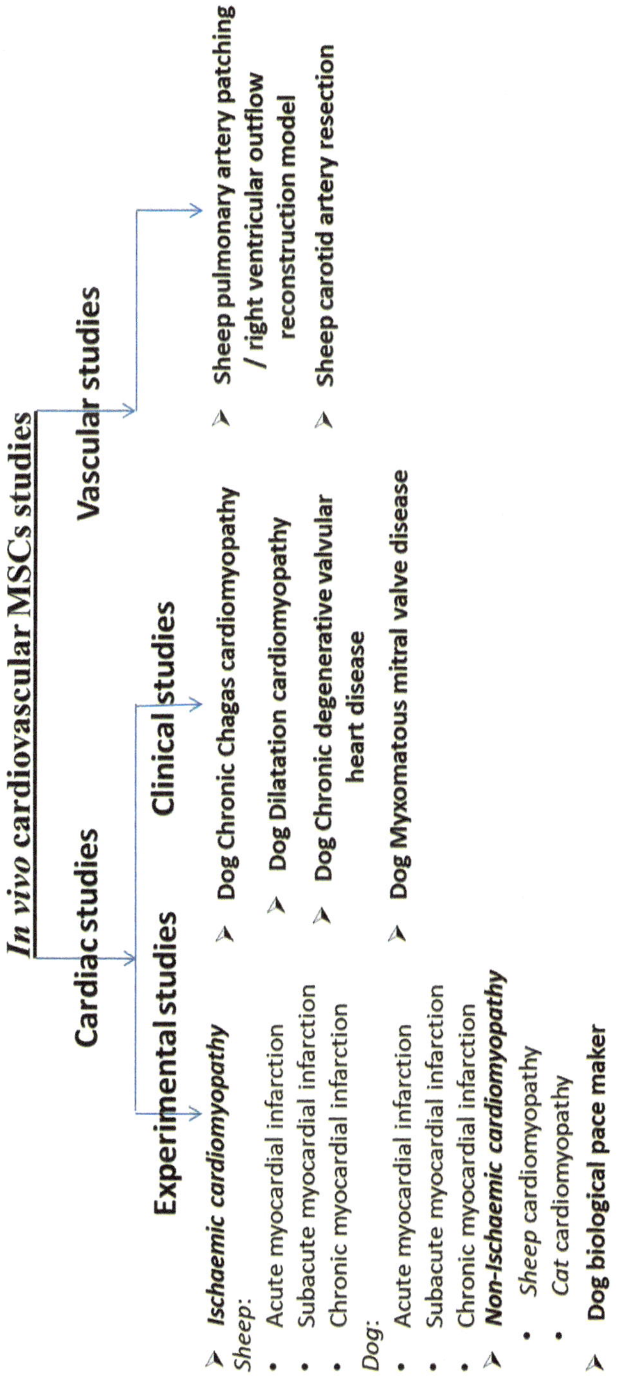

Fig. 6.2 Mesenchymal stem cell potential therapeutic effects on various in vivo experimental and clinical cardiogenic studies

non-transfected MSCs, although significantly lower at day 30th and 60th than at day 1 (Hnatiuk et al. 2016). This might be achieved through overexpression of pro-survival genes (ANGP-1 and EPO) (Cerrada et al. 2013). Similarly in subacute MI model, vascular endothelial growth factor (VEGF)-transfected BM-MSCs (20 million) leads to the reduction in infarct size at around 31%. Myocardial salvage is mostly attributed to the level of sub-endocardium with ejection fraction to recover to the baseline value (Locatelli et al. 2015). HIF1-α and VEGF (a downstream gene of HIF1-α) are associated with the increase in the vascularity and induction of adult cardiomyocyte mitosis and thus, may provide additional therapeutic benefits. The therapeutic potential of endothelial cells against that of the MSCs in acute MI model (diagonal coronary artery ligation) is also made. Transplantation of both the cell types effectuates comparable cardiac improvement depicted by increase in vascular density and subsequent functional improvement. It is worth mentioning here that cardiomyocytes in either group are in variable stages of development in the border zone (Rabbani et al. 2017). As MSCs endothelial differentiation is considered to prevent an increase in the ischemic area, direct endothelial cell transplantation may help to improve the outcome. However, MSCs may additionally offer immunomodulatory/anti-inflammatory effects, normalize calcium uptake, prevent compensatory cardiomyocyte hypertrophy and above all can be used as allogeneic therapy.

6.4.1.1.2 Dog Studies

Dogs are also one of the commonly utilized translational model animals for the human cardiovascular system. Like sheep, dog MI models are created with the blockade of coronary arteries and divided into acute, subacute and chronic MI as per the delay in cell transplantation (Table 6.1 and Fig. 6.2). The cell implantation is made either as direct intramyocardial (Silva et al. 2005; Bartunek et al. 2007; Mathieu et al. 2009) or intracoronary (Hao et al. 2015; Chang et al. 2015; Sun et al. 2016). The subject number ranges from $n = 10$ (Silva et al. 2005; Hao et al. 2015) to $n = 30$ (Bartunek et al. 2007). Therapeutic effects are evaluated for a period of 28 days (Chang et al. 2015) to 17 weeks (Mathieu et al. 2009) through variable diagnostic techniques like histopathology, immunohistochemistry, speckle-tracking echocardiography, magnetic resonance imaging (MRI), immunohistochemistry, macro-morphometry and polymerase chain reaction (PCR).

In acute MI models, two of the studies transplant MSCs directly intracoronary (Chang et al. 2015; Hao et al. 2015) while one makes coronary venous retroperfusion (Sun et al. 2016). The cellular concentration is mostly under 20 million (Chang et al. 2015; Hao et al. 2015; Sun et al. 2016) while in case of subacute and chronic models, higher cell doses (> 100 million) are transplanted. All these studies are conducted as simple randomized placebo-controlled or as randomized placebo-controlled comparative studies.

In a chronic MI model, allogeneic MSCs transplanted intramyocardially 30 days later transdifferentiate into smooth muscle cells (positive for smooth muscle actin) and MSCs located in the luminal face of the endothelium of several vessels transdifferentiate into endothelial cells (expressed factor VIII). This is correlated with the higher capillary density in the anterolateral wall of MSC-treated MI hearts. The cell-treated cardia has significantly better cardiac functions (left ventricular ejection fraction) (Silva et al. 2005). Considering the small animal number and ability of the dog heart to develop collateral circulation, it may not be genuine to conclude the results decisively. There are other studies that fail to demonstrate such positive results and even demonstrate MSCs potential to develop fibroblasts rather than cardiomyocytes (Wang et al. 2000) and lead to myocardial calcifications (Yoon et al. 2004).

To waive off such lacunas and potential threats, MSCs preconditioning with the myocardial factors is also evaluated. A study that transplants preconditioned MSCs 8 weeks after ligation demonstrates myocyte-like differentiation of specified MSCs by second week post cell transplantation. Cardiac myocyte-specific markers expressed are observed till the end of the study (12-week follow-up). Transplanted cells form new islands of cardiac-like cells within the infarcted tissue. These cells have striations, parallel orientation and expression of connexin 43 similar to the myocytes. Furthermore, no calcification is reported. Despite such promising results, the functional improvement is not that significant. Furthermore, a caspase-8 inhibitor added to prevent apoptotic cell death might additionally impart some beneficial effects (Bartunek et al. 2007). Similarly, another study that makes retrograde coronary venous infusion MSCs along with bFGF demonstrates enhanced neovascularization of the infarcted myocardium and inhibited apoptosis. However, there is only a slight strain improvement without a substantial increase in global cardiac functions (Sun et al. 2016). In the subacute MI model, retrograde coronary venous infusion of bFGF augments engraftment and differentiation capacity of transplanted MSCs. The engrafted MSCs (dual labelled with EGFP and cardiac-specific markers) primarily differentiate into smooth muscle cells and endothelial cells (co-expression of EGFP with FVIII and α-SMA). This combination recovers cardiac function and prevents compensatory remodelling (Wang et al. 2015).

In comparative studies, MSCs transplantation is supplemented with ultrasound-targeted microbubble destruction (UTMD) (Chang et al. 2015) or basic fibroblast growth factor (bFGF) (Wang et al. 2015; Sun et al. 2016). Although simple cell transplantation improves cardiac function and reduces infarct size, supplementary additions enhance further improvement. With UTMD, BM-MSCs densely deposit in the coronary artery and infarcted area with maximum number of α-actin positive cells (Chang et al. 2015). UTMD effectuates microvascular ruptures and widening of the endothelial interspaces and thus might enhance permeability (Wang et al. 2013). It can additionally stimulate secretion of endogenous vascular growth factors and thus might promote angiogenesis (Yoshida et al. 2005). MSCs transplantation followed by UTMD markedly improves myocardial perfusion that in-turn improves heart function and the wall motion score index (Ling et al. 2013). bFGF widely distributed in higher concentrations in the heart is considered to enhance neovascularization in case of myocardial infarction (Scheinowitz et al. 1997). Comparing the therapeutic effect of cardiac mononuclear cells with that of the BM-MSCs (transplanted 11 weeks later), the former cell types are more effective in chronic MI models. BM-MNCs lead to the greater improvement in cardiac contractility and regional left ventricular systolic function, reduction of infarct size and abrogated adverse remodelling as compared to the autologous cultured MSCs (Mathieu et al. 2009). BM-MNCs might improve neovascularization as the cellular fraction carries larger amounts of CD14+ cells (Urbich and Dimmeler 2004). Besides, the cellular crosstalk between multiple cells present in the MNCs fraction might also lead to the better results. Such findings need further validation taking into consideration type of model, cell dose and route of transplantation.

A single study on goat left anterior descending coronary artery ligation model, autologous goat scaffold (small intestinal submucosal film) loaded with BM-MSCs is evaluated. Transplantation of such a construct prevents left ventricular chamber dilatation, improves collateral perfusion besides the myocardial contractility and cardiac function (Liao et al. 2006).

MSCs appear promising in MI but fall short of typical myocardial tissue regeneration. Overall MSCs myocardial regeneration potential improves with the gene transfection, although currently evidence is limited enough to be conclusive on their definitive therapeutic efficacy and safety. There are numerous factors that affect their therapeutics. Cellular dose may be regulated as per the duration of therapy post infarction or the size of the infarct. Besides the route and technique of implantation, available microenvironment determined by time and extent of infarction needs to be evaluated. Since MSCs characteristics are affected by the available microenvironment, it may not be feasible to transplant the cells in the infarct area. Simultaneous incorporation of the scaffolds, however, remains to be studied in detail. Besides, the main impediment in their definitive use in clinical settings is their potential to develop arrhythmogenic foci.

6.4.1.2 Non-Ischaemic Cardiomyopathy

A very limited literature is available over non-ischaemic experimental animal studies (Table 6.1). A single sheep study available of a non-ischaemic model of cardiomyopathy is induced with anthracycline. This drug induces injury to cardiomyocytes, small vessels and Purkinje fibre leading to increased myocardial fibrosis. This injury is exhibited by moderate–severe left ventricular dysfunction (Psaltis et al. 2008). In such a model, trans-endocardially (catheter-based) injected allogeneic BM-MSCs (20 injections of 5.5 million cells, total dose @ 109 million, 10 weeks later) attenuates the end-systolic dilation and stabilizes ejection fraction. Affected cardiac tissue has less fibrosis with an increased density of karyokinetic cardiomyocytes and myocardial arterioles as compared to the control. As the cells are sparsely present in the myocardium by fourth week, their potential therapeutic benefits might occur through paracrine actions. Notable paracrine actions might involve the protection of the cardiomyocyte including those undergoing karyokinesis ($Ki67^+$) and stimulation of cardiac progenitor cells (Psaltis et al. 2010).

In cats, non-ischaemic cardiomyopathy (Table 6.1 and Fig. 6.2) is induced with L-isoproterenol. This injury depresses systolic and diastolic function concomitant with the profound atrial and ventricular dilatation. Also, a significant myocyte hypertrophy and replacement fibrosis occurs after injury (Angert et al. 2011). Intracoronary transplantation of autologous MSCs (@ one million) leads to the slightly better functional improvement in comparison to the similar dose of cardiac stem cells. In relation to prevention of pathological cardiac remodelling, comparable results are demonstrated. MSCs might have effectuated repair through their paracrine action over myocardial cells and or progenitor cells. Furthermore, progenitor cell proliferation has been greater in the atria than the ventricle, although without any known explanation. However, FACS analysis of the cells remains to be done (Taghavi et al. 2015).

6.4.1.2.1 Clinical Studies

6.4.1.2.1.1 Chronic Chagas Cardiomyopathy

In a single clinical study on chronic Chagas cardiomyopathy ($n = 5$), MSCs transplanted through the right and left coronary arteries (@100 million) improve the cardiac function in $n = 3$ although $n = 1$ dies. At 3 months, peak velocity of aortic flow is achieved while at 4-month pre-ejection period, isovolumic relaxation time, and Tei index (evaluates systolic and diastolic performances) of myocardial performance are significantly reduced (Sousa et al. 2011). The cause of single death in the study may be due to the supraventricular paroxysmal tachycardia that could have developed either due to the disease (fibrosis foci in the Chagas disease) or due to the lack of electromechanical integration of the transplanted MSCs with the recipient myocardium (Menasché 2009). Although no changes in electrocardiography and blood pressure are recorded, structural or functional changes cannot be ruled out considering the clinical condition. Further, continuous ECG (Holter) monitoring is not conducted to confirm ECG changes, if any. Also, myocardial cellular distribution is not known, and regenerated cardiac histological data to confirm actual tissue regeneration remains to be conducted. This study additionally has limited animal numbers and lacks control groups. In general, MSCs may improve cardiac function in Chagas disease cardiomyopathy but further studies are desired to confirm safety, feasibility, utility and efficacy of the cells (Table 6.2 and Fig. 6.2).

6.4.1.2.1.2 Dilatation Cardiomyopathy (DCM)

DCM may be genetic or acquired. In dog, oligo-genetic mode of inheritance together with the three single nucleotide polymorphisms (SNPs) explains most of the cases of the DCM (Vollmar et al. 2019). In cattle, genetic causes like Holstein gene in Holstein Friesian and curly hair coat in polled Hereford are considered to be one of the reasons for DCM. Besides, ingestion of feed (monensin, lasalocid, salinomycin) or deficiency of Vitamin E and selenium, copper or excessive molybdenum (Reef 1990; Boemo et al. 1991; Nart et al. 2004; Reef and Mcguirk 2015) may promote DCM.

There are no experimental animal studies over DCM, and MSCs therapeutic potential remains to be evaluated. A single dog pilot study ($n = 15$) is available that evaluates the safety of retrograde coronary venous delivery of MSCs (Table 6.2 and Fig. 6.2). Stromal-derived factor-1 transduced AD-MSCs (@ ten million) transplanted through retrograde coronary venous route appear safe but without any improvement in left ventricular size and function up to 2-month period. In this study,

Table 6.2 MSCs clinical studies on cardiomyopathies and valvular diseases in dog

Clinical condition	Study type/no. of animals/duration of study	Biomaterial	Evaluation methods	Result	References
Chronic Chagas cardiomyopathy	Case report/$n = 5$/ 6 months	Autologous BM-MSCs (100×10^6 implantation into right and left coronary artery)	Electrocardiography and echocardiography	Improved outcome	Sousa et al. (2011)
Dilated cardiomyopathy	Pilot study/$n = 15$/ 2 months	Stromal-derived factor-1 transduced AD-MSCs (10×10^6 transplanted retrograde coronary venous delivery)	Electrocardiograms (ECG), echocardiograms and Holter monitoring	No significant improvement	Pogue et al. (2013)
Chronic degenerative valvular heart disease	Comparative case controlled clinical study/$n = 20$/60 days	Puppy deciduous teeth stem cells (pDSCs) (IV 1×10^6 cells of pDSC per kg of body weight at day 0 and 14 after Dx)	Electrocardiography, complete transthoracic echocardiography, thoracic radiography, and blood pressure	Improved outcome	Petchdee and Sompeewong (2016)
Myxomatous mitral valve disease (MMVD)	Double-blind, placebo-controlled clinical trial/ $n = 10$/6 months	Allogeneic Wharton jelly–derived MSCs (2×10^6 cells/kg, IV), autologous serum for 3 injections 3 weeks apart	Echocardiographic data, ECG findings, serum cardiac biomarker concentrations, CBC, and serum biochemical analysis	No significant improvement	Yang et al. (2021)

MSCs mesenchymal stem cells, *AD* adipose tissue derived, *BM* bone marrow derived, *n* total number of animals in all the groups of the study; Positive outcome: indicates improved outcome with MSC transplantation even if only one parameter improves. Study period: Indicates maximum time period for which study is conducted direct intramyocardial transplantation of MSCs is avoided to prevent any kind of malignant arrhythmia, and thus intracoronary venous delivery is opted that decreases anaesthetization time as compared to coronary arterial transplantation (Pogue et al. 2013). However, as the control group is lacking, it becomes difficult to compare results. As DCM has multiple aetiologies, correlation of aetiology to MSCs therapeutic mechanisms may be made and accordingly MSCs are utilized. Besides, MSCs transfected with the desired genes, transplanted at specific dose, frequency and route, and additionally provided with growth factors may also be evaluated for possible therapeutic benefits. Furthermore, the long-term effects against the control may be studied.

6.4.1.2.1.3 Valvular Diseases

Myxomatous mitral valvular disease (MMVD) is the most common acquired cardiac disease in dogs to cause CHF. Around two-third of the cardiac failure cases are of congestive heart failure (CHF) in dogs (Parker and Kilroy-Glynn 2012). MMV degeneration leads to mitral insufficiency that in turn causes cardiac enlargement. In later stages, contractile dysfunction may set in due to the increase in inflammatory mediators and subsequent fibrosis (Borgarelli and Buchanan 2012; Dillon et al. 2012). Finally, volume overload causes the CHF. MMVD is being managed to delay CHF onset through medicine like pimobendan that enhances survival by around 5 months (Boswood et al. 2016). Surgical repair of mitral valve and chordae tendineae may reverse changes, but lack of expertise and cost of operation makes it least opted (Uechi 2012). As such, regenerative medicine employing MSCs is being evaluated.

In MMVD, no experimental study could locate that utilizes MSCs. Two available clinical studies are discussed here (Table 6.2). A placebo-controlled study fails to demonstrate any significant difference in the survival time, time to the first adjustment of diuretic drug dosage, echocardiographic variables and serum cardiac biomarker concentrations with intravenous allogeneic WJ-MSCs against the placebo control up to 6-month period. Post cell transplantation, however, decreases in blood lymphocyte, monocyte and eosinophil counts are recorded and might be due to transplanted MSCs anti-inflammatory effects (Yang et al. 2021). This study transplants single dose of the cells with no relevant data on MSCs engraftment and histological analysis of the affected heart. On the other hand, two intravenous doses of deciduous teeth-derived MSCs (cells are not characterized) at fortnight intervals show an increase in the left ventricular ejection fraction at day 30th in dogs. This left ventricular ejection fraction improvement, however, wanes at 60th day. This study did not consider detailed parameters as taken in the above study (Petchdee and Sompeewong 2016). Failure of early MSCs effects to make it to the long-term benefits is also reported in human study (Fu and Chen 2020). As there is variability in design, cell doses and concentration, further uniform studies may be conducted to get some conclusive results. Further, MSCs may be transfected or some growth factors may be added. Animal number and duration of study have to be on a higher side to compare and evaluate long-term benefits, if any.

6.4.2 Cardiac Pacemaker

In animals, pacemaker implantation may not be that common procedure but burgeoning ethical treatment of animals and pet animal culture ensures the paradigm shift towards improved animal care. Dog and horse clinical reports establish the need for pacemakers (Cunningham and Rush 2007; Johnson et al. 2007; Petchdee et al. 2018). Electronic pacemakers, used for complete heart block or sinus node dysfunction, pose challenges of biologic responsiveness compared to that of the native tissue, autonomic modulation, their battery life, electronic and magnetic interference and above all the need of intracardiac catheterization. As such biologic pacemakers are being evaluated (Jun et al. 2012). Among various cell-based biological pacemakers, MSCs appear promising due to their characteristic properties of enhancing the cardiac sympathetic hyperinnervation, VEGF-induced angiogenesis and ability to differentiate into the cardiomyocytes (Kim et al. 2010). Besides, their lentiviral vector transfection with hyperpolarization-activated cyclic nucleotide-gated cation channels (HCN) may additionally favour funny current (I_f) generation (Di Francesco 2010). As MSCs transfection allows long-term integration, thus, HCN may be expressed for a longer period suitable in chronic conditions like sick sinus syndrome (Seppen et al. 2001). Among various HCN isoforms, HCN2 (mainly in the ventricle) has widely been evaluated with HCN4 predominant in the SA node (Lu et al. 2013).

A few experimental studies on dogs as pacemaker model animals show promising results with simple or transfected MSCs, although not without apprehensions of fatal outcomes like arrhythmia (Table 6.3 and Fig. 6.2). To evaluate MSCs role as biological pacemaker after creating atrioventricular block, simple cells (Kim et al. 2010; Hou et al. 2011) or HCN-transfected MSCs (Plotnikov et al. 2007; Jun et al. 2012; Lu et al. 2013) are used. Intracardiac transplantation of simple xenogeneic (human) MSCs enhances sympathetic nerve responsiveness (increased nerve growth factor beta) and angiogenesis (increased VEGF), although the connexin 43 mRNA and connexion 43 positive gap junctions remain lower. Such a protein expression, however, can be enhanced with the concomitant electrical pacing (Kim et al. 2010). Sole existence of gap junctions may not be sufficient; their differentiation to the functional cardiomyocytes and expression of the sufficient ion channel may actually help in preventing arrhythmia. Additionally, sympathetic hyperinnervation improves ventricular function in a failed heart but may act as a potential substrate for ventricular arrhythmias (Zhou et al. 2004; Plotnikov et al. 2007). Furthermore, intramyocardial injection may lead to the highly localized density of transplanted MSCs and that might exaggerate the problems of cell-to-cell conduction and proarrhythmia (Fukushima et al. 2007). The above study has utilized multiple site injections but the actual effective cell dose that is devoid of any adverse effect remains to be seen.

To do away with limitations of the electronic pacemakers, right atrial and ventricular anastomosis followed by MSCs transplantation into the gap between the anastomotic surfaces is also evaluated. Such a surgical method along with transplanted MSCs is able to control the current within the physiological limits

Table 6.3 MSCs preclinical experimental studies on biological pacemaker in dog

Study type/number of animals/study period	Biomaterial used	Evaluation criteria	Overall result	References
Dog biological pace maker model				
Randomized controlled comparative study/$n = 8$/ 42 days	Xenogeneic mHCN2MSCs (low dose: >7,00,000; low dose: <7,00,000: injected subepicardially	Histology, immunohistochemistry, catecholamine responsiveness	Improved outcome	Plotnikov et al. (2007)
Randomized placebo-controlled comparative study/$n = 18$/ 4 weeks	Xenogeneic BM-MSCs (1×10^7 epicardial injection)	PCR, immunostaining	Improved outcome	Kim et al. (2010)
Randomized placebo-controlled study/$n = 24$/ 8 weeks	Allogeneic BM-MSCs (0.3 mL with 1×10^8/mL)	Immunohistochemistry, ECG	Improved outcome	Hou et al. (2011)
Randomized placebo-controlled study/$n = 24$/ 8 weeks	mHCN4BM-MSCs (3.14×10^6 (injected subepicardially)	ECG, histology	Improved outcome	Jun et al. (2012)
Randomized placebo-controlled study/$n = 14$/ 6 weeks	mHCN4-MSCs or MSCs (3.0 to 3.5×10^6 (injected subepicardially)	Heart rate variability (HRV), histology, WB	Improved outcome	Lu et al. (2013)

MSCs mesenchymal stem cells, *BM* bone marrow derived, *n* total number of animals in all the groups of the study; Positive outcome: indicates improved outcome with MSC transplantation even if only one parameter improves. Study period: Indicates maximum time period for which study is conducted

contrary to the surgical procedure (creating higher electric resistance). The surgico-cell combination leads to the development of the observable connexin 43 and troponin I. An important limitation in the study is keeping the normal conduction pathway intact, although a secondary conduction pathway is also developed. To make this technology clinically useful, various questions remain to be answered. Action potentials from normal atrium fail to propagate through the cardiac anastomosis region and cause ventricular capture. Anastomosis increases QRS complex duration as ventricles are activated through tissue other than the His-Purkinje system. There are chances of simultaneous systole of atria and ventricles. Therefore, MSCs must act like atrioventricular (AV) nodes or should get ex vivo differentiated before their in vivo transplantation (Hou et al. 2011).

In order to produce funny currents that are responsive to sympathetic flow, HCN2/HCN4-transfected MSCs (three million) transplanted subepicardially provide appropriate chronotropic response to physical exercise for up to 6 weeks (Lu et al. 2013). Moreover, MSCs (>7,00,000) injected cardiac areas neither show apoptosis nor cellular or humoral rejection at 2 weeks (Jun et al. 2012; Plotnikov et al. 2007). However, these studies are preliminary and HCN-transfected MSCs continued long-term function and animal survival benefits require further investigations.

6.4.3 Congenital or Acquired Cardiac or Vascular Defects

Cardiac defects may be extracardiac or intracardiac. In small animals (dogs and cats), extracardiac vascular deformities include anastomoses between the hepato-portal system and the systemic circulation (portosystemic shunts, including patent ductus venosus) (Rothuizen et al. 1982). Within the heart, valvular deformities like pulmonic or aortic stenosis and atrioventricular dysplasias, besides the ventricular septal defects and atrial septal defects, are reported (Darke 1986). In horse, atrial and ventricular septal defects are seen in tetralogy of Fallot and valvular stenosis (Scansen 2019). In cattle, fusion of the walls of the aorta and pulmonary artery, patent ductus arteriosus, tricuspid valve dysplasia, interventricular septal defect, intra-atrial septal defects, dextroposed aorta, transposition of the aorta and pulmonary artery is reported (West 1988). In relation to the acquired defects, infarcted areas, vessel aneurysms and stenotic vessels due to the trauma or infection, phlebitis, etc., require regenerative medicine.

Congenital heart diseases may require surgical reconstruction like right ventricular outflow tract (RVOT) and atrioventricular defects. Currently used conduits tend to lack regenerative potential, undergo calcification and stenose. Around 88% of these reconstructions require replacement around 5 years after transplantation (Sinzobahamvya et al. 2008). As such tissue-engineered valve construction is being studied. Among various tissue-engineered constructed valves, MSCs are loaded on various scaffolds like poly-4-hydroxybutyrate (P4HB)-coated polyglycolic acid (PGA) nonwoven biodegradable mesh scaffolds (Mendelson et al. 2007), polyglycolic acid (PGA)/poly-L-lactic acid (PLLA) (Engelmayr and Sacks 2006; Ramaswamy et al. 2010), modified polyethylene glycol hydrogels (Ouyang et al. 2008), polydioxanone, pericardial patches fitted with a polytetrafluoroethylene monocusp (Kalfa et al. 2010), polyvinyl alcohol and dextran (PVA/Dx) (Alexandre et al. 2014) or decellularized arterial matrices (Zhao et al. 2010; Zhang et al. 2016a, b). The type and processing of the tissue-engineered valves have bearing on their efficacy in the treatment of the congenital heart diseases (Ramaswamy et al. 2010, 2014).

Among various veterinary species, notable species for valve replacement studies is sheep considering similarity with human cardiovascular physiology (Row et al. 2017). As one of the common complications with bioprosthetic valves is its calcification, sheep models also develop rapid calcification (Schoen et al. 1994). In sheep pulmonary artery patching models (Table 6.4 and Fig. 6.2), MSC-seeded scaffold patches (poly-4-hydroxybutyrate (P4HB)-coated polyglycolic acid (PGA) nonwoven biodegradable mesh scaffolds) although show early (first 2 weeks) thrombus

Table 6.4 MSCs preclinical experimental studies on large vessel reconstruction in sheep

Study type/number of animals/study period	Biomaterial used	Evaluation criteria	Overall result	References
Sheep pulmonary artery patching/right ventricular outflow reconstruction model				
Randomized study/$n = 7$/ 6 weeks	Autologous MSCs (2×10^5 per cm^2 of scaffold) EPCs per square centimetre) and endothelial progenitor cells (EPCs, 2×10^6 per cm^2 of scaffold), poly-4-hydroxybutyrate (P4HB)-coated polyglycolic acid (PGA) nonwoven biodegradable mesh scaffolds (10×20 mm)	Histology and immunohistochemistry	Improved outcome	Mendelson et al. (2007)
Randomized study/$n = 8$/ 6 weeks	MSCs (2×10^5 per cm^2 and endothelial progenitor cells (2×10^6 per cm^2), nonwoven polyglycolic acid biopolymer	Ultrasonography and angiography	Improved outcome	Mettler et al. (2008)
Randomized comparative study/$n = 10$/8 weeks	UCB-MSCs (3×10^6 per cm^2), polydioxanone, pericardial patches fitted with a polytetrafluoroethylene monocusp	Echocardiography, magnetic resonance imaging (MRI), histology, immunohistochemistry and biochemical assays	Improved outcome	Kalfa et al. (2010)
Sheep carotid artery resection model				
Randomized placebo-controlled study/$n = 7$/ 5 months	BM-MSCs differentiated smooth muscle-like cells (SMCs, 2×10^7) and endothelial cells (ECs, 1×10^7), collagen gel, decellularized scaffold	Histology, immunohistochemistry, electron microscopy	Improved outcome	Zhao et al. (2010)
Randomized controlled comparative study/$n = 24$/ 24 weeks	Human WJ-MSCs (1×10^4) along with polyvinyl alcohol and dextran (PVA/Dx; 90/10 v/v)	Graft explanation, Doppler studies, histopathology, immunohistochemistry, scanning electron microscopy, digital image analysis	Improved outcome	Alexandre et al. (2014)

MSCs mesenchymal stem cells, *AD* adipose tissue derived, *BM* bone marrow derived, *WJ-MSCs* Wharton's jelly derived, *UCB* umbilical cord blood, *n* total number of animals in all the groups of the study; Positive outcome: indicates improved outcome with MSC transplantation even if only one parameter improves. Study period: Indicates maximum time period for which study is conducted formation and macrophage infiltration, but with time (fourth week) the loaded scaffolds develop GAGs, collagen. The luminal surface covers through the host artery-derived pannus that contains α-SMA-positive cells and laminated elastin. α-SMA-positive cells in the pannus are less organized compared to the native tissue (Mendelson et al. 2007). The engineered patches (polydioxanone, pericardial patches fitted with a polytetrafluoroethylene monocusp) along with the cells neither appear stenotic nor aneurysmal and are devoid of growth potential, have less fibrosis and calcifications and no thrombus (Kalfa et al. 2010). Even addition of endothelial progenitor cells might further enhance successful creation of an anatomically functional, autologous tissue-engineered pulmonary artery (Mettler et al. 2008). Mesenchymal stem cells may not potentially differentiate into the endothelial cells under in vivo system unless exposed to different scaffolds like arterial matrices or the vasculogenic factors like VEGF (Zhang et al. 2016a, b). It is worth mentioning that the studies conducted are of limited period with limited details on their absorption. Considering the importance of absorption of synthetic scaffolds, it is essential to consider properties of the scaffold.

Apart from the pulmonary patching, MSCs may be loaded on to the suitable scaffold (porcine urinary bladder) for closure of cardiac or vascular defects (Table 6.4). One of the dog studies shows increased regional mechanical function and exhibited myocyte-specific actinin staining with sarcomere spacing similar to that of normal myocytes in the right ventricular myocardial resection model (Potapova et al. 2008). Furthermore, in case of sheep, carotid artery resection models, collagen gel and/or decellularized scaffold loaded with BM-MSCs (20 million) or smooth muscle differentiated BM-MSCs (ten million) after implantation are demonstrated to be patent, anti-thrombotic and stable up to 5 months. These engineered constructs show a smooth muscle, endothelium and collagen and elastin up to 5 months. However, non-cell seeded grafts occlude within 2 weeks (Zhao et al. 2010). Xenogeneic WJ-MSCs loaded on polyvinyl alcohol and dextran too maintain similar patency rate and systolic/diastolic blood flow velocities compared to polytetrafluoroethylene grafts. Nearby sites to MSCs transplantation show lower fibrosis layer and detectable endothelial cells (Alexandre et al. 2014).

These studies appear promising but are limited enough to be conclusive. These tissue -engineered patches have limitations of potential of thrombus formation and are currently unreliable to achieve complete closure of the defects. Besides, the individual scaffold properties and the dose of cells loaded remain to be evaluated. The role of cells and of the internal adjacent tissue has to be understood as healing may progress from that as well. A comparative study on the basis of the scaffold types may be conducted. Clinical feasibility of these TE patches yet remains to be tested.

6.5 Conclusion(s)

Heart disease puts animal life at risk and an immediate death may occur depending on the type of cardiac ailment. The vascular diseases too affect the heart activity and tend to develop heart diseases. In animals, mostly, non-ischaemic HDs are reported.

Currently, the cardiovascular diseases are managed through palliative treatment. Regenerative medicine employing MSCs is considered to have potential to actually regenerate the ailing tissue of the heart. There is in vitro evidence that MSCs differentiate into the cardiomyocyte-like cells, although in vivo studies have failed to establish the fact. Animals have mostly been utilized as translational models for ischaemic conditions and as biological pacemaker. These studies although show an improved outcome although without any definitive conclusion. Under various experimental studies especially the ischaemic cardiomyopathies, MSCs therapeutic mechanisms like prevention of inflammation, cardiomyocyte hypertrophy, reduction in infarct size and scar formation and thereby improvement in physiological functions have been reported. Lack of clinical studies on MIs demands insights through clinical studies. In case of non-ischaemic conditions, little to no improvement has been reported. MSCs therapeutic evaluation in non-ischaemic HDs of animal is very limited. The reports involve the cardiomyopathy or clinical DCM and MMVD or chronic degenerative valvular heart disease although without any observable improvement. MSCs have also been utilized for vascular patching loaded on to the scaffolds with an improved outcome. However, these studies are preliminary, and thus, further studies in the area are desired to reach to any conclusion. In general, MSCs therapeutic applications for cardiovascular ailments are in preliminary stages and extensive studies are desired to reach to any conclusion.

References

Ahmad T, Mukherjee S, Pattnaik B, Kumar M, Singh S, Kumar M, Rehman R, Tiwari BK, Jha KA, Barhanpurkar AP et al (2014) Miro1 regulates intercellular mitochondrial transport & enhances mesenchymal stem cell rescue efficacy. EMBO J 33:994–1010

Airey JA, Almeida-Porada G, Colletti EJ, Porada CD, Chamberlain J, Movsesian M, Sutko JL, Zanjani ED (2004) Human mesenchymal stem cells form Purkinje fibers in fetal sheep heart. Circulation 109:1401–1407

Aktas MS, Ozkanlar Y, Oruc E, Sozdutmaz I, Kirbas A (2015) Myocarditis associated with foot-and-mouth disease in suckling calves. Vet Arhiv 85(3):273–282

Alexakis C, Partridge T, Bou-Gharios G (2007) Implication of the satellite cell in dystrophic muscle fibrosis: a self-perpetuating mechanism of collagen overproduction. Am J Physiol Cell Physiol 293:C661–C669

Alexandre N, Ribeiro J, Gärtner A, Pereira T, Amorim I, Fragoso J, Lopes A, Fernandes J, Costa E, Santos-Silva A, Rodrigues M, Santos JD, Maurício AC, Luís AL (2014) Biocompatibility and hemocompatibility of polyvinyl alcohol hydrogel used for vascular grafting—in vitro and in vivo studies. Biomed Mater Res 102(12):4262–4275

Amable PR, Teixeira MV, Carias RB, Granjeiro JM, Borojevic R (2014) Protein synthesis and secretion in human mesenchymal cells derived from bone marrow, adipose tissue and Wharton's jelly. Stem Cell Res Ther 5:53

Angert D, Berretta RM, Kubo H, Zhang H, Chen X, Wang W, Ogorek B, Barbe M, Houser SR (2011) Repair of the injured adult heart involves new myocytes potentially derived from resident cardiac stem cells. Circ Res 108:1226–1237

Baisan RA, Condurachi E, Turcu CA, Vu V (2021) Prevalence of cardiac diseases in small animals: a five-year single-centre retrospective study. Rev Rom Med Vet 31(2):35–40

Bartunek J, Croissant JD, Wijns W, Gofflot S, de Lavareille A, Vanderheyden M, Kaluzhny Y, Mazouz N, Willemsen P, Penicka M, Mathieu M, Homsy C, De Bruyne B, McEntee K, Lee IW,

Heyndrickx GR (2007) Pretreatment of adult bone marrow mesenchymal stem cells with cardiomyogenic growth factors and repair of the chronically infarcted myocardium. Am J Physiol Heart Circ Physiol 292:H1095–H1104

Beigi F, Schmeckpeper J, Pow-Anpongkul P, Payne JA, Zhang L, Zhang Z, Huang J, Mirotsou M, Dzau VJ (2013) C3orf58, a novel paracrine protein, stimulates cardiomyocyte cell-cycle progression through the PI3K-AKT-CDK7 pathway. Circ Res 113:372–380

Bian S, Zhang L, Duan L, Wang X, Min Y, Yu H (2014) Extracellular vesicles derived from human bone marrow mesenchymal stem cells promote angiogenesis in a rat myocardial infarction model. J Mol Med (Berl) 92:387–397

Boemo CM, Tucker JC, Huntington PJ et al (1991) Monensin toxicity in horses: an outbreak resulting in the deaths of ten horses. Aust Vet J 9:103

Bonagura JD (2019) Overview of equine cardiac disease. Vet Clin North Am Equine Pract 35(1): 1–22. https://doi.org/10.1016/j.cveq.2019.01.001

Borgarelli M, Buchanan JW (2012) Historical review, epidemiology and natural history of degenerative mitral valve disease. J Vet Cardiol 14:93–101

Bortolotti F, Ruozi G, Falcione A, Doimo S, Dal Ferro M, Lesizza P, Zentilin L, Banks L, Zacchigna S, Giacca M (2017) In vivo functional selection identifies Cardiotrophin-1 as a cardiac engraftment factor for mesenchymal stromal cells. Circulation 136:1509–1524

Boswood A, Haggstrom J, Gordon SG et al (2016) Effect of pimobendan in dogs with preclinical myxomatous mitral valve disease and cardiomegaly: the EPIC study—a randomized clinical trial. J Vet Intern Med 30:1765–1779

Buczinski S, Fecteau G, Francoz D et al (2005) Les affections cardiaques congénitales du veau : une approche clinique diagnostique simple. Med Vet Québec 35:79–85

Buczinski S, Fecteau G, DiFruscia R (2006) Ventricular septal defects in cattle: 25 cases. Can Vet J 47:246–252

Carrade DD, Lame MW, Kent MS, Clark KC, Walker NJ, Borjesson DL (2012) Comparative analysis of the immunomodulatory properties of equine adult-derived mesenchymal stem cells. Cell Med 4(1):1–11

Cassano JM, Fortier LA, Hicks RB, Harman RM, Van de Walle GR (2018) Equine mesenchymal stromal cells from different tissue sources display comparable immune-related gene expression profiles in response to interferon gamma (IFN)-γ. Vet Immunol Immunopathol 202:25–30

Cerrada I, Ruiz-Saurí A, Carrero R, Trigueros C, Dorronsoro A, Sanchez-Puelles JM, Diez-Juan A, Montero JA, Sepúlveda P (2013) Hypoxia-inducible factor 1 alpha contributes to cardiac healing in mesenchymal stem cells-mediated cardiac repair. Stem Cells Dev 22:501–511

Cervio E, Barile L, Moccetti T, Vassalli G (2015) Exosomes for intramyocardial intercellular communication. Stem Cells Int 2015:482171

Chang X, Liu J, Liao X, Liu G (2015) Ultrasound-mediated microbubble destruction enhances the therapeutic effect of intracoronary transplantation of bone marrow stem cells on myocardial infarction. Int J Clin Exp Pathol 8(2):2221–2234

Cheng Y, Yi G, Conditt GB, Sheehy A, Kolodgie FD, Tellez A, Polyakov I, Gu A, Aboodi MS, Wallace-Bradley D, Schuster M, Martens T, Itescu S, Kaluza GL, Basu S, Virmani R, Granada JF, Sherman W (2013) Catheter-based endomyocardial delivery of mesenchymal precursor cells using 3D echo guidance improves cardiac function in a chronic myocardial injury sheep model. Cell Transplant 22(12):2299–2309

Chiu CP, Blau HM (1985) 5-Azacytidine permits gene activation in a previously noninducible cell type. Cell 40:417–424

Chung ES, Miller L, Patel AN, Anderson RD, Mendelsohn FO, Traverse J, Silver KH, Shin J, Ewald G, Farr MJ et al (2015) Changes in ventricular remodelling and clinical status during the year following a single administration of stromal cell-derived factor-1 non-viral gene therapy in chronic ischaemic heart failure patients: the STOP-HF randomized phase II trial. Eur Heart J 36: 2228–2238

Cunningham M, Rush JE (2007) Transvenous pacemaker placement in a dog with atrioventricular block and persistent left cranial vena cava Suzanne. J Vet Cardiol 9:129–134

Darke PGG (1985) Myocardial disease in small animals. Br Vet J 141:342

Darke PGG (1986) Congenital heart defects in small animals. Br Vet J 142:203

Davis JL, Gardner SY, Schwabenton B, Breuhaus BA (2002) Congestive heart failure in horses: 14 cases (1984–2001). JAVMA 220(10):1512–1515

Dayan V, Sotelo V, Delfina V, Delgado N, Rodriguez C, Suanes C, Langhain M, Ferrando R, Armand K, Benech A, Touriño C (2016) Human mesenchymal stromal cells improve cardiac perfusion in an sheep immunocompetent animal model. J Invest Surg 29(4):1–8

De Lisio M, Jensen T, Sukiennik RA, Huntsman HD, Boppart MD (2014) Substrate and strain alter the muscle-derived mesenchymal stem cell secretome to promote myogenesis. Stem Cell Res Ther 5:74

Di Francesco D (2010) The role of the funny current in pacemaker activity. Circ Res 106:434–446

Dill SG, Moise NS, Meschter CL (1986) Cardiac failure in a stallion secondary to metastasis of an anaplastic pulmonary carcinoma. Equine Vet J 18:414–417

Dillon AR, Dell'Italia LJ, Tillson M et al (2012) Left ventricular remodeling in preclinical experimental mitral regurgitation of dogs. J Vet Cardiol 14:73–92

DiVincenti LJ, Westcott R, Lee C (2014) Sheep (*Ovis aries*) as a model for cardiovascular surgery and management before, during, and after cardiopulmonary bypass. J Am Assoc Lab Anim Sci 53(5):439–448

Dixon JA, Gorman RC, Stroud RE, Bouges S, Hirotsugu H, Gorman JH, Martens TP, Itescu S, Schuster MD, Plappert T, St John-Sutton MG, Spinale FG (2009) Mesenchymal cell transplantation and myocardial remodeling after myocardial infarction. Circulation 120(11):S220–S229

Donders R, Bogie JFJ, Ravanidis S, Gervois P, Vanheusden M, Marée R, Schrynemackers M, Smeets HJM, Pinxteren J, Gijbels K et al (2018) Human Wharton's jelly-derived stem cells display a distinct immunomodulatory and proregenerative transcriptional signature compared to bone marrow-derived stem cells. Stem Cells Dev 27:65–84

Driehuys S, Van Winkle TJ, Sammarco CD, Drobatz KJ (1998) Myocardial infarction in dogs and cats: 37 cases (1985–1994). J Am Vet Med Assoc 213(10):1444–1448

Emmert MY, Weber B, Wolint P, Frauenfelder T, Zeisberger SM, Behr L, Sammut S, Scherman J, Brokopp CE, Schwartländer R, Vogel V, Vogt P, Grünenfelder J, Alkadhi H, Falk V, Boss A, Hoerstrup SP (2013) Intramyocardial transplantation and tracking of human mesenchymal stem cells in a novel intra-uterine pre-immune fetal sheep myocardial infarction model: a proof of concept study. PLoS One 8(3):e57759

Engelmayr GC, Sacks MS (2006) A structural model for the flexural mechanics of nonwoven tissue engineering scaffolds. J Biomech Eng 128:610–622

Ford J, McEndaffer L, Renshaw R, Molesan A, Kelly K (2017) Parvovirus infection is associated with myocarditis and myocardial fibrosis in young dogs. Vet Pathol 54(6):964–971

Fu H, Chen Q (2020) Mesenchymal stem cell therapy for heart failure: a meta-analysis. Herz 45(6): 557–563

Fukushima S, Varela-Carver A, Coppen SR et al (2007) Direct intramyocardial but not intracoronary injection of bone marrow cells induces ventricular arrhythmias in a rat chronic ischemic heart failure model. Circulation 115:2254–2261

Golpanian S, Wolf A, Hatzistergos KE, Hare JM (2016) Rebuilding the damaged heart: mesenchymal stem cells, cell-based therapy, and engineered heart tissue. Physiol Rev 96:1127–1168

Grieve SM, Bhindi R, Seow J, Doyle A, Turner AJ, Tomka J, Lay W, Gill A, Hunyor SN, Figtree GA (2010) Microvascular obstruction by intracoronary delivery of mesenchymal stem cells and quantification of resulting myocardial infarction by cardiac magnetic resonance. Circ Heart Fail 3:e5–e6

Gugjoo MB, Hoque M, Saxena AC, Zama MMS (2013) Radiographic, electrocardiographic and echocardiographic features of dilatation cardiomyopathy. Indian Vet J 90:19–22

Gugjoo MB, Saxena AC, Hoque M, Zama MMS (2014) M–mode echocardiographic study in dogs. Afr J Agr Res 9(3):387–396

Gugjoo MB, Amarpal CV, Wani MY, Dhama K, Sharma GT (2018) Mesenchymal stem cell research in veterinary medicine. Curr Stem Cell Res Ther 13(8):645–657

Gugjoo MB, Amarpal I, Makhdoomi DM, Sharma GT (2019) Equine mesenchymal stem cells: properties, sources, characterization and potential therapeutic applications. J Equine Vet Sci 72: 16–27

Gugjoo MB, Hussain S, Amarpal SRA, Dhama K (2020) Mesenchymal stem cell-mediated immuno-modulatory and anti-inflammatory mechanisms in immune and allergic disorders. Recent Pat Inflamm Allergy Drug Discov 14:000–000

Gugjoo MB, Dar SH, Ahamd RA, Hussain HA, Dar MD, Parrah JD, Dhama K (2021) Traumatic reticuloperitonitis in cattle and buffalo: recent advances in understanding of etiopathogenesis, diagnosis and treatment. Indian J Anim Sci 91(11):895–905

Hamamoto H, Gorman JH, Ryan LP, Hinmon R, Martens TP, Schuster MD, Plappert T, Kiupel M, St John-Sutton MG, Itescu S, Gorman RC (2009) Allogeneic mesenchymal precursor cell therapy to limit remodeling after myocardial infarction: the effect of cell dosage. Ann Thorac Surg 87(3):794–801

Hao LJ, Hao J, Fang W, Han C, Zhang K, Wang X (2015) Dual isotope simultaneous imaging to evaluate the effects of intracoronary bone marrow-derived mesenchymal stem cells on perfusion and metabolism in canines with acute myocardial infarction. Biomed Rep 3:447–452

Hatzistergos KE, Quevedo H, Oskouei BN, Hu Q, Feigenbaum GS, Margitich IS, Mazhari R, Boyle AJ, Zambrano JP, Rodriguez JE et al (2010) Bone marrow mesenchymal stem cells stimulate cardiac stem cell proliferation and differentiation. Circ Res 107:913–922

Hatzistergos KE, Saur D, Seidler B, Balkan W, Breton M, Valasaki K, Takeuchi LM, Landin AM, Khan A, Hare JM (2016) Stimulatory effects of mesenchymal stem cells on cKit+ cardiac stem cells are mediated by SDF1/CXCR4 and SCF/cKit signaling pathways. Circ Res 119:921–930

Healy AM (1996) Endocarditis in cattle: a review of 22 cases. Ir Vet J 49:43–48

Hnatiuk AP, Ong S-G, Olea FD, Locatelli P, Riegler J, Lee WH, Jen CH, De Lorenzi A, Gimenez CS, Laguens R, Wu JC, Crottogini A (2016) Allogeneic mesenchymal stromal cells overexpressing mutant human hypoxia-inducible factor 1-a (HIF1-a) in an sheep model of acute myocardial infarction. J Am Heart Assoc 5:e003714

Hou Y-B, Zou C-W, Wang Y, Li D-C, Li Q-B, Li H-X, Zhang H-Z, Zhang Q, Fan Q-X (2011) Establishing a new electrical conduction pathway by anastomosis of the right auricle and right ventricle assisted by mesenchymal stem cells in a canine model. Transplant Proc 43:3980–3986

Houtgraaf JH, de Jong R, Kazemi K, de Groot D, van der Spoel TI, Arslan F, Hoefer I, Pasterkamp G, Itescu S, Zijlstra F, Geleijnse ML, Serruys PW, Duckers HJ (2013) Intracoronary infusion of allogeneic mesenchymal precursor cells directly after experimental acute myocardial infarction reduces infarct size, abrogates adverse remodeling, and improves cardiac function. Circ Res 113(2):153–166

Jain AK, Smith EJ, Rothman MT (2006) The coronary venous system: an alternative route of access to the myocardium. J Invasive Cardiol 18(11):563–568

Jin Y, Kong H, Stodilka RZ, Wells RG, Zabel P, Merrifield PA, Sykes J, Prato FS (2005) Determining the minimum number of detectable cardiac-transplanted 111In-tropolone-labelled bone-marrow-derived mesenchymal stem cells by SPECT. Phys Med Biol 50:4445–4455

Johnson MS, Martin MWS, Henley W (2007) Results of pacemaker implantation in 104 dogs. J Small Anim Pract 48:4–11

Jun C, Zhihui Z, Lu W, Yaoming N, Lei W, Yao Q, Zhiyuan S (2012) Canine bone marrow mesenchymal stromal cells with lentiviral mHCN4 gene transfer creates cardiac pacemakers. Cytotherapy 14:529–539

Kalfa D, Bel A, Chen-Tournoux A, Martina AD, Rochereau P, Coz C, Bellamy V, Bensalah M, Vanneaux V, Lecourt S, Mousseaux E, Bruneval P, Larghero J, Menasche P (2010) A polydioxanone electrospun valved patch to replace the right ventricular outflow tract in a growing lamb model. Biomaterials 31:4056–4063

Kim SK, Pak H-N, Park JH, Fang YF, Kim GI, Park YD, Hwang C, Kim Y-H, Kim BS (2010) Cardiac cell therapy with mesenchymal stem cell induces cardiac nerve sprouting, angiogenesis, and reduced connexin43-positive gap junctions, but concomitant electrical pacing increases connexin43-positive gap junctions in canine heart. Cardiol Young 20(3):308–317

Kim U, Shin D-G, Park J-S, Kim Y-J, Park S-I, Moon Y-M, Jeong K-S (2011) Homing of adipose-derived stem cells to radiofrequency catheter ablated canine atrium and differentiation into cardiomyocyte-like cells. Int J Cardiol 146:371–378

Lee WS, Suzuki Y, Graves SS, Iwata M, Venkataraman GM, Mielcarek M et al (2011) Canine bone marrow-derived mesenchymal stromal cells suppress alloreactive lymphocyte proliferation in vitro but fail to enhance engraftment in canine bone marrow transplantation. Biol Blood Marrow Transplant 17(4):465–475

Liao B, Deng L, Wang F (2006) Effects of bone marrow mesenchymal stem cells enriched by small intestinal submucosal films on cardiac function and compensatory circulation after myocardial infarction in goats. Zhongguo Xiu Fu Chong Jian Wai Ke Za Zhi 20(12):1248–1252

Ling Z-Y, Shu S-Y, Zhong S-G, Luo J, Su L, Liu Z-Z, Lan X-B, Yuan GB, Zheng Y-Y, Ran H-T, Wang Z-G, Y-H YIN (2013) Ultrasound targeted microbubble destruction promotes angiogenesis and heart function by inducing myocardial microenvironment change. Ultrasound Med Biol 39(11):2001–2010

Liu Y, Song J, Liu W et al (2003) Growth and differentiation of rat bone marrow stromal cells: does 5-azacytidine trigger their cardiomyogenic differentiation? Cardiovasc Res 58:460–468

Locatelli P, Olea FD, Hnatiuk A, De Lorenzi A, Cerdá M, Giménez CS, Sepúlveda D, Laguens R, Crottogini A (2015) Mesenchymal stromal cells overexpressing vascular endothelial growth factor in sheep myocardial infarction. Gene Ther 22(6):449–457

Loffredo FS, Steinhauser ML, Gannon J, Lee RT (2011) Bone marrow derived cell therapy stimulates endogenous cardiomyocyte progenitors and promotes cardiac repair. Cell Stem Cell 8:389–398

Lu W, Yaoming N, Boli R, Jun C, Changhai Z, Yang Z, Zhiyuan S (2013) mHCN4 genetically modified canine mesenchymal stem cells provide biological Pacemaking function in complete dogs with atrioventricular block. PACE 36:1138–1149

Ma N, Ding F, Zhang J, Bao C, Zhong H, Mei J (2013) Myocardial structural protein expression in umbilical cord blood mesenchymal stem cells after myogenic induction. Cell Biol Int 37:899–904

Margaryan R, Assanta N, Menciassi A, Burchielli S, Matteucci M, Agostini S, Lionetti V, Luchi C, Cariati E, Pucci A, Coceani F, Murzi B (2020) Selective perfusion of coronary vasculature in preterm sheep: a methodological innovation undermined by unfavourable operation of the foramen ovale. Can J Physiol Pharmacol 98:211–218

Marr CM, Reef VB, Brazil TJ et al (1998) Aorto-cardiac fistulas in seven horses. Vet Radiol Ultrasound 39:22–31

Mathieu M, Bartunek J, El Oumeiri B, Touihri K, Hadad I, Thoma P, Metens T, da Costa AM, Mahmoudabady M, Egrise D, Blocklet D, Mazouz N, Naeije R, Heyndrickx G, McEntee K (2009) Cell therapy with autologous bone marrow mononuclear stem cells is associated with superior cardiac recovery compared with use of nonmodified mesenchymal stem cells in a canine model of chronic myocardial infarction. J Thorac Cardiovasc Surg 138(3):646–653

Menasché P (2009) Stem cell therapy for heart failure: are arrhythmias a real safety concern? Circulation 119(20):2735–2740

Mendelson K, Aikawa E, Mettler BA, Sales V, Martin D, Mayer JE, Schoen FJ (2007) Healing and remodeling of bioengineered pulmonary artery patches implanted in sheep. Cardiovasc Pathol 16:277–282

Mettler BA, Sales VL, Stucken CL, Anttila V, Mendelson K, Bischoff J, Mayer JE (2008) Stem cell–derived, tissue-engineered pulmonary artery augmentation patches in vivo. Ann Thorac Surg 86:132

Mias C, Lairez O, Trouche E, Roncalli J, Calise D, Seguelas MH, Ordener C, Piercecchi-Marti MD, Auge N, Salvayre AN et al (2009) Mesenchymal stem cells promote matrix metalloproteinase secretion by cardiac fibroblasts and reduce cardiac ventricular fibrosis after myocardial infarction. Stem Cells 27:2734–2743

Miller PJ, Holmes JR (1985) Observations on seven cases of mitral insufficiency in the horse. Equine Vet J 17:181–190

Minguell JJ, Florenzano FM, Ramírez MR, Martínez RF, Lasala GP (2010) Intracoronary infusion of a combination of bone marrow-derived stem cells in dogs. Exp Clin Cardiol 15(2):17–20

Mirotsou M, Zhang Z, Deb A, Zhang L, Gnecchi M, Noiseux N, Mu H, Pachori A, Dzau V (2007) Secreted frizzled related protein 2 (Sfrp2) is the key Akt-mesenchymal stem cell-released paracrine factor mediating myocardial survival and repair. Proc Natl Acad Sci U S A 104: 1643–1648

Nakanishi C, Nagaya N, Ohnishi S, Yamahara K, Takabatake S, Konno T, Hayashi K, Kawashiri MA, Tsubokawa T, Yamagishi M (2011) Gene and protein expression analysis of mesenchymal stem cells derived from rat adipose tissue and bone marrow. Circ J 75:2260–2268

Nart P, Thompson H, Barrett DC et al (2004) Clinical and pathological features of dilated cardiomyopathy in Holstein-Friesian cattle. Vet Rec 155:355–361

Nout YS, Hinchcliff KW, Bonagura JD, Meurs KM, Papenfuss TL (2003) Cardiac amyloidosis in a horse. J Vet Intern Med 17:588–592

Ouyang H, Zhang JB, Liu Y, Li Q, Peng YH, Kang XJ, Wang YM, Wei XF, Yi DH, Liu WY (2008) Research on application of modified polyethylene glycol hydrogels in the construction of tissue engineered heart valve. Zhonghua Wai KeZaZhi 46(22):1723–1726

Parker HG, Kilroy-Glynn P (2012) Myxomatous mitral valve disease in dogs: does size matter? J Vet Cardiol 14(1):19–29

Perin EC, Silva GV, Assad JA, Vela D, Buja LM, Sousa AL et al (2008) Comparison of intracoronary and transendocardial delivery of allogeneic mesenchymal cells in a canine model of acute myocardial infarction. J Mol Cell Cardiol 44(3):486–495

Petchdee S, Sompeewong S (2016) Intravenous administration of puppy deciduous teeth stem cells in degenerative valve disease. Vet World 9(12):1429–1434

Petchdee S, Chanda M, Cherdchutham W (2018) Pacemaker implantation in horse with bradycardia-tachycardia syndrome. Asian J Anim Vet Adv 13:35–42

Plotnikov AN, Shlapakova I, Szabolcs MJ, Danilo P, Lorell BH, Potapova IA, Lu Z, Rosen AB, Mathias RT, Brink PR, Robinson RB, Cohen IS, Rosen MR (2007) Xenografted adult human mesenchymal stem cells provide a platform for sustained biological pacemaker function in canine heart. Circulation 116:706–713

Pogue B, Estrada AH, Sosa-Samper I, Maisenbacher HW, Lamb KE, Mincey BD, Erger KE, Conlon TJ (2013) Stem-cell therapy for dilated cardiomyopathy: a pilot study evaluating retrograde coronary venous delivery. J Small Anim Pract 54(7):361–366

Potapova IA, Doronin SV, Kelly DJ, Rosen AB, Schuldt AJT, Lu Z, Kochupura PV, Robinson RB, Rosen MR, Brink PR, Gaudette GR, Cohen IS (2008) Enhanced recovery of mechanical function in the canine heart by seeding an extracellular matrix patch with mesenchymal stem cells committed to a cardiac lineage. Am J Physiol Heart Circ Physiol 295:H2257–H2263

Psaltis PJ, Carbone A, Nelson A et al (2008) An ovine model of toxic, nonischemic cardiomyopathy—assessment by cardiac magnetic resonance imaging. J Card Fail 14:785–795

Psaltis PJ, Carbone A, Nelson AJ, Lau DH, Jantzen T, Manavis J, Williams K, Itescu S, Sanders P, Gronthos S, Zannettino ACW, Worthley SG (2010) Reparative effects of allogeneic mesenchymal precursor cells delivered transendocardially in experimental nonischemic cardiomyopathy. J Am Coll Cardiol Intv 3(9):974–983

Qian Q, Qian H, Zhang X et al (2012) 5-Azacytidine induces cardiac differentiation of human umbilical cord-derived mesenchymal stem cells by activating extracellular regulated kinase. Stem Cells Dev 21:67–75

Rabbani S, Soleimani M, Sahebjam M et al (2017) Effects of endothelial and mesenchymal stem cells on improving myocardial function in a sheep animal model. J Tehran Heart Cent 12(2): 65–71

Ramaswamy S, Gottlieb D, Engelmayr GC Jr, Aikawa E, Schmidt DE, Gaitan-Leon DM, Sales VL, Mayer JE Jr, Sacks MS (2010) The role of organ level conditioning on the promotion of engineered heart valve tissue development in-vitro using mesenchymal stem cells. Biomaterials 31:1114–1125

Ramaswamy S, Boronyak SM, Le T, Holmes A, Sotiropoulos F, Sacks MS (2014) A novel bioreactor for Mechanobiological studies of engineered heart valve tissue formation under pulmonary arterial physiological flow conditions. J Biomech Eng 136(12):121009
Reef VB (1990) A monensin outbreak in horses in the eastern United States: pathogenesis, clinical signs, and epidemiology. In: Proceedings of the 8th Annual Vet Med Forum, May 11–13, Washington, DC, p 619
Reef VB (1995) Evaluation of ventricular septal defects in horses using two-dimensional and Doppler echocardiography. Equine Vet J Suppl 19:86–96
Reef VB (1999) Cardiomyopathy and myocarditis in horses. World Equine Vet Rev 4:18–22
Reef VB, Mcguirk SM (2015) Disease of cardiovascular system. In: Smith BP (ed) Large internal medicine, 5th edn. Elsevier, Amsterdam, pp 427–260
Reef VB, Bain FT, Spencer PA (1998) Severe mitral regurgitation in horses: clinical, echocardiographic and pathologic findings. Equine Vet J 30:18–27
Rothuizen J, Van Den Inch TSGAM, Voorholl G, Van Der Leur RJT, Wouda W (1982) Congenital porto-systemic shunts in sixteen dogs and three cats. J Small Anim Pract 23:67
Row S, Swartz DD, Andreadis ST (2017) Animal models of cardiovascular disease as test beds of bioengineered vascular grafts. Drug Discov Today Dis Model 24:37
Scansen BA (2019) Equine congenital heart disease. Vet Clin North Am Equine Pract 35(1): 103–117
Scheinowitz M, Abramov D, Eldar M (1997) The role of insulin-like and basic fibroblast growth factors on ischemic and infarcted myocardium: a mini review. Int J Cardiol 59(1):1–5
Schoen FJ, Hirsch D, Bianco RW, Levy RJ (1994) Onset and progression of calcification in porcine aortic bioprosthetic valves implanted as orthotopic mitral valve replacements in juvenile sheep. J Thorac Cardiovasc Surg 108(5):880–887
Seppen J, Barry SC, Harder B, Osborne WR (2001) Lentivirus administration to rat muscle provides efficient sustained expression of erythropoietin. Blood 98:594–596
Shabbir A, Zisa D, Lin H, Mastri M, Roloff G, Suzuki G, Lee T (2010) Activation of host tissue trophic factors through JAK-STAT3 signaling: a mechanism of mesenchymal stem cell-mediated cardiac repair. Am J Physiol Heart Circ Physiol 299:H1428–H1438
Silva GV, Litovsky S, Assad JAR, Sousa ALS, Martin BJ, Vela D, Coulter SC, Lin J, Ober J, Vaughn WK, Branco RVC, Oliveira EM, He R, Geng Y-J, Willerson JT, Perin EC (2005) Mesenchymal stem cells differentiate into an endothelial phenotype, enhance vascular density, and improve heart function in a canine chronic ischemia model. Circulation 111:150–156
Sinclair KA, Yerkovich ST, Hopkins PMA, Chambers DC (2016) Characterization of intercellular communication and mitochondrial donation by mesenchymal stromal cells derived from the human lung. Stem Cell Res Ther 7(1):91
Sinzobahamvya N, Boscheinen M, Blaschczok HC, Kallenberg R, Photiadis J, Haun C et al (2008) Survival and reintervention after neonatal repair of truncus arteriosus with valved conduit. Eur J Cardiothorac Surg 34:732–737
Soltani L, Rahmani HR, Joupari MD, Ghaneialvar H, Mahdavi AH, Shamsara M (2016) Ovine fetal mesenchymal stem cell differentiation to cardiomyocytes, effects of co-culture, role of small molecules; reversine and 5-azacytidine. Cell Biochem Funct 34:250–261
Sousa MG, Paulino-Junior D, Pascon JPE, Pereira-Neto GB, Carareto R, Champion T, Camacho AA (2011) Cardiac function in dogs with chronic Chagas cardiomyopathy undergoing autologous stem cell transplantation into the coronary arteries. Can Vet J 52(8):869–874
Sun Q-W, Zhen L, Wang Q, Sun Y, Yang J, Li Y-J, Li R-J, Ma N, Li Z-A, Wang L-Y, Nie S-P, Yang Y (2016) Assessment of retrograde coronary venous infusion of mesenchymal stem cells combined with basic fibroblast growth factor in canine myocardial infarction using strain values derived from speckle-tracking echocardiography. Ultrasound Med Biol 42(1):272–281
Suzuki G, Iyer V, Lee TC, Canty JM Jr (2011) Autologous mesenchymal stem cells mobilize cKit+ and CD133+ bone marrow progenitor cells and improve regional function in hibernating myocardium. Circ Res 109:1044–1054

Taghavi S, Sharp TE, Duran JM, Makarewich CA, Berretta RM, Starosta T, Kubo H, Barbe M (2015) Autologous c-kit+ mesenchymal stem cell injections provide superior therapeutic benefit as compared to c-kit+ cardiac-derived stem cells in a feline model of isoproterenol-induced cardiomyopathy. Clin Transl Sci 8(5):425–431

Uccelli A, Pistoia V, Moretta L (2007) Mesenchymal stem cells: a new strategy for immunosuppression? Trends Immunol 28:219–226

Uechi M (2012) Mitral valve repair in dogs. J Vet Cardiol 14:185–192

Urbich C, Dimmeler S (2004) Endothelial progenitor cells functional characterization. Trends Cardiovasc Med 14:318–322

Vela DC, Silva GV, Assad JAR, Sousa ALS, Coulter S, Fernandes MR, Perin EC, Willerson JT, Buja LM (2009) Histopathological study of healing after allogenic mesenchymal stem cell delivery in myocardial infarction in dogs. J Histochem Cytochem 57:167–176

Vollmar C, Vollmarb A, Keenec BW, Foxd PR, Reesee S, Kohn B (2019) Dilated cardiomyopathy in 151 Irish wolfhounds: characteristic clinical findings, life expectancy and causes of death. Vet J 245:15–21

Vulliet PR, Greeley M, Halloran SM, MacDonald KA, Kittleson MD (2004) Intra-coronary arterial injection of mesenchymal stromal cells and microinfarction in dogs. Lancet 363:783–784

Wang JS, Shum-Tim D, Galipeau J, Chedrawy E, Eliopoulos N, Chiu RC (2000) Marrow stromal cells for cellular cardiomyoplasty: feasibility and potential clinical advantages. J Thorac Cardiovasc Surg 120:999–1005

Wang G, Zhuo Z, Xia H, Zhang Y, He Y, Tan W, Gao Y (2013) Investigation into the impact of diagnostic ultrasound with microbubbles on the capillary permeability of rat hepatomas. Ultrasound Med Biol 39:628–637

Wang X, Nie S-P, Zhen L, Miao H-T, Wu X-X, Ren H-M, Shi S-T (2014) Retrograde coronary venous infusion provides targeted cell engraftment into infarcted myocardium. Int J Cardiol 172:e279–e281

Wang X, Zhen L, Miao H, Sun Q, Yang Y, Que B, Lopes Lao EP, Wu X, Ren H, Shi S, Lau WB, Ma X, Ma C, Nie S (2015) Concomitant retrograde coronary venous infusion of basic fibroblast growth factor enhances engraftment and differentiation of bone marrow mesenchymal stem cells for cardiac repair after myocardial infarction. Theranostics 5(9):995–1006

West HJ (1988) Congenital anomalies of the bovine heart. Br Vet J 144:123–130

Whitlock RH, White NA, Rowland GN et al (1978) Monensin toxicosis in horses: clinical manifestations. In: Proceedings. Annual Meetings of the American Association of Equine Practitioners, pp 473–486

Worth LT, Reef VB (1998) Pericarditis in horses: 18 cases (1986–1995). J Am Vet Med Assoc 212: 248–253

Xiao H, Yang Y-J, Lin Y-Z, Peng S, Lin S, Song Z-Y (2018) Transcription factor Tbx18 induces the differentiation of c-kit$^+$ canine mesenchymal stem cells (cMSCs) into SAN-like pacemaker cells in a co-culture model in vitro. Am J Transl Res 10(8):2511–2528

Yang VK, Meola DM, Davis A, Barton B, Hoffman AM (2021) Intravenous administration of allogeneic Wharton jelly–derived mesenchymal stem cells for treatment of dogs with congestive heart failure secondary to myxomatous mitral valve disease. Am J Vet Res 82(6):487–493

Yoon YS, Park JS, Tkebuchava T, Luedeman C, Losordo DW (2004) Unexpected severe calcification after transplantation of bone marrow cells in acute myocardial infarction. Circulation 109: 3154–3157

Yoshida J, Ohmori K, Takeuchi H, Shinomiya K, Namba T, Kondo I, Kiyomoto H, Kohno M (2005) Treatment of ischemic limbs based on local recruitment of vascular endothelial growth factor-producing inflammatory cells with ultrasonic microbubble destruction. J Am Coll Cardiol 46:899–905

Zhang W, Huo Y, Wang X, Jia Y, Su L, Wang C, Li Y, Yang Y, Liu Y (2016a) Decellularized ovine arteries as biomatrix scaffold support endothelial of mesenchymal stem cells. Heart Vessels 31: 1874–1881

Zhang Z, Yang J, Yan W, Li Y, Shen Z, Asahara T (2016b) Pretreatment of cardiac stem cells with exosomes derived from mesenchymal stem cells enhances myocardial repair. J Am Heart Assoc 5:e002856

Zhao Y, Zhang S, Zhou J, Wang J, Zhen M, Liu Y, Chen J, Qi Z (2010) The development of a tissue-engineered artery using decellularized scaffold and autologous sheep mesenchymal stem cells. Biomaterials 31(2):296–307

Zhao Y, Li T, Wei X, Bianchi G, Hu J, Sanchez PG, Xu K, Zhang P, Pittenger MF, WU J, Griffith BP. (2012) Mesenchymal stem cell transplantation improves regional cardiac remodeling following ovine infarction. Stem Cells Transl Med 1:685–695

Zheng Y, Sampaio LC, Li K, Silva GV, Cabreira-Hansen M, Vela D, Segura AM, Bovè C, Perin EC (2013) Safety and feasibility of mapping and stem cell delivery in the presence of an implanted left ventricular assist device: a preclinical investigation in sheep. Tex Heart Inst J 40(3):229–234

Zhou S, Chen LS, Miyauchi Y et al (2004) Mechanisms of cardiac nerve sprouting after myocardial infarction in dogs. Circ Res 95:76–83

7 Mesenchymal Stem Cells Therapeutic Applications in Gastrointestinal Disorders

Abstract

Gastrointestinal system (GIT) encompasses various organs/tissues starting from mouth down to the anal area through oesophagus, stomach and intestines. One of the important associated organs is the liver. These organs have a specific role and any imbalance in the function tends to hamper GIT physiological and functional balance. Although the healing power in such a system is good, there are many ailments that currently pose a challenge for the clinician. To address such ailments, mesenchymal stem cells (MSCs) are being evaluated for their ability to regenerate the tissue. MSCs characteristic properties make them potential candidates for the regeneration of GIT. In oral cavity, MSCs therapeutic effects on periodontal defects, mucosal ulcers, vocal fold injury and laryngotracheal fistula have been evaluated with a good outcome. Oesophageal defects treated with scaffolds loaded with MSCs appear promising. Intestinal affections like anal furunculosis and inflammatory bowel disease (IBS) treated with MSCs also show better outcome. Besides, MSCs appear to have a potential therapeutic role in liver affections like fibrosis/acute liver injury. However, all these studies are very limited and carried out mostly under experimental settings. There is still a long way to go to apply MSCs clinically as evidence-based medicine in GIT affections. The current chapter focuses on MSCs potential therapeutic applications in GIT affections.

Keywords

Anal furunculosis · Inflammatory bowel disease · Liver fibrosis · Mesenchymal stem cells · Oesophageal defects · Oral mucosal ulcers · Periodontal defects · Vocal fold injury

M. B. Gugjoo, *Therapeutic Applications of Mesenchymal Stem Cells in Veterinary Medicine*, https://doi.org/10.1007/978-981-19-3277-9_7

7.1 Introduction

Gastrointestinal (GIT) system starts with the oral cavity down through the oesophagus, stomach, intestines and finally ends at the anal region. The liver also makes part of the system due to the associated functionality. It is a large system that has diverse functions with different components having specific role and accordingly the physiological mechanism. These organs undergo various diseases that may also involve other organs and are currently less curable or incurable. Oral cavity structures like periodontium, alveolar mucosa, sublingual mucosa, lip and cheek mucosa, tongue, palatal mucosa, vocal cords and pharyngeal mucosa undergo various ailments depending upon the aetiology. Oesophagus lacks a serosal layer and has segmental circulation/innervation that together with constant movement makes healing a complicated matter (Long and Orlando 2002). There may be developmental problems like oesophageal atresia or injury or pill-induced strictures, besides functional impairment like gastroesophageal reflux disorder (GERD)-induced ulcers. Pill- and GERD-induced ulcers has a prevalence of around 2–7% (Splechler 1995; Vito et al. 2007). Stomach or intestines that have good healing power also suffer with ulcers and may undergo leaks after any surgical repair or injuries. These may also have problems of blocked circulation in volvulus/obstructions or intussusceptions that lead to tissue damage. Further, the inflammatory bowel disease and perianal fistula are one of the common ailments seen in dogs. In some of these conditions, MSCs potential role is evaluated to a little extent in veterinary species. The current chapter focuses on various conditions in which MSCs potential therapeutics is evaluated.

7.2 MSCs in GIT Affections

GIT is one of the organ systems that has additional concern of microbes as it directly comes in contact with the external environment. Thus, protection against external antigens and maintenance of metabolic and immunological homeostasis are of prime importance (Luissint et al. 2016). MSCs due to their properties are considered to have the potential to address the issues with GIT affections (Fig. 7.1). MSCs may 'home in' to the inflamed areas in response to the secreted factors like SDF-1 and CX3CL1. These factors act on the receptors like CXCR4 and CX3CR1 or intercellular adhesion molecules (Chen et al. 2018; Li et al. 2019; Fu et al. 2020; Gugjoo et al. 2020a). These cells tend to differentiate into the epithelial and endothelial cells under ex vivo, although such a mechanism under in vivo system remains to be considered therapeutically important. MSCs do not survive for long at the implantation site and their engraftment and differentiation into tissue-specific cells remain to have enough evidence (Gugjoo et al. 2020a). MSCs offer regenerative effects through immune regulation/anti-inflammatory and anti-oxidative actions and promote epithelial cell homeostasis (Lazebnik et al. 2010; Sémont et al. 2010; Ciccocioppo et al. 2018; Zhang et al. 2019; Gugjoo et al. 2020b; Liu and Jiang 2021). MSCs mediate immune regulation either through paracrine actions (soluble factors and microvesicles) or cell–cell-mediated interactions (Gugjoo et al. 2020a;

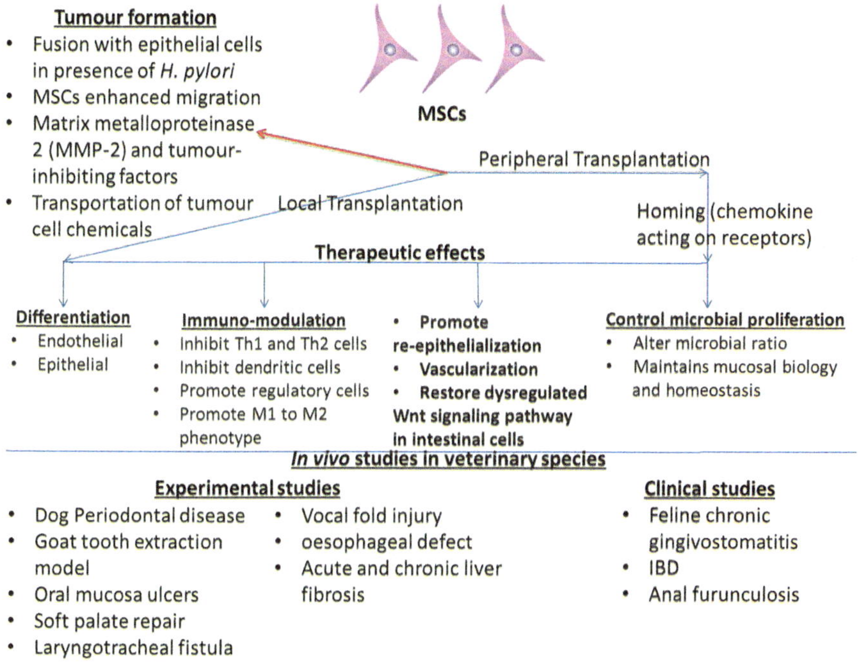

Fig. 7.1 Mesenchymal therapeutic effects and their potential applications in various gastrointestinal affections

Gugjoo and Pal 2020). Immune regulation required especially in inflammatory conditions like anal fistulae and inflammatory bowel disease may occur through macrophage reprogramming, inhibition of T-helper (Th1, Th2) and dendritic cells, and promotion of the regulatory cells (Lopez-Santalla and Garin 2021). The cells partially restore dysregulated Wnt-signalling pathways in intestinal cells in inflammatory conditions like IBD (Gao et al. 2020). Immunomodulatory/anti-inflammatory effects may occur through various secretory factors like indoleamine 2,3-dioxygenase (IDO) and prostaglandin E2 (PGE2), transforming growth factor beta (TGF-β1), nitrous oxide (NO) and interleukin (IL)10. Their immunomodulatory role through polarization of M1 to M2 phenotype is also well established (Lopez-Santalla and Garin 2021; Zhai et al. 2021). MSCs may enhance fibroblast proliferation and migration to enhance re-epithelialization in oral lesions (Tarnawski et al. 1991; El-Menoufy et al. 2010). These cells may promote angiogenesis through vascular endothelial growth factor (VEGF) secretion (Gerritsen et al. 2003). MSCs interaction with antibiotics and host immune response may control bacterial infections (Johnson et al. 2017). These cells may alter microbial ratio in the intestines which tend to help in maintaining the mucosal biology and homeostasis (Dong et al. 2019). However, a growing concern is of tumour formation that in specific infections like *Helicobacter pylori* may occur. MSCs fuse with the epithelial cells and may transform into any phenotype. Such a fusion may promote MSCs migration through tumour necrosis factor-α (TNF-α) via Nuclear Factor-kappa B-dependent pathway (Ferrand et al. 2011). MSCs secretome contains the matrix metalloproteinase 2 (MMP-2) and tumour-inhibiting factors that may enhance cancer cell growth and metastasis (Lee et al. 2013; Yang et al. 2015). Additionally, these cells may transport proteins, miRNA and metabolites from tumour cells (Vallabhaneni et al. 2015). All these features are the causes of concern with the stem cell application and thus, needs thorough understanding.

Below are discussed the potential therapeutic applications of MSCs in various GIT affections of the veterinary species (Fig. 7.1).

7.3 Oral Cavity

The oral mucosal and sometimes the submucosal lesions initially appear as an inflammatory condition commonly known as stomatitis. Oral mucosal ulceration may arise as a sequela to trauma, microbial infections, recurrent aphthous stomatitis or through systemic disorders. These ulcers are painful as oral mucosa is richly innervated (Porter and Scully 2000; Field and Longman 2003). Simple ulcers are usually self-limiting unlike that of the chronic oral ulcers that usually last from 4 to 6 weeks (Alamoudi et al. 2014). As these ulcers create painful oral conditions, prevent food intake, increase salvation and tend to recur, novel therapeutics like stem cells are being evaluated for their full addressal. In relation to animal mouth diseases, mainly periodontal or mucosal lesions have been studied for mesenchymal stem cell therapeutic potential. However, the potential of these cells goes beyond and may help in addressing the nerve damages or other fibrous conditions of the mouth as well. It is worth mentioning that the oral ulceration sometimes arises through the underlying systemic diseases like renal failure or intestinal/stomach affections that need to be focused as well.

7.3.1 Experimental Studies

7.3.1.1 Periodontal Defects

Periodontal disease (PD) is a group of inflammatory diseases varying from gingivitis (early stage) to periodontitis (advanced stage). It is commonly reported in dog (Albuquerque et al. 2012), cat (Perry and Tutt 2015) and horse (Klugh 2006) clinics, although other species are not immune to these diseases. It is usually caused by the formation of bacterial plaque in the periodontium in dog and cat (Albuquerque et al. 2012), while in horse mostly feed accumulation initiates the periodontitis (Colyer 1906). PD in dogs affects around 80% of the population ageing 2 years or more (Niemiec 2008a, b). It may be correlated with the renal, hepatic and cardiac disorders (Pavlica et al. 2008; Glickman et al. 2009), although currently without any direct link (Peddle et al. 2009a, b). Cat PD occurs in around 0.7%–10% of the population (Arzi et al. 2010; Alpsoy et al. 2015; Jennings et al. 2015) with only 7 out of 10 cases responding to the available care. Feline chronic gingivostomatitis (FCGS) is treated with the full-mouth or near full-mouth tooth extraction, and the refractory cases (about 30%) require frequent therapies of antibiotics, corticosteroids and analgesics (Jennings et al. 2015). Current regenerative therapies like guided tissue regeneration and enamel matrix derivatives lack clinical predictability (Esposito et al. 2009) and as such investigations for other therapeutics are undergoing. Autologous bone implantation therapies show better outcomes in alveolar or furcation defects but have limited availability. Osseointegration of the transplant that closely resembles the natural tooth lacks features in relation to the periodontal ligament and the connective tissue link via cementum. Lack of these structures in the implant fails to adapt to the occlusal forces especially in relation to the biomechanical forces. Additionally, periodontal ligament managed physiologic migration of natural teeth to maintain interdental and interocclusal relationships, sensitive proprioceptive and cellular defence mechanisms is also absent (Marei et al. 2009).

To evaluate regenerative medicine applications in periodontal affections, dogs have been used as a common model animal for understanding the diseases (Albuquerque et al. 2012). To improve clinical predictability, stem cell (due to their immunomodulatory and/or anti-inflammatory role, in addition to the paracrine properties and transdifferentiation capability) therapy is being evaluated. A recent study shows that AD-MSCs from calicivirus-positive feline chronic gingivostomatitis (FCGS) cats vary in their features and pinpoint their response against the diseases. AD-MSCs from FCGS cats may have lower zeta potential (higher zeta potential provides colloidal stability and reduces chances of aggregation) as compared to normal cats. AD-MSCs secretome from FCGS cats lacks proteins involved in metabolic functions. Exosomal proteins involved in metabolic processes, immune responses, regulation of biological processes and gene expression may exclusively be secreted (Villatoro et al. 2021). Thus, further studies in this area are desired to understand stem cell response and its use for clinical applications. Considering the poor responsive MSCs from chronic FCGS cat donors, transplantation of extrinsic allogeneic MSCs may maintain normal homeostasis.

In dog, MSCs role in varied defects have been studied ranging from alveolar bone defect (Han et al. 2011; Pourebrahim et al. 2013; Paknejad et al. 2015; Yuanzheng et al. 2015) to the furcation defects of various degrees (Simsek et al. 2012; Suaid et al. 2012; Tobita et al. 2013; Yu et al. 2013; Nagahara et al. 2015; Takewaki et al. 2017). The bone defects created are either one-wall infrabony (Zang et al. 2016; Tsumanuma et al. 2011) or three-walled (Khorsand et al. 2013) or simple periodontal fenestration (Li et al. 2009) (Table 7.1). In these models, MSCs from various sources including that of the periodontal structures, bone marrow, peripheral blood and adipose tissue and at variable doses (ranging from two million to the 20 million) are evaluated for regenerative potential with or without the scaffolds. Most of the studies are conducted as comparative ones although some are placebo-controlled. The evaluations are made through histometry, histopathology, scanning electron microscopy (SEM) and micro-CT, although the first two techniques have most commonly been used. Locally transplanted MSCs engraft the periodontal structures

Table 7.1 MSCs preclinical experimental studies in periodontal disease in goat and dog

Study type/animal number/study period	Biomaterial used	Evaluation criteria	Overall result	References
Canine tooth extraction (bilateral, titanium fixture) in goat				
Randomized placebo-controlled study/$n = 10$/ 1 month	Autologous BM-MSCs (9.4×10^6 per scaffold), poly (DL-Lactide-co-glycolide)	Radiography, histology, histophotometric and scanning electron microscopy	Improved outcome	Marei et al. (2009)
Furcation/periodontal bone defects models in dog				
Randomized comparative study/$n = 5/8$ weeks	Freshly isolated and cryopreserved BM-MSCs (5×10^6) seeded on collagen scaffold	Histological and histomorphometric parameters	Improved outcome	Li et al. (2009)
Randomized comparative study/$n = 4/3$ months	Blood MSCs (2×10^6/mL), nano-hydroxyapatite/collagen (nHAC) and calcium sulphate hemihydrate (CSH)	Histology	Improved outcome	Han et al. (2011)
Randomized placebo-controlled study/$n = 7$/ 6 months	Autologous PDL-MSCs	Histometric parameters, scanning electron microscopy (SEM)	Improved outcome	Suaid et al. (2011)
Randomized comparative study/$n = 4/8$ weeks	Autologous PDL-MSCs, alveolar periosteum MSCs (AP) vs. BM-MSCs (3–5 $\times 10^4$ cell sheets)	Histological and histometric parameters	Improved outcome with PDL-MSCs than others	Tsumanuma et al. (2011)
Randomized comparative study/$n = 3/8$ weeks	Autologous BM-MSCs (1×10^7), PRP	Histomorphometric analysis and histological analysis	Improved outcome	Simsek et al. (2012)
Randomized comparative study/$n = 7/3$ months	Autologous PDL-MSCs (3×10^5), carrier sponge	Histomorphometrically	Improved outcome	Suaid et al. (2012)
Randomized placebo-controlled study/$n = 10$/ 8 weeks	Autologous dental pulp MSCs (2×10^7) and bio-Oss granules after 4 weeks	Histologically and histomorphometrically	Improved outcome	Khorsand et al. (2013)
Randomized comparative study/$n = 4/8$ weeks	AD-MSCs, hydroxyapatite/beta-tricalcium phosphate scaffold	Histomorphometry	Improved outcome but less than autograft	Pourebrahim et al. (2013)

Randomized comparative study/*n* = 8/2 months	Autologous AD-MSCs (1.5×10^7), PRP	Histologic, immuno-histologic and radiography	Improved outcome	Tobita et al. (2013)
Randomized placebo-controlled study/*n* = 4/8 weeks	Xenogeneic Gingiva MSC cell sheets	Histology, histometry, immunohistochemistry	Improved outcome	Yu et al. (2013)
Randomized comparative study/*n* = 12/8 weeks	BM-MSCs (2×10^7), atelocollagen, beta tricalcium phosphate	Histology	Improved outcome	Nagahara et al. (2015)
Randomized comparative study/*n* = 12/8 weeks	BM-MSCs (2×10^7/mL), anorganic bovine bone mineral	Histometry, histology	Improved outcome	Paknejad et al. (2015)
Randomized comparative study/*n* = 20/6 months	Autologous bone graft, BM-MSCs, platelet-rich fibrin	Histology	Improved outcome	Yuanzheng et al. (2015)
Randomized comparative study/*n* = 6/8 weeks	BM-MSCs (1×10^7/mL), chitosan/β-glycerophosphate (β-GP)	Histology, histometry and immunohistochemistry, micro-CT	Improved outcome	Zang et al. (2016)
Randomized comparative-controlled study/*n* = 7/12 weeks	Autologous BM-MSCs and osteogenic-induced BM-MSCs (7×10^4 cell sheet)	Micro-computed tomography, histological analysis and morphometric analysis	Improved outcome with either cell type	Takewaki et al. (2017)

MSCs mesenchymal stem cells, *AD* adipose tissue derived, *BM* bone marrow derived, *PDL* periodontal ligament derived, *PRP* platelet rich plasma, *n* total number of animals in all the groups of the study; Improved outcome: Indicates improved outcome with MSC transplantation even if only one parameter improves; Study period: Indicates maximum time period for which study is conducted and differentiate into the osteoblasts, fibroblasts and odontoblasts (Shi and Gronthos 2003; Akizuki et al. 2005; Wei et al. 2010).

Transplantation of injectable tissue-engineered bone construct (composed of nano-hydroxyapatite/collagen (nHAC) and calcium sulphate hemihydrate and osteogenic-induced peripheral blood MSCs) leads to significantly more bone-implant contact and bone density as compared to the scaffold and the control at 3 months period (Han et al. 2011). Calcium sulphate hemihydrate (CSh) upon degradation releases calcium and sulphur of which the former degradation product combines with the phosphorus producing the calcium phosphate. CSh degradation may lower pH that enhances angiogenesis and local release of growth factors to enhance bone formation. Resorption of the whole assembly provides an osteoconductive surface activating the osteoblast recruitment and thus bone formation. Add-on MSCs might differentiate into osteoblast like cells, stimulate angiogenesis, secrete growth factors and tissue-specific matrices like collagen I, besides, prevent healed tissue degradation (Murashima et al. 2002). Loading of anorganic bovine bone mineral (ABBM) granules along with BM-MSCs 5 weeks after intrabony periodontal defect (non-placebo controlled) creation enhances repair. In cell-scaffold transplanted defects, more new cementum and periodontal ligament (PDL) with higher alveolar bone height occurs as compared to the simple transplantation of ABBM granules, although new bone formation was comparable (Paknejad et al. 2015). Similar results were also reported for a similar model that used autologous dental pulp MSCs and Bio-oss granules after 4 weeks of defect creation (Khorsand et al. 2013) and BM-MSCs along with the chitosan/anorganic bovine bone carrier (Zang et al. 2016). Implantation of autologous bone along with the autologous BM-MSCs and PRP in unilateral alveolar clefts enhances bone formation that connects with the underlying maxilla at 6-month period. Important to note is the maximum loss of the healed tissue in the control bone graft treated clefts while such a loss was minimal and the bone was denser in the cell/scaffold/PRP treatment (Yuanzheng et al. 2015). As there is a high absorption rate of the bone graft in the alveolar clefts, implantation of the cells along with the PRP might have prevented such a loss and enhanced the tissue integration. MSCs differentiate into osteoblasts and also secrete growth factors and factors that prevent matrix degradation. In contrast, transplantation of BM-MSCs sheets combined with β-tricalcium phosphate (β-TCP) and collagen in one-wall periodontal defect fails to increase the thickness of newly formed cementum while periodontal ligament improves (Hasegawa et al. 2006). Thus, MSCs loaded with specific scaffolds may enhance cementum length and periodontal ligament formation, although results may vary with respect to the actual bone formation. This may need additional understanding in the role of cells and scaffolds in relation to the defect type.

In class III furcation defects, transplantation of the gingival MSCs (Yu et al. 2013), autologous periodontal ligament MSCs along with the collagen I (Suaid et al. 2012), AD-MSCs along with the PRP (Tobita et al. 2013) and BM-MSCs in atelocollagen (Kawaguchi et al. 2004) enhances periodontal ligament, cementum and bone regeneration. Although graft requirement is considered important, however, one of these studies demonstrates healing with MSCs even without the graft

(Yu et al. 2013). Thus, almost all the studies of furcation defects report periodontal tissue formation involving periodontal ligament, cementum and new bone formation, while in case of the unilateral bony defect models, variable results with respect to the new bone formation may occur. Such differences may occur with the model type, cell dose, scaffold incorporated and tissue-engineered construct transplantation method and time, besides the evaluation criteria and period of evaluation.

There are various comparative studies conducted on the cell source and scaffold-type basis. In class II furcation defects, MSCs along with the PRP or autogenous bone graft with PRP and autogenous bone graft may show comparable periodontal tissue repair. All these biomaterials transplanted at 5 weeks post defect formation enhance periodontal regeneration. The class II furcation defects completely fill with the cementum, alveolar bone and periodontal ligament at eighth week after treatment (Simsek et al. 2012). As non-significant bone formation occurs with the simple use of PRP, it may be considered that PRP might help in periodontal ligament regeneration but may not be sufficient enough to enhance bone formation for which scaffolds like autogenous graft or viable cells like MSCs may be required. MSCs loaded with the graft (tricalcium phosphate) may not show as good periodontal regeneration as MSCs loaded with the collagen I and tricalcium phosphate (Nagahara et al. 2015). Likewise, MSCs along with their self-produced extracellular matrix enhance periodontal tissue regeneration (Takewaki et al. 2017). In comparison to the collagen scaffold, transplantation of MSCs (cryopreserved and/or fresh) provides good healing in periodontal fenestrations (Li et al. 2009). Among various cell types (periodontal ligament (PDL)-MSCs vs. Alveolar periosteum MSCs (AP) vs. BM-MSCs), transplantation of PDL-MSCs may lead to the significant improvement in the cementum formation and PDL fibre orientation. Additionally nerve and alveolar bone regeneration may occur in one-wall bony defect models with PDL-MSC-transplantation (Tsumanuma et al. 2011). MSCs, fresh or cryopreserved, may have comparable periodontal tissue regeneration (Li et al. 2009). Considering the implantation of osteogenic-induced MSCs, a better outcome may occur, although healing in relation to other non-bony structures has to be confirmed (Takewaki et al. 2017). Considering the regenerative ability of MSCs together with the scaffold (hydroxyapatite/beta-tricalcium phosphate scaffold), autograft appears more fruitful in alveolar bone defect models (Pourebrahim et al. 2013).

All these studies demonstrate that MSCs with or without the grafts potentially improve the periodontal regeneration but with some variability as per the model, graft and cells. The results of these studies are based on the small animal size and limited evaluations that too for shorter durations. As such further extensive studies considering such limitations are desired. Above all, the clinical applications remain to be made in dogs.

In addition to dogs, a single goat study evaluates MSCs role in tooth extraction model (Marei et al. 2009). Defects implanted with the autologous BM-MSCs loaded on to the poly (DL-Lactide-co-glycolide) and stabilized with titanium fixtures grow periodontal tissue-like bundles of connective tissue fibres that extend from wound periphery towards the surface of the fixed titanium as early as 10 days and further increase up to the last evaluation period (1 month). In control, such type of healing at

10 days may also occur, although much poor by 30th day (Marei et al. 2009). This comparable healing in the control at an early period might have been ensured by the presence of progenitor cells from the remnant periodontal ligament. These progenitor cells have tendency to differentiate into formative cells like cementoblasts and fibroblasts till the peri-implant wound area is close to the periphery of the extraction socket (Cho et al. 2016). The lack of complete regeneration in the control may be either due to the limited capacity of the remnant periodontal progenitor cells or these cells are insufficient to form and maintain the healed tissue (Marei et al. 2009). Thus, MSCs appear useful, although long duration studies with sufficient numbers of animals are desired. Furthermore, MSCs potential therapeutic mechanisms remain to be evaluated.

Overall regenerative medicine employing MSCs along with scaffolds/grafts offers a proof-of-therapeutics in the periodontal diseases. However, there is a significant lack of the literature on the exact role of the MSCs in such conditions. There is a need to evaluate MSCs engraftment and ability to differentiate, if any. Besides, the paracrine effects like immunomodulatory potential together with the other mechanisms like matrix secretion and prevention of degradation remain to be elucidated. These studies have employed limited animals to offer significant statistical analysis of the results. The long-term studies remain to be conducted and finally the clinical applications remain to be conducted.

7.3.1.2 Oral Lesions and Mucosal Ulcers

MSCs may play a significant role in healing of the oral lesions including the soft cleft palate (palatoschisis), although such studies remain to be conducted in veterinary species. Exogenous BM-MSCs orient and integrate along the axis of the skeletal myocytes post-surgical repair (mandibular symphysiotomy) of soft palate in horses. These cells remain viable under the soft palate epithelium (Carstanjen et al. 2006).

MSCs role in oral ulcers in veterinary species has exclusively been evaluated in the canine experimental studies (Table 7.2), although clinical applications remain to be conducted. These chemically (formocresol) induced chronic and recurrent oral ulcers upon transplantation of BM- or AD-MSCs 3 days later improve mucosal regeneration earliest by first and second week (El-Menoufy et al. 2010; Alamoudi et al. 2014; Aly et al. 2014). Formocresol induces coagulation necrosis-promoting inflammatory processes associated with the mononuclear cells, vascular dilatation and new fibril formation (AnaTeresa et al. 2008). Thus, a condition more relevant to the chronic ulcers is produced. Available studies are mostly comparative in nature especially against the current available treatments or the route and dose of the cells.

Considering MSCs against the dexamethasone or control, the induced ulcers treated with the cells show significantly accelerated healing. These cell-treated ulcers show increased expression of epithelialization and fibroplasia (EGF and collagen genes) and the neovascularization genes (VEGF and PDGF genes) comparatively (Alamoudi et al. 2014). The local transplantation of the cells in comparison to the intravenous injection may be preferred as early healing occurs in the former case (Aly et al. 2014). The cell transplantation right from the ten million doses offers clinical improvement (better histology than control); however, higher (20 million)

Table 7.2 MSC's preclinical experimental studies in oral non-periodontal, laryngotracheal (larynx and vocal fold) stenosis and oesophageal resection in dog

Study type/ animal number/ duration of study	Biomaterial used	Evaluation criteria	Overall result	References
Induced oral ulcers in dog				
Randomized placebo-controlled comparative study/$n = 18$/ 15 days	BM-MSCs 1×10^7 (low dose) 2×10^7 (high dose)	Clinical evaluation (Wound Assessment Parameter Scoring Tool) and polymerase chain reaction for collagen gene	Improved outcome	El-Menoufy et al. (2010)
Randomized placebo-controlled comparative study/$n = 18$/ 15 days	AD-MSCs (1×10^7), dexamethasone	Histopathologically, gene expression analysis	Improved outcome with cells	Alamoudi et al. (2014)
Randomized placebo-controlled comparative study/$n = 18$/ 15 days	BM-MSCs (2×10^7), intravenous and local implantation	Clinical and histopathological	Improved outcome with local as compared to IV	Aly et al. (2014)
Vocal fold injury				
Randomized comparative study/$n = 8$/ 2 months	BM-MSCs (0.5 to 2.0×10^6), atelocollagen	Morphology, histology	Improved outcome	Kanemaru et al. (2003)
Randomized comparative study/$n = 10$/ 24 weeks	AD-MSCs ($1–3 \times 10^6$), Atelocollagen	Morphology, histology	Improved outcome	Lee et al. (2006)
Randomized comparative study/$n = 17$/ 6 months	Differentiated AD-MSCs (3–4 mL, 1.0×10^5 cells/mL), AD-MSCs, vocal fold fibroblasts	Histology, immunohistochemistry	Improved outcome higher with differentiated AD-MSCs	Hu et al. (2014)
Laryngotracheal (larynx and vocal fold) stenosis in dog				
Randomized placebo-controlled comparative study/$n = 7$/ 6 weeks	BM-MSCs (2×10^6), conditioned medium (2 mL)	Histology	Improved outcome	Iravani et al. (2017)
Oesophageal resection model				

(continued)

Table 7.2 (continued)

Study type/animal number/duration of study	Biomaterial used	Evaluation criteria	Overall result	References
Randomized comparative study/$n = 12$/ 8 weeks	Autologous oral mucosal epithelial cells (OMECs, 1×10^7), acellular porcine small intestinal submucosa	Radiography, histology, immunohistochemistry	Improved outcome	Wei et al. (2009)
Randomized comparative study/$n = 6$/ 12 weeks	BM-MSCs (1×10^7), acellular porcine small intestinal submucosa	Radiography, histology	Improved outcome	Tan et al. (2013)
Randomized experimental study/$n = 8$/ 19 months	Autologous AD-MSCs (15×10^6), acellular porcine small intestinal submucosa	Endoscopic assessment of mucosal regrowth, histology	Improved outcome	La Francesca et al. (2018)

MSCs mesenchymal stem cells, *AD* adipose tissue derived, *BM* bone marrow derived, *n* Total number of animals in all the groups of the study; Improved outcome: Indicates improved outcome with MSC transplantation even if only one parameter improves; Study period: Indicates maximum time period for which study is conducted

cell dose may additively improve neovascularization (enhanced VEGF expression) (El-Menoufy et al. 2010). MSCs properties are affected by acute inflammation, and a slightly delayed local transplantation in these studies might have avoided such a phase for a better outcome. Considering the intravenous route, sufficient cells might not reach the lesion or the cells might not be activated as per the local environment. Furthermore, sufficient cell concentration may be imperative to offer better outcomes.

These preliminary studies bet for MSCs therapeutic efficacy in oral ulcers; however, considering the paucity of information required for evidence-based medicine is very much lacking. It is also worth mentioning that recent studies could not be located in large animals. Clinical trials are desired for evaluation of the MSCs safety, feasibility and finally efficacy under natural conditions.

7.3.1.3 Vocal Fold Injury and Laryngotracheal Fistula

Vocal fold is a delicate structure that upon an injury undergoes hyperplasia, scarring and to contracture (Rousseau et al. 2003). This causes permanent dysphonia and/or dysphagia. Thus, MSC-based therapeutics are being evaluated to offer scarless and actual regeneration of the tissue. In dog vocal fold cauterized resection models, transplantation of MSCs along with the scaffolds like 3D atelocollagen (Kanemaru et al. 2003; Lee et al. 2006) or fibroblasts differentiated from MSCs with collagen-hyaluronic (Hu et al. 2014) prevents vocal fold scarring and atrophy by 24 weeks after the treatment. The slightly delayed (4–7 days) cell transplantation may mimic natural courses and also prevent the effects of an extreme inflammatory environment. These cells remain viable and aligned along the vocal fold tissue up to 8 weeks. The cells prevent tissue atrophy and granuloma formation (Lee et al. 2006). Comparing the role of fibroblast derived from AD-MSCs against the AD-MSCs and vocal fold fibroblasts, the former cell type lead to the smooth tissue surfaces as compared to the later. The elastin concentration remains higher with the differentiated fibroblasts as compared to the other cells but collagen secretion may be higher with both the differentiated AD-MSCs and simple AD-MSCs. In relation to matrices (hyaluronic acid, decorin and fibronectin), all the cell types may have comparable roles (Hu et al. 2014). Immature cells (strongly stained nuclei) in thyroarytenoid muscle fibres (Kanemaru et al. 2003) and immature mesenchymal cells may be localized in the connective tissue of the subepithelial layer (Hu et al. 2014). However, transdifferentiation of MSCs into vocal fold cells remains to be seen.

The larynx and trachea upon injury may heal by concentric scar contraction creating narrowed lumen, laryngotracheal stenosis (LTS). For smooth and effective swallowing and phonation, the laryngeal and tracheal healing has to be scar free. In laryngotracheal fistula (mechanical trauma caused by Tru-cut punch forceps) model, BM-MSCs (two million) or their conditioned medium (CM, 2 mL) enhance healing although with little fibrosis in the vocal folds as compared to the control up to 6 weeks. Epithelialization, degree and type of inflammation between the cell- or CM-treated and the control may remain comparable. BM-MSCs may have a slight edge over the CM in reduction of the fibrosis although statistically non-significant (Iravani et al. 2017). These cells may give rise to the stable and rational extracellular matrix protein arrangement without any formation of the pseudomembrane or scar (Kumai et al. 2010). Considering the limited animal number, statistical differences may not be seen.

7.3.2 Clinical Studies

Among veterinary species, feline is the only species clinically treated with MSCs for chronic gingivostomatitis (CGS), a chronic, debilitating idiopathic oral mucosal inflammatory disease (Arzi et al. 2016) (Table 7.3). In feline chronic gingivostomatitis, intravenous transplantation of autologous AD-MSCs or allogeneic AD-MSCs (@ 20 million; 2 doses 30 days apart,) prove useful. Around 77% of the cases show significant improvement (Arzi et al. 2016, 2020). Among $n = 9$, only $n = 7$ cases complete the study out of which $n = 5$ respond to treatment by either complete clinical remission ($n = 3$) or substantial clinical improvement ($n = 2$). Rest of the $n = 2$ remain non-responders. Same authors in another study report that $n = 4$ out of $n = 7$ respond to the treatment. $n = 2$ show complete clinical remission, while other $n = 4$ show substantial clinical improvement (Arzi et al. 2017). No clinical

Table 7.3 MSC's clinical studies on chronic gingivostomatitis, inflammatory bowel disease and perianal fistula in dog and cat

Clinical study design/ animal number/ duration	Biomaterial used	Evaluations	Overall result	References
Feline chronic gingivostomatitis				
Case series/ $n = 9$/ 24 weeks	Autologous AD-MSCs (20×10^6; 2 doses 30 days apart, intravenously)	Histology and immunohistochemistry, blood immune cell subsets, serum protein, and cytokine levels	Improved outcome	Arzi et al. (2016)
Case series/ $n = 7$/ 20 months	Allogeneic AD-MSCs (20×10^6; 2 doses 30 days apart, intravenously)	Histology and immunohistochemistry, bloodwork and lymphocyte phenotyping	Improved outcome	Arzi et al. (2017)
Multicentric cohort study (case series)/ $n = 18$/ 3 months	Autologous ($n = 13$) and allogeneic ($n = 5$) AD-MSCs (20×10^6, 2 doses 30 days apart, intravenously)	Histology and immunohistochemistry, bloodwork and lymphocyte phenotyping	Improved outcome	Arzi et al. (2020)
Dog inflammatory bowel disease				
Randomized placebo-controlled study/$n = 7$/ 6 weeks	Allogeneic AD-MSCs (2×10^6/kg b. wt. intravenously)	Clinical Inflammatory Bowel Disease Activity Index (CIBDAI) and Canine Chronic Enteropathy Clinical Activity Index (CCECAI), and C-reactive protein (CRP)	Improved outcome	Perez-Merino et al. (2015b)
Randomized open trial study/$n = 11$/ 120 days	AD-MSCs (2×10^6 cells/kg b. wt., intravenous)	Endoscopic index (CIBDEI), histological score (HS)	Improved outcome	Pérez-Merino et al. (2015a)
Randomized controlled study/$n = 32$/ 6 months	Allogeneic AD-MSCs (2×10^6/kg b. wt. intravenously, slow infusion)	Clinical inflammatory bowel disease activity index (CIBDAI) and the Canine Chronic Enteropathy Clinical Activity Index (CCECAI)	Improved outcome	Cristóbal et al. (2021)
Cat idiopathic inflammatory bowel disease				
Randomized placebo-controlled study/$n = 14$/ 2 months	Allogeneic AD-MSCs (2×10^6/kg b wt.; two doses 2 weeks apart)	Clinical evaluation based on questionnaire	Improved outcome	Webb and Webb (2015)
Dog perianal fistula				

(continued)

Table 7.3 (continued)

Clinical study design/ animal number/ duration	Biomaterial used	Evaluations	Overall result	References
Randomized open label trial/$n = 6$/ 6 months	Human embryonic stem cell derived MSCs + cyclosporine	Clinical evaluation	Improved outcome	Ferrer et al. (2016)

MSCs mesenchymal stem cells, *AD* adipose tissue derived, *n* total number of animals in all the groups of the study; Improved outcome: Indicates improved outcome with MSC transplantation even if only one parameter improves; Study period: Indicates maximum time period for which study is conducted

recurrence occurs post complete clinical remission up to 2 years (Arzi et al. 2016). Responders exhibit systemic immunomodulation as revealed by decrease in numbers of circulating CD8+ T cells, a normalization of the CD4/CD8 ratio, decrease in CD3+ T cells and CD20+ B cells, decrease in neutrophil counts and pro-inflammatory mediators [interferon-γ and interleukin (IL)-1β] concentration (Arzi et al. 2016, 2020). The full clinical recovery may occur immediately and can be expected in 2–4 months post intravenous cell transplantation. This delay in response may arise due to their systemic mechanism of activation inhibition of the new T cells and/or providing regulatory signals to the activated cells. This takes time to show effects as activated pathogenic cells must first undergo apoptosis or reach an exhausted state for recovery to occur. In relation to the autologous MSCs, allogeneic MSCs transplantation may show further delayed response. This might be due to their trapping in the capillary vasculature of the lungs to a greater extent (Arzi et al. 2017, 2020). The full clinical recovery remains concomitant with an increase in IL-6. However, cats that cure completely show higher TNF-α levels. Clinical response to AD-MSCs therapy correlates with the reduction of the CD8 T cells ($<15\%$) with a simultaneous low expression of CD8 ($CD8^{lo}$) cells. The cats having $>15\%$ $CD8^{lo}$ cells may remain non-responders (Arzi et al. 2016). Thus, CD8+ cells might be the main inflammatory cells involved in CGS in cats. As in these studies, IL-6 along with the TGF-β was also increased; further studies are desired especially in controlling the chronic autoimmune diseases. IL-6 mainly regulates balance between regulatory T cells (Tregs) and effector Th17 cells (Kimura and Kishimoto 2010). Th17 plays an important role in autoimmune diseases (Kimura and Kishimoto 2010) which is induced through the IL-6 and the TGF-β (Bettelli et al. 2006; Veldhoen et al. 2006; Dominitzki et al. 2007).

7.4 Oesophagus

Oesophagal healing is a bit of a challenge due to its anatomy and physiological activity as mentioned above. Usually, affections are either as the tissue loss or arise as the functional disorder. In animals, both small and large, the common ailment seen is the oesophageal choke, a medical emergency often requiring the surgical intervention (Haven 1990). In delayed cases of oesophageal obstruction, the tissue often becomes debilitated, necrosed while in many other cases, complications arise due to the strictures, diverticula, leakage of contents or broncho-oesophageal fistula post-surgical repair or trauma (Bezdekova 2012). In cases of neuro-physiological ailments like megaesophagus, concord in its peristalsis loses (dysmotility), and thus, difficulty in deglutition occurs leading to vomiting or reverse peristalsis (Bexfield et al. 2006). In this case, the neurological ailments need to be addressed, although such stem cell studies in animals are currently lacking. One of the commonly seen conditions is gastrointestinal reflux disease including the gastroesophageal reflux disease (GERD) seen in dogs and cats. Further, the oesophageal submucosal dissection (ESD) is now-a-days becoming a common procedure to remove oesophageal cancers (Soetikno et al. 2003). Such conditions either physiologically or through endoscopic dissection develop oesophageal ulcers (Rodríguez-Alarcón et al. 2015), and thus, as discussed above MSCs may play a promising role. There are stem cell studies that particularly focus on the tissue development through engineered cell/ scaffolds in oesophageal ulcer or the resection models (Table 7.2).

Oral epithelial cells transplanted post ESD engraft to the defect site and enhance epithelialization (Takagi et al. 2010). These cells lead to complete wound healing, with no observable stenosis at fourth week after endoscopic lower oesophageal mucosal dissection (Ohki et al. 2006). This shows that endoscopy along with the stem cell transplantation may be an effective way to prevent complications of the stricture and fibrosis. However, there is a need to evaluate long-term results for the ulceration including the tissue matrices formed and inflammation regulation. These findings may then be correlated with the cause of the ulceration.

Dog cervical oesophageal defect (5 cm in length and 50% in circumference) treated with acellular porcine small intestinal submucosa loaded with BM-MSCs or oral mucosal epithelial cells show improved outcome. Oesophageal lumen surface appears smoother in the cell−/scaffold-treated animals as compared to that of the scaffold-treated animals at fourth week. Histologically, inflammatory cells are well controlled and a greater similarity of morphology and function between BM-MSCs and the oesophageal myoblasts occurs. These cell−/scaffold-treated oesophageal defects develop long bundles of skeletal muscles with greater micro-vessel density by 12th week. However, in such a shorter duration, nerve fibre regeneration is mostly lacking (Wei et al. 2009; Tan et al. 2013). Nerve fibre regeneration may occur in the long run and therefore extended studies are desired. Besides, the studies conducted and animals incorporated are small enough to offer any conclusion(s).

In pig full-thickness oesophageal resection model (6 cm, mid-lower thoracic oesophagus) regeneration of the tissue with a feasible and reproducible tissue-engineered construct occurs. The synthetic polyurethane electro-spun grafts seeded with autologous AD-MSCs aid in the gradual structural regrowth of endogenous oesophageal tissue. This includes layers like squamous mucosa, submucosa and smooth muscle layers along with the blood vessels (La Francesca et al. 2018). Addition of copper to the scaffold (acellular porcine small intestinal submucosa) promotes such a re-epithelialization, re-vascularization and muscular regeneration in canine model (Tan et al. 2014), and thus, further studies are desired to evaluate the role of MSCs. Such types of studies are desired to be conducted in a detailed manner to confirm MSCs possible therapeutic mechanisms for tube-like structures like oesophagus, stomach and intestines. Comparative studies are desired to confirm efficacy against the currently available therapeutic options.

7.5 Intestines

In veterinary species, MSCs have been evaluated for specific types of enteropathies that too in small animals like dogs and cats. The studies include inflammatory bowel disease (IBD) and anal fistula. However, the potential may go beyond involving the neurological ailments, digestive disorders or in general the overall tissue regeneration.

7.5.1 Inflammatory Bowel Disease (IBS)

In small animals like dogs and cats and large ones like horses, IBD is a common chronic enteropathy resulting from the interactions of the complex multiple factors like microbes, environment and immunological complications, besides the genetic predisposition. However, the notable factor understood so far is the associated immunological complications (Janeczko et al. 2008; Trepanier 2009; Jergens et al. 2010). On the basis of histopathology, IBD is grouped as granulomatous enteritis, multisystemic eosinophilic epitheliotropic disease, lymphocytic-plasmacytic enterocolitis, diffuse eosinophilic enterocolitis and proliferative enteritis (Schumacher et al. 2000).

Currently, no specific treatment is available and is mostly managed by palliative therapeutics (corticosteroids or cyclosporine) to reduce intestinal mucosal inflammation (Dye et al. 2013). All these treatment options may not show desired clinical response and condition often relapses (Allenspach et al. 2006) with possible adverse effects (Stroup et al. 2006). IBD is associated with the compromise in the normal mucosal barrier increasing the gut permeability and exposure to commensal microbes leading to the mucosal immunity and hyper-tolerance to intestinal antigens (Janeczko et al. 2008; Trepanier 2009; Zimmermann et al. 2012). To understand the disease and develop effective therapeutics, MSCs are being evaluated. In dog IBD cases, higher binding of IgG to gut bacteria occurs with greater TNF-α production from macrophage (Soontararak et al. 2019). As chronic inflammatory diseases like IBD in dog and horse are triggered by Th17- or other IL17-secreting cells, MSCs may offer an effective therapy through regulation of inflammatory cytokines and M1 macrophages (Singer and Caplan 2011; Olofsson et al. 2015; Kol et al. 2016; Songa et al. 2019). In general, MSCs have the ability to produce immune modulators, antimicrobials after stimulation (Kol et al. 2014), besides other pro-healing factors, and these cells are considered to provide effective therapeutics in IBD. In cats, AD-MSC-derived PGE2 may prevent the inflammatory response by increasing FOXP3+ Tregs (An et al. 2018). Compared to humans, animal studies are very limited with numerous questions to be unanswered (Pérez-Merino et al. 2015a). Thus, MSCs are mainly looked at relieving the inflammation and symptoms (MacDonald and Barrett 2020).

In dogs, MSCs clinical efficacy for the inflammatory bowel disease is evaluated (Table 7.3). IBD treated with the similar doses of the AD-MSCs with or without the prednisone shows improved results. Improvement in clinical scores [Clinical Inflammatory Bowel Disease Activity Index (CIBDAI), Canine Chronic Enteropathy Clinical Activity Index (CCECAI), and C-reactive protein (CRP)] occurs up to 6 months (Pérez-Merino et al. 2015a, b; Cristóbal et al. 2021). Prednisone treatment along with MSCs leads to the sharp fall in the clinical signs as compared to the cells only that show gradual effect. No relapse in the cases occurs within the 6-month period (Cristóbal et al. 2021). Scores for endoscopy, gross examination and histology improve with the intravenous transplantation of MSCs. An average reduction of histological score of 27.2% may be achieved with no histologic remission in any of these dogs. Further, CRP levels reduce with the transplantation of the MSCs, a significant finding from other studies using other treatments (Pérez-Merino et al. 2015b). Immunosuppressants often elicit side effects that are not seen with the MSCs. As immunosuppressive drugs effect obtunds few years later, such unresponsiveness against MSCs remains to be evaluated. Thus, long-term studies at least for a few years are desired.

In feline chronic enteropathy clinical study (lymphocytic-plasmacytic enteritis; conducted blinded and non-blinded) (Table 7.3), two intravenous doses of allogeneic AD-MSCs at fortnight period appears effective. Out of $n = 7$ (blinded cases), $n = 5$ animal owners report significant improvement amounting to the complete clinical recovery, while $n = 2$ reports show persistently modest improvement. In non-blinded cases, $n = 1$ shows marked improvement with its equal number showing no change in symptoms or animal is lost to follow-up. Contrarily, placebo-treated cases report no change or even worsening of the symptoms with the time (Webb and Webb 2015).

Most of these studies are non-blinded and non-comparative with limited animals. Additional supplementary therapeutics may make results inadvertent. Although clinical improvement has been reported, it is worth noting that the possible therapeutic effects of MSCs need to be evaluated and whether all the cases are treated equally. This will help in elucidating the possible aetiological factors involved in dog IBD and their therapeutics. The exact mechanism although remains to be evaluated; however, MSCs potential immunosuppressive features like secretion of immunomodulatory factors like NO, TGF-β, and others like ability to secrete matrices and prevent its degradation may enhance mucosal integrity (Krampera et al. 2003; Beyth et al. 2005; Ren et al. 2008; Stappenbeck and Miyoshi 2009; Gugjoo and Pal 2020;

Gugjoo et al. 2020c). These cells may additionally improve small intestinal wall integrity by regulating the endogenous epithelial cell homeostasis (Semont et al. 2010).

7.5.2 Anal Furunculosis/Fistula

Anal furunculosis is one of the most common clinical inflammatory conditions seen in dog especially German shepherd (Day 1993). It arises as a result of deranged immune response to the intestinal or skin commensal organisms and shares histopathological features with the human counterpart (Ferrer et al. 2016). Currently, specific treatment that could effectively heal it once for all is lacking and the condition often relapses following the withdrawal of treatment.

A single open-label clinical trial incorporating refractory cases of perianal fistula ($n = 6$) shows that xenogeneic embryonic stem cell-derived MSCs provide an effective treatment strategy (Table 7.3). Transplantation of MSCs leads to the reduction in expression levels of serum inflammatory mediators (IL-2 and IL-6) post 1 month. A local injection of the cells along with cyclosporine leads to the closure of fistula by the third month with only half of the dose of cyclosporine required. In 33% of the cases, the condition relapses by sixth month (Ferrer et al. 2016). Serum cytokine levels are required to be evaluated beyond 1 month as cyclosporine dosage derange beyond this period. As MSCs are considered to inhibit both CD8+ and CD4+ cells and decrease inflammatory cytokines like IL-2 and IL-8, their application might help to resolve the signs. In comparison to humans, these studies are very limited in veterinary species and thus may be undertaken to evaluate potential therapeutic effects of MSCs in dogs.

7.5.3 Endotoxaemia

Endotoxaemia is not an uncommon problem in animals. In dogs, sheep and goats usually infections like Parvoviral enteritis or Clostridial enteritis lead to the endotoxaemia, while in horse colic leads to endotoxaemia caused by gram-negative infection (Harlan et al. 1983; Seekamp et al. 1992; Otto et al. 1997; Tinker et al. 1997; Thoefner et al. 2001; Senior et al. 2011). Clinically, endotoxaemia is associated with the release of the NO, production of thromboxane A2 and prostaglandin F2α, prostaglandin (PG) I2 and PGE2 that lead to hypotension. Release of various chemokines like ICAM-1 and VCAM-1 along with β2 integrins also enhances endothelial binding of neutrophils (Morris 1991; Jones 2015; MacDonald and Barrett 2020).

MSCs impart immunomodulation through cell–cell contact or through a secretome that harbours soluble factors or microvesicles. MSCs or their secretome tend to inhibit the T- and B- cell proliferation and simultaneously enhance regulatory T cell progression (Peroni and Borjesson 2011). These cells inhibit proliferation of the natural killer cells and block monocyte differentiation into immature dendritic cells and thereby their cytokine release like TNF-α production (Peroni and Borjesson

2011). In general, MSCs reduce circulating pro-inflammatory cytokines like TNF-α, IL-6, IL-1β and IFN-γ (Xu et al. 2007; Gonzalez-Rey et al. 2009; Mei et al. 2010; Gugjoo et al. 2020c). Currently, no study on veterinary species is available to demonstrate MSCs therapeutic effect against endotoxaemia. As the cells have potential immunomodulatory/anti-inflammatory effects, their potential role in endotoxaemia may be evaluated (MacDonald and Barrett 2020).

7.6 Liver

Liver performs metabolic and synthetic functions of the body (Miyajima et al. 2014). It has higher regeneration potential but does undergo various diseases that need special attention. The common diseases of liver in small animals are inflammatory (hepatitis), fibrotic, hepato-portal shunts and tumours (Watson 2004; Geraldine 2004; Bexfield et al. 2012; Eulenberg and Lidbury 2018). In ruminants, parasitic diseases like hydatidosis and fascioliasis, among others, are commonly reported that lead to liver dysfunction (Aminzare et al. 2018). To address these ailments and make the liver a fully functional organ, regenerative medicine employing MSCs are being evaluated. This may potentially provide translational research on stem cell regenerative therapy for hepatic diseases of humans.

7.6.1 In Vitro MSC Hepatocyte Differentiation

In liver, hepatocytes arise from endoderm-based hepatic progenitor cells that are available in limited concentrations (Ijzer et al. 2010). These progenitor cells have good regeneration power but fall short in some cases. To provide a viable and ready to use therapeutic agent, MSCs are being evaluated considering their extended differentiation potential into endoderm lineages including the hepatocyte-like cells. There are numerous studies that demonstrate their ability to differentiate into hepatocyte-like cells. MSCs differentiate into hepatocyte-like cells with cocktail of chemicals or growth factors. Employing growth factors like hepatocyte growth factor (HGF), fibroblast growth factor-4 (FGF-4) and basic fibroblast growth factor (bFGF), besides other culture ingredients differentiate cattle UC-MSCs (Xiong et al. 2014; Dueñas et al. 2014) and feline AD-MSCs (Fujimoto et al. 2020) into the hepatocyte-like cells. Both early progenitor genes (AFB) and a late mature hepatocyte genes (ALB) are expressed in such induced cells (Xiong et al. 2014). FGF-4 enhances hepatocyte growth and development, HGF enhances their glycogen synthesis and simultaneously inhibit their apoptosis (Jiang et al. 2002), while bFGF regulates their proliferation and motility (Kin et al. 1997). Other cocktails of chemicals that involve proteins like α-feroprotein, albumin, α1-antitrypsin and connexin 32 along with enzymes like tyrosine aminotransferase and cytochrome P450 differentiate cattle BM-MSCs into the hepatocyte-like cells. Extensive characterization shows the upregulation of ALB, CNX32 and cytochrome P34A4 (CYP3A4) and late markers (α1-AT and TAT) (Dueñas et al. 2014). Hepatocyte late markers (α1AT and TAT) degrade tyrosine and inhibit proteases (Rehman et al. 2004; Jedicke et al. 2014), while gap junction protein (CNX32) accounts for major liver connexin proteins (90%) (Nakashima et al. 2004). Apart from the growth factors or other proteins, hepatocyte-like differentiation of dog BM-MSCs may be effectuated through their transfection with retroviruses expressing Foxa1 and Hnf4a. Such a retroviral transfection of these cells changes their morphology towards circular to equilaterally circular. The hepatocyte-like cells show significantly higher expression of albumin (ALB) and cadherin (CDH) with an ability to perform hepatic functions (Nitta et al. 2020). Thus, MSCs have a tendency to differentiate towards endodermal lineage including hepatocyte-like cells.

7.6.2 In Vivo Liver Studies

7.6.2.1 Dog Liver Injury Model

In vivo regenerative potential of MSCs in hepatic models has mainly been evaluated in dogs among various veterinary species (Table 7.4). In dogs, both acute and chronic hepatic injury models have been evaluated. In acute or chronic injury model (carbon tetrachloride induced), autologous or allogeneic AD-MSCs (Matsuda et al. 2017; Teshima et al. 2017) or BM-MSCs (Nishimura et al. 2019) of variable doses improve the hepatic injury. MSCs are transplanted either in the peripheral veins (Matsuda et al. 2017) or the hepatic artery (Nishimura et al. 2019). MSCs after peripheral or intrahepatic artery transplantation home to the liver, although the cells transplanted through the former route may also get trapped in the lungs and thus are required in higher doses. An improvement in liver function enzymes (AST, AT and ALB) and parenchyma occurs with an ability to synthesize glycogen (Matsuda et al. 2017; Teshima et al. 2017; Nishimura et al. 2019; Yan et al. 2019).

To evaluate the effective route of cell transplantation, splenic vein transplanted MSCs significantly home to the liver and improve serum liver enzymes effectively as compared to the MSCs transplanted through peripheral vein. MSCs transplanted through either route lead to the significant reduction in inflammatory cytokines (IL-1, IL-6, IL-8 and IFNγ) and promote anti-inflammatory cytokines (IL-4 and IL-10, HGF and VEGFA). Although splenic vein transplanted cells are better engrafted to the liver than the peripheral vein, the cell concentration remains low pointing to their paracrine actions as the main therapeutic mechanism (Teshima et al. 2017). In chronic liver injury (MSCs transplanted after 10 weeks of injury), BM-MSCs transplanted through hepatic artery lead to the better outcome including the excretion of dye (indocyanine green) and reduction in liver infarction as compared to the peripheral vein transplanted MSC that shows better results as compared to the non-transplanted animals. Further higher dose of the cells to the extent of 1.2 million/kg b. wt. through intrahepatic artery does not elicit any adverse reaction (depicted by normal clinical parameters together with no arterial embolization and necrotic biliary areas in the liver tissue) (Nishimura et al. 2019). Thus, MSCs may be better implanted through splenic vein or hepatic artery, although for easy application, peripheral vein may also be opted.

Table 7.4 MSC's preclinical experimental studies on liver injuries and hepatocutaneous syndrome

Study design/ period	Biomaterial	Evaluation criteria	Overall result	References
Dog liver injury models				
Randomized placebo-controlled study ($n = 16$)/4 weeks	*Autologous BM-MSCs* (4×10^5/ kg)	CE-CT IMAGING, liver biopsy, histology, histomorphometry, DNA microarray analysis, PCR	Improved outcome	Matsuda et al. (2017)
Randomized placebo-controlled study ($n = 9$)/ 15 days	AD-MSCs [(2 doses of 2×10^6/kg wt at day 3 and 8 after induction through peripheral vein (PV) and splenic vein (SV)]	Histology, immunohistochemistry, blood examination and PCR	Improved outcome	Teshima et al. (2017)
Randomized placebo-controlled study ($n = 22$)/ 12 weeks	Autologous BM-MSCs (4×10^5 cells/kg, in left hepatic artery and peripheral vein after 10 weeks injury)	Computed tomography (CT), blood examination, LFT, liver biopsy, Histomorphometry	Improved outcome	Nishimura et al. (2019)
Randomized placebo-controlled study ($n = 6$)/ 5 days	AD-MSCs (1×10^7 peripheral vein 10 h after injury)	Serology, histopathology	Improved outcome	Yan et al. (2019)
Hepatocutaneous syndrome				
Case report ($n = 1$)/ 30 months	Allogeneic AD-MSCs (46 doses of $5X10^7$ amongst 8 intrahepatic, 3 weeks after diagnosis), hepato-protective treatment	Clinical evaluation and post mortem analysis	Improved outcome	Nam et al. (2017)

MSCs mesenchymal stem cells, *AD* adipose tissue derived, *BM* bone marrow derived, *n* total number of animals in all the groups of the study; Improved outcome: Indicates improved outcome with MSC transplantation even if only one parameter improves; Study period: Indicates maximum time period for which study is conducted

7.6.2.2 Hepatocutaneous Syndrome

In a clinical case of dog hepato-cutaneous syndrome (Table 7.4), multiple doses of allogeneic AD-MSCs (50 million) along with hepatoprotective drugs enhances the survival period to 30 months. However, simultaneous supplementary treatment with amino acid and other standard therapy may also have played a role in increased survival. Considering the normal survival period of 6 months with standard therapy, it may be said MSCs along with the standard therapy improves survival period. Post mortem analysis reveals the liver fibrosis and superficial necrolytic dermatitis (Nam et al. 2017). Thus, MSCs may prove effective and other options may be evaluated to further improve the animal condition and survival. However, the post mortem findings of liver fibrosis and superficial necrolytic dermatitis ask for cautious use of the MSCs.

MSCs might therapeutically reduce liver cell damage, prevent oxidative stress, downregulates inflammatory cytokines (TNF-α and IFN-γ), offers anti-inflammatory effects and immunosuppressive/immunoregulatory actions (Wang et al. 2009; Quintanilha et al. 2014). MSCs anti-inflammatory effects through regulation of macrophages and dendritic cells may indirectly inhibit stellate cells involved in fibrosis (Shin et al. 2016).

7.6.2.3 Hepatocellular Carcinoma

The commonest liver tumour that accounts for more than 50% tumour cases is hepatocellular carcinoma (HCC) (Liptak 2007; Zhao et al. 2021). Prognosis varies with the type of HCC, for massive, but focal HCC, it is good but in case of nodular and diffuse HCC it is poor (Liptak et al. 2004; Liptak 2007). AD-MSCs may significantly enhance proliferation (enhanced expression of transforming growth factor-beta 1, epidermal growth factor A, hepatocyte growth factor, platelet-derived growth factor-beta, vascular endothelial growth factor, and insulin-like growth factor) and invasion (enhanced expression of matrix metalloproteinase) of the dog HCC (Teshima et al. 2018). These studies make a note of caution for utilization of MSCs in tumours.

7.7 Conclusion(s)

GIT affections are characterized on the basis of the particular organ involvement, although combined affections of many organs may occur simultaneously. There is generally good epithelial regeneration potential of GIT; however, it may get compromised in some cases. To deal with such conditions, regenerative medicine employing MSCs appears promising. Regenerative medicine involving MSCs and scaffolds/grafts plays a significant therapeutic role in periodontal diseases. Scaffold/graft has a significant role to play and thus, due care is desired in its selections. Clinical studies remain to be conducted. In experimental oral ulcer, vocal fold and laryngotracheal injury models in dogs, MSCs although appear promising but standardization of cell source, dose, route of transplantation and frequency of dosing remains to be done. Besides, its actual clinical applicability and feasibility remains to be confirmed. In case of oesophageal defects, MSCs loaded on scaffolds also appear promising but the studies are currently very limited to reach to the conclusion. In relation to GERD and stomach affections, no MSCs study could be located. Intestinal affections like IBD, chronic enteropathy and anal furunculosis have been entertained clinically with MSCs and the results have improved. These studies are very limited, and in anal furunculosis, relapse of clinical signs may occur. Thus, further studies are desired in the area including the effect of MSCs frequency to provide evidence-based medicine for intestinal affections. MSCs differentiate into the hepatocyte-like cells although such a differentiation remains to be confirmed under the in vivo system. In the liver, experimental acute and chronic liver injury studies conducted in dogs show positive outcomes with MSCs. Thus, further extensive studies in all these areas and in other GIT affections like enterotoxaemia, hepato-cutaneous syndrome and hepatocellular carcinomas are desired. These studies should be uniform employing the cells of specific sources, passage, concentrations, route and frequency with sufficient numbers of animals to evaluate results statistically. These experimental studies may thereby be extended to the clinical settings for a comprehensive evaluation of safety, feasibility and finally efficacy of MSCs.

References

Akizuki T, Odam S, Komaki M, Tsuchioka H, Kawakatsu N, Kikuchi A, Yamato M, Okano T, Ishikawa I (2005) Application of periodontal ligament cell sheet for periodontal regeneration: a pilot study in beagle dogs. J Periodontal Res 40:245–251

Alamoudi NM, El Ashiry EA, Farsi NM, El Derwi DA, Atta HM (2014) Treatment of oral ulcers in dogs using adipose tissue-derived mesenchymal stem cells. J Clin Pediatr Dent 38(3):215–222

Albuquerque C, Morinha F, Requicha J, Martins T, Dias I, Guedes- Pinto H, Bastos E, Viegas C (2012) Canine periodontitis: the dog as an important model for periodontal studies. Vet J 191(3): 299–305

Allenspach K, Rüfenacht S, Sauter S, Gröne A, Steffan J, Strehlau G, Gaschen F (2006) Pharmacokinetics and clinical efficacy of cyclosporine treatment of dogs with steroid-refractory inflammatory bowel disease. J Vet Intern Med 20:239–244

Alpsoy E, Akman-Karakas A, Uzun S (2015) Geographic variations in epidemiology of two autoimmune bullous diseases: pemphigus and bullous pemphigoid. Arch Dermatol Res 307: 291–298

Aly LA, El-Menoufy H, Sadeq HS, Ragae A, Sabry D (2014) Efficiency of systemic versus intralesional bone marrow derived stem cells in regeneration of oral mucosa after induction of formocresol induced ulcers in dogs. Dent Res J (Isfahan) 11(2):212–221

Aminzare M, Hashemi M, Faz SY, Raeisi M, Hassanzadazar H (2018) Prevalence of liver flukes infections and hydatidosis in slaughtered sheep and goats in Nishapur, Khorasan Razavi, Iran. Vet World 11(2):146–150

An J-H, Song W-J, Li Q, Kim S-M, Yang J-I, Ryu M-O, Nam AR, Bhang DH, Jung Y-C, Youn H-Y (2018) Prostaglandin E2 secreted from feline adipose tissue-derived mesenchymal stem cells alleviate DSS-induced colitis by increasing regulatory T cells in mice. BMC Vet Res 14: 354

AnaTeresa S, Luis C, Lizeti T, Oliveira R (2008) Histological analysis of the association between formocresol and endotoxin in the subcutaneous tissue of mice. Braz Dent J 19:33–39

Arzi B, Murphy B, Cox DP et al (2010) Presence and quantification of mast cells in the gingival of cats with tooth resorption, periodontitis and chronic stomatitis. Arch Oral Biol 55:148–154

Arzi B, Mills-Ko E, Verstraete FJM, Kol A, Walker NJ, Badgley MR, Fazel N, Murphy WJ, Vapniarsky N, Borjesson DL (2016) Therapeutic efficacy of fresh, autologous mesenchymal stem cells for severe refractory gingivostomatitis in cats. Stem Cells Transl Med 5:75–86

Arzi B, Clark KC, Sundaram A, Spriet M, Verstraete FJM, Walker NJ, Loscar MR, Fazel N, Murphy WJ, Vapniarsky N, Borjesson DL (2017) Therapeutic efficacy of fresh, allogeneic mesenchymal stem cells for severe refractory feline chronic Gingivostomatitis. Stem Cells Transl Med 6(8):1710–1722

Arzi B, Peralta S, Fiani N, Vapniarsky N, Taechangam N, Delatorre U, Clark KC, Walker NJ, Loscar MR, Lommer MJ, Fulton A, Battig J, Borjesson DL (2020) A multicenter experience using adipose derived mesenchymal stem cell therapy for cats with chronic, non-responsive gingivostomatitis. Stem Cell Res Ther 11:115

Bettelli E, Carrier Y, Gao W, Korn T, Strom TB, Oukka M, Weiner HL, Kuchroo VK (2006) Reciprocal developmental pathways for the generation of pathogenic effector TH17 and regulatory T cells. Nature 441(7090):235–238

Bexfield NH, Watson PJ, Herrtage ME (2006) Esophageal dysmotility in young dogs. J Vet Intern Med 20(6):1314–1318

Bexfield NH, Buxton RJ, Vicek TJ, Day MJ, Bailey SM, Haugland SP et al (2012) Breed, age and gender distribution of dogs with chronic hepatitis in the United Kingdom. Vet J 193(1):124e8

Beyth S, Borovsky Z, Mevorach D, Liebergall M, Gazit Z, Aslan H, Galun E, Rachmilewitz J (2005) Human mesenchymal stem cells alter antigen-presenting 376 cell maturation and induce T-cell unresponsiveness. Blood 105:2214–2219

Bezdekova B (2012) Esophageal disorders in horses—a review of literature. Pferdeheilkunde 2: 187–192

Carstanjen B, Desbois C, Hekmati M, Behr L (2006) Successful engraftment of cultured autologous mesenchymal stem cells in a surgically repaired soft palate defect in an adult horse. Can J Vet Res 70:143–147

Chen Z, Chen Q, Du H, Xu L, Wan J (2018) Mesenchymal stem cells and CXC chemokine receptor 4 overexpression improved the therapeutic effect on colitis via mucosa repair. Exp Ther Med 16: 821–829

Cho H, Tarafder S, Fogge M, Kao K, Lee CH (2016) Periodontal ligament stem/progenitor cells with protein-releasing scaffolds for cementum formation and integration on dentin surface. Connect Tissue Res 57(6):488–495

Ciccocioppo R, Fioravante M, Frulloni L (2018) Mesenchymal stromal cell therapy in intestinal diseases. Curr Opin Organ Transplant 23:679–689

Colyer JF (1906) Variations and diseases of the teeth of horses. Trans Odontol Soc Great Br New Ser 38:42–47

Cristóbal JI, Duque FJ, Usón-Casaús JM, Ruiz P, Nieto EL, Pérez-Merino EM (2021) Effects of allogeneic mesenchymal stem cell transplantation in dogs with inflammatory bowel disease treated with and without corticosteroids. Animals 11:2061

Day M (1993) Immunopathology of anal furunculosis in the dog. J Small Anim Pract 34(8): 381–389

Dominitzki S, Fantini MC, Neufert C et al (2007) Cutting edge: trans-signaling via the soluble IL-6R abrogates the induction of FoxP3 in naive CD4+CD25 T cells. J Immunol 179:2041–2045

Dong X, Feng X, Liu J, Xu Y, Pan Q, Ling Z, Yu J, Yang J, Li L, Cao H (2019) Characteristics of intestinal microecology during mesenchymal stem cell-based therapy for mouse acute liver injury. Stem Cells Int 2019:2403793

Dueñas F, Becerra V, Cortes Y, Vidal S, Sáenz L, Palomino J, De Los Reyes M, Peralta OA (2014) Hepatogenic and neurogenic differentiation of bone marrow mesenchymal stem cells from abattoir-derived bovine fetuses. BMC Vet Res 10:154

Dye T, Diehl K, Wheeler S, Westfall D (2013) Randomized, controlled trial of budesonide and prednisone for the treatment of idiopathic inflammatory bowel disease in dogs. J Vet Intern Med 27:1385–1391

El-Menoufy H, Aly LA, Aziz MT, Atta HM, Roshdy NK, Rashed LA, Sabry D (2010) The role of bone marrow-derived mesenchymal stem cells in treating formocresol induced oral ulcers in dogs. J Oral Pathol Med 39:281–289

Esposito M, Grusovin MG, Papanikolaou N, Coulthard P, Worthington HV (2009) Enamel matrix derivative (Emdogain(R)) for periodontal tissue regeneration in intrabony defects. Cochrane Database Syst Rev 2009:CD003875

Eulenberg VM, Lidbury JA (2018) Hepatic fibrosis in dogs. J Vet Intern Med 32:26e41

Ferrand J, Lehours P, Schmid-Alliana A, Mégraud F, Varon C (2011) *Helicobacter pylori* infection of gastrointestinal epithelial cells *in vitro* induces mesenchymal stem cell migration through an NF-κB-dependent pathway. PLoS One 6(12):e29007

Ferrer L, Kimbrel EA, Lam A, Falk EB, Zewe C, Juopperi T, Lanza R, Hoffman A (2016) Treatment of perianal fistulas with human embryonic stem cell-derived mesenchymal stem cells: a canine model of human fistulizing Crohn's disease. Regen Med 11(1):33–43

Field EA, Longman LP (2003) Tyldesley's oral medicine, 5th edn. Oxford University Press, Oxford

Fu Y, Ni J, Chen J, Ma G, Zhao M, Zhu S, Shi T, Zhu J, Huang Z, Zhang J, Chen J (2020) Dual-functionalized MSCs that express CX3CR1 and IL-25 exhibit enhanced therapeutic effects on inflammatory bowel disease. Mol Ther 28(4):1214–1228

Fujimoto Y, Yokozeki T, Yokoyama A, Tabata Y (2020) Basic fibroblast growth factor enhances proliferation and hepatocyte growth factor expression of feline mesenchymal stem cells. Regener Ther 15:10–17

Gao J-G, Yu M-S, Zhang M-M, Gu X-W, Ren Y, Zhou X-X, Chen D, Yan T-L, Li Y-M, Jin X (2020) Adipose-derived mesenchymal stem cells alleviate TNBS-induced colitis in rats by influencing intestinal epithelial cell regeneration, Wnt signaling, and T cell immunity. World J Gastroenterol 26(26):3750–3766

Geraldine BH (2004) Effect of breed on anatomy of Porto-systemic shunts resulting from congenital diseases in dogs and cats: a review of 242 cases. Aust Vet J 82(12):746e9

Gerritsen M, Tomlinson J, Zlot C, Ziman M, Hwang S (2003) Using gene expression profiling to identify the molecular basis of the synergistic actions of hepatocyte growth factor and vascular endothelial growth factor in human endothelial cells. Br J Pharmacol 140:595–610

Glickman LT, Glickman NW, Moore GE, Goldstein GS, Lewis HB (2009) Evaluation of the risk of endocarditis and other cardiovascular events on the basis of the severity of periodontal disease in dogs. J Am Vet Med Assoc 234:486–494

Gonzalez-Rey E, Anderson P, Gonzalez MA, Rico L, Buscher D, Delgado M (2009) Human adult stem cells derived from adipose tissue protect against experimental colitis and sepsis. Gut 58: 929–939

Gugjoo MB, Pal A (2020) Mesenchymal stem cell immuno-modulatory and/anti-inflammatory properties. In: Gugjoo MB, Pal A (eds) Mesenchymal stem cells in veterinary sciences. Springer Nature, Berlin, pp 47–66

Gugjoo MB, Amarpal I, Fazili MR, Shah RA, Mir MS, Sharma GT (2020a) Goat mesenchymal stem cell basic research and potential applications. Small Rum Res 183:106045

Gugjoo MB, Hussain S, Shah RA, Dhama K (2020b) Mesenchymal stem cell-mediated immuno-modulatory and anti-inflammatory mechanisms in immune and allergic disorders. Inflamm Allergy Drug Targets 14(1):3–14

Gugjoo MB, Pal A, Sharma GT (2020c) Mesenchymal stem cell and its properties. In: Gugjoo MB, Pal A (eds) Mesenchymal stem cell in veterinary sciences. Springer Nature, Berlin, pp 13–26

Han X, Liu H, Wang D, Su F, Zhang Y, Zhou W, Li S, Yang R (2011) Alveolar bone regeneration around immediate implants using an injectable nHAC/CSH loaded with autogenic blood-acquired mesenchymal progenitor cells: an experimental study in the dog mandible. Clin Implant Dent Relat Res 15:390–401

Harlan J, Winn R, Hildebrandt J, Harker L (1983) Selective inhibition of thromboxane synthesis during experimental endotoxemia in the goat: effects on pulmonary haemodynamics and lung lymph flow. Br J Clin Pharmacol 15(1):123S–126S

Hasegawa N, Kawaguchi H, Hirachi A et al (2006) Behavior of transplanted bone marrow-derived mesenchymal stem cells in periodontal defects. J Periodontol 77:1003–1007

Haven ML (1990) Bovine esophageal surgery. Vet Clin North Am Food Anim Pract 6(2):359–369

Hu R, Ling W, Xu W, Han D (2014) Fibroblast-like cells differentiated from adipose-derived mesenchymal stem cells for vocal fold wound healing. PLoS One 9(3):e92676

Ijzer J, Schotanus BA, Vander Borght S, Roskams TA, Kisjes R, Penning LC, Rothuizen J, van den Ingh TS (2010) Characterisation of the hepatic progenitor cell compartment in normal liver and in hepatitis: an immunohistochemical comparison between dog and man. Vet J 184(3):308–314

Iravani K, Sobhanmanesh A, Ashraf MJ, Hashemi SB, Mehrabani D, Zare S (2017) The healing effect of conditioned media and bone marrow-derived stem cells in Laryngotracheal stenosis: a comparison in experimental dog model. World J Plastic Surg 6(2):190–197

Janeczko S, Atwater D, Bogel E et al (2008) The relationship of mucosal bacteria to duodenal histopathology, cytokine mRNA, and clinical disease activity in cats with inflammatory bowel disease. Vet Microbiol 128:178–193

Jedicke N, Struever N, Aggrawal N, Welte T, Manns MP, Malek NP, Zender L, Janciauskiene S, Wuestefeld T (2014) Alpha-1-antitrypsin inhibits acute liver failure in mice. Hepatology 59: 2299–2308

Jennings MW, Lewis JR, Soltero-Rivera MM et al (2015) Effect of tooth extraction on stomatitis in cats: 95 cases (2000–2013). J Am Vet Med Assoc 246:654–660

Jergens A, Crandell J, Morrison J, Deitz K, Pressel M, Ackermann M, Suchodolski J, Steiner J, Evans R (2010) Comparison of Oral prednisone and prednisone combined with metronidazole for induction therapy of canine inflammatory bowel disease: a randomized-controlled trial. J Vet Intern Med 24:269–277

Jiang Y, Jahagirdar BN, Reinhardt RL, Schwartz RE, Keene CD, OrtizGonzalez XR, Reyes M, Lenvik T, Lund T, Blackstad M (2002) Pluripotency of mesenchymal stem cells derived from adult marrow. Nature 418:41–49

Johnson V, Webb T, Norman A, Coy J, Kurihara J, Regan D, Dow S (2017) Activated mesenchymal stem cells interact with antibiotics and host innate immune responses to control chronic bacterial infections. Sci Rep 7:9575

Jones S (2015) Diseases of the equine alimentary tract. Large animal internal medicine, 5th edn. Elsevier, St. Louis, MO, pp 638–842

Kanemaru S, Nakamura T, Omori K, Kojima H, Magrufov A, Hiratsuka Y, Hirano S, Ito J, Shimizu Y (2003) Regeneration of the vocal fold using autologous mesenchymal stem cells. Ann Otol Rhinol Laryngol 112:915–920

Kawaguchi H, Hirachi A, Hasegawa N et al (2004) Enhancement of periodontal tissue regeneration by transplantation of bone marrow mesenchymal stem cells. J Periodontol 75:1281–1287

Khorsand A, Eslaminejad MB, Arabsolghar M, Paknejad M, Ghaedi B, Rokn AR, Moslemi N, Nazarian H, Jahangir S (2013) Autologous dental pulp stem cells in regeneration of defect created in canine periodontal tissue. J Oral Implantol 39(4):433–443

Kimura A, Kishimoto T (2010) IL-6: regulator of Treg/Th17 balance. Eur J Immunol 40(7): 1830–1835

Kin M, Sata M, Ueno T, Torimura T, Inuzuka S, Tsuji R, Sujaku K, Sakamoto M, Sugawara H, Tamaki S, Tanikawa K (1997) Basic fibroblast growth factor regulates proliferation and motility of human hepatoma cells by an autocrine mechanism. J Hepatol 27(4):677–687

Klugh DO (2006) A review of equine periodontal disease. AAEP Proc 52:551–558

Kol A, Foutouhi S, Walker NJ, Kong NT, Weimer BC, Borjesson DL (2014) Gastrointestinal microbes interact with canine adipose-derived mesenchymal stem cells in vitro and enhance immunomodulatory functions. Stem Cells Dev 23:1831–1843

Kol A, Walker NJ, Nordstrom M, Borjesson DL (2016) Th17 pathway as a target for multipotent stromal cell therapy in dogs: implications for translational research. PLoS One 11(2):e0148568

Krampera M, Glennie S, Dyson J, Scott D, Laylor R, Simpson E, Dazzi F (2003) Bone marrow mesenchymal stem cells inhibit the response of naive and memory antigen-specific T cells to their cognate peptide. Blood 101:3722–3729

Kumai Y, Kobler JB, Park H, Galindo M, Herrera VL, Zeitels SM (2010) Modulation of vocal fold scar fibroblasts by adipose-derived stem/stromal cells. Laryngoscope 120:330–337

La Francesca S, Aho JM, Barron MR, Blanco EW, Soliman S, Kalenjian L, Hanson AD, Todorova E, Marsh M, Burnette KL, DerSimonian H, Odze RD, Wigle DA (2018) Long-term regeneration and remodeling of the pig esophagus after circumferential resection using a retrievable synthetic scaffold carrying autologous cells. Sci Rep 8(1):4123

Lazebnik LB, Konopliannikov AG, Kniazev OV, Parfenov AI, Tsaregorodtseva TM, Ruchkina IN, Khomeriki SG, Rogozina VA, Konopliannikova OA (2010) Use of allogeneic mesenchymal stem cells in the treatment of intestinal inflammatory diseases. Ter Arkh 82(2):38–43

Lee BJ, Wang SG, Lee JC, Jung JS, Bae YC, Jeong HJ, Kim HW, Lorenz RR (2006) The prevention of vocal fold scarring using autologous adipose tissue-derived stromal cells. Cells Tissues Organs 184:184–204

Lee JK, Park SR, Jung BK, Jeon YK, Lee YS, Kim MK, Kim YG, Jang JY, Kim CW (2013) Exosomes derived from mesenchymal stem cells suppress angiogenesis by down-regulating VEGF expression in breast cancer cells. PLoS One 8:e84256

Li H, Yan F, Lei L, Li Y, Xiao Y (2009) Application of autologous cryopreserved bone marrow mesenchymal stem cells for periodontal regeneration in dogs. Cells Tissues Organs 190:94–101

Li X, Wang Q, Ding L, Wang Y-X, Zhao Z-D, Mao N, Wu C-T, Wang H, Zhu H, Ning S-B (2019) Intercellular adhesion molecule-1 enhances the therapeutic effects of MSCs in a dextran sulfate sodium-induced colitis models by promoting MSCs homing to murine colons and spleens. Stem Cell Res Ther 10:267

Liptak JM (2007) Cancer of the gastrointestinal tract. In: Withrow & MacEwen's small animal clinical oncology, 4th edn. Saunders Elsevier, St. Louis, MO, pp 483–491

Liptak JM, Dernell WS, Monnet E, Powers BE, Bachand AM, Kenney JG et al (2004) Massive hepatocellular carcinoma in dogs: 48 cases (1992–2002). J Am Vet Med Assoc 225:1225–1230

Liu J, Jiang D (2021) Role of mesenchymal and other stem cell therapy in intestinal diseases. In: Stem cell innovation in health & disease. Associated Press, New York, NY, pp 147–163

Long JD, Orlando RC (2002) Anatomy, histology, embryology, and developmental abnormalities of the esophagus. In: Feldman M, Fieldman LS, Sleisenger MH et al (eds) Gastrointestinal and liver diseases. WB Saunders, Philadelphia, pp 551–560

Lopez-Santalla M, Garin MI (2021) Improving the efficacy of mesenchymal stem/stromal-based therapy for treatment of inflammatory bowel diseases. Biomedicine 9(11):1507

Luissint AC, Parkos CA, Nusrat A (2016) Inflammation and the intestinal barrier: leukocyte–epithelial cell interactions, cell junction remodeling, and mucosal repair. Gastroenterology 151:616–632

MacDonald ES, Barrett JG (2020) The potential of mesenchymal stem cells to treat systemic inflammation in horses. Front Vet Sci 6:507

Marei MK, Saad MM, El-Ashwah AM, El-Backly RM, Al-Khodary MA (2009) Experimental formation of periodontal structure around titanium implants utilizing bone marrow mesenchymal stem cells: a pilot study. J Oral Implantol 35(3):106–129

Matsuda T, Takami T, Sasaki R, Nishimura T, Aibe Diaz Paredes B, Quintanilha LF, Matsumoto T, Ishikawa T, Yamamoto N, Tani K et al (2017) A canine liver fibrosis model to develop a therapy for liver cirrhosis using cultured bone marrow–derived cells. Hepatol Commun 1(7):691–703

Mei SH, Haitsma JJ, Dos Santos CC, Deng Y, Lai PF, Slutsky AS et al (2010) Mesenchymal stem cells reduce inflammation while enhancing bacterial clearance and improving survival in sepsis. Am J Respir Crit Care Med 182:1047–1057

Miyajima A, Tanaka M, Itoh T (2014) Stem/progenitor cells in liver development, homeostasis, regeneration, and reprogramming. Cell Stem Cell 14(5):561e74

Morris DD (1991) Endotoxemia in horses. A review of cellular and humoral mediators involved in its pathogenesis. J Vet Intern Med 5:167–181

Murashima Y, Yoshikawa G, Wadachi R, Sawada N, Suda H (2002) Calcium sulfate as a bone substitute for various osseous defects in conjunction with apicectomy. Int Endod J 35:768–774

Nagahara T, Yoshimatsu S, Shiba H, Kawaguchi H, Takeda K, Iwata T, Mizuno N, Fujita T, Kurihara H (2015) Introduction of a mixture of β-tricalcium phosphate into a complex of bone marrow mesenchymal stem cells and type I collagen can augment the volume of alveolar bone without impairing cementum regeneration. J Periodontol 86(3):456–464

Nakashima Y, Ono T, Yamanoi A, El-Assal ON, Kohno H, Nagasue N (2004) Expression of gap junction protein connexin32 in chronic hepatitis, liver cirrhosis, and hepatocellular carcinoma. J Gastroenterol 39:763–768

Nam A, Han M-S, Go D-M, Kim D-Y, Seo K-W, Youn H-Y (2017) Long-term management with adipose tissue-derived mesenchymal stem cells and conventional treatment in a dog with hepatocutaneous syndrome. J Vet Intern Med 31:1514–1519

Niemiec BA (2008a) Periodontal disease. Top Companion Anim Med 23:72–80
Niemiec BA (2008b) Periodontal therapy. Top Companion Anim Med 23:81–90
Nishimura T, Takami T, Sasaki R, Aibe Y, Matsuda T, Fujisawa K, Matsumoto T, Yamamoto N, Tani K, Taura Y, Sakaida I (2019) Liver regeneration therapy through the hepatic artery-infusion of cultured bone marrow cells in a canine liver fibrosis model. PLoS One 14(1): e0210588
Nitta S, Kusakari Y, Yamada Y, Kubo T, Neo S, Igarashi H, Hisasue M (2020) Conversion of mesenchymal stem cells into a canine hepatocyte-like cells by Foxa1 and Hnf4a. Regener Ther 14:165–176
Ohki T, Yamato M, Murakami D, Takagi R, Yang J, Namiki H, Okano T, Takasaki K (2006) Treatment of oesophageal ulcerations using endoscopic transplantation of tissue-engineered autologous oral mucosal epithelial cell sheets in a canine model. Gut 55(12):1704–1710
Olofsson KM, Hjertner B, Fossum C, Press CM, Lindberg R (2015) Expression of T helper type 17 (Th17)-associated cytokines and toll-like receptor 4 and their correlation with Foxp3 positive cells in rectal biopsies of horses with clinical signs of inflammatory bowel disease. Vet J 206: 97–104
Otto CM, Drobatz KJ, Soter C (1997) Endotoxemia and tumor necrosis factor activity in dogs with naturally occurring parvoviral enteritis. J Vet Intern Med 11(2):65–70
Paknejad M, Eslaminejad MB, Ghaedi B, Rokn A-R, Khorsand A, Etemad-Moghadam S, Alaeddini M, Dehghan MM, Moslemi N, Nowzari H (2015) Isolation and assessment of mesenchymal stem cells derived from bone marrow: histologic and histomorphometric study in a canine periodontal defect. J Oral Implantol 41:284–291
Pavlica Z, Petelin M, Juntes P, Erzen D, Crossley DA, Skaleric U (2008) Periodontal disease burden and pathological changes in organs of dogs. J Vet Dent 25:97–105
Peddle GD, Drobatz KJ, Harvey CE, Adams A, Sleeper MM (2009a) Association of periodontal disease, oral procedures, and other clinical findings with bacterial endocarditis in dogs. J Am Vet Med Assoc 234:100–107
Peddle GD, Sleeper MM, Ryan MJ, Kittleson MD, Pion P (2009b) Questions validity of study on periodontal disease and cardiovascular events in dogs. J Am Vet Med Assoc 234:1525–1528
Pérez-Merino EM, Uson-Casaus JM, Duque-Carrasco J, Zaragoza-Bayle C, Marinas-Pardo L, Hermida-Prieto M, Vilafrnca-Compte M, Barrera-Chacon R, Gualtieri M (2015a) Safety and efficacy of allogeneic adipose tissue-derived mesenchymal stem cells for treatment of dogs with inflammatory bowel disease: endoscopic and histological outcomes. Vet J 206:391–397
Pérez-Merino EM, Uson-Casaus JM, Zaragoza-Bayle C, Duque-Carrasco J, Marinas-Pardo L, Hermida-Prieto M, Barrera-Chacon R, Gualtieri M (2015b) Safety and efficacy of allogeneic adipose tissue-derived mesenchymal stem cells for treatment of dogs with inflammatory bowel disease: clinical and laboratory outcomes. Vet J 206:385–390
Peroni JF, Borjesson DL (2011) Anti-inflammatory and immunomodulatory activities of stem cells. Vet Clin North Am Equine Pract 27:351–362
Perry R, Tutt C (2015) Periodontal disease in cats: back to basics--with an eye on the future. J Feline Med Surg 17(1):45–65
Porter SR, Scully C (2000) Adverse drug reactions in the mouth. Clin Dermatol 18:525–532
Pourebrahim N, Hashemibeni B, Shahnaseri S, Torabinia N, Mousavi B, Adibi S, Heidari F, JafaryAlavi M (2013) A comparison of tissue engineered bone from adipose derived stem cell with autogenous bone repair in maxillary alveolar cleft model in dogs. Int J Oral Maxillofac Surg 42:562–568
Quintanilha LF, Takami T, Hirose Y, Fujisawa K, Murata Y, Yamamoto N et al (2014) Canine mesenchymal stem cells show antioxidant properties against thioacetamide-induced liver injury in vitro and in vivo. Hepatol Res 44:E206–E217
Rehman KK, Ayesha Q, Khan AA, Ahmed N, Habibullah CM (2004) Tyrosine aminotransferase and gamma-glutamyl transferase activity in human fetal hepatocyte primary cultures under proliferative conditions. Cell Biochem Funct 22:89–96

Ren G, Zhang L, Zhao X, Xu G, Zhang Y, Roberts AI, Zhao RC, Shi Y (2008) Mesenchymal stem cell-mediated immunosuppression occurs via concerted action of chemokines and nitric oxide. Cell Stem Cell 2:141–150

Rodríguez-Alarcón CA, Beristain-Ruiz DM, Rivera-Barreno R, Díaz G, Usón-Casaús JM, García-Herrera R, Pérez-Merino EM (2015) Gastroesophageal reflux in anesthetized dogs: a review. Rev Colomb Cienc Pecu 28:144–155

Rousseau B, Hirano S, Scheidt TD, Welham NV, Thibeault SL, Ford CN, Bless DM (2003) Characterization of vocal fold scarring in a canine model. Laryngoscope 113:620–627

Schumacher J, Edwards JF, Cohen ND (2000) Chronic idiopathic inflammatory bowel diseases of the horse. J Vet Intern Med 14:258–265

Seekamp A, Dwenger A, Weidner M, Regel G, SturmJ A (1992) Phagocytic cell function in recurrent endotoxemia in sheep. Circ Shock 37(4):263–269

Sémont A, Mouiseddine M, François A, Demarquay C, Mathieu N, Chapel A, Saché A, Thierry D, Laloi P, Gourmelon P (2010) Mesenchymal stem cells improve small intestinal integrity through regulation of endogenous epithelial cell homeostasis. Cell Death Differ 17(6):952–961

Semont A, Mouiseddine M, François A, Demarquay C, Mathieu N, Chapel A, Saché A, Thierry D, Laloi P, Gourmelon P (2010) Mesenchymal stem cells improve small intestinal integrity through regulation of endogenous epithelial cell homeostasis. Cell Death Differ 17:952–961

Senior JM, Proudman CJ, Leuwer M, Carter SD (2011) Plasma endotoxin in horses presented to an equine referral hospital: correlation to selected clinical parameters and outcomes. Equine Vet J 43:585–591

Shi S, Gronthos S (2003) Perivascular niche of postnatal mesenchymal stem cells in human bone marrow and dental pulp. J Bone Miner Res 18:696–704

Shin TH, Kim HS, Kang TW, Lee BC, Lee HY, Kim YJ et al (2016) Human umbilical cord blood-stem cells direct macrophage polarization and block inflammasome activation to alleviate rheumatoid arthritis. Cell Death Dis 7:e2524

Simsek SB, Keles GC, Baris S, Cetinkaya BO (2012) Comparison of mesenchymal stem cells and autogenous cortical bone graft in the treatment of class II furcation defects in dogs. Clin Oral Investig 16:251–258

Singer NG, Caplan AI (2011) Mesenchymal stem cells: mechanisms of inflammation. Annu Rev Pathol 6:457–478

Soetikno RM, Gotoda T, Nakanishi Y, Soehendra N (2003) Endoscopic mucosal resection. Gastrointest Endosc 57(4):567–579

Songa W-J, Lia Q, Ryua M-O, Nama A, Ana J-H, Jungb YC, Ahnc J-O, Youn H-Y (2019) Canine adipose tissue-derived mesenchymal stem cells pre-treated with TNF-alpha enhance immunomodulatory effects in inflammatory bowel disease in mice. Res Vet Sci 125:176–184

Soontararak S, Chow L, Johnson V, Coy J, Webb C, Wennogle S et al (2019) Humoral immune responses against gut bacteria in dogs with inflammatory bowel disease. PLoS One 14: e0220522

Splechler SJ (1995) Complications of gastroesophageal reflux disease. In: Castell DO (ed) The esophagus. Little Brown, New York, pp 533–545

Stappenbeck TS, Miyoshi H (2009) The role of stromal stem cells in tissue regeneration 488 and wound repair. Science 324:1666–1669

Stroup ST, Behrend EN, Kemppainen RJ, Smith-Carr S (2006) Effects of oral administration of controlled-ileal-release budesonide and assessment of pituitary-adrenocortical axis suppression in clinically normal dogs. Am J Vet Res 67:1173–1178

Suaid FF, Ribeiro FV, E Silva Gomes TRL, Silvério KG, Carvalho MD, Nociti FH Jr, Casati MZ, Sallum EA (2011) Autologous periodontal ligament cells in the treatment of class III furcation defects: a study in dogs. J Clin Periodontol 39:377–384

Suaid FF, Ribeiro FV, Silva Gomes TRLE, Gonzales Silvério K, Carvalho MD, Nociti FH Jr, Casati MZ, Antônio SE (2012) Autologous periodontal ligament cells in the treatment of class III furcation defects: a study in dogs. J Clin Periodontol 39(4):377–384

Takagi R, Murakami D, Kondo M, Ohki T, Sasaki R, Mizutani M, Yamato M, Nishida K, Namiki H, Yamamoto M, Okano T (2010) Fabrication of human oral mucosal epithelial cell sheets for treatment of esophageal ulceration by endoscopic submucosal dissection. Gastrointest Endosc 72(6):1253–1259

Takewaki M, Kajiya M, Takeda K, Sasaki S, Motoike S, Komatsu N, Matsuda S, Ouhara K, Mizuno N, Fujita T et al (2017) MSC/ECM cellular complexes induce periodontal tissue regeneration. J Dent Res 96(9):984–991

Tan B, Wei R-Q, Tan M-Y, Luo J-C, Deng L, Chen X-H, Hou J-L, Li X-Q, Yang Z-M, Xie H-Q (2013) Tissue engineered esophagus by mesenchymal stem cell seeding for esophageal repair in a canine model. J Surg Res 182(1):40–48

Tan B, Wang M, Chen X, Hou JL, Chen XH, Wang Y, Li-Ling J, Xie HQ (2014) Tissue engineered esophagus by copper—small intestinal submucosa graft for esophageal repair in a canine model. Sci China Life Sci 57(2):248–255

Tarnawski A, Hollander D, Stachura J, Gergely H, Krause W, Sarfeh J (1991) Role of angiogenesis in healing of experimental gastric ulcer. In: Halter F, Garner A, Tytgat GNJ (eds) Mechanisms of peptic ulcer healing. Kluwer Academic, Dordrecht/Boston/London, pp 165–171

Teshima T, Matsumoto H, Michishita M, Matsuoka A, Shiba M, Nagashima T, Koyama H (2017) Allogenic adipose tissue-derived mesenchymal stem cells ameliorate acute hepatic injury in dogs. Stem Cells Int 2017:3892514

Teshima T, Matsumoto H, Koyama H (2018) Soluble factors from adipose tissue-derived mesenchymal stem cells promote canine hepatocellular carcinoma cell proliferation and invasion. PLoS One 13(1):e0191539

Thoefner MB, Ersboll AK, Jensen AL, Hesselholt M (2001) Factor analysis of the interrelationships between clinical variables in horses with colic. Prev Vet Med 48:201–214

Tinker MK, White NA, Lessard P, Thatcher CD, Pelzer KD, Davis B et al (1997) Prospective study of equine colic incidence and mortality. Equine Vet J 29:448–453

Tobita M, Uysal CA, Guo X, Hyakusoku H, Mizuno H (2013) Periodontal tissue regeneration by combined implantation of adipose tissue-derived stem cells and platelet-rich plasma in a canine model. Cytotherapy 15(12):1517–1526

Trepanier L (2009) Idiopathic inflammatory bowel disease in cats. Rational treatment selection. J Feline Med Surg 11:32–38

Tsumanuma Y, Iwata T, Washio K, Yoshida T, Yamada A, Takagi R, Ohno T, Lin K, Yamato M, Ishikawa I et al (2011) Comparison of different tissue-derived stem cell sheets for periodontal regeneration in a canine 1-wall defect model. Biomaterials 32:5819–5825

Vallabhaneni KC, Penfornis P, Dhule S, Guillonneau F, Adams KV, Mo YY, Xu R, Liu Y, Watabe K, Vemuri MC, Pochampally R (2015) Extracellular vesicles from bone marrow mesenchymal stem/stromal cells transport tumor regulatory microRNA, proteins, and metabolites. Oncotarget 6:4953–4967

Veldhoen M, Hocking RJ, Atkins CJ, Locksley RM, Stockinger B (2006) TGFbeta in the context of an inflammatory cytokine milieu supports de novo differentiation of IL-17-producing T cells. Immunity 24(2):179–189

Villatoro AJ, Martín-Astorga MC, Alcoholado C, Cárdenas C, Fariñas F, Becerra J, Visser R (2021) Altered proteomic profile of adipose tissue-derived mesenchymal stem cell exosomes from cats with severe chronic gingivostomatitis. Animals (Basel) 11(8):2466

Vito D, Lidia D, Ermira Z et al (2007) A case of oesophageal ulcer developed after taking homeopathic pill in a young woman. World J Gastroenterol 13(14):2132–2134

Wang Y, Zhang A, Ye Z, Xie H, Zheng S (2009) Bone marrow-derived mesenchymal stem cells inhibit acute rejection of rat liver allografts in association with regulatory T-cell expansion. Transplant Proc 41:4352–4356

Watson PJ (2004) Chronic hepatitis in dogs: a review of current understanding of the aetiology, progression, and treatment. Vet J 167:228–241

Webb TL, Webb CB (2015) Stem cell therapy in cats with chronic enteropathy: a proof-of-concept study. J Feline Med Surg 17(10):901–908

Wei R-Q, Tan B, Tan M-Y, Luo J-C, Deng L, Chen X-H, Li X-Q, Zuo X, Zhi W, Yang P, Xie H-Q, Yang Z-M (2009) Grafts of porcine small intestinal submucosa with cultured autologous oral mucosal epithelial cells for esophageal repair in a canine model. Exp Biol Med (Maywood) 234(4):453–461

Wei N, Gong P, Liao D, Yang X, Li X, Liu Y, Yuan Q, Tan Z (2010) Auto transplanted mesenchymal stromal cell fate in periodontal tissue of beagle dogs. Cytotherapy 12:514–521

Xiong H, Bai C, Wu S, Gao Y, Lu T, Hu Q, Guan W, Ma Y (2014) Biological characterization of mesenchymal stem cells from bovine umbilical cord. Anim Cell Syst 1(1):59–67

Xu J, Woods CR, Mora AL, Joodi R, Brigham KL, Iyer S et al (2007) Prevention of endotoxin-induced systemic response by bone marrow derived mesenchymal stem cells in mice. Am J Physiol Lung Cell Mol Physiol 293:L131–L141

Yan Y, Fang J, Wen X, Teng X, Li B, Zhou Z, Peng S, Arisha AH, Liu W, Hua J (2019) Therapeutic applications of adipose-derived mesenchymal stem cells on acute liver injury in canines. Res Vet Sci 126:233–239

Yang Y, Bucan V, Baehre H, von der Ohe J, Otte A, Hass R (2015) Acquisition of new tumor cell properties by MSC-derived exosomes. Int J Oncol 47:244–252

Yu X, Ge S, Chen S, Xu Q, Zhang J, Guo H, Yang P (2013) Human gingiva-derived mesenchymal stromal cells contribute to periodontal regeneration in beagle dogs. Cells Tissues Organs 198: 428–437

Yuanzheng C, Yan G, Ting L, Yanjie F, Peng W, Nan B (2015) Enhancement of the repair of dog alveolar cleft by an autologous iliac bone, bone marrow–derived mesenchymal stem cell, and platelet-rich fibrin mixture. Plast Reconstr Surg 135:1405–1412

Zang S, Jin L, Kang S, Hu X, Wang M, Wang J, Chen B, Peng B, Wang Q (2016) Periodontal wound healing by transplantation of jaw bone marrow derived mesenchymal stem cells in chitosan/an organic bovine bone carrier into one-wall infrabony defects in beagles. J Periodontol 87:971–981. https://doi.org/10.1902/jop.2016.150504

Zhai Q, Dong J, Zhang X, He X, Fei D, Jin Y, Li B, Jin F (2021) Mesenchymal stem cells enhance therapeutic effect and prevent adverse gastrointestinal reaction of methotrexate treatment in collagen-induced arthritis. Stem Cells Int 2021:8850820

Zhang F, Peng W, Zhang J, Dong W, Yuan D, Zheng Y, Wang Z (2019) New strategy of bone marrow mesenchymal stem cells against oxidative stress injury via Nrf2 pathway: oxidative stress preconditioning. J Cell Biochem 120(12):19902–19914

Zhao LX, Zhang K, Shen BB, Li JN (2021) Mesenchymal stem cell-derived exosomes for gastrointestinal cancer. World J Gastrointest Oncol 13(12):1981–1996

Zimmermann K, Haas A, Oxenius A (2012) Systemic antibody responses to gut microbes in health and disease. Gut Microb 3:42–47

Mesenchymal Stem Cells Therapeutic Applications in Lung Disorders

8

Abstract

Lung is one of the essential organs of the live animal whose function is to oxygenate blood for proper cellular respiration. Lung innate resistance to varied assaults at times falls short with resultant varied types of ailments. In early stage, lung function may be reversible but in advanced cases current therapeutics are insufficient for complete recovery. As such role of the stem cells especially mesenchymal stem cells (MSCs) is being evaluated. Their therapeutic effects have been demonstrated to occur through direct cell–cell contact or through secretome leading to immunomodulation, anti-oxidant and antimicrobial actions. These cells may also give rise to epithelial cells in lungs and effectuate pro-secretory actions and prevention of matrix degradation. However, a growing concern is that MSCs may undergo mesenchyme transition developing myo-fibroblasts and leading to lung fibrosis. In veterinary species, MSCs have been evaluated in various experimental models of lung ailments. These studies, although very limited, appear promising. The current chapter describes the MSCs potential role in lung affections of veterinary species.

Keywords

Asthma · Acute respiratory distress syndrome · Lung · Mesenchymal stem cell · Recurrent airway obstruction

8.1 Introduction

Lung is an essential organ that ensures blood oxygen saturation necessary for cellular respiration. It has a typical architecture that ensures maximum possible gaseous exchange. In doing so, it simultaneously gets exposed to the environment through inhalation of air and the contaminants therein. The close association of the

M. B. Gugjoo, *Therapeutic Applications of Mesenchymal Stem Cells in Veterinary Medicine*, https://doi.org/10.1007/978-981-19-3277-9_8

capillaries with the lung alveoli keeps it in almost direct contact with the haematogenous etiological agents, if any. To face such challenges, lung has naturally developed better regeneration power but falls short against numerous infectious or non-infectious aetiological agents (Sharma et al. 2016). Any kind of lung affection initially elicits an inflammatory response that impairs its function and if it continues a more chronic condition develops leading to lung parenchymal damage. There are therapeutic options with limited potential to inhibit or reverse the advanced lung pathologies of destructive diseases (Sharma et al. 2016).

In order to address lung affections, more advanced regenerative therapeutic options are being evaluated. Cell based regenerative therapy is one of the options and in lung perspective undifferentiated cell may be more preferable option due to the existence of diverse cellular phenotypes. Among numerous stem cell types, mesenchymal stem cells (MSCs) are mainly being evaluated. In the recent past, lung MSCs have been harvested and characterized in various species including that of veterinary species like sheep (Ingenito et al. 2012; Ma et al. 2017) and cattle (Hu et al. 2015). The extensive characterization of lung MSCs still lag behind the MSCs from other organs (Sharma et al. 2016). Harvesting MSCs from lung, predominantly located in vascular zone of the lung arteries, is the tedious and risky job as compared to that of the other sources (Steens et al. 2021). In animals lung affections are not uncommon and vary from slight inflammatory condition to the chronic fibrosis including cat idiopathic pulmonary fibrosis (Cohn et al. 2004), interstitial lung disease (Reinero 2019) and asthma (Quimby and Borjesson 2018); canine infectious respiratory disease ((Lappin et al. 2017), interstitial lung disease (Reinero 2019); cattle acute respiratory distress syndrome, bronchiolitis obliterans, fibrosing alveolitis, and pneumonitis (Breeze 1985); horse inflammatory airway disease, interstitial pneumonia and recurrent airway obstruction (Rush and Mair 2004), among others. These diseases may currently be treated in initial stages but chronic disease conditions progress towards an irreversible stage (fibrosis). As such regenerative medicine is being evaluated to ensure therapeutics that can cure both acute as well as the advanced lung diseases for which cellular therapy involving stem cells are considered as the potential therapeutic candidates.

8.2 MSCs Therapeutic Effects in Lung Pathologies

There are wider type of lab animal lung studies ranging from emphysema (Ingenito et al. 2012; Guan et al. 2013), asthma (Zeng et al. 2015), fibrosis (Cargnoni et al. 2009; Moodley et al. 2009), acute respiratory distress syndrome (Németh et al. 2009; Mei et al. 2010) to chronic obstructive pulmonary disease (Zhang et al. 2014). All these studies approve MSCs potential therapeutic effect. MSCs effect ranges from reduction of inflammation and fibrotic effects (Moodley et al. 2010) to the pluripotency differentiation into varied cell types including the bronchioles and alveoli, besides the neovascularization (Anversa et al. 2011) (Fig. 8.1). However, application of stem cells in lung regenerative medicine may not be as simple as it seems. Translation of lab animal study results into clinical veterinary species and

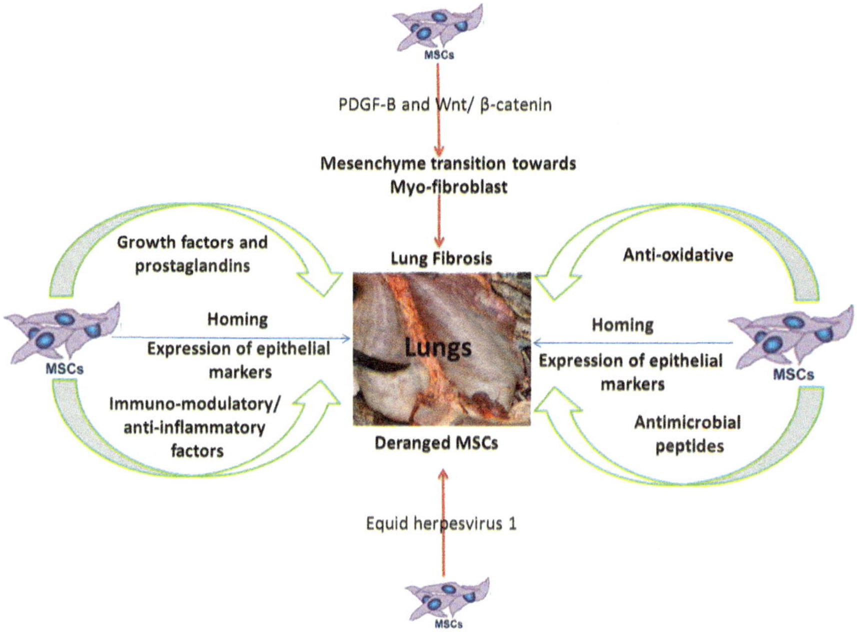

Fig. 8.1 Pictorial representation of mesenchymal stem cells (MSCs) therapeutic effects in pulmonary affections

human demands prior safety and efficacy confirmation. The current lack of literature pertaining to MSCs in these species makes clinical application a distant future. Add-on recently MSCs have been attributed to take part in the development of lung fibrotic changes through myofibroblast differentiation. This is considered to be triggered by local pathological state of the tissue (Veevers-Lowe et al. 2011; Sharma et al. 2016; El Agha et al. 2017; Cao et al. 2018) and wherein resident MSCs more likely participate into myofibroblast formation rather than their external sources like bone marrow-derived ones (El Agha et al. 2017). Local pathological changes like hypobaric oxygen and mechanical stress may trigger such a mesenchymal transition (Schermuly et al. 2005; Berg et al. 2010). The notable factors involved in such mesenchyme transition towards fibrosis are platelet-derived growth factor subunit B and Wnt/β-catenin (Heldin and Westermark 1999; Barst 2005; Cao et al. 2018). Some of the respiratory infectious agents (Equid herpesvirus 1) additionally affect MSCs properties including their surface proteins (Claessen et al. 2015). In lung disorders, primarily activation and replenishment of MSCs should be aimed through rejuvenation of the local micro-environment.

MSCs tend to home, engraft, transdifferentiate and secrete a plethora of biological factors in a tissue/organ including the lungs (Kang and Park 2014; El Agha et al. 2017). Various chemokines and their receptors like SDF1 and its receptor (CXR4), cytokine receptor (CCR2) and its intracellular adapter besides, biological factors like growth factors and prostaglandin (TGF-β, IGF-1, iNOS, PGE2, VEGF) ensure

MSCs homing and their regenerative effects (Belema-Bedada et al. 2008; Schipani and Kronenberg 2008). MSCs express lung epithelial markers under in vitro conditions, and thus, their transdifferentiation might potentially help in epithelial regeneration (Macpherson et al. 2005; Van Arnam et al. 2005; Fang et al. 2019). These cells, including lung MSCs, secrete extracellular matrices that potentially promote epithelial and endothelial cell haptotaxis. Such an extracellular matrix promotes epithelial cell attachment comparable to that of the collagen type I and Matrigel (Ingenito et al. 2012). The cell secretome contains microvesicles or their conditioned medium that can potentially bring about the regeneration through trophic factors, immunomodulation/anti-inflammatory, anti-oxidant and antimicrobial factors (Gugjoo et al. 2020). Lung MSCs may express FGF2 and FGF10 proteins and the transcripts like VEGF-A, VEGF-C and angiopoetin-1 (Ingenito et al. 2012). In acute lung injuries, MSCs may help in recruitment and maturation of lung dendritic cells through activation of Notch signalling (Lu et al. 2020). Their secretion of stanniocalcin-1 (STC1) can potentially rectify the inadequacy of communication between mesenchymal and epithelial cells. STC1 effectuates the anti-fibrosis through the reduction of oxidative stress, endoplasmic reticulum stress, besides the alveolar epithelial cell TGF-β1 production (Ono et al. 2015). MSCs prevent hyaluronic acid (HA)-induced TGF-β-mediated Smad pathways that potentially induce pulmonary fibrosis (Hao et al. 2018). AD-MSCs extracellular vesicles may also hamper apoptosis and necrosis of type II alveolar epithelial cells, thereby preventing lung injury and pulmonary fibrosis (Ingenito et al. 2012; Gao et al. 2020). Furthermore, immunomodulatory/anti-inflammatory actions may be one of the most important mechanisms in ensuring ailing lung rejuvenation (Lu et al. 2020). MSCs from almost all the animal species modulate innate and acquired immunity (Dias et al. 2019; Gugjoo et al. 2020; Park et al. 2021; Sung et al. 2021). This anti-inflammatory or immunomodulatory power may be innate in their secretome (Qiang et al. 2016; Hao et al. 2018; Sung et al. 2021) and enhances upon stimulation with inflammatory mediators like IFN-γ as in horse BM-MSCs (Caffi et al. 2020; Park et al. 2021). Am-MSCs conditioned medium (10%) or microvesicles (50×10^6/mL) modulate alveolar macrophage and thus may be effective in equine lung diseases (Zucca et al. 2016). MSCs ability to secret hepatocyte growth factor may hamper lung fibrosis (Lan et al. 2017). MSCs or their microvesicles/conditioned medium contain antimicrobial peptides that may be effective in preventing microbial infection (Harman et al. 2017; Cortés-Araya et al. 2018; Bujňáková et al. 2020; Marx et al. 2020). MSCs also inhibit neutrophil reactive oxygen species (Espinosa et al. 2020); however, these cells may enhance CD8+ T cells cytotoxic effects through granzyme B upregulation (Taechangam et al. 2021).

Therefore, MSCs carry properties potentially able to enhance regeneration although caution has to be made to prevent such innate properties not diverge towards the pathological condition with the availability of hostile microenvironment. Therefore, MSCs role has to be evaluated and their characteristic cellular and molecular features need to be thoroughly understood as per the clinical affection before their in vivo applications (Sharma et al. 2016).

8.3 In Vivo Mesenchymal Stem Cell Studies in Veterinary Species

In veterinary species, lung regenerative studies employing MSCs are very limited, while human clinical trials limited to phases I and II are being undertaken extensively. In animals, lung regenerative studies are mostly restricted to the experimental translational models or experimental models for direct clinical applications. Among various animal species, sheep is considered as a potential model animal for numerous human respiratory diseases (Scheerlinck et al. 2008; Guillamón and Luis 2010), while horse may be an effective model animal for human asthma (Lange-Consiglio et al. 2019). Stem cell research in respiratory ailments of veterinary species may provide clinical basis of MSCs therapeutics for veterinary and human medicine and also shows discourse of MSCs response against lung affections including the infectious ones.

8.3.1 MSCs Regenerative Effects

MSCs regenerative effects have been evaluated in various veterinary species like sheep, dog, cat and horse, but overall, the literature is too limited to offer any conclusion and has been detailed in Tables 8.1 and 8.2. Below are discussed, these studies in all the mentioned species.

8.3.2 Sheep Studies

MSCs lung repair potential has been evaluated in experimental models of emphysema (Ingenito et al. 2012), acute respiratory distress syndrome (Asmussen et al. 2014; Kocyildirim et al. 2017; Cardenes et al. 2019; Chaleshtori et al. 2020; Millar et al. 2020) and smoke inhalation (Baljinnyam et al. 2020) (Table 8.1).

In lung experimental emphysema study (randomized placebo control), endobronchial implantation of autologous lung-derived MSCs along with fibrinogen, fibronectin and poly-L-lysine is well tolerated with simultaneous increase in lung perfusion as compared to control. Increase in lung mass is characterized by cell retention, and increased cellularity and extracellular matrix at 1-month period. The cellular density in some lung areas may not show appearance consistently comparable to that of the normal. Small cell fraction engrafted to the alveolar septum conveys possible MSCs paracrine actions as the main role of therapeutic effects. An interesting finding of expression of α-SMA by lung MSCs under ex vivo cultures does not concord to the collagen type I in the lung tissue engrafted with these extrinsic lung MSCs (Ingenito et al. 2012). However, contrary to this finding, α-SMA+ CD45+ cells of bone marrow have been concurred with the fibrosis through secretion of collagen type I (Ishii et al. 2005, 2007; Liebler et al. 2008). In the above study, lung MSCs phenotype is α-SMA+ CD45− contrary to the BM-MSCs α-SMA+ CD45+. Besides, the microenvironment (inflammation and cytokine expression) and/or the

Table 8.1 MSCs preclinical experimental studies in various lung affection models in sheep, dog and cat

Study design/ period	Biomaterial used	Evaluation criteria	Overall result	Reference
Sheep lung emphysema model				
Randomized placebo control ($n = 10$)/ 4 weeks	Autologous lung MSCs (5–10 × 10^6 endo-bronchially implanted), scaffold (1% fibrinogen, 20 μg/mL fibronectin and 20 μg/mL poly-L-lysine)	Clinical toxicity, tissue mass (CT), lung physiology, scintigraphy, histology	Improved outcome	Ingenito et al. (2012)
Sheep acute respiratory distress syndrome (ARDS) model				
Randomized placebo control ($n = 19$)/24 h	Human BM-MSCs [(low dose: 5 × 10^6/ kg) (high dose: 10 × 10^6/kg)]	Post-mortem lung wet-to-dry weight ratio, bronchoalveolar lavage, serum biochemical parameters	Improved outcome	Asmussen et al. (2014)
Randomized placebo control ($n = 14$)/6 h	Human BM-MSCs (4 × 10^6, 10 × 10^6, 40 × 10^6) intrabronchially 30 min later	Wet-to-dry weight ratio, histopathology	Improved outcome	Rojas et al. (2014)
Randomized placebo control ($n = 11$)/6 h	BM-MSCs (multipotent adult progenitor cells @ 40 × 10^6 intra-tracheally to left lower lobe)	Blood gas results, radiography, histopathology	Improved outcome	Kocyildirim et al. (2017)
Randomized placebo control ($n = 6$)/5 h	BM-MSCs (1 × 10^6/ kg by intravenous route and one million cells/kg by endobronchial route)	PET/CT	Improved outcome	Cardenes et al. (2019)
Randomized placebo control ($n = 10$)/ 168 h	Autologous BM-MSCs (5 × 10^7 intrapulmonary injection at 24 h post LPS injection)	Clinical, complete blood count, analysis of arterial blood gases and the concentrations of IL6, IL10, TNF-α, total protein, IgM and albumin BAL	Improved outcome	Chaleshtori et al. (2020)
Randomized placebo control ($n = 14$)/24 h	iPSC-derived human MSCs (hMSCs, 3 × 10^8 Endo-bronchially)	Respiratory and hemodynamic variables, histopathology and lung injury, inflammatory cytokines, haematological, biochemical parameters	Improved outcome	Millar et al. (2020)

(continued)

Table 8.1 (continued)

Study design/ period	Biomaterial used	Evaluation criteria	Overall result	Reference
Sheep smoke inhalation injury				
Randomized placebo control ($n = 10$)/48 h	Allogeneic AD-MSCs (20×10^7 1-h post injury intravenously)	Lung lymph flow, extravascular lung water content, protein content in plasma and lung lymph fluid	Improved outcome	Ihara et al. (2017)
Randomized placebo control ($n = 11$)/48 h	Allogeneic AD-MSCs (20×10^7 1-h post injury intravenously)	Immunohistochemistry, immunoblotting	Non-significant effect	Baljinnyam et al. (2020)
Goat bronchopleural fistula (upper tracheal lobectomy) model				
Randomized control ($n = 9$)/ 28 days	BM-MSCs (2×10^6/ mL, 5 mL injected bronchoscopically), fibrin glue	Histopathology, CT and MRI	Improved outcome	Petrella et al. (2014)
Cat asthma model				
Randomized placebo control ($n = 9$)/ 12 months	Allogeneic, AD-MSCs (6 doses of 1.44×10^7 bimonthly intravenous)	Bronchoalveolar lavage fluid cytology, pulmonary mechanics and clinical scoring, computed tomographic (CT)	Improved outcome	Trzil et al. (2014)
Randomized placebo control ($n = 15$)/ 9 months	Allogeneic AD-MSCs (5 doses of 2×10^6, 4×10^6, 4.7×10^6, 1×10^7 and 1×10^7 in each cat at 0, 14, 28, 98, 130 days)	Airway eosinophilia, pulmonary mechanics, computed tomography, immunologic assays	Improved outcome	Trzil et al. (2016)
Dog cardiopulmonary bypass model				
Randomized placebo control ($n = 6$)/3 h	Human Am-MSCs (xenogeneic 2×10^7)	Measurement of oxygenation and respiratory index, ELISA	Improved outcome	Qiang et al. (2016)
Randomized placebo control ($n = 8$)/ 180 days	Human UC-MSCs (xenogeneic 1×10^6 per kg b. wt. intratracheal)	Blood gas analysis, CT, ELISA, histology, WB, immunofluorescence	Improved outcome	Hao et al. (2018)

MSCs mesenchymal stem cells, *AD* adipose tissue derived, *Am* amniotic membrane derived, *BM* bone marrow derived, *UC* umbilical cord derived, *iPSCs* induced pluripotent stem cells, *n* total number of animals in all the groups of the study; Improved outcome: Indicates improved outcome with MSC transplantation even if only one parameter improves; Study period: Indicates maximum time period for which study is conducted scaffold used in this study might also lead to such differential outcomes. MSCs surface proteins like syndecan-2 and $\alpha5\beta1$ integrins bind with the poly-L-lysine through its heparin sulphate chain and fibronectin-binding site, respectively, that activates mitogen-activated protein (MAP) kinase leading to the initiation of cell cycling (Cohen et al. 1998; Choi et al. 2005). The above study evaluates results for the shorter duration (28 days) during which the fibrosis does not usually become obvious. Chances of fibrosis increases with the time and thus, further longer duration studies are highly desired. Currently, the exact mechanism behind such results remains to be elucidated.

Sheep acute respiratory distress syndrome (ARDS) model has been created through injection of microbes (*P. aeruginosa*) (Asmussen et al. 2014) or endotoxin (LPS) (Rojas et al. 2014; Kocyildirim et al. 2017; Cardenes et al. 2019; Chaleshtori et al. 2020; Millar et al. 2020). In ARDS, MSCs intravenous or the endobronchial route may show comparable results as the cells have good bio-distribution to the lungs and also get trapped in the microvasculature. Inflammation in the lungs may further promote such a distribution (Cardenes et al. 2019). However, in multiple systemic affections, MSCs may have to be given in higher doses upon intravenous route and/or through the endobronchial route for lung affections.

Autologous or xenogeneic MSCs have been evaluated in a randomized placebo control ARDS model (Table 8.1). In the ARDS model induced by intrapulmonary LPS, reduction in severity of respiratory signs and hemodynamic stability occurs post 24 h delayed xenogeneic BM-MSCs transplantation. The improvement (in comparison to the control lungs and initial lungs at 24 h of LPS injection) is exhibited by significant increase in partial pressure of oxygen and its saturation, besides the interleukin (IL-10). There is a significant decrease in PCO2, total protein, IgM, IL-6, TNF-α, albumin, WBCs, neutrophils and macrophages (Rojas et al. 2014; Chaleshtori et al. 2020). MSCs concentration might also play some role especially when transplanted intravenously as median lung water content (pulmonary oedema) significantly decreases in lungs treated with higher cells (ten million per kg b. wt.) than the lower cell dose (five million per kg b. wt.) 1 h post injury. However, oxygenation increases with the lower cell dose as well by 24-h period. The xenogeneic cells show no adverse effect on the systemic haemodynamics (pulmonary haemodynamics and systemic arterial pressure) (Asmussen et al. 2014). These studies have been conducted only for a shorter duration and may not offer greater details. Besides, the treatment in these studies is initiated immediately or early after injury, it may not be quite feasible in clinical settings.

Further, intratracheal MSCs or iPSC-derived human MSCs implantation to left lower lobe along with the extracorporeal membrane oxygenation (ECMO) may improve lung parenchymal appearance, although improvement in oxygenation and pulmonary function remains to be observed (Kocyildirim et al. 2017; Millar et al. 2020). Lack of observable improvement in pulmonary function or oxygenation might lead to the significant elevation in transmembrane oxygenator pressure gradients and could be accompanied by pulmonary artery thrombosis. This questions the endobronchial use of xenogeneic MSCs during veno-venous-ECMO (Millar et al. 2020). ECMO may develop systemic inflammatory responses exhibited in the form of cytokine storm, leukocyte activation and multi-organ dysfunction

(Adrian et al. 1998). MSCs may improve sheep ARDS with special significance to be given to their route of implantation, dose and whether ECMO should be ensued or not. In the above studies, early MSCs transplantation made is not possible clinically and the co-morbidities that occur in clinical cases cannot be replicated. Therefore, further extensive studies are desired before any conclusion can be drawn.

In case of smoke inhalation injury model, intravenous transplantation of AD-MSCs within 1 h post injury attenuates the adverse changes in the lungs. Reduced hyperpermeability of the pulmonary microvascular evidenced by attenuation of increased lung lymph flow and increased pulmonary gas exchange occurs. Such an improvement remains devoid of the changes in the expression of lipoxin A4 (LXA4) and angiopoietin-2 (ANG2) at 48 h period (Ihara et al. 2017). MSCs are considered to attenuate lung injury through LXA4 (activates alveolar epithelial sodium channel, Na, K-ATPase, and thereby increases alveolar fluid clearance) and ANG2 increases vascular leaks to lead to ARDS (Parikh et al. 2006; Wang et al. 2013). This discrepancy in the findings may be due to the variable time of evaluation as the evaluation in this study has been a bit delayed (48 h) compared to that demonstrated in earlier studies (Lange et al. 2012). However, in combined burn and smoke inhalation injury, intravenous xenogeneic AD-MSCs transplantation fail to attenuate the pulmonary dysfunctions. This might be due to the fact that higher cell concentration in lungs is achieved in smoke injury cases while in combined cases, cells may preferably localize burn areas (Baljinnyam et al. 2020).

In all these studies, therapeutic benefits of MSCs appear although limited in one way or other. There is currently lack of sufficient mechanistic evidence of MSCs therapeutics including their ability to engraft, immunomodulate, matrix secretion. This is further aggravated by the interspecies differences especially for the large veterinary species and the rodents in which study results do not correlate.

8.3.3 Cat Studies

In feline practice, asthma is one of the commonly seen respiratory problems characterized by an airway eosinophilia and airway hyperresponsiveness (AHR) (Norris Reinero et al. 2004). This condition is considered to have hallmark features similar to humans (Norris Reinero et al. 2004). Commonly utilized drugs like glucocorticoids and bronchodilators fail to provide effective results consistently (Cocayne et al. 2011; Durrani et al. 2011). In order to address this problem, regenerative medicine employing MSCs is being evaluated (Mirershadi et al. 2020).

Cat asthma studies have been conducted as randomized placebo controlled experimental models induced with Bermuda grass (Trzil et al. 2014, 2016) (Table 8.1). Intravenous injection of multiple doses (6 doses) of allogeneic AD-MSCs (14 million) fail to show any improvement in airway eosinophilia or AHR. However, significant finding of attenuated CT indices of airway remodelling (common feature in asthma) occurs at 6 months post last cell dose. This change, however, may not sustain till 12 months. As cell treated and control lungs show non-significant differences in airway eosinophilia, allergen-specific IgE and T lymphocyte proliferation besides, the stimulated IL-10 production. It points that decrease in airway remodelling through MSCs is less likely related to the attenuation of the TH_2-predominating environment. Furthermore, non-significant differences in relation to allergen responsiveness also fail to explain such a remodelling with MSCs (Trzil et al. 2014). MSCs failure in improvement in airway eosinophilia or AHR is contrary to that seen in rodents and thus underscores the translation of rodent studies for feline asthma. Such differences may be due to the model differences, evaluation criteria and use of different bronchoprovocants (Park et al. 2010; Goodwin et al. 2011). Another study from the same group demonstrates that multiple intravenous doses (5 doses) of allogeneic AD-MSCs fail to show any improvement in airway eosinophilia up to 133 although AHR does reduce. By the ninth-month period, eosinophilia percentage reach to normal ranges along with the significant reduction in lung attenuation and bronchial wall thickening in the cell-treated animals (Trzil et al. 2016). The findings are significant but are marred by the lack of effective and balanced sample sizes in this pilot study. Furthermore, as control lungs also show reduced airway eosinophilia, allergen challenges may decrease in the chronic study period. In general, MSCs lung disease modulation may not occur in an immediate period but possibly in a delayed period; however, further studies are warranted to confirm the efficacy of MSCs in cat asthma. As variable cell concentration within the study at different time intervals and between the studies was utilized with variable results, it is important to evaluate the minimal dose required. Above all the clinical trials may be conducted to confirm actual safety, feasibility and safety of the cells.

8.3.4 Broncho Pleural Fistula

Pulmonary cancers are one of the common cancers seen worldwide (Wakelee et al. 2007) and a common procedure performed is its surgical resection. However, such a procedure develops bronchopleural fistula (BPF) in 1–4% patients with a mortality rate of 12.5% to 71.2% (Sonobe et al. 2000). Thus, regenerative medicine is being evaluated for effectiveness (Table 8.1). In a goat bronchopleural fistula model, allogeneic BM-MSCs (two million in fibrin glue) via submucosal bronchoscopic injection heal fistula by 28th day as evidenced by proliferation of fibroblast and their collagenous matrix secretion. All these animals survive without any evidence of pleural empyema (Petrella et al. 2014). This study has limited animal size with no scaffold-based control and as the cells are transplanted in acute phase as compared to the clinical scenario where these cases are usually brought later. Further chances of leakage remain to be seen in the long run. Thus, this single study on BPF is promising but is insufficient enough to provide any conclusive evidence.

8.3.5 Dog Studies

Like other veterinary species, MSCs studies are also very limited in dogs. In a single experimental study each on ischaemia and reperfusion injury (Qiang et al. 2016) and radiation injury (Hao et al. 2018), MSCs have proven to be effective in attenuating lung damage (Table 8.1). In ischaemia and reperfusion injury models following cardiopulmonary bypass (CBP), MSCs transplanted into the femoral vein are detected in lung tissues within 3 h. These cells tend to elevate arterial blood oxygen index and reduce respiratory index. Further, these cells ameliorate elevated levels of inflammatory mediators like TNF-α and IL-8 levels and improve IL-10 starting from the 15 min post cell transplantation as compared to the control. MSCs transplantation may prevent increase in MMP-9 (degrades capillary basement membrane and increases capillary permeability) correlates well with reduction in capillary permeability and reduction in lung pathology (Qiang et al. 2016). In the radiation injury model, partial pressure of oxygen in human UC-MSC transplanted (@ one million per kg b. wt. endotracheally) lungs increases gradually over time. Levels of hydroxyproline and TGF-β levels in the peripheral blood reduce while TGF-α gradually increases much significantly in comparison to the control. Also, phospho-Smad2 and phospho-Smad3 reduce significantly in the cell-transplanted lung tissues. Overall, the lung injury remains milder in the cell treated as confirmed by CT and expression of interleukins and superoxide dismutase. In consent with the previous CBP-induced ischaemia and reperfusion injury, MSCs prevent inflammatory mediators like TNF-α and IL-1 and additionally the superoxide dismutase (SOD) and HA in the radiation injury (Hao et al. 2018). HA is considered to induce pulmonary fibrosis through TGF-β (Midgley et al. 2013) that follows Smad-dependent pathways (Willis and Borok 2007). Thus, inhibition of these factors may markedly prevent pulmonary fibrosis. Therefore, MSCs through the immunomodulatory and/or regenerative effects may ameliorate lung injuries induced through cardiopulmonary bypass and radiation. However, these studies are preliminary and should be carried extensively to confirm MSCs safety and efficacy.

8.3.6 Horse Studies

8.3.6.1 Recurrent Airway Obstruction

Equine asthma syndrome encompasses various inflammatory diseases of the airways including recurrent airway obstruction (RAO) and inflammatory airway disease (IAD). The former condition is a more chronic and severe form while the latter one is a mild inflammatory response (Couëtil et al. 2016). Horses tend to develop RAO that may occur either genetically or through exposure to moldy hay, dusty straw and pollens activating the innate and adaptive immune system. Prevalence of hypersensitivity to moulds or dust in horses ranges from 2 to 80% (Leguillette 2003). The clinical condition is characterized by persistent airway inflammation exhibited by neutrophil and non-eosinophil infiltration as also seen in the cats (Leclere et al. 2011; Couëtil et al. 2016) with no effective therapy (Dauvillier et al. 2011). These conditions are often managed by glucocorticoid treatment. Glucocorticoid receptors (GCR) are seen in all the tissues, of which GCR-α is widely expressed and prime glucocorticoid effector (Mainguy-Seers and Lavoie 2021). Its interactions lead to the anti-inflammatory response (activation of IL-10) and simultaneous transrepression of pro-inflammatory genes [nuclear factor-kappa B (NF-κB) and activator protein-1] (Vandewalle et al. 2018). Apart from genomic responsiveness, non-genomic responsive therapeutic effects like intracellular calcium level alteration, reduction in smooth muscle contraction and reactive oxygen species modulation also occur with glucocorticoids (Panettieri et al. 2019). A big challenge with the glucocorticoids is the development of resistance to its therapeutic effects and associated side effects. A limited understanding especially in relation to the cellular mechanisms involved in the resistance to GCs currently makes it difficult to overcome the challenge. Glucocorticoid receptor, GCR-β, does not bind to the glucocorticoid rather its interaction with glucocorticoid responsive genes modulates the steroid responsiveness leading to the treatment resistance (Payne and Freishtat 2012; Barnes 2013). This resistance is also associated with the genetic make-up, neutrophilic inflammation (Green et al. 2002) and associated morbidities like obesity (Telenga et al. 2012). Besides, the GC administration techniques and management may also affect the outcome in horse (Mainguy-Seers and Lavoie 2021). Thus, regenerative medicine is being evaluated to act as a potential replacement option. MSCs due to their immunomodulation properties and regenerative effects may offer some therapeutic benefits.

A randomized controlled chronic respiratory disease (RAO) study compares intratracheal BM-MNCs (@ 500 million to 1000 million added to 5 mL of fresh autologous serum to prepare a 20 mL volume) transplantation to that of the oral dexamethasone treatment (Table 8.2). Cell therapy shows respiratory function improvement comparable to that of the dexamethasone. The neutrophil concentration is much reduced while macrophage percentage improves in bronchioalveolar lavage fluid (BALF) with both the treatments normally seen in healthy horses. A significant finding of the increase in IL-10 concentration in cell-treated animals as compared to the dexamethasone is seen. IL-10 is a potent anti-inflammatory factor with the added feature of polarization of macrophage (M)1 to M2 phenotype. Thus, BM-MNCs may ameliorate the inflammatory response of recurrent airway obstruction (Barussi et al. 2016). Thus, compared to the conventional therapy that majorly

Table 8.2 MSCs preclinical experimental and clinical study in horse

Study design/ period	Biomaterial used	Evaluation criteria	Overall result	Reference
Equine recurrent airway obstruction				
Case-controlled comparative study ($n = 8$)/ 21 days	BM-MNCs (5×10^8 to 1×10^9 intra-tracheal injection), dexamethasone	Bronchoalveolar lavage fluid (BALF) collection and analysis, interleukin analysis	Improved outcome	Barussi et al. (2016)
Severe equine asthma				
Randomized-controlled comparative trial ($n = 20$)/ 1 year	Autologous AD-MSCs (100×10^6 intrabronchial application)	Endoscopic examination, BALF collection, and cytology, cytokine quantification	Improved outcome	Adamič et al. (2022)

MSCs mesenchymal stem cells, *AD* adipose tissue derived, *MNCs* mononuclear cells, *n* total number of animals in all the groups of the study; Improved outcome: Indicates improved outcome with MSC transplantation even if only one parameter improves; Study period: Indicates maximum time period for which study is conducted acts through regression of inflammatory processes, the cell therapy may additionally repair affected lungs and improve the tolerance to allergens. BM-MNCs contain MSCs and growth factors and thus may offer better results in immune reactions or inflammatory environments. In a similar way, AD-MSCs intrabronchial transplantation leads to the early anti-inflammatory effects (reduced expression of IL-17 and reduced concentrations of IL-1β, IL-4 and TNFα) and long-term (1 year) clinical improvement. This study shows possible pathways of immunomodulatory effects that MSCs offer. This includes both anti-inflammatory cytokines and the cytokines that represent particular immune responses including type 1 (T1: IFNγ), type 2 (T2: IL-4) and type 17 (T17: IL-17) (Adamič et al. 2022) (Table 8.2).

Barring few studies, the pulmonary affection-related clinical studies in veterinary species to present evidence-based therapeutic effect of MSCs remain unavailable. As these studies are conducted as translational studies, their potentiality may be extended through uniform case-controlled studies. MSCs source, passage number, concentration, route and repetition need to be standardized for any conclusive statement. Further, more animal size and duration are highly desired to evaluate results statistically and for a longer duration.

8.3.7 MSCs and Respiratory Viral Infection

In respiratory syncytial virus infection, lung-resident MSCs get infected and undergo cytopathy. The viral dissemination from these cells may occur through cell-to-cell contact, although cell replication increases the cell concentrations. Lung-resident MSCs release type I and III interferon (IFNs) and stimulate genes, an antiviral state. This antiviral response is also combined with the potential to repair and regenerate the lung tissue through the secretion of various growth factors like VEGF and TGF-β. This response tends to be different from that of the alveolar epithelial cells (Brügger et al. 2021). To understand and make effective applications of MSCs, further extensive studies are desired.

8.4 Conclusion(s)

Lung ailments are not uncommon in animals and in severe cases become life threatening. There are many lung ailments that progress towards the irreversible conditions and become the cause of morbidity, miserable life and finally death. One of the causes for development of lung fibrosis is the involvement of MSCs whose role thus needs to be evaluated. Various preliminary experimental studies utilizing MSCs in lung ailments like emphysema, acute respiratory distress syndrome, smoke inhalation, asthma, lung injury and RAO conducted on veterinary species appear promising. However, there is very little literature on such studies to draw any conclusive evidence. MSCs prevent immune/inflammatory reactions, show anti-oxidant actions and differentiate into epithelial lineage and prevent cell apoptosis besides other pro-healing factors. These may inhibit pulmonary fibrosis through inhibition of various associated pathways.

References

Adamič N, Mihevc SP, Blagus R, Kramarič P, Krapež U, Majdič G, Viel L, Hofman AM, Bienzle D, Vengust M (2022) Effect of intrabronchial administration of autologous adipose-derived mesenchymal stem cells on severe equine asthma. Stem Cell Res Ther 13:23

Adrian K, Mellgren K, Skogby M, Friberg LG, Mellgren G, Wadenvik H (1998) Cytokine release during long-term extracorporeal circulation in an experimental model. Artif Organs 22:859–863

Anversa P, Kajstura J, Leri A, Loscalzo J (2011) Tissue-specific adult stem cells in the human lung. Nat Med 17:1038–1039

Asmussen S, Ito H, Traber DL, Lee JW, Cox RA, Hawkins HK, McAuley DF, McKenna DH, Traber LD, Zhuo H, Wilson J, Herndon DN, Prough DS, Liu KD, Matthay MA, Enkhbaatar P (2014) Human mesenchymal stem cells reduce the severity of acute lung injury in a sheep model of bacterial pneumonia. Thorax 69(9):819–825

Baljinnyam T, Radnaa E, Yosuke N, Fukuda S, Prough DS, Enkhbaatar P (2020) Cutaneous burn diminishes beneficial effect of intravenously administered mesenchymal stem cells on acute lung injury induced by smoke inhalation in sheep. Burns 46(8):1914–1923

Barnes PJ (2013) Corticosteroid resistance in patients with asthma and chronic obstructive pulmonary disease. J Allergy Clin Immunol 131:636–645

Barst RJ (2005) PDGF signaling in pulmonary arterial hypertension. J Clin Invest 115:2691–2694

Barussi FCM, Bastos FZ, Leite LMB, Fragoso FYI, Senegaglia AC, Brofmanb PRS, Nishiyamac A, Pimpãoa CT, Michelotto PVJ (2016) Intratracheal therapy with autologous bone marrow-derived mononuclear cells reduces airway inflammation in horses with recurrent airway obstruction. Respir Physiol Neurobiol 232:35–42

Belema-Bedada F, Uchida S, Martire A, Kostin S, Braun T (2008) Efficient homing of multipotent adult mesenchymal stem cells depends on FROUNT-mediated clustering of CCR2. Cell Stem Cell 2:566–575

Berg T et al (2010) β-Catenin regulates mesenchymal progenitor cell differentiation during hepatogenesis. J Surg Res 164:276–285

Breeze R (1985) Respiratory diseases in cattle. Veterinary clinics of North America. Food Anim Pract 1(2):311–346

Brügger M, Démoulins T, Barut GT, Zumkehr B, Oliveira Esteves BI, Mehinagic K, Haas Q, Schögler A, Rameix-Welti M-A, Eléouët J-F, Moehrlen U, Marti TM, Schmid RA, Summerfield A, Posthaus H, Ruggli N, Hall SRR, Alves MP (2021) Pulmonary mesenchymal stem cells are engaged in distinct steps of host response to respiratory syncytial virus infection. PLoS Pathog 17(7):e1009789

Bujňáková D, Cuvalov A, Ižek MC, Humenik F, Salzet M, Ižková DC (2020) Canine bone marrow mesenchymal stem cell conditioned media affect bacterial growth, biofilm-associated Staphylococcus aureus and AHL-dependent quorum sensing. Microorgan 8:1478

Caffi V, Espinosa G, Gajardo G, Morales N, Durán MC, Uberti B, Morán G, Plaza A, Henríquez C (2020) Pre-conditioning of equine bone marrow-derived mesenchymal stromal cells increases their immunomodulatory capacity. Front Vet Sci 7:318

Cao H, Wang C, Chen X, Hou J, Xiang Z, Shen Y, Han X (2018) Inhibition of Wnt/β-catenin signaling suppresses myofibroblast differentiation of lung resident mesenchymal stem cells and pulmonary fibrosis. Sci Rep 8:13644

Cardenes N, ArandaValderrama P, Carney JP, Torres JS, Alvarez D, Kocydirim E, Smith WJA, Ting AE, Lagazzi L, Yu Z, Mason S, Santos E, Lopresti BJ, Rojas M (2019) Cell therapy for ARDS: efficacy of endobronchial versus intravenous administration and biodistribution of MAPCs in a large animal model. BMJ Open Respir Res 6:e000308

Cargnoni A et al (2009) Transplantation of allogeneic and xenogeneic placenta-derived cells reduces bleomycin-induced lung fibrosis. Cell Transplant 18:405–422

Chaleshtori SS, Dezfouli MRM, Abbasi J, Dehghan MM, Fakhr MJ, SYadollahi H, Mirabad MM. (2020) Prevention of LPS-induced acute respiratory distress syndrome in sheep by bone marrow-derived mesenchymal stem/stromal cells. Life Sci 263:118600

Choi S, Lee E, Kwon S, Park H, Yi JY, Kim S, Han IO, Yun Y, Oh ES (2005) Transmembrane domain induced oligomerization is crucial for the functions of syndecan-2 and syndecan-4. J Biol Chem 280:42573–42579

Claessen C, Favored H, Ma H, Osterrieder N, De Schauwer C, Piepers S, Van de Walle GR (2015) Equid herpesvirus 1 (EHV1) infection of equine mesenchymal stem cells induces a pUL56-dependent downregulation of select cell surface markers. Vet Microbiol 176(1–2):32–39

Cocayne CG, Reinero CR, DeClue AE (2011) Subclinical airway inflammation despite high-dose oral corticosteroid therapy in cats with lower airway disease. J Feline Med Surg 13:558–563

Cohen AR, Woods DF, Marfatia SM, Walther Z, Chishti AH, Anderson JM (1998) Human CASK/LIN-2 binds syndecan-2 and protein 4.1 and localizes to the basolateral membrane of epithelial cells. J Cell Biol 142:129–138

Cohn LA, Norris CR, Hawkins EC, Dye JA, Johnson CA, Williams KJ (2004) Identification and characterization of an idiopathic pulmonary fibrosis-like condition in cats. J Vet Intern Med 18:632–641

Cortés-Araya Y, Amilon K, Rink BE, Black G, Lisowski Z, Donadeu FX, Esteves CL (2018) Comparison of antibacterial and immunological properties of mesenchymal stem/stromal cells from equine bone marrow, endometrium, and adipose tissue. Stem Cells Dev 27:1518–1525

Couëtil LL, Cardwell JM, Gerber V, Lavoie J-P, Léguillette R, Richard EA (2016) Inflammatory airway disease of horses—revised consensus statement. J Vet Intern Med 30:503–515

Dauvillier J, Felippe MJB, Lunn DP, Lavoie-Lamoureux A, Leclère M, Beauchamp G, Lavoie JP (2011) Effect of long-term fluticasone treatment on immune function in horses with heaves. J Vet Intern Med 25:549–557

Dias IE, Pinto PO, Barros LC, Viegas CA, Dias IR, Carvalho PP (2019) Mesenchymal stem cells therapy in companion animals: useful for immune-mediated diseases? BMC Vet Res 15:358

Durrani SR, Viswanathan RK, Busse WW (2011) What effect does asthma treatment have on airway remodeling? Current perspectives. J Allergy Clin Immunol 128:439–448

El Agha E, Kramann R, Schneider RK, Li X, Seeger W, Humphreys BD, Bellusci S (2017) Mesenchymal stem cells in fibrotic disease. Cell Stem Cell 21(2):166–177

Espinosa G, Plaza A, Schenffeldt A, Alarcón P, Gajardo G, Uberti B, Morán G, Henríquez C (2020) Equine bone marrow-derived mesenchymal stromal cells inhibit reactive oxygen species production by neutrophils. Vet Immunol Immunopathol 221:109975

Fang S, Zhang S, Dai H, Hu X, Li C, Xing Y (2019) The role of pulmonary mesenchymal cells in airway epithelium regeneration during injury repair. Stem Cell Res Ther 10(1):366

Gao Y, Sun J, Dong C, Zhao M, Hu Y, Jin F (2020) Extracellular vesicles derived from adipose mesenchymal stem cells alleviate PM2.5-induced lung injury and pulmonary fibrosis. Med Sci Monit 26:e922782

Gugjoo MB, Hussain S, Amarpal, Shah RA, Dhama K (2020) Mesenchymal stem cell-mediated immuno-modulatory and anti-inflammatory mechanisms in immune and allergic disorders. Recent Patents Inflamm Allergy Drug Discov 14(1):3–14

Goodwin M, Sueblinvong V, Eisenhauer P, Ziats NP, LeClair L, Poynter ME, Steele C, Rincon M, Weiss DJ (2011) Bone marrow-derived mesenchymal stromal cells inhibit Th2-mediated allergic airways inflammation in mice. Stem Cells 29(7):1137–1148

Green RH, Brightling CE, Woltmann G, Parker D, Wardlaw AJ, Pavord ID (2002) Analysis of induced sputum in adults with asthma: identification of subgroup with isolated sputum neutrophilia and poor response to inhaled corticosteroids. Thorax 57:875–879

Guan X-J et al (2013) Mesenchymal stem cells protect cigarette smoke-damaged lung and pulmonary function partly via VEGF-VEGF receptors. J Cell Biochem 114:323–335

Guillamón MDLH, Luis BC (2010) The sheep as a large animal experimental model in respiratory diseases research. Arch Bronconeumol 46(10):499–501

Hao Y, Ran Y, Lu B, Li J, Zhang J, Feng C, Fang J, Ma R, Qiao Z, Dai X, Xiong W, Liu J, Zhou Q, Hao J, Li R, Dai J (2018) Therapeutic effects of human umbilical cord derived mesenchymal stem cells on canine radiation-induced lung injury. Int J Radiat Oncol Biol Phys 102(2):407–416

Harman RM, Yang S, He MK, Van de Walle GR (2017) Antimicrobial peptides secreted by equine mesenchymal stromal cells inhibit the growth of bacteria commonly found in skin wounds. Stem Cell Res Ther 8:157

Heldin CH, Westermark B (1999) Mechanism of action and in vivo role of platelet-derived growth factor. Physiol Rev 79:1283–1316

Hu P, Pu Y, Li X, Zhu Z, Zhao Y, Guan W, Ma Y (2015) Isolation, in vitro culture and identification of a new type of mesenchymal stem cell derived from fetal bovine lung tissues. Mol Med Rep 12:3331–3338

Ihara K, Fukuda S, Enkhtaivan B, Trujillo R, Perez-Bello D, Nelson C et al (2017) Adipose-derived stem cells attenuate pulmonary microvascular hyperpermeability after smoke inhalation. PLoS One 12(10):e0185937

Ingenito EP, Tsai L, Murthy S, Tyagi S, Mazan M, Hoffman A (2012) Autologous lung-derived mesenchymal stem cell transplantation in experimental emphysema. Cell Transplant 21:175–189

Ishii G, Sangai T, Sugiyama K, Ito T, Hasebe T, Endoh Y, Magae J, Ochiai A (2005) In vivo characterization of bone marrow-derived fibroblasts recruited into fibrotic lesions. Stem Cells 23:699–706

Ishii G, Ito TK, Aoyagi K, Fujimoto H, Chiba H, Hasebe T, Fujii S, Nagai K, Sasaki H, Ochiai A (2007) Presence of human circulating progenitor cells for cancer stromal fibroblasts in the blood of lung cancer patients. Stem Cells 25:1469–1477

Kang MH, Park HM (2014) Evaluation of adverse reactions in dogs following intravenous mesenchymal stem cell transplantation. Acta Vet Scand 56:16

Kocyildirim E, Cárdenes N, Ting A, Cáceres E, BermUdez C, Rojas M (2017) The use of GMP-produced bone marrow-derived stem cells in combination with extracorporeal membrane oxygenation in ARDS: an animal model. ASAIO J 63(3):324–332

Lan Y-W, Theng S-M, Huang T-T, Choo K-B, Chen C-M, Kuo H-P, Chong K-Y (2017) Oncostatin M-preconditioned mesenchymal stem cells alleviate bleomycin-induced pulmonary fibrosis through paracrine effects of the hepatocyte growth factor. Stem Cells Transl Med 6(3): 1006–1017

Lange M, Szabo C, Traber DL, Horvath E, Hamahata A, Nakano Y, Traber LD, Cox RA, Schmalstieg FC, Herndon DN, Enkhbaatar P (2012) Time profile of oxidative stress and neutrophil activation in ovine acute lung injury and sepsis. Shock 37(5):468–472

Lange-Consiglio A, Stucchi L, Zucca E, Lavoie JP, Cremonesi F, Ferrucci F (2019) Insights into animal models for cell-based therapies in translational studies of lung diseases: is the horse with naturally occurring asthma the right choice? Cytotherapy 21(5):525–534

Lappin MR, Blondeau J, Boothe D, Breitschwerdt EB, Guardabassi L, Lloyd DH, Papich MG, Rankin SC, Sykes JE, Turnidge J, Weese JS (2017) Antimicrobial use guidelines for treatment of respiratory tract disease in dogs and cats: Antimicrobial Guidelines Working Group of the International Society for Companion Animal Infectious Diseases. J Vet Intern Med 31(2):279–294

Leclere M, Lavoie-Lamoureux A, Lavoie JP (2011) Heaves, an asthma-like disease of horses. Respirology 16:1027–1046

Leguillette R (2003) Recurrent airway obstruction–heaves. Vet Clin North Am Equine Pract 19:63–86

Liebler JM, Lutzko C, Banfalvi A, Senadheera D, Aghamohammadi N, Crandall ED, Borok Z (2008) Retention of human bone marrow-derived cells in murine lungs following bleomycin-induced lung injury. Am J Physiol Lung Cell Mol Physiol 295:L285–L292

Lu Z, Meng S, Chang W, Fan S, Xie J, Guo F, Yang Y, Qiu H, Liu L (2020) Mesenchymal stem cells activate notch signaling to induce regulatory dendritic cells in LPS-induced acute lung injury. J Transl Med 18(1):241

Ma C, Liu C, Li X, Lu T, Bai C, Fan Y, Guan W, Guo Y (2017) Cryopreservation and multipotential characteristics evaluation of a novel type of mesenchymal stem cells derived from Small Tailed Han Sheep fetal lung tissue. Cryobiology 75:7–14

Macpherson H, Keir P, Webb S, Samuel K, Boyle S, Bickmore W, Forrester L, Dorin J (2005) Bone marrow derived SP cells can contribute to the respiratory tract of mice in vivo. J Cell Sci 118: 2441–2450

Mainguy-Seers S, Lavoie J-P (2021) Glucocorticoid treatment in horses with asthma: a narrative review. J Vet Intern Med 35:2045–2057

Marx C, Gardner S, Harman RM, de Walle GRV (2020) The mesenchymal stromal cell secretome impairs methicillin-resistant Staphylococcus aureus biofilms via cysteine protease activity in the equine model. Stem Cells Transl Med 9:746–757

Mei SHJ et al (2010) Mesenchymal stem cells reduce inflammation while enhancing bacterial clearance and improving survival in sepsis. Am J Respir Crit Care Med 182:1047–1057

Midgley AC, Rogers M, Hallett MB et al (2013) Transforming growth factor-b1 (TGF-b1)-stimulated fibroblast to myofibroblast differentiation is mediated by hyaluronan (HA)-facilitated epidermal growth factor receptor (EGRF) and CD44 co-localization in lipid rafts. J Biol Chem 288:14824–14838

Millar J, Bartnikowski N, Passmore MR, Obonyo N, Malfertheiner MV, von Bahr V, Redd MA, Hoe LS, Ki KK, Pedersen S, Boyle AJ, Baillie JK, Shekar K, Palpant N, Suen JY, Matthay MA, McAuley DF, Fraser JF (2020) Combined mesenchymal stromal cell therapy and ECMO in ARDS: a controlled experimental study in sheep. Am J Respir Crit Care Med 202(3):383–392

Mirershadi F, Ahmadi M, Rezabakhsh A, Rajabi H, Rahbarghazi R, Keyhanmanesh R (2020) Unraveling the therapeutic effects of mesenchymal stem cells in asthma. Stem Cell Res Ther 11: 400

Moodley Y, Atienza D, Manuelpillai U, Samuel CS, Tchongue J et al (2009) Human umbilical cord mesenchymal stem cells reduce fibrosis of bleomycin-induced lung injury. Am J Pathol 175:303–313

Moodley Y et al (2010) Human amnion epithelial cell transplantation abrogates lung fibrosis and augments repair. Am J Respir Crit Care Med 182:643–651

Németh K et al (2009) Bone marrow stromal cells attenuate sepsis via prostaglandin E(2)-dependent reprogramming of host macrophages to increase their interleukin-10 production. Nat Med 15:42–49

Norris Reinero CR, Decile KC, Berghaus RD et al (2004) An experimental model of allergic asthma in cats sensitized to house dust mite or bermuda grass allergen. Int Arch Allergy Immunol 135:117–131

Ono M, Ohkouchi S, Kanehira M, Tode N, Kobayashi M, Ebina M, Nukiwa T, Irokawa T, Ogawa H, Akaike T, Okada Y, Kurosawa H, Kikuchi T, Ichinose M (2015) Mesenchymal stem cells correct inappropriate epithelial-mesenchyme relation in pulmonary fibrosis using stanniocalcin-1. Mol Ther 23(3):549–560

Panettieri RA, Schaafsma D, Amrani Y, Koziol-White C, Ostrom R, Tliba O (2019) Non-genomic effects of glucocorticoids: an updated view. Trends Pharmacol Sci 40:38–49

Parikh SM, Mammoto T, Schultz A, Yuan HT, Christiani D, Karumanchi SA, Sukhatme VP (2006) Excess circulating angiopoietin-2 may contribute to pulmonary vascular leak in sepsis in human. PLoS Med 3(3):e46

Park HK, Cho KS, Park HY, Shin DH, Kim YK, Jung JS, Park SK, Roh HJ (2010) Adipose-derived stromal cells inhibit allergic airway inflammation in mice. Stem Cells Dev 19(11):1811–1818

Park SG, An JH, Li Q, Chae HK, Park SM, Lee JH, Ahn JO, Song WJ, Youn HY (2021) Feline adipose tissue-derived mesenchymal stem cells pretreated with IFN-γ enhance immunomodulatory effects through the PGE2 pathway. J Vet Sci 22(2):e16

Payne AS, Freishtat RJ (2012) Conserved steroid hormone homology converges on nuclear factor kappaB to modulate inflammation in asthma. J Invest Med 60:13–17

Petrella F, Toffalorio F, Brizzola S, Martino De Pas T, Rizzo S, Barberis M, Pelicci P, Spaggiari L, Acocella F (2014) Stem cell transplantation effectively occludes bronchopleural fistula in an animal model. Ann Thorac Surg 97:480–483

Qiang Y, Liang G, Yu L (2016) Human amniotic mesenchymal stem cells alleviate lung injury induced by ischemia and reperfusion after cardiopulmonary bypass in dogs. Lab Invest 96:537–546

Quimby JM, Borjesson DL (2018) Mesenchymal stem cell therapy in cats: current knowledge and future potential. J Feline Med Surg 20:208–216

Reinero C (2019) Interstitial lung diseases in dogs and cats part II: known cause and other discrete forms. Vet J 243:55–64

Rojas M, Cárdenes N, Kocyildirim E, Tedrow JR, Cáceres E, Deans R, Ting A, Bermúdez C (2014) Human adult bone marrow-derived stem cells decrease severity of lipopolysaccharide-induced acute respiratory distress syndrome in sheep. Stem Cell Res Ther 5:42

Rush B, Mair T (2004) Non-infectious pulmonary diseases and diagnostic techniques. In: Equine respiratory diseases. Blackwell Science, Hoboken, NJ, pp 189–231

Scheerlinck JP, Snibson KJ, Bowles VM, Sutton P (2008) Biomedical applications of sheep models: from asthma to vaccines. Trends Biotechnol 26:259–266
Schermuly RT et al (2005) Reversal of experimental pulmonary hypertension by PDGF inhibition. J Clin Invest 115:2811–2821
Schipani E, Kronenberg HM (2008) Adult mesenchymal stem cells. In: StemBook. The stem cell research community. Harvard Stem Cell Institute, Cambridge, MA, pp 2008–2009
Sharma K, Husain SY, Das P, Hussain M, Syed MA (2016) Regenerative potential of mesenchymal stem cells: therapeutic applications in lung disorders. In: Pham P (ed) Liver, lung and heat regeneration. Stem cells in clinical applications. Springer, Cham, pp 77–117
Sonobe M, Nakagawa M, Ichinose M, Ikegami N, Nagasawa M, Shindo T (2000) Analysis of risk factors in bronchopleural fistula after pulmonary resection for primary lung cancer. Eur J Cardiothoracic Surg 18:519–523
Steens J, Klar L, Hansel C, Slama A, Hager T, Jendrossek V, Aigner C, Klein D (2021) The vascular nature of lung-resident mesenchymal stem cells. Stem Cells Transl Med 10:128–143
Sung S-E, Seo M-S, Kang K-K, Choi J-H, Lee S, Sung M, Kim K, Lee GW, Lim J-H, Yang SY, Yim S-G, Kim S-K, Park S, Kwon Y-S, Yun S (2021) Mesenchymal stem cell exosomes derived from feline adipose tissue enhance the effects of anti-inflammation compared to fibroblasts-derived exosomes. Vet Sci 8(9):182
Taechangam N, Walker NJ, Borjesson DL (2021) Feline adipose-derived mesenchymal stem cells induce effector phenotype and enhance cytolytic function of CD8+ T cells. Stem Cell Res Ther 12(1):495
Telenga ED, Tideman SW, Kerstjens HA et al (2012) Obesity in asthma: more neutrophilic inflammation as a possible explanation for a reduced treatment response. Allergy 67:1060–1068
Trzil JE, Masseau I, Webb TL, Chang CH, Dodam JR, Cohn LA, Liu H, Quimby JM, Dow SW, Reinero CR (2014) Long term evaluation of mesenchymal stem cell therapy in a feline model of chronic allergic asthma. Clin Exp Allergy 44(12):1546–1557
Trzil JE, Masseau I, Webb TL, Chang C-H, Dodam JR, Liu H, Quimby JM, Dow SW, Reinero CR (2016) Intravenous adipose-derived mesenchymal stem cell therapy for the treatment of feline asthma: a pilot study. J Feline Med Surg 18(12):981–990
Van Arnam JS, Herzog E, Grove J, Bruscia E, Ziegler E, Swenson S, Krause DS (2005) Engraftment of bone marrow-derived epithelial cells. Stem Cell Rev 1:21–27
Vandewalle J, Luypaert A, De Bosscher K et al (2018) Therapeutic mechanisms of glucocorticoids. Trends Endocrinol Metab 29:42–54
Veevers-Lowe J, Ball SG, Shuttleworth A, Kielty CM (2011) Mesenchymal stem cell migration is regulated by fibronectin through α5β1-integrin-mediated activation of PDGFR-β and potentiation of growth factor signals. J Cell Sci 124:1288–1300
Wang Q, Lian QQ, Li R, Ying BY, He Q, Chen F, Zheng X, Yang Y, Wu DR, Zheng SX, Huang CJ, Smith FG, Jin SW (2013) Lipoxin A(4) activates alveolar epithelial sodium channel, Na,K-ATPase, and increases alveolar fluid clearance. Am J Respir Cell Mol Biol 48(5):610–618
Wakelee HA, Chang ET, Gomez SL, Keegan TH, Feskanich D, Clarke CA, Holmberg L, Yong LC, Kolonel LN, Gould MK, West DW (2007) Lung cancer incidence in never smokers. J Clin Oncol 25:472–478
Willis BC, Borok Z (2007) TGF-beta-induced EMT: mechanisms and implications for fibrotic lung disease. Am J Physiol Lung Cell Mol Physiol 293:L525–L534
Zeng S-L, Wang L-H, Li P, Wang W, Yang J (2015) Mesenchymal stem cells abrogate experimental asthma by altering dendritic cell function. Mol Med Rep 12:2511–2520
Zhang W-G, He L, Shi X-M, Wu S-S, Zhang B, Mei L, Xu Y-J, Zhang Z-X, Zhao J-P, Zhang H-L (2014) Regulation of transplanted mesenchymal stem cells by the lung progenitor niche in rats with chronic obstructive pulmonary disease. Respir Res 15:33
Zucca E, Corsini E, Galbiati V, Lange-Consiglio A, Ferrucci F (2016) Evaluation of amniotic mesenchymal cell derivatives on cytokine production in equine alveolar macrophages: an in vitro approach to lung inflammation. Stem Cell Res Ther 7(1):137

Mesenchymal Stem Cells in Reproductive System: Reproductive and Therapeutic Applications

9

Abstract

Veterinary species like farm animals and pet animals are being reproduced to develop progeny of elite germplasm that can ensure high production and reproductive potential. Assisted reproductive technologies (ARTs) are being utilized to produce elite and endangered germplasm. One of the common ARTs is in vitro fertilization of embryos by nuclear transfer technology, although the success rate to provide viable offspring is quite low. Also, animals suffer from various reproductive disorders that call for their culling. To address these problems in animal reproduction, mesenchymal stem cells (MSCs) are being evaluated. MSCs have the characteristic properties of being primitive with minimal epigenetic changes and show paracrine actions through soluble factors or microvesicles. Soluble factors include growth factors and other immunomodulatory and/or anti-inflammatory factors while microvesicles harbour genetic materials like RNA, DNA and mitochondria, among others. These features can potentially help in better success rate in ARTs like nuclear transfer and in alleviating the reproductive disorders of animals. In veterinary species, MSCs role is evaluated in promoting growth of follicle and embryo, as nuclear donor in nuclear transfer technology, and as adjuncts to the semen cryoprotectants, besides the therapeutic agents in ovarian, uterine and testicular affections. MSCs in many studies offer promising outcomes although with some intricacies and are currently limited to provide enough evidence for their standard utilization. The current chapter focuses on the MSCs sources, especially the reproductive/foetal membranes, besides their potential reproductive applications.

Keywords

Endometrial disorders · Mesenchymal stem cells · Nuclear donor · Ovarian disorder · Testicular affection · Cryoprotectant

M. B. Gugjoo, *Therapeutic Applications of Mesenchymal Stem Cells in Veterinary Medicine*, https://doi.org/10.1007/978-981-19-3277-9_9

9.1 Introduction

Farm and pet animals are usually being reared for production of animal origin food and other products and as companions, respectively. This asks for much attention over their reproduction. Apart from elite germplasm, transgenic cattle are resistant to mastitis (Donovan et al. 2005) and can produce milk rich in desired proteins like human α-lactalbumin (Yang et al. 2011). This helps in boosting both the economy and health of the human. A wide range of reproductive advancements in veterinary reproduction occurs with each passing day. One of the common fields is assisted reproductive technologies (ARTs). Such technologies like nuclear transfer are aimed at producing elite germplasm. However, the main component of the technology, the nuclear donor, remains an efficiency-limiting step. In successful animal cloning (one of the ARTs), epigenetic errors incorporated in adult fibroblasts (nuclear donor) are considered as main impediments (Hochedlinger and Jaenisch 2002; Wang et al. 2011). Recent focus towards in vitro follicle and embryo development poses the challenge of ensuring efficient supporting factors. Furthermore, veterinary species commonly undergo numerous genital affections. To address these problems in veterinary species, an all-in-one solution in the form of stem cells is being considered.

To waive off the epigenetic issues of current nuclear donor cells, stem cells including the mesenchymal stem cells (MSCs) theoretically make a good candidate for nuclear transfer. These cells have little epigenetic modification as depicted by their lower levels of methylated DNA and/or higher levels of histone acetylation together with the higher telomerase activity. MSCs can undergo easy reprogramming and result in efficient cloning (Enright et al. 2003; Cardoso et al. 2012). Current research on reproduction demonstrates that stem cells can potentially be utilized for in vitro embryo and germ cell production, besides the regeneration of the ailing reproductive organs (West et al. 2013; Asgharzadeh et al. 2015). MSCs may differentiate into the germ cell-like cells (Yan et al. 2015) and show paracrine activity to promote growth and alleviate the reproductive ailments (Peng et al. 2020). The paracrine actions may occur through the soluble factors or the exosomes. The former containing various immunomodulatory/anti-inflammatory factors, growth factors, antioxidants, antimicrobial peptides (AMPs) while the latter harbours the genetic materials like miRNA, and proteins and fats (Gugjoo et al. 2020). Through the cell–cell contact, MSCs may transfer mitochondria via the gap junctions and tunnelling nanotubes (Sinclair et al. 2016; Yu et al. 2016; Mead and Tomarev, 2017; He et al. 2018; Feng et al. 2019). Thus, a lot of thrust is being given to the reproductive applications of MSCs for elite animal genetic dissemination, transgenic animal production, in vitro follicle and embryo production, besides their therapeutics in various genital disorders (Hill and Dobrinsky 2006). This chapter discusses MSCs reproductive/foetal membrane sources, in vitro ability to differentiate into putative germ cells and finally their potential reproductive applications in veterinary species (Fig. 9.1).

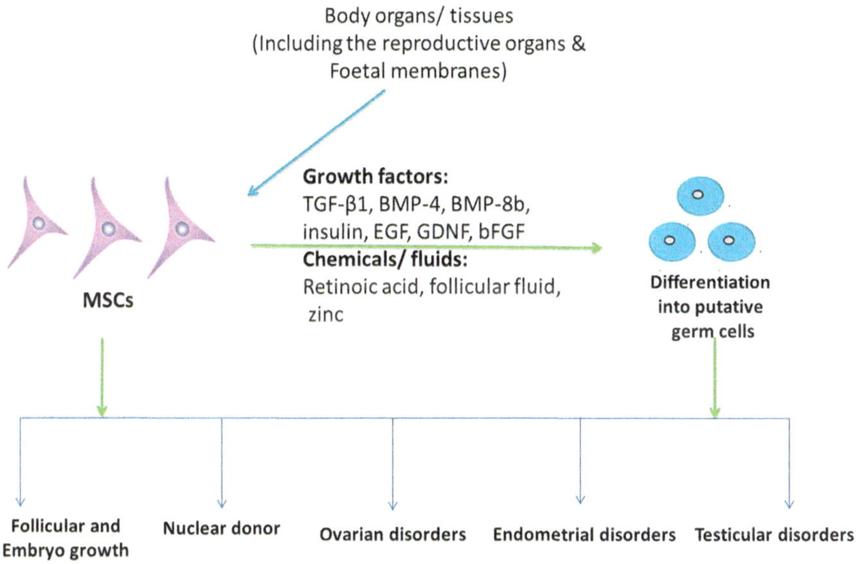

Fig. 9.1 Mesenchymal stem cells (MSCs) paracrine effects and their potential applications in assisted reproductive technologies and therapeutic effects

9.2 MSCs from Reproductive Organs and Foetal Membranes

MSCs have been harvested from a wide variety of body tissues including those from reproductive organs and foetal membranes as detailed in Table 9.1. Among various animal species, dog or cat foetal adnexa-derived MSCs may be limited as the pet owner elects castration of their animals before sexual maturity. Further, dog umbilical cord harbours limited blood (umbilical cord blood, UCB) volume and thereby limiting availability of MSCs (Seo et al. 2009; Sultana et al. 2018). An alternative to such pet sources is the tissues removed in castration like uterus (endometrium) and ovaries (Sultana et al. 2018), although dog endometrium-derived MSCs (End-MSCs) show limited proliferation as compared to their adipose tissue derived (AD)-MSCs (De Cesaris et al. 2017).

MSCs properties vary with source, age and physiological status of the donor. Cattle endometrium-derived MSCs (End-MSCs) harvested at various stages of the oestrus cycle show comparable proliferation potential (de Moraes et al. 2016). These cells in the same stage but at different times (early vs. late luteal stage) show similar characteristics but might become more amenable towards mesodermal lineage in late luteal stage (Cabezas et al. 2014). However, cyclicity affects End-MSCs characteristics. End-MSCs from anestral goats show higher proliferation potential as compared to those from cyclic goats (Tamadon et al. 2017). Non-cyclic (anestrus) endometrium might accumulate various growth factors (bFGF, TGF-β) potentially enhancing their cellular proliferation (Tamada et al. 2000; Tamadon et al. 2017).

Table 9.1 MSCs derived from placental or reproductive organs in veterinary species and their characterization

Source	Positive markers/negative markers/differentiation	Reference
Dog MSCs from foetal membranes		
AF-MSC	Oct-4, CD44, DLA-DRA1 and DLA-79/CD34, CD45/osteogenic chondrogenic, adipogenic, neurogenic	Uranio et al. (2011)
AM-MSC	CD90, CD105, Oct-4, CD44, CD184, and CD29/CD3, CD11c, CD28, CD38, CD62L, CD34, CD45, CD41a/osteogenic chondrogenic, adipogenic, neurogenic	Uranio et al. (2011) and Park et al. (2012)
UCB-MSC	CD29, CD33, CD44, CD105, CD184, Oct4/CD4, CD8a, CD10, CD14, CD20, CD24, CD31, CD34, CD38, CD41a, CD45, CD49b, CD41/61, CD62p, CD73, CD90, CD133, HLA-DR/ osteogenic, chondrogenic, neurogenic	Seo et al. (2009)
UCT-MSC	CD44, CD54, CD61, CD80, CD90, CD105, Flk1, HMGA2, SOX2, NANOG, OCT3/4, OCT-4, CD44, CD184, and CD29/CD4, CD8A, CD25, CD33, CD34 and CD117, CD45/ osteogenic chondrogenic, adipogenic, neurogenic	Uranio et al. (2011) and Lee et al. (2013)
UCV-MSC	CD44, CD73, CD29, CD90, CD105/CD14, CD34, CD45, CD117/osteogenic, adipogenic, chondrogenic, neurogenic	Zucconi et al. (2010) and Ryu et al. (2009)
WJ-MSC	CD44, CD73, CD90, CD105/ CD3, CD14, CD11c, CD28, CD38, CD62L, CD34, CD45, CD41a/osteogenic, chondrogenic, neurogenic	Ryu et al. (2009) and Seo et al. (2012)
Dog MSCs from reproductive organs		
End-MSC	CD44, CD29, CD90, CD13, CD133, CD73, CD105, Oct4/ CD45, CD34, CD31 (low to nil)/ osteogenic chondrogenic, adipogenic	De Cesaris et al. (2017)
Ov-MSC	CD90, CD44, NANOG, TERT, SOX2, OCT4/CD34, CD45/ osteogenic chondrogenic, adipogenic, neurogenic, endoderm, primordial germ cells	Trindade Hill et al. (2017)

(continued)

Table 9.1 (continued)

Source	Positive markers/negative markers/differentiation	Reference
Cattle MSCs from foetal membranes (AF: amniotic fluid derived, Am: amniotic membrane derived, UCB: Umbilical cord blood derived, WJ: Wharton's jelly derived; YS: Yolk sac derived)		
AF-MSCs	CD29, CD44, CD73, CD90, CD105, CD166, MHC-I, Oct-4, C-Myc, CD34 (controversial), β-integrin/CD14, CD34, CD44, CD45, CD90, MHC-II, Sox2, cytokeratin, E-cadherin, CD73 (controversial)/adipogenic, chondrogenic, osteogenic, neurogenic	Corradetti et al. (2013), Gao et al. (2014), Rossi et al. (2014), Gonzales da Silva et al. (2016b) and Chang et al. (2018)
Am-MSCs	CD29, CD44, CD73, CD105, CD166, MHC-I, Nanog, Sox-2, Oct-4, C-Myc/CD14, CD34, CD45, MHC-II/Adipogenic, Chondrogenic, osteogenic, neurogenic	Corradetti et al. (2013), Mann et al. (2013) and Campos et al. (2017)
UCB-MSCs	CD29, CD44, CD73, CD90, CD166, Oct-4, Nanog/CD34, CD45, BLA-DR, MHC-II/ adipogenic, chondrogenic, osteogenic, hepatocyte-, islet cell-, neurocyte-like cells	Raoufi et al. (2011), Xiong et al. (2014) and Campos et al. (2017)
WJ-MSCs	CD29, CD73, CD90, CD105, POUF51, ITSN1/CD34, CD44 (controversial), CD45/adipogenic, chondrogenic, osteogenic, neural-like cells	Cardoso et al. (2012), Cardoso et al. (2016) and da Silva et al. (2016a)
YS-MSCs	CD90, CD105, and CD79 /CD45, CD44, and CD79/adipogenic, chondrogenic, osteogenic	Mançanares et al. (2015)
Cattle MSCs from endometrium		
End-MSCs	CD29, CD44, CD117, STAT-3, Oct-4, Sox-2, c-KIT, vimentin/ CD34 (low expression), MHC-II Nanog (controversial)/adipogenic, chondrogenic, osteogenic	Cabezas et al. (2014), de Moraes et al. (2016) and Lara et al. (2017)
Buffalo MSCs from foetal membranes (UC: Umbilical cord derived)		
Am-MSCs	CD29, CD44, CD90, CD73, CD105, CD166, OCT4, SOX2, NANOG, TERT, REX-1, SSEA-1, SSEA-4 and TRA-1-81, cytokeratin 18 (epithelial)/CD34/ adipogenic, chondrogenic, osteogenic, neurogenic	Dev et al. (2012), Ghosh et al. (2015), Sadeesh et al. (2016) and Deng et al. (2018)
UC-MSCs	OCT4, NANOG and SOX2/−/ adipogenic, osteogenic	Singh et al. (2013)

(continued)

Table 9.1 (continued)

Source	Positive markers/negative markers/differentiation	Reference
WJ-MSCs	CD-73, CD-105 and CD-90, Stro-1, Oct-4, Nanog, Sox-2/CD34/adipogenic, chondrogenic, osteogenic	Sreekumar et al. (2014)
Sheep MSCs from foetal membranes		
AF-MSC	CD13, CD29, CD44, CD58, CD90, CD106, CD166, OCT4, CD29, CD58 /CD14, CD31, CD45, CD49f/adipogenic, chondrogenic, osteogenic	Shaw et al. (2011), Colosimo et al. (2013) and Tian et al. (2016)
Goat MSCs from foetal membranes (Cot: Cotyledonary)		
AF, WJ and CB	CD73, CD90 and CD105, Oct4, Sox2, Nanog, KLF, cMyc, FoxD3/CD34/adipogenic, chondrogenic, osteogenic	Somal et al. (2016)
Cot-MCs	CD29, CD33, and CD166/CD31, CD45/adipogenic (weak), chondrogenic, osteogenic	Ribitsch et al. (2017)
UC-MSCs	–/CD44/CD38, CD45, CD41/61	Fadel et al. (2011)
Goat MSCs from endometrium		
End-MSCs	CD271/CD49f/adipogenic, chondrogenic, osteogenic, smooth muscle	Letouzey et al. (2015)
Cat MSCs from foetal membranes		
Am-MSCs	CD73, CD90, CD44, CD105, OCT4, SOX2/CD30, CD34, CD14, CD45, CD73, CD79/adipogenic, chondrogenic and osteogenic	Iacono et al. (2012), Vidane et al. (2014), Cardoso et al. (2016) and Vidane et al. (2016)

Further, gestational stage might also affect foetal membrane MSCs proliferation potential or concentration consequent to the foetal development. First-trimester cattle amniotic fluid (AF)-MSCs are more in concentration than second and third trimesters (Rossi et al. 2014). Apart from reproductive sources, term sheep AD-MSCs proliferate significantly faster than male sheep AD-MSCs (Dar et al. 2021). Thus, further details in such variations need to be understood to make definitive use of these cells.

9.3 MSCs Promote Follicular and Embryo Growth

In vitro follicular, oocyte and embryo culture environment is aimed at mimicking the female reproductive tract (Tan and Tan 2003; Benkhalifa et al. 2012). However, there are limitations in the current in vitro culture environments to mimic the typical reproductive features (Baldoceda-Baldeon et al. 2014; Campelo et al. 2014; Salzano et al. 2014). To improve such a system, the main focus remains on culture media and cell-based co-culture systems. MSCs due to their characteristic secretory and cell–cell-based interactions are considered as potential candidates to waive off such limitations.

MSCs may support and promote growth and activation of preantral follicles (Xia et al. 2015; Sousa et al. 2021). Bovine ovary MSCs co-culture is able to support growth of the bovine small preantral follicles (Itoh and Hoshi 2000). MSCs may not only support growth but also promote proliferation and activation of preantral follicles. Goat preantral follicles enclosed in ovarian fragments and co-cultured with the Wharton's jelly (Wj)-MSCs for 7 days increase in percentage as compared to those cultured in FBS and basal medium. MSCs beneficial effect on proliferation of preantral follicles occurs irrespective of the FBS as its absence does not hamper growth. MSCs may also increase preantral follicle activation that usually occurs by coordination of suppressive and activating biomolecules (Sousa et al. 2021). BM-MSCs may offer insufficient follicular or oocyte development in goats but critically maintain survival of preantral follicles during 7 days of culture (Arrivabene Neves et al. 2019). The differences in results with MSCs might occur due to the differences in their concentrations (only 1000 cells in the latter study), sources (Wj-MSCs vs. BM-MSCs) and technique of co-culture (ovarian fragment enclosure).

Apart from preantral follicles, MSCs may offer a suitable alternative to foetal calf serum on sheep oocyte maturation and successive in vitro embryo development. These cells additionally perform better than the oviduct epithelial cells. MSCs show comparable effects on oocyte maturation (MII stage) but better indirect effects on blastocyst development stages as compared to the oviduct epithelial cells (Asgharzadeh et al. 2015). In ovines, the direct effect of MSCs co-culture improves embryo hatching rate and total cell number especially in the first 3 days of culture in comparison to the oviduct epithelial cells (Heidari et al. 2013). An extensive evaluation on such co-culture study in bovines shows that bovine AD-MSCs enhance blastocyst rate and embryo quality as confirmed through POU5F1 and G6PDH expressions in comparison to the cumulus cells. Considering the effect of cell concentration, higher cell concentration @ 10,000 cells per mL in comparison to lower cell concentration @ 1000 per mL appears more effective. Furthermore, the direct AD-MSCs-based co-culture has more beneficial effects than their conditioned medium (Miranda et al. 2016). Considering the lack of any specific consensus on optimal conditioning period, variable results with conditioned medium may be seen. As in the above study, a 24-h culture-based conditioned medium (although 100%) is used; it may not have optimal concentration of different paracrine factors. POU5F1 maintains pluripotency of developing mammalian embryo and embryonic stem cells, while G6PDH ensures steroid and fatty acid biosynthesis, nucleotide and nucleic acid synthesis through carbohydrate metabolism and NADPH availability, and increasing the ribose 5-phosphate supply. Both these markers, although not sole markers, are considered as markers of embryo quality (Rodríguez-Alvarez et al. 2013). However, equine BM-MSCs ability to enhance in vitro bovine embryo production falls short of the potential of equine amniotic epithelial cells (Lange-Consiglio et al. 2012). Although numerous similarities have been demonstrated between the equine BM-MSCS and amniotic epithelial cells, there are still differences in these cells, notably the presence of the epidermal growth factor (Lange Consiglio et al. 2010).

MSCs secrete a variety of substances comprising of cytokines/growth factors and exosomes containing microRNAs that can promote growth and neutralize ROS (Baglio et al. 2012; Park et al. 2013). Notable growth factors secreted by MSCs are the transforming growth factor β1, GDF9, BMP15 and hepatocyte growth factor (Uccelli et al. 2008; Kim 2012; Xia et al. 2015). Further, MSCs ability to secrete stem cell factor (upregulates c-kit mRNA expression and stimulates the phosphatidylinositol 3-kinase/AKT pathway) may have a more pronounced effect on follicular development (Thuwanut et al. 2017). Thus, MSCs secreted factors may stimulate oocyte gene expression and play an important role in the development of preantral follicles. MSCs may neutralize toxic components and reactive oxygen species (ROS) in a co-culture system and additionally might secrete various paracrine factors to bring early embryonic development (Benkhalifa et al. 2012).

These studies effectively favour co-culture of MSCs as compared to the oviduct epithelial cells. These studies are insufficient to provide evidence for their standard utilization as there is lack of information on the MSCs source, their concentration, possible mechanism involved and above all the specific period for which it can be maintained.

9.4 MSCs as Nuclear Donor

Nuclear transfer technology (NTT) is aimed at producing transgenic animals. Currently, NTT has a little efficiency of 0–5% in producing the young ones successfully. In bovine, a minute percentage (1.7%) of reconstructed oocytes and transferred NT embryos (11.5%) develop to term. New born may additionally exhibit abnormalities at birth (Keefer 2015). Among various factors, a notable limiting factor is considered to be the nucleus-donor cell (Kwong et al. 2014). Employing MSCs as a donor karyoplast may lead to better blastocyst formation in goats (Kwong et al. 2014), cattle (da Silva et al. 2016b; Nazari et al. 2016) and buffalo (Sadeesh et al. 2016). Few cattle studies, however, fail to demonstrate much improvement with MSCs, and the results are comparable to that with the fibroblasts (Kato et al. 2004; Colleoni et al. 2005). One of the goat studies shows better cleavage rate with AD-MSCs, although blastocyst formation remains comparable to the fibroblasts (Ren et al. 2014). The possible reasons for the differences in the result need further insights.

Extensive studies on equine show better cleavage and blastocyst formation with MSCs. BM-MSCs and UC-MSCs as nuclear donors lead to the higher cleavage and blastocyst rate as compared to the fibroblasts (Olivera et al. 2016, 2018). However, the pregnancy loss of the NT-derived embryos is significant leading to the comparable delivery rates with both the MSCs and fibroblasts but lower than the artificial insemination (AI). This pregnancy loss of the NT-derived embryo irrespective of the donor type currently remains the cause of concern especially during the first trimester. In addition to the higher cleavage and blastocyst rate, the births seen with the MSCs as nuclear donor remained viable and healthy in contrast to the fibroblast where half of the live births remained viable. Furthermore, the pregnancy duration remains same in MSCs births and the AI (310–365 days), while it increases in case of fibroblast births (367–382 days). This might also lead to the delayed hospital discharge of fibroblast delivered foals than the MSCs (Olivera et al. 2018). There are possible chances of placental insufficiency leading to reduced nutrient supply in extended periods. Additionally, enhanced creatinine levels in the neonatal foals are also recorded revealing their dysfunctional placenta (Cummins et al. 2008). All these findings point towards limited viability of the foals in the neonatal phase.

MSCs source and species involved might play a significant role in overall NTT success. Cattle AF-MSCs in comparison to the AD-MSCs as donor karyoplast show better fusion rate (85.89% vs. 50.06%) but comparable cleavage rate (39.08% vs. 38.26%). The percent of calves born are comparable (12.5% vs. 25%) (da Silva et al. 2016b). Another study from same authors demonstrates a slightly lower fusion rate (69.4 $\pm$ 10.2%) with cattle Wj-MSCs but better than fibroblasts (44.8 $\pm$ 0.0%). The percent pregnancies achieved with these cells as nucleus donors are 20% (da Silva et al. 2016a). With larger cells, fusion rate might increase due to increased contact surface as is the case with cattle AF-MSCs/WJ-MSCs in comparison to the AD-MSCs/fibroblasts (da Silva et al. 2016b). It is worth mentioning that calves or young ones born to large nuclear donor cells may also be larger in size leading to large 'offspring syndrome'. Such young ones have lower survival chances due to the foetal abnormalities arising from limited uterine space and dystocia (da Silva et al. 2016), in addition to the epigenetic dysregulation (Farin et al. 2006). As some of the studies show variable results, species differences might also play a role. High levels of H4K5 histone acetylation in embryos derived from goat AD-MSCs may be one of the factors in their high cloning efficiency (Ren et al. 2014). Besides, the recipient oocyte also plays an important role in determining the fusion and cleavage rates and finally to the percent pregnancies. Fresh cattle BM-MSCs yield better blastocysts with fresh oocytes as compared to the cryopreserved oocytes (Kato et al. 2004).

MSCs are considered to have a relatively undifferentiated genome and thus can be reprogrammed more efficiently by recipient oocytes in order to reactivate expression of early embryonic genes. The relative abundance of target transcripts in MSCs-NT embryos may closely follow IVF-derived embryos as compared with fibroblast-NT embryos. However, not all the transcripts show a normal path even in the MSCs-NT clones. Variations in expression levels of these genes may have serious implications and might be contrary to the normal development of the foetus (Sadeesh et al. 2016; Nazari et al. 2016). To achieve better success, MSCs may further be modulated through some chemicals. TSA (non-selective HDAC inhibitor) might enhance their in vitro developmental capacity without compromising the cytological quality for intraspecies (porcine) and/or interspecies (porcine/bovine) cloned embryos (Opiela et al. 2017).

9.5 MSC Therapeutic Applications

9.5.1 Female Reproductive System

In the female reproductive system, MSCs are being evaluated for their ability to waive off various affections of the ovary and the endometrium as detailed below.

9.5.1.1 Bilateral Ovarian Dystrophy in Cattle

Varied available studies demonstrate MSCs role in ovarian disorders, although only a single study evaluates the potential therapeutic effect of MSCs. In clinical infertility cases failed with protocols like ovulation synchronization (OVUSYNCH) and Progesterone Releasing Intravaginal Device (PRID) (Pacala et al. 2010), MSCs may offer therapeutic effects. In cattle placebo-controlled study, cattle AF-MSCs injected into ovary induce oestrus in bilateral ovarian dystrophic cattle. The cell-treated ovaries have observable follicles and preovulatory follicles observable at day 15 after cell injection, while placebo-treated ovaries show the absence of any ovarian follicle. Out of 8 ancestral cows, 4 animals (50%) entered into the oestral phase in which 50% (2 animals) after ovulation have a successful pregnancy. This demonstrates that MSCs may help in inducing oestrus provided other unknown factors are also checked (Chang et al. 2018).

9.5.1.1.1 Ovum Pick-Up-Induced Lesions

In cattle, follicular puncture is a common procedure in ovum pick-up (OPU) technique. This procedure however, often leads to the inflammation, ovarian lesion of thickening of tunica albuginea, stromal fibrosis and adhesions (Petyim et al. 2001; Viana et al. 2003). Even repeated procedures lead to the endocrine abnormalities like increased plasma FSH/LH concentrations, altered follicle growth (Viana et al. 2010) and may potentially develop cystic ovarian disease (Faria et al. 2018). This progressively declines the cumulus–oocyte complex and ultimately the embryo production.

One of the studies demonstrates that intraovarian injection of AD-MSCs (@2.5 million) in acute cases of OPU ($n = 5$) may favour production of superior oocytes and embryos through amelioration of the negative effects of follicular puncture. However, in chronic cases, improvement in in vitro embryo production (IVEP) may not be significant (Malard et al. 2020). As the beneficial results occur within the shorter duration (after 15 days), the possible beneficial effect of MSCs is on the growing antral follicles but whether on tertiary or primordial or growing primary or secondary follicles, it remains to be seen. Therefore, MSCs may show antiapoptotic effects thereby increasing the antral follicle population as reported in humans (Elfayomy et al. 2016; Mohamed et al. 2018). An increased expression of SLC2A3 in the embryos, produced from MSC-treated oocytes, is positively correlated with the glucose uptake and viability of preimplantation embryos (Aller et al. 2012). Glucose being hydrophilic, its cellular transportation occurs through solute carrier family 2 (SLC2A) transporters, among which SLC2A3 is particularly important for embryos. One of the mice studies shows that SLC2A3 deletion may not impact early embryo development until the blastocyst stage but shows lethality after implantation (Schmidt et al. 2009).

Chronic cases (multiple OPU sessions and progressive failure in producing embryos and pregnancies over the years) may develop scar tissue in ovarian stroma resulting in its hardening that affects the preantral follicular development (Petyim et al. 2010; Brito et al. 2014). As in the above study, one-third of antral follicle count is calculated as compared to the healthy heifers, multiple OPU procedures might have damaged the actual ovarian tissue and thus inefficacy of MSCs in enhancing their function.

9.5.1.2 Follicular Cyst

Ovarian follicular cyst (OFC) is one of the commonly reported reproductive problems which arises due to the hormonal imbalances (Peng et al. 2020). The commonly used hormonal treatment protocol gets resistance and additionally leads to other side effects and can come into the food chain through meat or milk. In order to avoid these constraints, other therapeutic options are being evaluated including the MSCs. Placenta-derived MSCs transplantation (one million or six million) along with drainage offers higher recovery rates comparable or better than established hormonal protocol. Such a transplantation enhances the progesterone concentration although without any effect on estradiol. MSCs also induce higher percent oestrus and conception rates in inseminated animals. In relation to the cell concentration, no significant detail is available (animal number less, $n = 4$) but the calf was born to cattle which was injected with lower cell dose. It is concluded that MSCs injection may achieve higher recovery, and oestrus, comparable to that of the hormones (GnRH-PG–GnRH). Overall conception rate remains lower in this study (Peng et al. 2020) than previously reported (Probo et al. 2011). This poor outcome might be due to the use of poor subject animals culled by the local farmers.

In an equine prospective randomized study, intraovarian transplantation of allogeneic BM-MSCs (2 doses of ten million cells, 6 weeks apart) in aged mares ($n = 8$) fails to effectuate significant changes in follicle number and oocyte recovery (on follicle aspiration) and maturation rates up to 6 months. Besides, blastocyst rate also remains same after intracytoplasmic sperm injection (ICSI) in these mares. Similarly, in young mares ($n = 6$), non-significant changes in follicle number also occur. Even no change in anti-Mullerian hormone (AMH) concentrations is recorded indicating no effect on growing follicles. Also MSCs engraftment is lacking even after the second dose of the cells. However, a significant finding of upregulated mRNA expression [testis-specific 10 (*TSGA10*), Ras-association domain family member 9 (*RASSF9*), tachykinin precursor (*TAC1*)] in MSC-treated ovaries as compared to the non-treated ovaries demands further insights (Grady et al. 2019). The results in relation to the aged mare ovaries concord with the cattle study (Malard et al. 2020) but non-significant results in young mares demand further insights. In aged or chronic cases, it may be necessary to induce injury for the cells to engraft. Although the cells do not localize in the ovarian tissue, significant change in the expression of some genes may occur through their paracrine effects.

The possible differences in results across cattle and equine studies may be attributed to the species variability or with the source of MSCs. The positive effect in acute cases could be due to MSCs homing to the inflammatory lesions and increasing follicle recruitment and reduced atresia of follicles. As in the case of chronic lesions, such a store may not be available and thus observable beneficial results could not be achieved. Further extensive studies are therefore desired to confirm MSCs efficacy, if any in clinical settings.

9.5.1.3 Endometrial Affections

Endometritis, inflammation and/or infection of the endometrium, represents one of the main causes of subfertility. It is the third most common disease affecting horses (Canisso et al. 2020), while in cattle it accounts for about 23.6 to 42.6% postpartum cases (Okawa et al. 2017). Cattle End-MSCs properties may be affected in clinical conditions of endometritis (Lara et al. 2017). To ensure effective and sufficient MSCs availability in the affected tissues, extrinsic supply of MSCs may provide some respite. However, MSCs from a particular source may bear characteristic features more relevant to the particular condition and as such may be more effective. Amniotic membrane under in vitro study expresses genes involved in early pregnancy, preimplantation conceptus development and their regulators. Even MSCs from such a tissue may additionally show better expression of anti-inflammatory mediators like prostaglandin-endoperoxide synthase 2 (*PTGS2*) than the diestrus endometrium MSCs (Corradetti et al. 2014). Autologous equine End-MSCs transplantation similar to artificial insemination is made in the case of diestrus phase for endometritis. These cells localize in the uterine lumen for a period of 24 h and attenuate self-afflicted inflammatory response, although the cells are not engrafted to the endometrium (Rink et al. 2018; de Oliveira Tongu et al. 2021). Allogeneic BM-MSCs (@ 20 million) injected 24 h post-breeding modulate local inflammatory response in mares. These cells reduce uterine neutrophil number by over 50% at 6 h post-breeding (inflammatory response to spermatozoa) and increase IL-1Ra protein (Ferris et al. 2014) and IL-10 (de Oliveira Tongu et al. 2021). Apart from the self-inflicted inflammatory reaction, allogeneic AD-MSCs or End-MSCs (each @ 20 million) also show anti-inflammatory action (reduction in IL-6 and TNFα) against an induced equine persistent breeding-induced endometritis (PBIE). The cells, however, are available only for a limited period predicting their more predominant paracrine activity (Navarrete et al. 2020).

9.5.1.3.1 Endometriosis

A chronic fibrosed condition of endometrium known as endometriosis arises due to the repeated endometrial inflammation and can affect reproductive health (LeBlanc and Causey 2009). MSCs characteristic features may have some alleviating role to play. An interesting finding in an in vitro study demonstrates that allogeneic AD-MSCs home peri-glandular and glandular tissue of endometriosis-affected tissue, while such homing is limited in normal endometrium. A mixture of response towards the inflammatory mediators (downregulated IL1B, IL10, TNFA and IL1RN and upregulated expression levels of IL6 and IL8) and matrix-degrading enzymes (decreased MMP2 and TIMP2 and increased MMP9) is seen both in the healthy and endometriozed tissues (Falomo et al. 2015). The engraftment of the allogeneic AD-MSCs (20 million) in endometriosis-affected mares ($n = 6$) post artificial insemination-like intrauterine transplantation is also confirmed in an in vivo study.

The cells effectively home into the endometrium including glandular and non-glandular endometrium (Mambelli et al. 2013). Histologically, an improvement in glandular epithelia of endometriosis uterus occurs after transplantation of these cells. Post allogeneic MSCs transplantation, co-expression of markers like vimentin, laminin, smooth-muscle-α-actin and CK18 does not occur (Mambelli et al. 2014). In endometriosis, altered protein expression patterns of these markers occur, and thus MSCs may be useful in conditions of pregnancy failure caused by deranged endometrial cells.

Apart from AI-like intrauterine implantation of the cells, autologous BM-MSCs (@ 12 million) directly transplanted (at 12 sites with individual dose of one million) into the uterus of a mare fail to offer any respite in subfertility and degenerated endometrium (grades II-b and III) in the mares ($n = 6$). Barring localized inflammatory reactions at MSCs injection on day 15 and absent at day 30, no observable clinical changes occur up to day 60 (Alvarenga et al. 2016). These limited available studies show promising role of MSCs in endometrial affections; however, the route of transplantation of the cells, their source may have a role to play and thus, further extensive studies involving large case numbers with details of the mechanism of action need to be undertaken. Further, the actual replenishment of the endometrial cells, if any need to be evaluated and for longer duration.

9.5.2 Male Reproductive System

In male animals, stem cell applications are being aimed at ensuring the semen banks with high recovery, in addition to their therapeutic applications. Mostly in vitro studies are currently available in veterinary species that evaluate MSCs ability to transdifferentiate and help in recovering the semen with high recovery rate of spermatozoa and minimal abnormality.

9.5.2.1 Transdifferentiation of MSCs into Putative Germ Cells

MSCs, if successfully transdifferentiated, may act as a bank of spermatozoa. These cells although show potential to express the germ cell-like genes and change morphology but current procedures fail to demonstrate their actual transdifferentiation into both early and late germ cell-like cells. The details of the in vitro transdifferentiation of these cells in various veterinary species are discussed below.

MSCs express male germ cell-like characteristics upon their priming with some growth factors and chemicals, and ectopic overexpression. Growth factors like TGF-β1, BMP-4 and BMP-8b influence sheep and cattle BM-MSCs to change morphology from spindle shape to oval or rounded and express male germ cell-specific markers, although there is variability in morphology change (oval or rounded) and gene expression (Ghasemzadeh-Hasankolaei et al. 2014a, b, 2015; Cortez et al. 2018). There is variability in germ cell markers for each specific germ cell; priming of ovine BM-MSCs with TGF-β1 initiates spermatogonial stem cells and spermatogonia-like cells contrary to the BMP-4 and BMP-8 that initiate primordial germ cell-like properties. Treatment with any of these factors to MSCs shows no upregulated expression of Oct4 (Ghasemzadeh-Hasankolaei et al. 2014a, b; 2016). Growth factor priming of MSCs fails to effectuate expression of all the germ cell-specific genes (early and late markers). The transdifferentiated germinal cell-like cells express pluripotency markers like Nanog, Sox-4 and some specific germinal markers like deleted in azoospermia-like (Dazl), while simultaneously failing to express other germinal cell-specific genes [Vasa, Fragilis, Stella and male germinal cell genes (Stra8 and Piwil2)], signifying expression of mostly early germ cell-like differentiation (Cortez et al. 2018). Also dog study shows that AD-MSCs can be induced into primordial germ cell-like cells (PGCLCs) and germ-like cells as exhibited by their expression of PR domain-containing-1 (*Prdm1*), *PRDM14*, double sex and mab-3-related transcription factor (*Dmrt1*), as well as promyelocytic leukaemia zinc finger (*Plzf*) under BMP-4 (Wei et al. 2016). Other growth factors [insulin, epidermal growth factor (EGF), Glial Derived Neurotrophic Factor (GDNF) and basic fibroblast growth factor (bFGF)] used in combination with other chemicals favour dog ovarian MSCs to express primordial germ cell-like and male germ cell-like features (Trindade Hill et al. 2017). Morphology change to the round and expression of Oct4 and DDX4 confirms dog MSCs transdifferentiation into primordial germ cell lineage (PGCLs) (Trindade Hill et al. 2017). Overall, MSCs priming with growth factors transiently increases the expression of Oct4, Nanog and *Dazl*.

Apart from growth factors, other chemicals like retinoic acid (RA), follicular fluid and zinc may potentiate MSCs germ cell-like expression. Retinoic acid mostly increases *Dazl* expression in bovine foetal AD-MSCs and BM-MSCs while simultaneously repressing their Oct4 (Cortez et al. 2018; Cordero et al. 2021) and Nanog expression in AD-MSCs (Cordero et al. 2021). RA, a multipurpose factor, controls meiotic entrance of PGCs by exerting expression control of Cyp26b1 (Koubova et al. 2006). Reproductive tissue matrix or its fluids may affect MSCs growth and differentiation potential. At lower follicular concentration (2–10%, mean 5%), goat UC-MSCs proliferate at higher rate while at higher concentration (20%), differentiation into oocyte-like cells (large round cell masses and positive for specific germ cell markers—synaptonemal complex protein 3, *Scp3* and Vasa) occurs. These induced cells show higher *Zp2, Zp3, Scp3, Stra8 and GDF9* expressions as compared to the control (Qiu et al. 2012). Follicular fluids contain hormones (follicle-stimulating hormone, oestrogen and progesterone), *GDF9* and BMP15. The level of the *GDF9* correlates well with the oocyte nuclear maturation and embryo quality (Gode et al. 2011). Even the proteins and cytokines (EGF and insulin like growth factor, IGF) modulate oocytes and embryos (Dyce et al. 2011; Gode et al. 2011). Thus, dose-dependent effect of follicular fluid might make the difference in MSCs germ cell-like differentiation. Zinc supplementation to ram BM-MSCs upregulates their *VASA* and *ITGb1* expressions and downregulates PIWIL2 and Oct4. Genes like *DAZL* and *PGP 9.5* remain unexpressed. Thus, zinc does not stimulate MSCs transdifferentiation into germ cells rather modulates their expression and could be a possible reason for zinc-associated effects on spermatogenesis (Ghasemzadeh-Hasankolai et al. 2012). BMP-4 and RA that induce partial activation of germ cell gene expression possess common targets but induce different gene expression effects (Cordero et al. 2021).

Ectopic expression of various genes (*DAZL, STRA8* and *BOULE*) more efficiently reprograms the MSCs fate (Rosa and Brivanlou 2010). In rodents, *DAZL* is an important factor for preliminary primordial germ cell formation and exhibits it through restriction of the pluripotency and other somatic differentiation (Chen et al. 2014; Niu et al. 2014). Besides that its expression is also seen in the prenatal and postnatal testes (Brekhman et al. 2000). In comparison, *BOULE* mostly functions in the late stage of meiosis (Kee et al. 2009). *STRA8* is prominently known to act as gatekeeper for germ cell entry into meiosis, indispensably required for premeiotic DNA replication and its subsequent entry into prophase of meiosis I (Mark et al. 2008). Ectopic overexpression of germ cell-specific genes like *STRA8, BOULE and DAZL* attributes to goat MSCs characteristics of male germ cells, in both early and late stages. The cells show primordial germ cell (express *STELLA*, *C-KIT* and *MVH)*, premeiotic (*DAZL*, *BOULE*, *STRA8*, *PIWIL2* and *RNF17) and* meiotic marker (synaptonemal complex protein 3) expression. Further, ectopic co-expression of all the three genes by these cells shows more resemblance to the germ cells than those expressing only one of the genes. However, percent of transdifferentiated cells remain quite low (Li et al. 2017; Zhang et al. 2019). Overexpression of *DAZL* and *STRA8* in bicistronic and *DAZL, STRA8 and BOULE* in tricistronic vectors upregulates Oct4, Nanog and *PIWIL2* in bovine foetal AD-MSCs (Cordero et al. 2021). In goat BM-MSCs, ectopic expression of *Dazl* gene appears superior followed in order by culture priming of retinoic acid (RA) and BMP-4 (Yan et al. 2015). Goat MSCs co-transfection with mRNA of *STRA8, BOULE and DAZL* also results in upregulation of Oct4 (Li et al. 2017; Zhang et al. 2019). *DAZL* expression overlaps with that of the Oct4 and Nanog till Oct4 downregulates and Nanog levels decrease. Oct4 downregulation occurs with their entry into meiosis at spermatogonial stage B and Nanog levels decrease when migrating primordial germ cells (PGCs) reach the gonad (Saitou and Miyauchi 2016). Oct4 being a pluripotency marker with good correlation with germinal markers is considered to have a role in germ cell development.

From all these studies, it becomes clear that MSCs respond to specific growth factors and chemicals, whether priming or through overexpression, to change morphology and express germ cell-like markers. However, the priming with growth factors or chemicals may not be as efficient as that initiated with the overexpression. These inducing agents may upregulate pluripotency markers that overlap with the germ cell markers expression and downregulation of the former downregulates the germ cell-like markers. The mechanisms involved under such processes largely remain unclear. Furthermore, the germ cell-specific marker's expression (under RT-PCR) has also been seen in the uninduced MSCs raising queries over their actual differentiation into germ cell lineage (Ghasemzadeh-Hasankolai et al. 2012). Thus, further studies in the area are desired for clear and controlled evidence of MSCs differentiation into germ cell-like cells and finally for clinical applications.

To effectuate normal functioning of testis, MSCs or pre-differentiated MSCs may be used to rejuvenate the testis. In an equine study, allogeneic BM-MSCs

(20 million) intratesticular injection appears safe clinically (no effect on testis or semen) (de MelloPapa et al. 2020). The in vitro differentiated ovine BM-MSCs after in vivo testicular transplantation survive in the ram testis and home to the basement membrane of seminiferous tubules wherein spermatogonia-like cell colonies (expressed PGP9.5) form. Considering the role of in vitro differentiation factors, cells differentiated under TGF-β1 show higher propensity towards homing and colony formation in comparison to those differentiated under retinoic acid (Ghasemzadeh-Hasankolaei et al. 2015). Thus, further studies in the area are desired to use cells clinically.

9.5.2.2 Semen Cryopreservation

Semen cryopreservation is a common protocol in veterinary sciences irrespective of the species. However, cryopreserved sperms undergo damage of variable kind and thus lead to limited efficiency. Despite sperm cell's ability to adapt to osmotic changes (Mughal et al. 2018), these cells still undergo some osmotic stress during the freeze–thaw cycle. This leads to the increased production of reactive oxygen species (ROS), apoptosis-like changes, DNA damage and an increase in tail defects in the sperm cells (Ball 2008). Apart from sperm cryopreservation, elite germplasm male animals too undergo testicular disorders that often become the cause of their culling (Turner 2007). MSCs due to their characteristics may be employed to help in preserving the germplasm and to cure the disorders of testis effectively.

Ex-vivo studies demonstrate improved post-thaw sperm characteristics with dog Am-MSC-conditioned medium (CM) (Mahiddine et al. 2020) or dog AD-MSCs (Qamar et al. 2020) or their exosomes (Qamar et al. 2019). Based on the MSCs population and CM percentage, low concentration of cells (2.5 million vs. 5.0 million) (Qamar et al. 2020) and around 10% of the conditioned medium (Mahiddine et al. 2020) may enhance the quality-related parameters of frozen–thawed sperm cells. The percentage of coiled tail sperm does not improve but significant improvement in the bent tail sperm occurs with the incorporation of CM (Mahiddine et al. 2020). Higher cell concentrations of five million might lead to the higher cell packing, thereby leading to less cell survivability (Qamar et al. 2020). Limited improved results or even negative impact with the higher concentration of conditioned medium (>15%) might have been due to the change in osmolarity of the spermatozoa milieu (Mahiddine et al. 2020). Like conditioned medium, AD-MSCs exosome incorporation (@25, 50 or 100 μg/mL) in cryomedium also leads to the improvement in sperm motility, live sperm percentage, their membrane and acrosome integrity in dogs. This is supported by the statistically higher expression of plasma membrane repair and chromatin material-related genes in exosome-treated spermatozoa (Qamar et al. 2019). MSCs secrete or their CM contain growth factors, antioxidants, intermediate filaments, extracellular matrix components (ECM), growth enzymes, besides other proteins modulating cell metabolism. Furthermore, dog amniotic membrane (Am)-MSC conditioned medium contains apolipoprotein A-1 (sperm capacitation and motility), thioredoxin and apolipoprotein E (antioxidant activity). Also the serum albumin present in CM may help in fertility improvement and sperm cryopreservation (Mahiddine et al. 2020).

9.6 Conclusion(s)

MSCs have numerous sources including that of the genital organs and foetal membranes. These cells have been evaluated in numerous reproductive techniques or ailments and appear promising. In case of NTT, MSCs appear useful, but currently limited success demands further studies to effectuate efficient and fruitful results. Limited available literature evidences MSCs potential to support growth of follicles or embryo but due to the lack of sufficient literature on their mechanism of action, dose titration and above all the long-term effects currently restrict their clinical or standard applications. These cells are also being evaluated in various reproductive ailments like ovarian, uterine and testicular disorders with fruitful results. But lack of comprehensive literature on their safety, mechanism of action and variable results with respect to the efficacy limits their utilization in female reproduction ailments. In male reproductive system, their ability to transdifferentiate is currently incomplete. These cells can be differentiated into the germ cell-like cells but mostly of early-stage male progenitor germ like-cells and their transdifferentiation potential varies with the source and differentiation techniques. However, the success rate and standardization of the therapeutics remain to be established.

References

Aller JF, Mucci NC, Kaiser GG, Callejas SS, Alberio RH (2012) Effect of repeated eCG treatments and ovum pick-up on ovarian response and oocyte recovery during early pregnancy in suckling beef cows. Anim Reprod Sci 133:10–15

Alvarenga MA, do Carmo MT, Segabinazzi LG, Guastali MD, Maia L, Landim-Alvarenga FC (2016) Feasibility and safety of endometrial injection of autologous bone marrow mesenchymal stem cells in mares. J Equine Vet Sci 42:12–18

Arrivabene Neves C, dos Santos SL, Sousa de Carvalho CE, Carvalho MS, Sarmento JLR, Cavalcante TV, Arrivabene M, Arrivabene Neves T, de Sousa Bezerra MÉ, Gomes Júnior AL, Gomes da Silva CM, Martins de Carvalho MA (2019) Culture of goat preantral follicles in situ associated with mesenchymal stem cell from bone marrow. Zygote 28:65–71

Asgharzadeh S, Mirshokraei P, Hassanpour H, Ahmadi E, Nazari H (2015) The effect of mesenchymal stem cells as co-culture in in vitro nuclear maturation of ovine oocytes. Anim Sci Paper Rep 33(3):223–232

Baglio SR, Pegtel DM, Baldini N (2012) Mesenchymal stem cell secreted vesicles provide novel opportunities in (stem) cell-free therapy. Front Physiol 3:359

Baldoceda-Baldeon LM, Gagné D, Vigneault C, Blondin P, Robert C (2014) Improvement of bovine in vitro embryo production by vitamin K2 supplementation. Reproduction 148:489–497

Ball BA (2008) Oxidative stress, osmotic stress and apoptosis: impacts on sperm function and preservation in the horse. Anim Reprod Sci 107:257–267

Benkhalifa M, Demirol A, Sari T, Balashova E, Tsouroupaki M, Giakoumakis Y et al (2012) Autologous embryo-cumulus cells co-culture and blastocyst transfer in repeated implantation failures: a collaborative prospective randomized study. Zygote 20:173–180

Brekhman V, Itskovitz-Eldor J, Yodko E, Deutsch M, Seligman J (2000) The DAZL1 gene is expressed in human male and female embryonic gonads before meiosis. Mol Hum Reprod 6(5): 465–468

Brito IR et al (2014) Alginate hydrogel matrix stiffness influences the *in vitro* development of caprine preantral follicles. Mol Reprod Dev 81:636–645

Cabezas J, Lara E, Pacha P, Rojas D, Veraguas D, Saravia F, Rodríguez-Alvarez L, Castro FO (2014) The endometrium of cycling cows contains populations of putative mesenchymal progenitor cells. Reprod Domest Anim 49(4):550–559

Campelo IS, Pereira AF, Alcântara-Neto AS, Canel NG, SouzaFabjan JMG, Teixeira DIA et al (2014) Effect of crotamine, a cell penetrating peptide, on blastocyst production and gene expression of in vitro fertilized bovine embryos. Zygote 24:1–10

Campos LL, Landim-Alvarenga FC, Ikeda TL, Monteiro BA, Maia L, Freitas-Dell'Aqua CP, De Vita B (2017) Isolation, culture, characterization and cryopreservation of stem cells derived from amniotic mesenchymal layer and umbilical cord tissue of bovine fetuses. Pesqui Vet Bras 37(3):278–286

Canisso IF, Segabinazzi LGTM, Fedorka CE (2020) Persistent breeding-induced endometritis in mares—a multifaceted challenge: from clinical aspects to immunopathogenesis and pathobiology. Int J Mol Sci 21:1432

Cardoso TC, Ferrari HF, Garcia AF, Novais JB, Silva-Frade C, Ferrarezi MC, Andrade AL, Gamiero R (2012) Isolation and characterization of Whartons jelly-derived multipotent mesenchymal stromal cells obtained from bovine umbilical cord and maintained in defined serum free three-dimensional system. BMC Biotechnol 12:18

Cardoso MT, Pinheiro AO, Vidane AS, Casals JB, de Oliveira VC, Gonçalves NJN, Martins DS, Ambrósio CE (2016) Characterization of teratogenic potential and gene expression in canine and feline amniotic membrane-derived stem cells. Reprod Domest Anim 51(Suppl. 3):1–7

Chang L-B, Peng S-Y, Chou C-J, Chen Y-J, Shiu J-S, Tu P-A, Gao S-X, Chen Y-C, Lin T-K, Wu S-C (2018) Therapeutic potential of amniotic fluid stem cells to treat bilateral ovarian dystrophy in dairy cows in a subtropical region. Reprod Domest Anim 53(4):1024–1024

Chen HH, Welling M, Bloch DB et al (2014) DAZL limits pluripotency, differentiation, and apoptosis in developing primordial germ cells. Stem Cell Rep 3(5):892–904

Colleoni S, Donofrio G, Lagutina I, Duchi R, Galli C, Lazzari G (2005) Establishment, differentiation, electroporation, viral transduction, and nuclear transfer of bovine and porcine mesenchymal stem cells. Cloning Stem Cells 7(3):154–166

Colosimo A, Russo V, Mauro A, Curini V, Marchisio M, Bernabò N, Alfonsi M, Mattioli M, Barboni B (2013) Prolonged in vitro expansion partially affects phenotypic features and osteogenic potential of sheep amniotic fluid-derived mesenchymal stromal cells. Cytotherapy 15:930–950

Cordero P, Guerrero-Moncayo A, De los Reyes M, Varas-Godoy M, Cortez J, Torres CG, Parraguez VH, Peralta OA (2021) Overexpression of DAZL, STRA8, and BOULE genes and treatment with BMP4 or retinoic acid modulate the expression of MSC overexpressing germ cell genes. Front Vet Sci 8:667547

Corradetti B, Meucci A, Bizzaro D, Cremonesi F, Lange Consiglio A (2013) Mesenchymal stem cells from amnion and amniotic fluid in the bovine. Reproduction 145:391–400

Corradetti B, Correani A, Romaldini A, Marini MG, Bizzaro D, Perrini C, Cremonesi F, Lange-Consiglio A (2014) Amniotic membrane-derived mesenchymal cells and their conditioned media: potential candidates for uterine regenerative therapy in the horse. PLoS One 9:e111324

Cortez J, Bahamonde J, De los Reyes M, Palomino J, Torres CG, Peralta OA (2018) In vitro differentiation of bovine bone marrow-derived mesenchymal stem cells into male germ cells by exposure to exogenous bioactive factors. Reprod Domest Anim 53(3):700–709

Cummins C, Carrington S, Fitzpatrick E, Duggan V (2008) Ascending placentitis in the mare: a review. Ir Vet J 61(5):307–313

da Silva CG, Martins CF, Cardoso TC, da Cunha ER, Bessler HC, Martins GHL, Pivato I, Bao SN (2016a) Production of bovine embryos and calves cloned by nuclear transfer using Mesenchymal stem cells from amniotic fluid and adipose tissue. Cell Reprogram 18(2):127–135

da Silva CG, Martins CF, Cardoso TC, da Cunha ER, Bessler HC, MacManus CM, Pivato I, Bao SN (2016b) Isolation and characterization of mesenchymal stem cells derived from bovine

Wharton's jelly and their potential use in cloning by nuclear transfer. Ciência Rural 46(10): 1830–1837

Dar ER, Gugjoo MB, Javaid M, Hussain S, Fazili MR, Dhama K, Alqahtani T, Alqahtani AM, Shah RA, Emran TB (2021) Adipose tissue- and bone marrowderived mesenchymal stem cells from Sheep: culture characteristics. Animals (Basel) 11(8):2153

De Cesaris V, Grolli S, Bresciani C, Conti V, Basini G, Parmigiani E, Bigliardi E (2017) Isolation, proliferation and characterization of endometrial canine stem cells. Reprod Domest Anim 52(2): 235–242

de MelloPapa P, Guasti PN, De Vita B, Nakazato NG, Maia L, de Freitas Dell'Aqua CP, da Scheeren VFC, Segabinazzi LGTM, de Andrade Junior LRP, Silva LFMC et al (2020) Clinical safety of intratesticular transplantation of allogeneic bone marrow multipotent stromal cells in stallions. Reprod Domest Anim 55:429–437

de Moraes CN, Maia L, Dias MC, Dell'Aqua CP, da Mota LS, Chapwanya A, Landim-Alvarenga FD, Oba E (2016) Bovine endometrial cells: a source of mesenchymal stem/progenitor cells. Cell Biol Int 40(12):1332–1339

de Oliveira Tongu EA, Segabinazzi LGTM, Alvarenga ML, Monteiro A, Papa FO, Alvareng MA (2021) Allogenic mesenchymal stem cell-conditioned medium does not affect sperm parameters and mitigates early endometrial inflammatory responses in mares. Theriogenology 169:1–8

Deng Y, Huang G, Zou L, Nong T, Yang X, Cui J, Wei Y, Yang S, Shi D (2018) Isolation and characterization of buffalo (*bubalus bubalis*) amniotic mesenchymal stem cells derived from amnion from the first trimester pregnancy. J Vet Med Sci 80(4):710–719

Dev K, Giri S, Kumar A, Yadav A, Singh B, Gautam S (2012) Derivation, characterization and differentiation of buffalo (Bubalus bubalis) amniotic fluid derived stem cells. Reprod Domest Anim 47(5):704–711

Donovan DM, Kerr DE, Wall RJ (2005) Engineering disease resistant cattle. Transgenic Res 14: 563–567

Dyce P, Liu J, Tayade C, Betts D, Li J (2011) In vitro and in vivo germ line potential of stem cells derived from newborn mouse skin. PLoS One 6:e20339

Elfayomy AK, Almasry SM, El-Tarhouny SA, Eldomiaty MA (2016) Human umbilical cord blood-mesenchymal stem cells transplantation renovates the ovarian surface epithelium in a rat model of premature ovarian failure: possible direct and indirect effects. Tissue Cell 48:370–382

Enright BP et al (2003) Epigenetic characteristics of bovine donor cells for nuclear transfer: levels of histone acetylation. Biol Reprod 69:1525–1530

Fadel L, VianaII BR, Feitosa MLT, ErcolinII ACM, RoballoII KCS, Casals JB, Pieri NCG, Meirelles FV, dos Santos MD, Miglino MA, Ambrósio CE (2011) Protocols for obtainment and isolation of two mesenchymal stem cell sources in sheep. Acta Cir Bras 26(4):267–273

Falomo ME, Ferroni L, Tocco I, Gardin C, Zavan B (2015) Immunomodulatory role of adipose-derived stem cells on equine endometriosis. Biomed Res Int 2015:1–6

Faria OAC et al (2018) Effects of active immunization against GnRH in oocyte donors with cystic ovarian disease. Reprod Fertil Dev 30:190

Farin PW, Piedrahita JA, Farin CE (2006) Errors in development of fetuses and placentas from in vitro produced bovine embryos. Theriogenology 65:178–191

Feng Y, Zhu R, Shen J, Wu J, Lu W, Zhang J, Zhang J, Liu K (2019) Human bone marrow mesenchymal stem cells rescue endothelial cells experiencing chemotherapy stress by mitochondrial transfer via tunneling nanotubes. Stem Cells Dev 28(10):674–682

Ferris RA, Frisbie DD, McCue PM (2014) Use of mesenchymal stem cells or autologous conditioned serum to modulate the inflammatory response to spermatozoa in mares. Theriogenology 82:36–42

Gao Y, Zhu Z, Zhao Y, Hua J, Ma Y, Guan W (2014) Multi-lineage potential research of bovine amniotic fluid mesenchymal stem cells. Int J Mol Sci 15:3698–3710

Ghasemzadeh-Hasankolaei M, Eslaminejad MB, Batavani R, Sedighi-Gilani M (2014a) Comparison of the efficacy of three concentrations of retinoic acid for transdifferentiation induction in sheep marrow-derived mesenchymal stem cells into male germ cells. Andrologia 46:24–35

Ghasemzadeh-Hasankolaei M, Sedighi-Gilani MA, Eslaminejad MB (2014b) Induction of ram bone marrow mesenchymal stem cells into germ cell lineage: using transforming growth factor-β superfamily growth factors. Reprod Domest Anim 49:588–598. https://doi.org/10.1111/rda.12327

Ghasemzadeh-Hasankolaei M, Eslaminejad MB, Batavani R, Ghasemzadeh-Hasankolaei M (2015) Male and female rat bone marrow-derived mesenchymal stem cells are different in terms of the expression of germ cell specific genes. Anat Sci Int 90(3):187–196

Ghasemzadeh-Hasankolaei M, Eslaminejad MB, Sedighi-Gilani M (2016) Derivation of male germ cells from ram bone marrow mesenchymal stem cells by three different methods and evaluation of their fate after transplantation into testis. In Vitro Cell Dev Biol Anim 52(1):49–61

Ghasemzadeh-Hasankolai M, Batavani R, Eslaminejad BM, Sedighi-Gilani M (2012) Effect of zinc ions on differentiation of bone marrow-derived mesenchymal stem cells to male germ cells and some germ cell-specific gene expression in rams. Biol Trace Elem Res 150:137–146

Ghosh K, Kumar R, Singh J, Gahlawat SK, Kumar D, Selokar NL, Yadav SP, Gulati BR, Yadav PS (2015) Buffalo (Bubalus bubalis) term amniotic-membrane-derived cells exhibited mesenchymal stem cells characteristics in vitro. In Vitro Cell Dev Biol Anim 51(9):915–921

Gode F, Gulekli B, Dogan E, Korhan P, Dogan S, Bige O, Cimrin D, Atabey N (2011) Influence of follicular fluid GDF9 and BMP15 on embryo quality. Fertil Steril 95:2274–2278

Grady ST, Watts AE, Thompson JA, Penedo MCT, Konganti K, Hinrichs K (2019) Effect of intra-ovarian injection of mesenchymal stem cells in aged mares. J Assist Reprod Genet 36:543–556

Gugjoo MB, Amarpal AAI, Aithal HP, Kinjavdekar P, Saikumar GS, Sharma GT (2020) Allogeneic Mesenchymal stem cells and growth factors in gel scaffold repair Osteochondral defect in rabbit. Regen Med 14:1261–1275

He GH, Zhang W, Ma YX, Yang J, Chen L, Song J, Chen S (2018) Mesenchymal stem cellsderived exosomes ameliorate blue light stimulation in retinal pigment epithelium cells and retinal laser injury by VEGF-dependent mechanism. Int J Ophthalmol 11:559–566

Heidari B, Shirazi A, Naderi M-M, Akhondi M-M, Hassanpour H, Sarvari A, Borjian S (2013) Effect of various co-culture systems on embryo development in ovine. Czech J Anim Sci 58(10):443–452

Hill JR, Dobrinsky I (2006) Male germ cell transplantation in livestock. Reprod Fertil Dev 18(1–2): 13–18. https://doi.org/10.1071/RD05123

Hochedlinger L, Jaenisch R (2002) Monoclonal mice generated by nuclear transfer from mature B and T donor cells. Nature 415:1035–1038

Iacono E, Cunto M, Zambelli D, Ricci F, Tazzari PL, Merlo B (2012) Could fetal fluid and membranes be an alternative source for mesenchymal stem cells (MSCs) in the feline species? A preliminary study. Vet Res Commun 36:107–118

Itoh T, Hoshi H (2000) Efficient isolation and long-term viability of bovine small preantral follicles in vitro. In Vitro Cell Dev Biol Anim 36(4):235–240

Kato Y, Imabayashi H, Mori T, Tani T, Taniguchi M, Higashi M, Matsumoto M, Umezawa A, Tsunoda Y (2004) Nuclear transfer of adult bone marrow mesenchymal stem cells: developmental totipotency of tissue-specific stem cells from an adult mammal. Biol Reprod 70(2): 415–418

Kee K, Angeles VT, Flores M, Nguyen HN, Reijo Pera RA, Human DAZL (2009) DAZ and BOULE genes modulate primordial germ-cell and haploid gamete formation. Nature 462(7270): 222–225

Keefer CL (2015) Artificial cloning of domestic animals. Proc Natl Acad Sci U S A 112:8874–8878

Kim JY (2012) Control of ovarian primordial follicle activation. Clin Exp Reprod Med 39(1):10–14

Koubova J, Menke DB, Zhou Q, Capel B, Griswold MD, Page DC (2006) Retinoic acid regulates sex-specific timing of meiotic initiation in mice. Proc Natl Acad Sci U S A 103(8):2474–2479

Kwong PJ, Nam HY, Wan Khadijah WE, Kamarul T, Abdullah RB (2014) Comparison of in vitro developmental competence of cloned caprine embryos using donor Karyoplasts from adult bone marrow mesenchymal stem cells vs. ear fibroblast cells. Reprod Domest Anim 49:249–253

Lange Consiglio A, Corradetti B, Bizzarro D, Cassano M, Cremonesi F (2010) Horse amnion: a source of mesenchymal (AMSCs) and epithelial stem cells (AESCs). 36th Annual Conference of International Embryo Transfer Society (IETS) Conference, 9–12 gennaio, Cordoba, Argentina. Reprod Fertil Dev 22:349

Lange-Consiglio A, Maggio V, Pellegrino L, Cremonesi F (2012) Equine bone marrow mesenchymal or amniotic epithelial stem cells as feeder in a model for the in vitro culture of bovine embryos. Zygote 20(1):45–51

Lara E, Velásquez A, Cabezas J, Rivera N, Pacha P, Rodríguez-Alvarez L, Saravia F, Castro FO (2017) Endometritis and in vitro PGE2 challenge modify properties of cattle endometrial mesenchymal stem cells and their transcriptomic profile. Stem Cells Int 2017:4297639

LeBlanc MM, Causey RC (2009) Clinical and subclinical endometritis in the mare: both threats to fertility. Reprod Domest Anim 44:10–22

Lee KS, Cha SH, Kang HW, Song JY, Lee KW, Ko KB, Lee HT (2013) Effects of serial passage on the characteristics and chondrogenic differentiation of canine umbilical cord matrix derived mesenchymal stem cells. Asian Australas J Anim Sci 26(4):588595

Letouzey V, Tan KS, Deane JA, Ulrich D, Gurung S, Ong YR, Gargett CE (2015) Isolation and characterisation of mesenchymal stem/stromal cells in the sheep endometrium. PLoS One 10(5): e0127531

Li PZ, Yan GY, Han L, Pang J, Zhong BS, Zhang GM, Wang F, Zhang YL (2017) Overexpression of STRA8, BOULE, and DAZL genes promotes goat bone marrow-derived mesenchymal stem cells in vitro Transdifferentiation toward putative male germ cells. Reprod Sci 24(2):300–312

Mahiddine FY, Kim JW, Qamar AY, Ra JC, Kim SH, Jung EJ, Kim MJ (2020) Conditioned medium from canine amniotic membrane-derived mesenchymal stem cells improved dog sperm post-thaw quality-related parameters. Animals 10:1899

Malard P, Peixer MAS, Grazia JG, dos Santos Sena Brunel H, Feres LF, Villarroel CL, Siqueira LGB, Dode MN, Pogue R, Viana JHM, Carvalho JL (2020) Intraovarian injection of mesenchymal stem cells improves oocyte yield and in vitro embryo production in a bovine model of fertility loss. Sci Rep 10:8018

Mambelli LI, Winter GHZ, Kerkis A, Malschitzky E, Mattos RC, Kerkis I (2013) A novel strategy of mesenchymal stem cells delivery in the uterus of mares with endometriosis. Theriogenology 79:744–750

Mambelli LI, Mattos RC, Winter GHZ, Madeiro DS, Morais BP, Malschitzky E, Miglino MA, Kerkis A, Kerkis I (2014) Changes in expression pattern of selected endometrial proteins following mesenchymal stem cells infusion in mares with endometrosis. PLoS One 9:e97889

Mançanares CA, Oliveira VC, Oliveira LJ, Carvalho AF, Sampaio RV, Mançanares AC, Souza AF, Perecin F, Meirelles FV, Miglino MA, Ambrósio CE (2015) Isolation and characterization of mesenchymal stem cells from the yolk sacs of bovine embryos. Theriogenology 84(6):887–898

Mann A, Yadav RP, Singh J, Kumar D, Singh B, Yadav PS (2013) Culture, characterization and differentiation of cells from buffalo (*Bubalus bubalis*) amnion. Cytotechnology 65(1):23–30

Mark M, Jacobs H, Oulad-Abdelghani M et al (2008) STRA8-deficient spermatocytes initiate, but fail to complete, meiosis and undergo premature chromosome condensation. J Cell Sci 121 (pt 19):3233–3242

Mead B, Tomarev S (2017) Bone marrow-derived mesenchymal stem cells-derived exosomes promote survival of retinal ganglion cells through miRNA-dependent mechanisms. Stem Cells Transl Med 6(4):1273–1285

Miranda MS, Nascimento HS, Costa MP, Costa NN, Brito KN, Lopes CT, Santos SS, Cordeiro MS, Ohashi OM (2016) Increasing of blastocyst rate and gene expression in co-culture of bovine embryos with adult adipose tissue-derived mesenchymal stem cells. J Assist Reprod Genet 33(10):1395–1403

Mohamed SA et al (2018) Human mesenchymal stem cells partially reverse infertility in chemotherapy-induced ovarian failure. Reprod Sci 25:51–63

Mughal DH, Ijaz A, Yousaf MS, Wadood F, Farooq U, Mahmood SA, Riaz A (2018) Effect of osmotic pressure on spermatozoa characteristics of cryopreserved buffalo bull (Bubalus bubalis) semen. J Appl Anim Res 46(1):274–277

Navarrete F, Saravia F, Cisterna G, Rojas F, Silva PP, Rodríguez-Alvarez L, Rojas D, Cabezas J, Mançanares ACF, Castro FO (2020) Assessment of the anti-inflammatory and engraftment potential of horse endometrial and adipose mesenchymal stem cells in an in vivo model of post breeding induced endometritis. Theriogenology 155:33–42

Nazari H, Shirazi A, Shams-Esfandabadi N, Afzali A, Ahmadi E (2016) The effect of amniotic membrane stem cells as donor nucleus on gene expression in reconstructed bovine oocytes. Int J Dev Biol 60:95–102

Niu Z, Hu Y, Liao M et al (2014) Conservation and function of Dazl in promoting the meiosis of goat male germline stem cells. Mol Biol Rep 41(5):2697–2707

Okawa H, Fujikura A, Wijayagunawardane MMP, Vos PLAM, Taniguchi M, Takagi M (2017) Effect of diagnosis and treatment of clinical endometritis based on vaginal discharge score grading system in postpartum Holstein cows. J Vet Med Sci 79(9):1545–1551

Olivera R, Moro LN, Jordan R, Luzzani C, Miriuka S, Radrizzani M, Donadeu FX, Vichera G (2016) In vitro and in vivo development of horse cloned embryos generated with iPSCs, mesenchymal stromal cells and fetal or adult fibroblasts as nuclear donors. PLoS One 11(10): e0164049

Olivera R, Moro LN, Jordan R, Pallarols N, Guglielminetti A, Luzzani C, Miriuka SG, Vichera G (2018) Bone marrow mesenchymal stem cells as nuclear donors improve viability and health of cloned horses. Stem Cells Cloning 11:13–22

Opiela J, Samiec M, Romanek J (2017) In vitro development and cytological quality of inter-species (porcine/bovine) cloned embryos are affected by trichostatin A-dependent epigenomic modulation of adult mesenchymal stem cells. Theriogenology 97:27–33

Pacala N, Corin N, Bencsik I, Dronca D, Cean A, Boleman A, Caraba V, Papp S (2010) Stimulation of the reproductive function at cyclic cows by ovsynch and PRID/eCG. Anim Sci Biotech 43: 317–320

Park SS, Lee YJ, Lee SH, Lee D, Choi K, Kim WH, Kweon OK, Han HJ (2012) Functional recovery after spinal cord injury in dogs treated with a combination of Matrigel and neural-induced adipose-derived mesenchymal stem cells. Cytotherapy 14:584–597

Park HY, Kim EY, Lee S-E, Choi H-Y, Moon JJ, Park M-J et al (2013) Effect of human adipose tissue-derived mesenchymal-stem-cell bioactive materials on porcine embryo development. Mol Reprod Dev 80:1035–1047

Peng SY, Wu TH, Lin TY, Hii LY, Chan KS, Fu TY, Chang SC, Shen PC, Liu KY, Shaw SW (2020) Application of cattle placental stem cells for treating ovarian follicular cyst. World J Stem Cells 12(11):1366–1376

Petyim S, Båge R, Forsberg M, Rodríguez-Martínez H, Larsson B (2001) Effects of repeated follicular punctures on ovarian morphology and endocrine parameters in dairy heifers. J Vet Med A Physiol Pathol Clin Med 48:449–463

Probo M, Comin A, Mollo A, Cairoli F, Stradaioli G, Veronesi MC (2011) Reproductive performance of dairy cows with luteal or follicular ovarian cysts after treatment with buserelin. Anim Reprod Sci 127(3–4):135–139

Qamar AY, Fang X, Kim MJ, Cho JK (2019) Improved post-thaw quality of canine semen after treatment with exosomes from conditioned medium of adipose-derived mesenchymal stem cells. Animals 9:865

Qamar AY, Fang X, Kim MJ, Cho JK (2020) Improved viability and fertility of frozen-thawed dog sperm using adipose-derived mesenchymal stem cells. Sci Rep 10:7034

Qiu P, Bai Y, Liu C, He X, Cao H, Li M, Zhu H, Hua J (2012) A dose-dependent function of follicular fluid on the proliferation and differentiation of umbilical cord mesenchymal stem cells (MSCs) of goat. Histochem Cell Biol 138:593–603

Raoufi MF, Tajik P, Dehghan MM, Eini F, Barin A (2011) Isolation and differentiation of mesenchymal stem cells from bovine umbilical cord blood. Reprod Domest Anim 46:95–99

Ren Y, Wu H, Ma Y, Yuan J, Liang H, Liu D (2014) Potential of adipose-derived mesenchymal stem cells and skeletal muscle-derived satellite cells for somatic cell nuclear transfer mediated Transgenesis in Arbas cashmere goats. PLoS One 9(4):e93583

Ribitsch I, Chang-Rodriguez S, Egerbacher M, Gabner S, Gueltekin S, Huber J, Schuster T, Jenner F (2017) Sheep placenta cotyledons: a noninvasive source of sheep mesenchymal stem cells. Tissue Eng 23(5):298–310

Rink BE, Beyer T, French HM, Watson E, Aurich C, Donadeu FX (2018) The fate of autologous endometrial mesenchymal stromal cells after application in the healthy equine uterus. Stem Cells Dev 27:1046–1052

Rodríguez-Alvarez L, Manriquez J, Velasquez A, Castro FO (2013) Constitutive expression of the embryonic stem cell marker OCT4 in bovine somatic donor cells influences blastocysts rate and quality after nucleus transfer. In Vitro Cell Dev Biol Anim 49(9):657–667

Rosa A, Brivanlou AH (2010) Synthetic mRNAs: powerful tools for reprogramming and differentiation of human cells. Cell Stem Cell 7:549

Rossi B, Merlo B, Colleoni S, Iacono E, Tazzari PL, Ricci F, Lazzari G, Galli C (2014) Isolation and in vitro characterization of bovine amniotic fluid derived stem cells at different trimesters of pregnancy. Stem Cell Rev Rep 10(5):712–724

Ryu HH, Lim JH, Byeon YE, Park JR, Seo MS, Lee YW, Kim WH, Kang KS, Kweon OK (2009) Functional recovery and neural differentiation after transplantation of allogenic adipose-derived stem cells in a canine model of acute spinal cord injury. J Vet Sci 10:273–284

Sadeesh EM, Shah F, Yadav PS (2016) Differential developmental competence and gene expression patterns in buffalo (Bubalus bubalis) nuclear transfer embryos reconstructed with fetal fibroblasts and amnion mesenchymal stem cells. Cytotechnology 68(5):1827–1848

Saitou M, Miyauchi H (2016) Gametogenesis from pluripotent stem cells. Cell Stem Cell 18:721–735

Salzano A, Albero G, Zullo G, Neglia G, Abdel-Wahab A, Bifulco G et al (2014) Effect of resveratrol supplementation during culture on the quality and cryotolerance of bovine in vitro produced embryos. Anim Reprod Sci 151:91–96

Schmidt S et al (2009) Essential role of glucose transporter GLUT3 for post-implantation embryonic development. J Endocrinol 200:23–33

Seo MS, Jeong YH, Park JR, Park SB, Rho KH, Kim HS, Yu KR, Lee SH, Jung JW, Lee YS, Kang KS (2009) Isolation and characterization of canine umbilical cord blood-derived mesenchymal stem cells. J Vet Sci 10:181–187

Seo MS, Park SB, Kang KS (2012) Isolation and characterization of canine Wharton's jelly-derived mesenchymal stem cells. Cell Transplant 21:1493–1502

Shaw SW, Bollini S, Nader KA, Gastaldello A, Mehta V, Filppi E, Cananzi M, Gaspar HB, Qasim W, De Coppi P, David AL (2011) Autologous transplantation of amniotic fluid-derived mesenchymal stem cells into sheep fetuses. Cell Transplant 20(7):1015–1031

Sinclair KA, Yerkovich ST, Hopkins PMA, Chambers DC (2016) Characterization of intercellular communication and mitochondrial donation by mesenchymal stromal cells derived from the human lung. Stem Cell Res Ther 7(1):91

Singh J, Mann A, Kumar D, Duhan JS, Yadav PS (2013) Cultured buffalo umbilical cord matrix cells exhibit characteristics of multipotent mesenchymal stem cells. In Vitro Cell Dev Biol Anim 49:408–416

Somal A, Bhat IA, Baiju I, Pandey S, Panda BSK, Thakur N, Sarkar M, Chandra V, Saikumar G, Sharma GT (2016) A comparative study of growth kinetics, in vitro differentiation potential and molecular characterization of fetal adnexa derived caprine mesenchymal stem cells. PLoS One 11(6):e0156821

Sousa RP, Duarte ABG, Pinto Y, Sá NAR, Alves BG, FWS C, Silva GC, Carvalho CES, Neto NMA, Rodrigues APR, Silva CMG, Figueiredo JR, Carvalho MAM (2021) Vitro activation and development of goat preantral follicles enclosed in ovarian tissue co-cultured with mesenchymal stem cells. Reprod Sci 28(6):1709–1717

Sreekumar TR, Ansari MM, Chandra VG, Sharma T (2014) Isolation and characterization of buffalo Wharton's jelly derived mesenchymal stem cells. J Stem Cell Res Ther 4:207

Sultana T, Lee S, Yoon H-Y, Lee JI (2018) Current status of canine umbilical cord blood-derived mesenchymal stem cells in veterinary medicine. Stem Cells Int 2018:8329174

Tamada H, Yoh C, Inaba T, Takano H, Kawate N, Sawada T (2000) Epidermal growth factor (EGF) in the goat uterus: immuno-histochemical localization of EGF and EGF receptor and effect of EGF on uterine activity in vivo. Theriogenology 54(1):159–169

Tamadon A, Mehrabani D, Zarezadeh Y, Rahmanifar F, Dianatpour M, Zare S (2017) Caprine endometrial mesenchymal stromal stem cell: multilineage potential, characterization, and growth kinetics in breeding and anestrous stages. Vet Med Int 2017:5052801

Tan X-W, Tan J-H (2003) Co-culture of embryos: influencing factors and mechanisms of action. Sheng Wu Gong Cheng XueBao 19:502–505

Thuwanut P, Comizzoli P, Wildt DE, Keefer CL, Songsasen N (2017) Stem cell factor promotes in vitro ovarian follicle development in the domestic cat by upregulating c-kit mRNA expression and stimulating the phosphatidylinositol 3-kinase/AKT pathway. Reprod Fertil Dev 29:1356–1368

Tian Y, Tao L, Zhao S, Tai D, Liu D, Liu P (2016) Isolation and morphological characterization of sheep amniotic fluid mesenchymal stem cells. Exp Anim 65(2):125–134

Trindade Hill AB, Therrien J, Garcia JM, Smith LC (2017) Mesenchymal-like stem cells in canine ovary show high differentiation potential. Cell Prolif 50(6):e12391

Turner RMO (2007) Pathogenesis, diagnosis, and management of testicular degeneration in stallions. Clin Tech Equine Pract 6(4):278–284

Uccelli A, Moretta L, Pistoia V (2008) Mesenchymal stem cells in health and disease. Nat Rev Immunol 8:726–736

Uranio MF, Valentini L, Lange-Consiglio A, Caira M, Guaricci AC, L'abbate A, Catacchio CR, Ventura M, Cremonesi F, Dell'aquila ME (2011) Isolation, proliferation, cytogenetic, and molecular characterization and in vitro differentiation potency of canine stem cells from Foetal adnexa: a comparative study of amniotic fluid, amnion, and umbilical cord matrix. Mol Reprod Dev 78:361–373

Viana JHM et al (2003) Caracterização de seqüelas subseqüentes à punção folicular em bovinos. Pesquisa Veterinária Brasileira 23:119–124

Viana JHM, Palhao MP, Siqueira LGB, Fonseca JF, Camargo LSA (2010) Ovarian follicular dynamics, follicle deviation, and oocyte yield in Gyr breed (Bos indicus) cows undergoing repeated ovum pick-up. Theriogenology 73:966–972

Vidane AS, Souza AF, Sampaio RV, Bressan FF, Pieri NC, Martins DS, Meirelles FV, Miglino MA, Ambrósio CE (2014) Cat amniotic membrane multipotent cells are nontumorigenic and are safe for use in cell transplantation. Stem Cells Cloning Adv Appl 7:71–78

Vidane AS, Pinheiro AO, Casals JB, Passarelli D, Hage MCFNS, Bueno RS, Martins DS, Ambrósio CE (2016) Transplantation of amniotic membrane-derived multipotent cells ameliorates and delays the progression of chronic kidney disease in cats. Reprod Domest Anim 51(Suppl. 3):1–11

Wang YS et al (2011) Production of cloned calves by combination treatment of both donor cells and early cloned embryos with 5-aza-2/−deoxycytidine and trichostatin a. Theriogenology 75:819–825

Wei Y, Fang J, Cai S, Lv C, Zhang S, Hua J (2016) Primordial germ cell-like cells derived from canine adipose mesenchymal stem cells. Cell Prolif 49(4):503–511

West FD, Shirazi R, Mardanpour P, Ozcan S, Dinc G, Hodges DH, Nayernia K et al (2013) In vitro-derived gametes from stem cells. Semin Reprod Med 31:33–38

Xia X, Wang T, Yin T, Yan L, Yan J, Cuilin L, Zhao L, Li M, Zhang Y, Jin H, Zhu X, Liu P, Li R, Qiao J (2015) Mesenchymal stem cells facilitate in vitro development of human preantral follicle. Reprod Sci 22(11):1367–1376

Xiong H, Bai C, Wu S, Gao Y, Lu T, Hu Q, Guan W, Ma Y (2014) Biological characterization of mesenchymal stem cells from bovine umbilical cord. Anim Cell Syst 1(1):59–67

Yan G, Fan Y, Li P, Zhang Y, Wang F (2015) Ectopic expression of DAZL gene in goat bone marrow-derived mesenchymal stem cells enhances the trans-differentiation to putative germ cells compared to the exogenous treatment of retinoic acid or bone morphogenetic protein 4 signalling molecules. Cell Biol Int 39(1):74–83

Yang B, Wang J, Tang B, Liu Y, Guo C, Yang P et al (2011) Characterization of bioactive recombinant human lysozyme expressed in milk of cloned transgenic cattle. PLoS One 6: e17593

Yu B, Shao H, Su C, Jiang Y, Chen X, Bai L, Zhang Y, Li Q, Zhang X, Li X (2016) Exosomes derived from MSCs ameliorate retinal laser injury partially by inhibition of MCP-1. Sci Rep 6:1–12

Zhang YL, Li PZ, Pang J, Wan YJ, Zhang GM, Fan YX, Wang ZY, Tao NH, Wang F (2019) Induction of goat bone marrow mesenchymal stem cells into putative male germ cells using mRNA for STRA8, BOULE and DAZL. Cytotechnology 71(2):563–572

Zucconi E, Vieira NM, Bueno DF, Secco M, Jazedje T, Ambrosio CE, Passos-Bueno MR, Miglino MA, Zatz M (2010) Mesenchymal stem cells derived from canine umbilical cord vein—a novel source for cell therapy studies. Stem Cells Dev 19:395–402

Mesenchymal Stem Cells Therapeutic Applications in Urinary System Disorders

10

Abstract

Urinary system is one of the important body systems for the survival of an individual. A majority of the non-obstructive affections of the kidney and bladder tend to progress towards irreversible conditions. Ailments like chronic renal failure and bladder atony or necrosis and tumour, besides the urinary incontinence currently remain unaddressed for which mesenchymal stem cells are being evaluated. In large animals, available limited literature shows MSCs potential therapeutic role for urinary affections like ischaemic renal injury and bladder tissue affections. However, the literature is limited enough to offer conclusive evidence for such a therapy. The current chapter focuses on the literature that details mesenchymal stem cell possible therapeutic mechanisms and in vivo applications.

Keywords

Acute renal failure · Bladder resection · Chronic renal failure · Mesenchymal stem cell · Veterinary species

10.1 Introduction

Urinary system comprises the kidneys, ureters, urinary bladder and urethra. Kidneys (bilateral) by and large perform a major function of excreting various toxic chemicals from blood while tissues like ureter, bladder and urethra act as route of excretion. Except for obstructive problems, the main structures that get affected in the urinary system are kidneys and bladder. Kidney is one of the most important organs of the body, the functioning of which determines the survivability of an individual. It not only maintains the body's internal osmotic pressure through filtration and absorption but also produces hormones like renin, erythropoietin,

M. B. Gugjoo, *Therapeutic Applications of Mesenchymal Stem Cells in Veterinary Medicine*, https://doi.org/10.1007/978-981-19-3277-9_10

kinins and prostaglandins (Chawla and Kimmel 2012). Renal tissues show remarkable regeneration ability being a less terminally differentiated organ. Depending on the type of insult, its persistence and changes induced, renal damage varies (Zubko and Frishman 2009). Kidney ailment is generally divided into acute and chronic form. Acute renal failure (ARF) involves the abrupt disruption in renal function, while chronic kidney disease (CKD) develops over a period of time wherein the actual renal parenchyma involving nephrons are affected (Polzin 2011; Lim et al. 2016; Thomson et al. 2019). Among different animal species, dog and cat renal ailments are frequently encountered in clinical settings while in sheep, goat, cattle and horse, urinary system ailments hardly get much attention except for the life-threatening emergency obstructive urolithiasis.

Sheep and cattle renal ailments are often undiagnosed in early stages and gets evident in a more generalized state progressing to renal failure (Mahouz et al. 2015). Despite the renal ailments in cattle/sheep, clinically they get little attention and might be due to the economic constraints. On post-mortem, approximately 32% and 17% kidney lesions are reported in sheep and cattle, respectively. Among various lesions, higher percentage of interstitial nephritis followed by glomerulonephritis and epithelionephritis are reported (Mahouz et al. 2015). In horse, renal diseases may not be common but their reversal is almost impossible. Renal ailments, especially the advanced cases often lead to higher mortality and morbidity for which currently no effective treatment is available (Vidane et al. 2016). In dog and cat clinical practice, ARF and CKD are frequently encountered characterized by functional and/or structural impairment (Lawson et al. 2015). In young dogs, ARF self-resolve contrary to the older dogs as aged kidneys usually have an underlying functional impairment (Feest et al. 1993; Tögel and Westenfelder 2010). In these older patients, such a condition might progress towards a chronic or end-stage renal disease (Goldberg and Dennen 2008). For chronic renal failure/end stage kidney disease, the only useful option is the kidney transplantation which currently remains very limited in veterinary practice. Apart from kidney ailments, urinary bladder frequently gets affected in animals. The bladder gets necrosed after rupture in case of obstructive urolithiasis and develops some functional affection, besides some tumorous growth. Bladder also develops atony or has other neurological affections leading to urinary incontinence. In all these cases, no specific treatment currently exists. In order to address such urinary ailments, regenerative medicine employing mesenchymal stem cells (MSCs) is being evaluated.

10.2 MSCs in Renal Diseases and Potential Applications

MSCs bear characteristic features potentially required for regeneration. These cells exist in almost all the body tissues including that of the urinary system like kidney, urethra and urine (Behr et al. 2007; Burdzinska et al. 2017; Zhang et al. 2014; Gugjoo et al. 2018, 2019, 2020). In kidney, renal papilla mainly harbours MSCs that tend to participate in tubular regeneration (Oliver 2004; Ji et al. 2016). MSCs have capability to migrate and home to the renal tubules and undergo differentiation as per the local microenvironment (Poulsom et al. 2001; Behr et al. 2007). MSCs home ischaemic, inflammatory or necrotic sites by migration across the endothelium (Kim and Cho 2013; Monsel et al. 2014). After tissue damage, various signalling molecules comprising chemokines, adhesion molecules and matrix metalloproteinases get released that interact with the chemokine receptors and integrins on MSCs and promote their migration and homing (Wei et al. 2013; Monsel et al. 2014). This makes it possible to transplant MSCs through routes other than the renal for therapeutics. MSCs differentiate into mesodermal lineages like osteogenic, chondrogenic, adipogenic or myogenic depending upon the type of available microenvironment (Kariminekoo et al. 2016). Such a differentiation has been demonstrated to extend to the endodermal and ectodermal lineages as well (Jeon et al. 2015). Despite ex vivo evidences, in vivo therapeutic effects through differentiation including the renal cells mostly remain inconceivable (Seo et al. 2005). This property can judiciously be utilized to develop extrinsic tissue for transplantation.

MSCs are immuno-evasive due to the lack of MHC-II and T-cell co-stimulatory molecules. Thus, MSCs from allogeneic or xenogeneic sources may be used for in vivo transplantations (Gugjoo et al. 2020). MSCs therapeutic role through non-differentiation mechanisms notably include the cell-to-cell interaction or through paracrine actions. The secretome contains soluble factors or microvesicles. Microvesicles contain lipids, proteins and miRNAs that have therapeutic effects (Bruno et al. 2009, 2012; Collino et al. 2010; Thebaud and Stewart 2012; Zhou et al. 2013) (Fig. 10.1). Renal growth may be promoted through soluble factors including growth factors and secreted extracellular matrix and cell adhesion proteins together with its remodelling enzymes, besides the angiogenic factors (Morigi et al. 2016; Gugjoo et al. 2020). The secretory factors like antioxidant, antiapoptotic, interleukins and growth factors are also considered important. Notable growth factors (GFs) like epithelial (EGF), vascular endothelial (VEGF), transforming (TGF) α and β, fibroblast (FGF), insulin-like type 1 (IGF-1), among others are considered as important ones (Murphy et al. 2013; Humenik et al. 2019). In general, MSCs promote growth of renal parenchyma and inhibit its degradation, prevent microbial growth, immune reaction and inflammation (Abumaree et al. 2012; Humenik et al. 2019; Marx et al. 2020; Luo et al. 2021). All these features offer an all-in-one regenerative solution in diverse urinary ailments. MSCs regenerative studies in animal species can offer therapeutics in veterinary urology and simultaneously provide proof-of-therapeutic principle for human medicine.

10.3 In Vivo Studies

10.3.1 Renal Insufficiency Studies

The literature on urinary regenerative medicine employing MSCs pertains to small ruminants like sheep and goat, and pet animals like cat and dog (Quimby et al. 2011, 2013, 2016; Lim et al. 2016; Rosselli et al. 2016; Vidane et al. 2016; Lee et al. 2017; Thomson et al. 2019). For renal efficacy, route of implantation may be an important

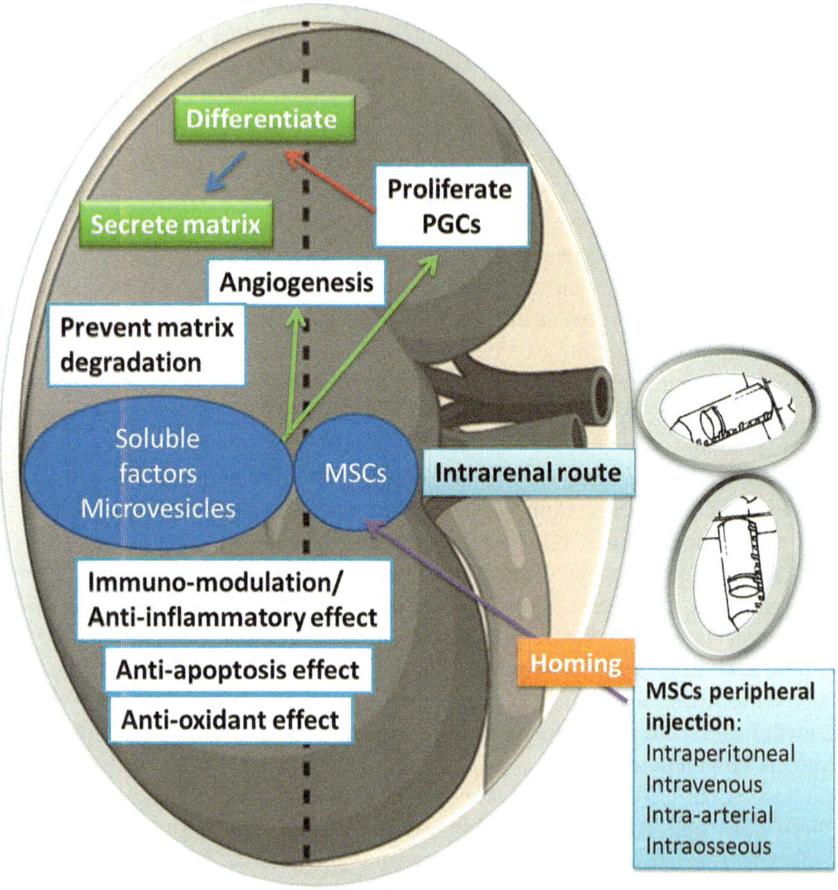

Fig. 10.1 Mesenchymal stem cell route of transplantation and their potential therapeutic effects in renal ailments

factor towards the result-oriented work. Direct intrarenal cellular infusion although may allow complete renal engraftment but requirement of anaesthetics together with its associated risk may make it less feasible. The intravenous injection of MSCs for CKD thus appears a preferable route but might require higher doses as the majority of the intravenously injected MSCs usually ends up in localization in lungs or liver (Vidane et al. 2016). Among various parenteral routes, autologous AD-MSCs renal arterial injection via femoral or carotid artery appears feasible and safe within a 3-month postoperative period (Thomson et al. 2019).

10.3.1.1 Sheep Studies

Sheep is utilized as a translational model animal for evaluation of MSCs renal regeneration potential in humans (Behr et al. 2007, 2009). The studies are conducted as ischemic renal injury (IRI) (randomized, placebo-controlled), created through balloon catheterization of renal artery to transiently occlude the vasculature. Such a blockade causes acute tubular and cortical necrosis (Table 10.1). MSCs therapeutic effect is evaluated in unilateral (Behr et al. 2007) or bilateral ischaemia (Behr et al. 2009) IRI models. Unilateral ischaemic study (biochemical parameters may remain unchanged as other kidney keeps function) evaluated for 3–5 weeks demonstrates that renal artery transplanted MSCs (@1.5 million per kg b. wt.) engraft to the glomeruli and tubules and express podocyte phenotype and tubular epithelial cell markers at around 3 weeks post transplantation. However, MSCs implanted at 1 h post-injury engraft significantly higher in comparison to those implanted at 2 weeks post injury (Behr et al. 2007). Thus, MSCs efficacy may be higher when implanted at earliest after injury. Contrarily in the bilateral IRI model, MSCs fail to engraft to the tubules, express glomerular cell markers and do not prevent biochemical changes of IRI (podocin and von Willebrand factor) at day 7. MSCs fail to promote kidney parenchymal repair and modulate cell death and cytokine release. It is being considered that MSCs post renal artery transplantation may remain localized in glomerular capillaries and renal vasculature during the first week followed by their migration to tubules and transdifferentiation after 2–3 weeks. As MSCs therapeutic effect is unlikely through the actual transdifferentiation, their paracrine effects may still show improvement (Behr et al. 2009). In comparison to the rodent or other lab animal models, significant differences in results can be observed. In lab animal models, a significant improvement in renal morphology and function occurs in IRI (Tögel et al. 2005, 2007; Herrera et al. 2007). This might arise mainly on the basis of differences in species and the cells. As MSCs from small animals have higher turnover than the large animals (Wagner and Storb 1996), the results may accordingly vary and thus, lab animals study results may not be recapitulated in large animals or humans. As the above-mentioned large animal studies are preliminary, studies in large animals that consider parameters at functional, morphological, cellular and molecular level may be evaluated.

10.3.1.2 Dog Studies

Like sheep, dogs are utilized as renal translational model animals for human stem cell studies. However, unlike sheep, the acute IRI in dog is mostly drug induced viz., higher dose of cisplatin (Lim et al. 2016) and gentamicin (Lee et al. 2017; He et al. 2021) with one study occluding the renal artery through clamping (Osman et al. 2020) (Table 10.1). Mostly these studies have been conducted as randomized, placebo-controlled studies. MSCs from bone marrow @ one million per kg b. wt. (Lim et al. 2016), umbilical cord blood @ 2 doses of one million (Lee et al. 2017) and adipose tissue @ ten million (He et al. 2021) and @ 5–15 million (Osman et al. 2020) are transplanted. The route of cell injection includes intravenous (Lim et al. 2016; He et al. 2021) and renal artery (Lee et al. 2017) or direct renal cortex (Osman et al. 2020). Considering the renal artery route in $n = 2$, significant allogeneic MSCs concentration localizes in the renal cortex by 24 h and is lost by day 8 post-implantation. However, such renal arterial implantation incites inflammation and leads to tubular necrosis, mineralization and fibrosis, although functional changes are not observed (Yoo et al. 2011). Thus, the route of cell implantation may be

Table 10.1 MSCs preclinical experimental studies on acute renal injury models in sheep, dog and cat

Study design/ period	Biomaterial	Evaluation criteria	Overall result	References
Sheep renal ischemic reperfusion injury model				
Randomized placebo-controlled study ($n = 24$)/ 5 weeks	Autologous BM-MSCs (appx 1.5 million/kg b. wt., renal artery)	Identification of engrafted cell, immune phenotype analysis	Cells engrafted to kidney	Behr et al. (2007)
Randomized placebo-controlled study ($n = 15$)/ 7 days	BM-MSCs (50×10^6, renal artery)	Immunophenotypic analysis, renal function, histology and urine analysis	No functional improvement	Behr et al. (2009)
Cat ischaemic renal injury model				
Randomized controlled study ($n = 18$)/ 6 days	Allogeneic AD-MSCs, BM-MSCs, fibroblasts (each cell @ 4×10^6 intravenous after 1 h of injury)	Serum and urine test (International Renal Interest Society, IRIS Grade)	Time, but not treatment, affects improvement	Rosselli et al. (2016)
Dog acute kidney injury model (drug induced)				
Randomized placebo controlled study ($n = 6$)/ 14 days	BM-MSCs (1×10^6 per kg b. wt. intravenously)	CBC, GFR, MRI	No improvement in the renal function	Lim et al. (2016)
Randomized placebo-controlled study ($n = 14$)/ 7 weeks	*UCB-MSCs* (1×10^6 bilaterally into renal cortico-medullary junction at days 6 and 23 post injury)	Renal function and histology	Improved renal tubular function with moderate tubular necrosis	Lee et al. (2017)
Randomized placebo-controlled study ($n = 12$)/ 8 days	*AD-MSCs* (2% FBS cultured @ 1×10^7 in forearm brachial vein)	Histology, immunohistochemistry, PCR, Western blotting	Kidney damage was alleviated with both allogeneic and xenogeneic cADSCs	He et al. (2021)

(continued)

Table 10.1 (continued)

Study design/ period	Biomaterial	Evaluation criteria	Overall result	References
Dog renal ischemia-reperfusion injury (vascular occlusion)				
Randomized placebo-controlled study ($n = 60$)/ 3 months	*BM-MSCs*, *AD-MSCs* (3 doses @ 5, 10, 15 × 10^6 into renal cortex)	Renogram, histopathology, and pro-inflammatory markers	Functional improvement to some extent	Osman et al. (2020)

MSCs mesenchymal stem cells, *AD* adipose tissue derived, *BM* bone marrow derived, *UCB* umbilical cord blood derived, *FBS* foetal bovine serum, *n* total number of animals in all the groups of the study; Improved outcome: Indicates improved outcome with MSCs transplantation even if only one parameter improves; Study period: Indicates maximum time period for which study is conducted

weighed as per the clinical condition. The regeneration is evaluated through variable parameters like renal function and histology, immunohistochemistry, gene and protein expression analysis for a period ranging from 8 days (He et al. 2021) to 7 weeks (Lee et al. 2017).

In all these dog studies, unlike sheep, an improvement with cellular transplantation is reported. However, it is worth mentioning that variable modes of IRI model preparation may not show comparable kidney damage, besides, gentamicin-induced renal necrosis leads to higher fatality (Lee et al. 2017). Intravenous transplantation of autologous BM-MSCs limits inflammatory and fibrotic changes and increases proliferation of tubular epithelial cells in injured kidneys as compared to the control. The cell transplantation also lowers tumour necrosis factor-α and transforming growth factor-β. Reduction in TNF-α is correlated with the increase in endotheliogenesis, contrary to the lab animal studies where VEGF mainly enhances peritubular capillaries (Asanuma et al. 2011). However, mononuclear inflammatory cell infiltration, tubular necrosis, dilated tubules and glomerular damage still remain discernible in both the cell treated and control at day 14. At 2 weeks, cell treatment does not improve clinical renal function in dogs. Further, BM-MSCs transplantation is correlated with leukopenia (Lim et al. 2016). Contrarily, UC-MSCs transplantation in corticomedullary junction is shown to improve renal function and limit tubular necrosis, tubular epithelial cell shedding and glomerular necrosis to moderate, mild to moderate and moderate levels, respectively against the severe level in the control (Lee et al. 2017). These findings are further correlated with reduced endoplasmic reticular stress upon intrarenal AD-MSCs transplantation (He et al. 2021). As is evident from above studies, MSCs from various sources and at variable doses may not show comparable results.

In renal artery occlusion model, renal parenchymal transplantation of AD-MSCs more effectively enhances creatinine clearance both at early (2 weeks) and late period (3 month) than BM-MSCs. Further, BM-MSC-implanted kidneys show significantly more tubular necrosis and delayed regeneration with slightly more pro-inflammatory mediator expression as compared to those implanted with AD-MSCs. In relation to dose efficacy, five million and ten million doses provide effective regeneration than 15 million cells (Osman et al. 2020). All these studies show good prospects for MSCs application in kidney diseases, although the current available literature conducted on limited animals is too limited to reach to the logical conclusion. There is a wide variability in the results among the species with the MSCs, their doses and route of transplantation and none of the study actually shows MSCs mechanisms. Further, as the cells viability remains to be evaluated for long, their actual pro-healing factors and microenvironment eliciting such a response remain largely understood.

10.3.1.3 Cat Studies

In cat, MSCs therapeutic role is evaluated in both experimental (Rosselli et al. 2016) and clinical (Quimby et al. 2011, 2013, 2016; Vidane et al. 2016; Thomson et al. 2019) renal diseases (Table 10.2). In an experimental cat renal ischaemic model, intravenous injection of four million allogeneic AD-MSCs, BM-MSCs and fibroblasts each in ($n = 5$) 1 h post injury fails to effectuate renal function recovery. Post cell transplantation, animals show hyperthermia that could not be considered to be anaesthesia induced as it is not resolved in a shorter duration normally seen in such cases. Animals that receive BM-MSCs show significantly fewer mitotic figures in the superficial cortex of ischemic kidneys as compared to ones that receive fibroblasts. The possible reason for such a positive effect could not be elucidated. However, tubular necrosis, granular casts, with mild to moderate inflammation developed after ischaemic injury are not regressed with any treatment. It is concluded that time but not treatment has a significant effect on renal function and factors that affect outcome are renal injury type, cell dose, frequency and allogenicity, duration, route and timing of administration (Rosselli et al. 2016). Further, species-specific effect in MSCs regeneration potential is seen as lab animals show better recovery as compared to the large animals like sheep and cat.

In Phase I clinical trial of cat CKD, renal arterial transplantation of autologous MSCs [dissolved in stromal vascular fraction (SVF) and made via the femoral or carotid artery under fluoroscopic guidance] appears safe and feasible (Thomson et al. 2019). The clinical application of MSCs in cats is conducted in chronic renal failure (CRF) cases of varied stages graded on the basis of International Renal Interest Society (IRIS). In an open-label baseline controlled pilot trial, intrarenal transplantation of autologous BM-MSCs and AD-MSCs is made at variable doses in grade II, III and IV IRIS cases. In IRIS grades II and III cats, ultrasound-guided intrarenal injection of MSCs is well tolerated with the potential to lead to slight improvement in renal function at 30 days interval. $n = 2$ out of $n = 4$ cats that receive AD-MSCs show modest improvement (statistically non-significant) in clinical parameters like creatinine and glomerular filtration rate (GFR) values. In IRIS stage IV cats, no clinical improvement occurs and could be due to the presence of crystals, consistent with melamine-cyanuric acid. It is concluded the intrarenal administration appears safe but may subvert the clinical benefits of MSCs as animal has to be anesthetized

Table 10.2 MSCs clinical studies on cat chronic kidney disease (CKD)

Clinical study design/period	Biomaterial used	Evaluation criteria	Overall result	References
Cat chronic kidney disease (CKD)				
Open-label, baseline-controlled trial ($n = 6$)/ 30 days	Autologous AD-MSCs, BM-MSCs ($1–40 \times 10^5$)	Nuclear scintigraphy glomerular filtration rate (GFR)	$N = 2$ experienced modest improvement in GFR	Quimby et al. (2011)
Three open-label, baseline-controlled, pilot clinical trials ($n = 16$)/ 8 weeks	Allogeneic AD-MSCs [3 doses of 2×10^6 (cryopreserved cells), 4×10^6 (cryopreserved cells), 4×10^6 (cells from cryopreserved adipose tissue)]	Serum biochemistry, complete blood count, urinalysis, urine protein, glomerular filtration rate, and urinary cytokine concentrations	No discernible clinically relevant improvement in renal functional parameters. Cryopreserved cells and not cryopreserved tissue tend to produced adverse effects	Quimby et al. (2013)
Randomized, placebo-controlled clinical trial ($n = 8$)/ 8 weeks	Allogeneic AD-MSCs (3 doses of 2×10^6 per kg b. wt. at 2, 4 and 6 weeks intravenously)	Blood and urinalysis, nuclear scintigraphy (GFR)	No significant change in lab tests	Quimby et al. (2016)
Open-label, baseline-controlled clinical trial ($n = 10$)/ 60 days	Allogeneic amniotic membrane MSCs [(2×10^6 in CKD cats 2 doses at 21 days interval) and healthy cat (5×10^5 bilateral intrarenal infusion)]	Ultrasonography and lab analysis	CKD cats treated with MSCs registered a significant improvement of renal function	Vidane et al. (2016)
Prospective cohort study (phase I clinical trial)/ $n = 5$/90 days	Autologous AD-MSCs (2 doses 2.4×10^6 and 3.7×10^6 into renal artery via the femoral or carotid artery)	Serum and urinalysis and quality of life assessments	Intra-arterial infusion of MSC into the renal artery in CKD cats is feasible and safe	Thomson et al. (2019)

MSCs mesenchymal stem cells, *AD* adipose tissue derived, *BM* bone marrow derived, *CKD* chronic kidney disease, *n* total number of animals in all the groups of the study; Improved outcome: Indicates improved outcome with MSCs transplantation even if only one parameter improves; Study period: Indicates maximum time period for which study is conducted and injected directly into the kidney. As the study follows dose escalation in the cases, the comparative results cannot be elucidated, although AD-MSCs show slightly better results. However, the control untreated group is not included (considering the study type as pilot feasibility and safety) and thus the slight improvement could have been spontaneously as well (Quimby et al. 2011).

The same researchers report three open-label, baseline-controlled, pilot clinical trials wherein allogeneic AD-MSCs either cryopreserved or derived from cryopreserved adipose tissue are injected thrice intravenously. Cryopreserved cells (two million and four million) develop adverse reactions especially those receiving higher doses (four million) with no reliable clinical improvement. Animals that receive four million cells harvested from cryopreserved AD although do not lead to any adverse effect but simultaneously no improvement in renal functional parameters occurs till 8-week follow-up (Quimby et al. 2013). Another study of the same group also reports MSCs safety but with non-significant improvement upon transplantation of intravenous cryopreserved AD-derived allogeneic MSCs (3 doses of @ two million per kg b. wt.) in cat CKD (Quimby et al. 2016). The possible adverse reactions especially with higher doses of cryopreserved cells may be explained by their possible internalization of the toxic compounds of cryopreservation media. It is concluded that long-term follow-up may be made to evaluate MSCs effect on cat CKD progression. Contrarily, another open-label, baseline-controlled clinical trial that made two doses of an intrarenal injection of allogeneic amniotic membrane MSCs (@ two million) demonstrates improved renal function clinically especially after the second dose. An improvement in serum creatinine and urine protein concentrations and urine-specific gravity is observed but without any change in kidney architecture and morphology after 2 months. The modest reduction in lymphocyte population is attributed to MSCs immunomodulatory effects. It is concluded that amniotic membrane (Am)-MSCs may prove renoprotective and might stabilize the clinical condition, although further extensive long-term studies are required (Vidane et al. 2016).

The current studies are very limited to give any conclusive result. There are contrasting reports on MSCs efficacy even in the case of the cat clinical CKD. Considering the results of various above-mentioned animal studies, it appears that MSCs application may not improve structural and morphological features of the kidney in an immediate follow-up period but the clinical renal functions may improve to some extent. Since these studies are preliminary, non-uniform and lack sequential and to the point focus, further extensive studies are desired. Experimental studies have to be specific with implantation of specific MSCs from specific sources having uniform culture technique, concentration, implantation, doses and finally evaluation criteria. The studies with large animal numbers to be conducted for longer duration to evaluate any structural or morphological improvement are highly desired. The comparative studies may be conducted to confirm the feasibility and efficacy of different cell sources, concentration, doses and routes. Side-by-side clinical trials may be conducted to evaluate actual feasibility and efficacy of the MSCs. Finally a conclusive statement can be drawn to arrive at evidence-based medicine, if any.

10.3.2 Urinary Bladder

Naturally, bladder comprises transitional epithelium that confers it with the ability to stretch or compress as per the demand. As such urinary bladder regenerative medicine is aimed at developing therapeutics to regenerate functional bladder affected congenitally or through acquired diseases (Gerbera and Rees 2009; Gugjoo et al. 2013; Martano et al. 2013; Shinada et al. 2020). The main aim remains to have a healthy, vascular and non-fibrotic bladder with functional smooth muscle cells and transitional epithelium (Drzewiecki et al. 2011). Functional losses in the bladder are exhibited by urinary incontinence in dog and cat (Silverman and Long 2000; James et al. 2015). For a large bladder loss, enterocystoplasty is being done, although it has associated complications like calculi formation, malignancies, urine leakage, infections and above all the metabolic disturbances (Oberpenning et al. 1999; Vishwakarma et al. 2020). Considering limited success and high rate of morbidity using current treatment modalities, more effective alternative strategies are desired to overcome the current challenges and provide better therapeutic outcome. In relation to muscle regenerative potential, in vitro studies demonstrate that MSCs acquire smooth muscle cell phenotype characterized by smooth muscle actin, myosin and calponin features (Kanematsu et al. 2005; Ross et al. 2006; Sharma et al. 2009). Besides, typical urothelium might help in smooth muscle induction from the mesenchyme (Baskin et al. 2001) and as such urinary bladder regeneration might be more imminent. An in vitro study demonstrates that a decellularized urinary bladder provides a suitable biomaterial for growth and proliferation of MSCs from human umbilical cord and as such may be evaluated in the in vivo settings (Vishwakarma et al. 2020).

In lab animals, MSCs have been evaluated in varied type of bladder ailments including cystitis interstitials (Adamowicz et al. 2014; Furuta et al. 2018), functional bladder regeneration (Snow-Lisy et al. 2015; Salehi-pourmehra et al. 2019), bladder and posterior urethral injuries (Demirel et al. 2017), fibrosis (Kim et al. 2016) and bladder tuberculosis (Yudintceva et al. 2018). All these studies show prospectus for MSCs applications in bladder affections. Studies on regenerative medicine employing stem cells in bladder affections may offer basis for their potential clinical applications in veterinary practice and may additionally provide proof-of-therapeutic principles for human medicine. Goat is a common model being used for studying the urinary incontinence in humans (Burdzinska et al. 2017). However, there is paucity of literature on MSCs application in bladder affections of large animals.

10.3.2.1 In Vivo Studies

In a dog urinary bladder reconstruction model (Table 10.3), human AD-MSCs seeded on bladder acellular matrix (BAM/hAD-MSCs, @ 1 lakh cells/cm^2 scaffold) in comparison to scaffold significantly increases maximum bladder volume and compliance evaluated through urodynamics. In comparison to the bladders transplanted with cell less scaffold, limited adhesion to the reconstructed bladder and a more organized smooth muscle structure and capillary density are noted in the bladder transplanted with cell-loaded scaffold (Hou et al. 2016). To consider the

Table 10.3 MSCs preclinical experimental studies on bladder reconstruction and urethral insufficiency

Study design/ period	Biomaterial	Evaluation criteria	Overall result	References
Bladder reconstruction model				
Randomized placebo-controlled study (n = 12)/ 6 months	Bladder acellular matrix (BAM), xenogeneic (human) AD-MSCs (1×10^5 cells per cm^2 scaffold)	Radiography, histopathology, urodynamics	Improved outcome	Hou et al. (2016)
Randomized controlled study (n = 20)/ 3 months	Bladder acellular matrix (BAM), AD-MSCs (15×10^6 cells/ cm^2)	Radiography, histopathology, immunofluorescence and molecular evaluation	Improved outcome	Pokrywczynska et al. (2018)
Urethral sphincter deficiency in goat				
Randomized placebo-controlled comparative study (n = 23)/ 84 days	Muscle-derived cells (MDCs), BM-MCs or co-transplantation of BM-MSCs and MDCs @ (29.6 ± 4.3 million, four locations)	Urethral pressure profile (UPP) [maximal urethral closure pressure (MUCP) and functional area (FA)]	Improved outcome	Burdzinska et al. (2018)

MSCs mesenchymal stem cells, *AD* adipose tissue derived, *BM* bone marrow derived, *n* total number of animals in all the groups of the study; Improved outcome: Indicates improved outcome with MSCs transplantation even if only one parameter improves; Study period: Indicates maximum time period for which study is conducted

minute but important factors, further extensive evaluation in the porcine model is evaluated. It is demonstrated that the use of BAM loaded with AD-MSCs develops a regenerated bladder mostly in the proximity to the native bladder wall while the centre of the scaffold forms scar at 3-month period. This might occur due to the endogenous regeneration beginning adjacent to the native bladder wall being activated after injury. In relation to myogenic regeneration of AD-MSCs, a little evidence (small fraction of αSMA-expressing cells) exists and might occur through cellular ingrowth from surrounding tissue. However, detrusor regrowth may be supported by MSCs (Pokrywczynska et al. 2017). Urothelial markers show divergence from the normal especially in relation to the uroplakin family that regulates superficial umbrella cells membrane permeability. As urine penetrates into the graft, it hampers the rebuilding of the neobladder wall including its nerves through cell death. However, AD-MSCs may be actively involved in angiogenesis considering the survivability of the cells (high PKH-26) in the areas with higher vascular and myogenic markers (Pokrywczynska et al. 2018). Modulation of the molecular pattern like Hedgehog signalling pathway in the healing process through MSCs in the tissue-engineered urinary bladder may ensure such prospects (Pokrywczynska et al. 2019).

Thus, current tissue engineering based urinary bladder regeneration falls short for the standard clinical practice. MSCs support may be incomplete in large bladder wall defect especially in the central graft region. There is a marginal MSCs population that are able to survive and develop smooth muscle and endothelial cells. Urine acts as an innate obstacle for their activity and thus, the scaffold that can prevent urine entry may be employed to allow improved results.

10.3.3 Urethral Sphincter Deficiency

Involuntary urine leakage occurs post external urinary sphincter damage (Walters 1989) and can occur as congenital anomaly or after injury. This leads to a condition known as stress urinary incontinence (SUI) (Nygaard and Heit 2004). Current treatment options like therapeutic agents, surgical and cellular therapy for SUI have one or the other kind of limitations. Thus, novel treatment is being evaluated under experimental models like dog (Eberli et al. 2009) and goat (Burdzinska et al. 2017, 2018). These models are created either through longitudinal sphincterotomy in rodents (Praud et al. 2007) or through the urethral distention in pig (Burdzińska et al. 2012) and partial sphincterotomy in dog (Eberli et al. 2009). In veterinary species, therapeutic studies involving MSCs are lacking barring a single study that evaluates the role of BM-MSCs and muscle-derived cells (MDCs) either alone or their co-transplantation in aged goats (>6 years, multiparous) (Table 10.3). In multiparous aged goats, urethral closure naturally weakens and thus acts a more relevant model than those induced with injuries. Autologous cells (29.6 ± 4.3 million) transplanted endoscopically into the urethra at four locations are located up to day 28, however, are lost by day 84 of the study. The co-transplantation of the cells in the external urethral sphincter contributes to striated muscle formation while such muscle formation is rarely seen in individual MSCs transplantations. Further, the cell co-transplantation in the submucosal region remains undifferentiated. In individual cell transplantations, improvement in urethral profile parameters (UPP) mostly remains non-significant. Further to the details, such comparative analysis offers interesting findings. UPP like maximal urethral closure pressure (MUCP) and functional area (FA) in MSCs transplanted urethra are better at day 28 than at day 84, while in case of MDCs, reverse trend is seen (Paczek 2017; Burdzinska et al. 2018). The current study fails to show much improved UPP than some of the previously reported large animal models (Eberli et al. 2012; Badra et al. 2013). The difference in the results among studies may be attributed to the variability in urethral condition prior to cell transplantation, cell dose and mode of cell transplantation (Eberli et al. 2012; Badra et al. 2013; Williams et al. 2015). The improved results in co-transplanted urethra may occur due to the MDCs ability to improve MSCs myogenic activity, while MSCs may improve MDCs survival through paracrine action. Considering the reversal effect with the individual cells, it may be that MSCs do not integrate with the host muscle to the extent of MDCs and thus only early improved results are seen with the former cell type.

Although the goat study discussed above appears promising with respect to cellular therapy, there are numerous questions to be answered that desire further extensive and organized studies.

10.4 Conclusion(s)

Renal diseases especially the chronic renal failure or end-stage kidneys remain unaddressed. To address the renal diseases, MSCs are being employed. Currently a limited literature on MSCs therapeutic effects on renal regeneration is available. Only a miniscule dog literature demonstrates MSCs potential therapeutic effects in the form of improvement in some clinical parameters while in other species the results are mostly disappointing. The studies have been evaluated only for the shorter durations although long-term results may vary quite considerably. Thus, further desired studies need to be conducted for longer duration with all the possible permutations and combinations to achieve desired therapeutic effects. In relation to the bladder, MSCs loaded on the scaffolds appear promising. However, current literature is limited enough to offer any conclusive evidence of tissue-engineered constructs employing MSCs for their clinical applications. In relation to the urethral incontinence, MSCs as a sole cellular therapeutic agent may not be quite significant but its combinations with other cell types and at specific does may improve the clinical condition that remains to be evaluated.

References

Abumaree M, Al Jumah M, Pace RA et al (2012) Immunosuppressive properties of mesenchymal stem cells. Stem Cell Rev 8:375–392

Adamowicz J, Pokrywczynska M, Drewa T (2014) Conditioned medium derived from mesenchymal stem cells culture as an intravesical therapy for cystitis interstitials. Med Hypotheses 82: 670–673

Asanuma H, Vanderbrink BA, Campbell MT et al (2011) Arterially delivered mesenchymal stem cells prevent obstruction-induced renal fibrosis. J Surg Res 168:e51–e59

Badra S, Andersson KE, Dean A, Mourad S, Williams JK (2013) Long-term structural and functional effects of autologous muscle precursor cell therapy in a nonhuman primate model of urinary sphincter deficiency. J Urol 190:1938–1945

Baskin L, DiSandro M, Li Y, Li W, Hayward S, Cunha G (2001) Mesenchymal-epithelial interactions in bladder smooth muscle development: effects of the local tissue environment. J Urol 165(4):1283–1288

Behr L, Hekmati M, Fromont G, Borenstein N, Noel L-H, Lelievre-Pegorier M, Laborde K (2007) Intra renal arterial injection of autologous mesenchymal stem cells in an ovine model in the postischemic kidney. Nephron Physiol 107:65–76

Behr L, Hekmati M, Houcinet K, Faussat A-M, Borenstein N, Noel L-H, Lelievre-Pegorier M, Laborde K (2009) Evaluation of the effect of autologous mesenchymal stem cell injection in a large-animal model of bilateral kidney ischemia reperfusion injury. Cell Prolif 42:284–297

Bruno S, Grange C, Deregibus MC et al (2009) Mesenchymal stem cell-derived microvesicles protect against acute tubular injury. J Am Soc Nephrol 20(5):1053–1067

Bruno S, Grange C, Collino F et al (2012) Microvesicles derived from mesenchymal stem cells enhance survival in a lethal model of acute kidney injury. PLoS One 7(3):e33115

Burdzińska A, Crayton R, Dybowski B, Koperski Ł, Idziak M, Fabisiak M, Pączek L, Radziszewski P (2012) Urethral distension as a novel method to simulate sphincter insufficiency in the porcine animal model. Int J Urol 19(7):676–682

Burdzinska A, Dybowski B, Zarychta-Wisniewska W, Kulesza A, Zagozdzon R, Gajewski Z, Paczek L (2017) The anatomy of caprine female urethra and characteristics of muscle and bone marrow derived caprine cells for autologous cell therapy testing. Anat Rec 300:577–588

Burdzinska A, Dybowski B, Zarychta-Wiśniewska W, Kulesza A, Butrym M, Zagozdzon R, Graczyk-Jarzynka A, Radziszewski P, Gajewski Z, Paczek L (2018) Intraurethral co-transplantation of bone marrow mesenchymal stem cells and muscle-derived cells improves the urethral closure. Stem Cell Res Ther 9(1):239

Chawla LS, Kimmel PL (2012) Acute kidney injury and chronic kidney disease: an integrated clinical syndrome. Kidney Int 82:516–524

Collino F, Deregibus MC, Bruno S et al (2010) Microvesicles derived from adult human bone marrow and tissue specific mesenchymal stem cells shuttle selected pattern of miRNAs. PLoS One 5(7):e11803

Demirel BD, Biçakci Ü, RizalaR R, Pinarli FA, Aydin O (2017) Histopathological effects of mesenchymal stem cells in rats with bladder and posterior urethral injuries. Turk J Med Sci 47:1912–1919

Drzewiecki BA, Thomas JC, Tanaka ST (2011) Bone marrow-derived mesenchymal stem cells: current and future applications in the urinary bladder. Stem Cells Int 2010:765167

Eberli D, Andersson K-E, Yoo JJ, Atala A (2009) A canine model of irreversible urethral sphincter insufficiency. BJU Int 103(2):248–253

Eberli D, Aboushwareb T, Soker S, Yoo JJ, Atala A (2012) Muscle precursor cells for the restoration of irreversibly damaged sphincter function. Cell Transplant 21:2089–2098

Feest TG, Round A, Hamad S (1993) Incidence of severe acute renal failure in adults: results of a community-based study. BMJ 306:481–483

Furuta A, Yamamoto T, Igarashi T, Suzuki Y, Egawa S, Yoshimura N (2018) Bladder wall injection of mesenchymal stem cells ameliorates bladder inflammation, overactivity, and nociception in a chemically induced interstitial cystitis-like rat model. Int Urogynecol J 29(11):1615–1622

Gerbera K, Rees P (2009) Urinary bladder botryoid rhabdomyosarcoma with widespread metastases in an 8-month-old Labrador cross dog. J S Afr Vet Assoc 80(3):199–203

Goldberg R, Dennen P (2008) Long-term outcomes of acute kidney injury. Adv Chronic Kidney Dis 15:297–307

Gugjoo MB, Zama MMS, Amarpal, Mohsina A, Saxena AC, Sarode IP (2013) Obstructive urolithiasis in buffalo calves and goats: incidence and management. J Adv Vet Res 3:109–113

Gugjoo MB, Amarpal I, Chandra V, Wani MY, Dhama K, Sharma GT (2018) Mesenchymal stem cell research in veterinary medicine. Curr Stem Cell Res Ther 13(8):645–657

Gugjoo MB, Amarpal I, Makhdoomi DM, Sharma GT (2019) Equine mesenchymal stem cells: properties, sources, characterization and potential therapeutic applications. J Equine Vet Sci 72: 16–27

Gugjoo MB, Amarpal I, Fazili MR, Shah RA, Mir MS, Sharma GT (2020) Goat mesenchymal stem cell basic research and potential applications. Small Rum Res 183:106045

He W, Qin D, Li B, Zhang H, Cheng X, Sun J, Hua J, Peng S (2021) Immortalized canine adipose-derived mesenchymal stem cells alleviate gentamicin-induced acute kidney injury by inhibiting endoplasmic reticulum stress in mice and dogs. Res Vet Sci 136:39–50

Herrera MB, Bussolati B, Bruno S, Morando L, Mauriello Romanazzi G, Sanavio F et al (2007) Exogenous mesenchymal stem cells localize to the kidney by means of CD44 following acute tubular injury. Kidney Int 72:430–441

Hou X, Shi C, Chen W, Chen B, Jia W, Guo Y, Ma C, Ye G, Kang J, Dai J (2016) Transplantation of human adipose-derived mesenchymal stem cells on a bladder acellular matrix for bladder regeneration in a canine model. Biomed Mater 11:031001

Humenik F, Cizkova D, Cikos S, Luptakova L, Madari A, Mudronova D, Kuricova M, Farbakova J, Spirkova A, Petrovova E, Cente M, Mojzisova Z, Aboulouard S, Murgoci A-N, Fournier I, Salzet M (2019) Canine bone marrow-derived mesenchymal stem cells: genomics, proteomics and functional analyses of paracrine factors. Mol Cell Proteomics 18:1824–1835

James DR, Collins D, Johnson PJ, Marchevsky AM (2015) Chronic urinary bladder torsion causing urinary incontinence in a cat. JFMS Open Rep 1:2055116915603384l

Jeon B-G, Jang S-J, Park J-S, Subbarao RB, Jeong G-J, Park B-W, Rho G-J (2015) Differentiation potential of mesenchymal stem cells isolated from human dental tissues into non-mesodermal lineage. Anim Cells Syst 19:321–331. https://doi.org/10.1080/19768354.2015.1087430

Ji M, Bai C, Li L, Fan Y, Ma C, Li X, Guan W (2016) Biological characterization of sheep kidney-derived mesenchymal stem cells. Exp Ther Med 12:3963–3971

Kanematsu A, Yamamoto S, Yamamoto S (2005) Induction of smooth muscle cell-like phenotype in marrow-derived cells among regenerating urinary bladder smooth muscle cells. Am J Pathol 166(2):565–573

Kariminekoo S, Movassaghpour A, Rahimzadeh A, Talebi M, Shamsasenjan K, Akbarzadeh A (2016) Implications of mesenchymal stem cells in regenerative medicine. Artif Cells Nanomed Biotechnol 44:749–757

Kim N, Cho S-G (2013) Clinical applications of mesenchymal stem cells. Korean J Intern Med 28: 387–402

Kim A, Yu HI, Heo J, Song M, Shin J-H, Lim J, Yoon S-J, Kim YH, Lee S, Kim SW, Oh W, Choi SJ, Shin D-M, Choo M-S (2016) Mesenchymal stem cells protect against the tissue fibrosis of ketamine-induced cystitis in rat bladder. Sci Rep 6:30881

Lawson J, Elliott J, Wheeler-Jones C, Syme H, Jepson R (2015) Renal fibrosis in feline chronic kidney disease: known mediators and mechanisms of injury. Vet J 203:18–26

Lee S-J, Ryu M-O, Seo M-S, Park S-B, Ahn J-O, Han S-M, Kang K-S, Bhang D-H, Youn H-Y (2017) Mesenchymal stem cells contribute to improvement of renal function in a canine kidney injury model. In Vivo 31(6):1115–1124

Lim C-Y, Han J-I, Kim S-G, Lee C-M, Park H-M (2016) Evaluation of autologous bone marrow–derived mesenchymal stem cells on renal regeneration after experimentally induced acute kidney injury in dogs. J Vet Res 77:208–217

Luo Y, Guo J, Zhang P, Cheuk YC, Jiang Y, Wang J, Xu S, Rong R (2021) Mesenchymal stem cell protects injured renal tubular epithelial cells by regulating mTOR-mediated Th17/Treg Axis. Front Immunol 12:684197

Mahouz F, Khoudja FB, Chikhaoui M (2015) Pathological study on renal diseases in cattle and sheep. J Anim Vet Adv 14(12):357–360

Martano M, Roperto F, de Cassia SR, Russo V, Borzacchiello G, Paciello O, Iovane V, Leonardi L, Maiolino P, Restucci B, Papparella S, Roperto S (2013) Bovine Papillomavirus Type 2 infection and a series of mesenchymal tumors of the urinary bladder in cattle. BioMed Res Int 2013: 814635

Marx C, Gardner S, Harman RM, de Walle GRV (2020) The mesenchymal stromal cell secretome impairs methicillin-resistant Staphylococcus aureus biofilms via cysteine protease activity in the equine model. Stem Cells Transl Med 9:746–757

Monsel A, Zhu Y-G, Gennai S, Hao Q, Liu J, Lee JW (2014) Cell-based therapy for acute organ injury: preclinical evidence and ongoing clinical trials using mesenchymal stem cells. Anesthesiology 121(5):1099–1121

Morigi M, Rota C, Remuzzi G (2016) Mesenchymal stem cells in kidney repair. Methods Mol Biol 1416:89–107

Murphy MB, Moncivais K, Caplan AI (2013) Mesenchymal stem cells: environmentally responsive therapeutics for regenerative medicine. Exp Mol Med 45:e54

Nygaard IE, Heit M (2004) Stress urinary incontinence. Obstet Gynecol 104:607–620

Oberpenning F, Meng J, Yoo JJ, Atala A (1999) De novo reconstitution of a functional mammalian urinary bladder by tissue engineering. Nat Biotechnol 17:149–155

Oliver JA (2004) Adult renal stem cells and renal repair. Curr Opin Nephrol Hypertens 13:17–22

Osman Y, Hamed SM, Barakat NM, Khater S, Gabr M, Mosbah A, Gaballah MA, Shaaban A (2020) Prophylaxis against renal ischemia-reperfusion injury in canine model: stem cell approach. Indian J Urol 36(1):44–49

Paczek L (2017) The anatomy of caprine female urethra and characteristics of muscle and bone marrow derived caprine cells for autologous cell therapy testing. Anat Rec 300(3):577–588

Pokrywczynska M, Jundzill A, Warda K, Buchholz L, Rasmus M, Adamowicz J, Bodnar M, Marszalek A, Helmin-Basa A, Michalkiewicz J, Gagat M, Grzanka A, Frontczak-Baniewicz M, Gastecka AM, Kloskowski T, Nowacki M, Ricordi C, Drewa T (2017) Does the mesenchymal stem cell source influence smooth muscle regeneration in tissue-engineered urinary bladders? Cell Transplant 26(11):1780–1791

Pokrywczynska M, Jundzill A, Rasmus M, Adamowicz J, Balcerczyk D, Buhl M, Warda K, Buchholz L, Gagat M, Grzanka D, Drewa T (2018) Understanding the role of mesenchymal stem cells in urinary bladder regeneration—a preclinical study on a porcine model. Stem Cell Res Ther 9:328

Pokrywczynska M, Rasmus M, Jundzill A, Balcerczyk D, Adamowicz J, Warda K, Buchholz L, Drewa T (2019) Mesenchymal stromal cells modulate the molecular pattern of healing process in tissue-engineered urinary bladder: the microarray data. Stem Cell Res Ther 10:176

Polzin DJ (2011) Chronic kidney disease in small animals. Vet Clin Small Anim 41:15–30

Poulsom R, Forbes SJ, Hodivala-Dilke K, Ryan E, Wyles S, Navaratnarasah S, Jeffery R, Hunt T, Alison M, Cook T, Pusey C, Wright NA (2001) Bone marrow contributes to renal parenchymal turnover and regeneration. J Pathol 195:229–235

Praud C, Sebe P, Biérinx AS, Sebille A (2007) Improvement of urethral sphincter deficiency in female rats following autologous skeletal muscle myoblasts grafting. Cell Transplant 16(7): 741–749

Quimby JM, Webb TL, Gibbons DS, Dow SW (2011) Evaluation of intrarenal mesenchymal stem cell injection for treatment of chronic kidney disease in cats: a pilot study. J Feline Med Surg 13: 418–426

Quimby JM, Webb TL, Habenicht LM, Dow SW (2013) Safety and efficacy of intravenous infusion of allogeneic cryopreserved mesenchymal stem cells for treatment of chronic kidney disease in cats: results of three sequential pilot studies. Stem Cell Res Ther 4:48

Quimby JM, Webb TL, Randall E, Marolf A, Valdes-Martinez A, Dow SW (2016) Assessment of intravenous adipose derived allogeneic mesenchymal stem cells for the treatment of feline chronic kidney disease: a randomized, placebo-controlled clinical trial in eight cats. J Feline Med Surg 18(2):165–171

Ross JJ, Hong Z, Hong Z (2006) Cytokine-induced differentiation of multipotent adult progenitor cells into functional smooth muscle cells. J Clin Investig 116(12):3139–3149

Rosselli DD, Mumaw JL, Dickerson V, Brown CA, Brown SA, Schmiedt CW (2016) Efficacy of allogeneic mesenchymal stem cell administration in a model of acute ischemic kidney injury in cats. Res Vet Sci 108:18–24

Salehi-pourmehra H, Rahbarghazib R, Mahmoudia J, Roshangar L, Chapplec CR, Hajebrahimid S, Abolhasanpoure N, Azghanif M-R (2019) Intra-bladder wall transplantation of bone marrow mesenchymal stem cells improved urinary bladder dysfunction following spinal cord injury. Life Sci 221:20–28

Seo MJ, Suh SY, Bae YC, Jung JS (2005) Differentiation of human adipose stromal cells into hepatic lineage in vitro and in vivo. Biochem Biophys Res Commun 328:258–264

Sharma AK, Fuller NJ, Fuller NJ (2009) Defined populations of bone marrow derived mesenchymal stem and endothelial progenitor cells for bladder regeneration. J Urol 182(4):1898–1905

Shinada M, Saeki K, Yoshitake R, Eto S, Tsuboi M, Chambers JK, Uchida K, Kato D, Yoshimoto S, Kamoto S, Ikeda N, Kinoshita R, Fujita N, Nishimura R, Nakagawa T (2020) Evaluation of epithelial and mesenchymal cell markers in canine urinary bladder transitional cell carcinoma. Vet J 66:105571

Silverman S, Long CD (2000) The diagnosis of urinary incontinence and abnormal urination in dogs and cats. Vet Clin N Am Small Anim Pract 30(2):427–448

Snow-Lisy DC, Diaz EC, Bury MI, Fuller NJ, Hannick JH, Ahmad N, Sharma AK (2015) The role of genetically modified mesenchymal stem cells in urinary bladder regeneration. PLoS One 10(9):e0138643

Thebaud B, Stewart DJ (2012) Exosomes: cell garbage can, therapeutic carrier, or Trojan horse? Circulation 126(22):2553–2555

Thomson AL, Berent AC, Weisse C, Langston CE (2019) Intra-arterial renal infusion of autologous mesenchymal stem cells for treatment of chronic kidney disease in cats: phase I clinical trial. J Vet Intern Med 33:1353–1361

Tögel FE, Westenfelder C (2010) Mesenchymal stem cells: a new therapeutic tool for AKI. Nat Rev Nephrol 6:179–183

Tögel F, Hu Z, Weiss K, Isaac J, Lange C, Westenfelder C (2005) Administered mesenchymal stem cells protect against ischemic acute renal failure through differentiation-independent mechanisms. Am J Physiol Renal Physiol 289:F31–F42

Tögel F, Weiss K, Yang Y, Hu Z, Zhang P, Westenfelder C (2007) Vasculotropic, paracrine actions of infused mesenchymal stem cells are important to the recovery from acute kidney injury. Am J Physiol Renal Physiol 292:F1626–F1635

Vidane AS, Pinheiro AO, Casals JB, Passarelli D, Hage MCFNS, Bueno RS, Martins DS, Ambrósio CE (2016) Transplantation of amniotic membrane-derived multipotent cells ameliorates and delays the progression of chronic kidney disease in cats. Reprod Domest Anim 51(Suppl. 3):1–11

Vishwakarma SK, Sarwar S, Adil MAM, Khan AA (2020) Biofabrication of cell-laden allografts of goat urinary bladder scaffold for organ reconstruction/regeneration. Tissue Cell 67:101443

Wagner JL, Storb R (1996) Preclinical large animal models for hematopoietic stem cell transplantation. Curr Opin Hematol 3:410–415

Walters MD (1989) Mechanisms of continence and voiding, with international continence society classification of dysfunction. Obstet Gynecol Clin North Am 16:773–785

Wei X, Yang X, Han Z-P, Qu F-F, Shao L, Shi Y-F (2013) Mesenchymal stem cells: a new trend for cell therapy. Acta Pharmacol Sin 34(6):747–754

Williams JK, Eckman D, Dean A, Moradi M, Allickson J, Cline JM, Yoo JJ, Atala A (2015) The dose-effect safety profile of skeletal muscle precursor cell therapy in a dog model of intrinsic urinary sphincter deficiency. Stem Cells Transl Med 4:286–294

Yoo J-H, Park C, Jung D-I, Lim C-Y, Kang B-T, Kim J-H, Park J-W, Kim J-H, Park H-M (2011) In vivo cell tracking of canine allogenic mesenchymal stem cells administration via renal arterial catheterization and Physiopathological effects on the kidney in two healthy dogs. J Vet Med Sci 73(2):269–274

Yudintceva NM, Bogolubova IO, Muraviov AN, Sheykhov MG, Vinogradova TI, Sokolovich EG, Samusenko IA, Shevtsov MA (2018) Application of the allogenic mesenchymal stem cells in the therapy of the bladder tuberculosis. J Tissue Eng Regen Med 12(3):e1580–e1593

Zhang D, Wei G, Li P, Zhou X, Zhang Y (2014) Urine-derived stem cells: a novel and versatile progenitor source for cell-based therapy and regenerative medicine. Gene Dis 1:8–17

Zhou Y, XuH XW et al (2013) Exosomes released by human umbilical cord mesenchymal stem cells protect against cisplatin-induced renal oxidative stress and apoptosis in vivo and in vitro. Stem Cell Res Ther 4:34

Zubko R, Frishman W (2009) Stem cell therapy for the kidney? Am J Ther 16(3):247–256

Mesenchymal Stem Cells Therapeutic Applications in Integumentary System Disorders

11

Abstract

Integumentary system in animals comprises skin, mammary gland and the hoof and horns. Among various skin affections, wounds usually represent one of the most common clinical presentations. Chronic wound healing poses a major challenge to the clinician and acute wound-healing enhancement is being aimed at reduction of patient morbidity. Mastitis, equine laminitis and dog atopic dermatitis are other important and most frequently reported clinical conditions in veterinary species. Advanced therapeutics in the form of regenerative medicine are currently being evaluated to enhance wound healing and cure other clinical dermatopathies, mastitis and other integumentary affections effectively and in a shorter duration. The main component in regenerative medicine is stem cells especially mesenchymal stem cell (MSC). The cell characteristic properties make them potential candidates to offer all-in-one therapeutics. Ex vivo and in vivo studies demonstrate MSCs to provide better results, although some reservations still exist to provide evidence-based medicine. MSCs are being evaluated for various therapeutic effects including the dedifferentiation, immunomodulation, antioxidant and antimicrobial effects, among others. In vivo studies conducted on various integumentary conditions like wounds (acute and chronic), atopic dermatitis, laminitis and mastitis show MSCs potential therapeutic effect. The current chapter provides comprehensive details of the therapeutic role of MSCs in various integumentary system affections in veterinary practice.

Keywords

Atopic dermatitis · Laminitis · Mastitis · Mesenchymal stem cell · Veterinary species · Wounds

M. B. Gugjoo, *Therapeutic Applications of Mesenchymal Stem Cells in Veterinary Medicine*, https://doi.org/10.1007/978-981-19-3277-9_11

11.1 Introduction

In animals, the integumentary system comprises the largest body organ (skin), hoof, horns and mammary glands. The skin comprises three major skin layers (epidermis, dermis and hypodermis), appendages like hair, claws or lamina (cornified layer of the epidermis) and subcutaneous muscles and fat. It constitutes about 12–24% of the body weight depending upon the species and age in question. Skin provides a barrier against all the external assaults and regulates temperature and gives a sense of touch (Lee et al. 2006). The specialized integumentary structures like the hoof that protects animal digital bones are formed of keratinized epithelium and modified underlying dermis. In the horns, keratinized epithelium of the skin covers the core bone. Mammary gland is one of the most important structures that produces milk.

Considering the role of skin and its direct exposure to external assaults, its wound-healing potential naturally remains very high through an organized and coordinated series of processes (Ojeh et al. 2015). This enhanced healing potential of skin may be attributed to the innate presence of different stem cells including the epidermal stem cells, dermal papilla and dermal sheath cells (Ojeh et al. 2015; Michler et al. 2018). However, in certain animal skin affections like wounds (chronic and extensive acute), burns, cattle teat fistula and equine proud flesh, such a potential may be insufficient or the cellular processes get deranged. Further, skin appendages like the equine hoof undergo devastating laminitis that currently lacks any effective treatment. Besides, skin allergic diseases like dog atopic dermatitis too remains to be effectively treated. Mammary gland that provides huge economic and nutritional benefits usually undergoes mastitis and fibrosis for which currently no effective and useful treatment is available. In order to address these clinical ailments of the skin, regenerative medicine employing adult stem cells like mesenchymal stem cells (MSCs) is being evaluated.

11.2 Why Mesenchymal Stem Cell Therapy?

Stem cells are one of the important components in regenerative medicine. Stem cell is a primitive undifferentiated cell with an ability to self-renew, multiply and differentiate. Among various stem cell types, MSCs are preferred over the others due to their broader availability and easy harvesting procedures. In addition, MSCs lack ethical and teratogenic issues usually associated with the embryonic and induced pluripotent stem cells (Gugjoo and Amarpal 2018; Gugjoo et al. 2018a, b, 2019, 2020a, b).

The donor tissue harbours very limited MSC concentration which makes it necessary to culture expand them for desired concentrations. These cells are available in variable concentrations and tend to show variable differentiation potential with respect to the tissue sources (Gugjoo et al. 2018b, 2019, 2020a, b). For their uniformity and typical stem cell nature as per the International Society for Cellular and Gene Therapy (ISCT), MSCs should be plastic adherent, express surface markers like CD73, CD90, CD105 and CD166 and lack hematopoietic markers like CD34 and CD45. Further, the cells should transdifferentiate into at least adipogenic, chondrogenic and osteogenic lineages (Dominici et al. 2006). Among the necessary prescribed criteria, MSCs in animal species show plastic adherence and differentiate into trilineages (adipogenic, chondrogenic and osteogenic) but fail to offer consensus on their surface marker expression (Gugjoo et al. 2019). The transcriptomic profile of adult tissue MSCs differs from that of the foetal membranes (Zhan et al. 2019), while canine dermal sheath cells express mesenchymal stem cell markers (Mercati et al. 2009). The surface marker variability may be explained by the variations in the tissues, methods of harvesting and surface marker antibodies utilized. Dynamism in MSCs immunophenotype additionally leads to such variabilities (Strioga et al. 2012; Cabezas et al. 2018). Utilization of detaching agents like trypsin may also impair cell surface receptor membrane proteins and thus, their detection (Gugjoo et al. 2019).

For regenerative effects, MSCs 'home' to the area affected and bring about the healing either through paracrine pro-healing secretions and/or through the transdifferentiation. Their typical anti-inflammatory and/or immunomodulatory, and antimicrobial properties are crucial for early and effective healing. In relation to the MSCs pro-healing properties, the local milieu (micro-environment) plays an important role and its complexity is considered as one of the important reasons for MSCs diverse stem cell characteristics. Such findings open the field of tissue engineering that evaluates the effect of scaffolds, cells, growth factors and mechanical factors. These factors have an essential role to mimic the environmental cues and affect MSCs secretome and differentiation (Gugjoo and Amarpal 2020; Tan et al. 2020). For a deranged healing environment, MSCs modulation of the inflammatory/hyperimmune environment and offering local antimicrobial effects might provide suitable conditions for wound healing to occur (Gugjoo et al. 2020a, b).

MSCs are considered to provide an all-in-one therapeutic solution for various ailments (Gugjoo et al. 2018a, b, 2019, 2020a, b) (Fig. 11.1). In general, MSCs regenerative role is considered to occur through the immunomodulatory/anti-inflammatory and antifibrotic activities, neutralizing action on the reactive oxygen species (ROS), angiogenesis through neovascularization, promoting fibroblasts/fibrocytes towards myofibroblast lineage and inhibiting microbial proliferation. Further, their direct differentiation into the dermal fibroblasts, keratinocytes and vascular endothelial cells (Fig. 11.1) (Jackson et al. 2012; Russell et al. 2020), in addition to the other cell lineages may also occur, although its currently questioned. Below are discussed the immunomodulatory/anti-inflammatory and antimicrobial roles of MSCs. The other therapeutic effects of MSCs are discussed in the relevant sections of the chapters.

11.2.1 MSCs Anti-Inflammatory/Immunomodulatory Role

Apart from the basic properties, MSCs show homing properties and locally offer immunomodulatory/anti-inflammatory effects, prevent tissue hypoxia/reperfusion injury and evade allogeneic rejection (Gugjoo et al. 2020a, b). The cells express lower levels of major histocompatibility complex (MHC) class I while MHC class II

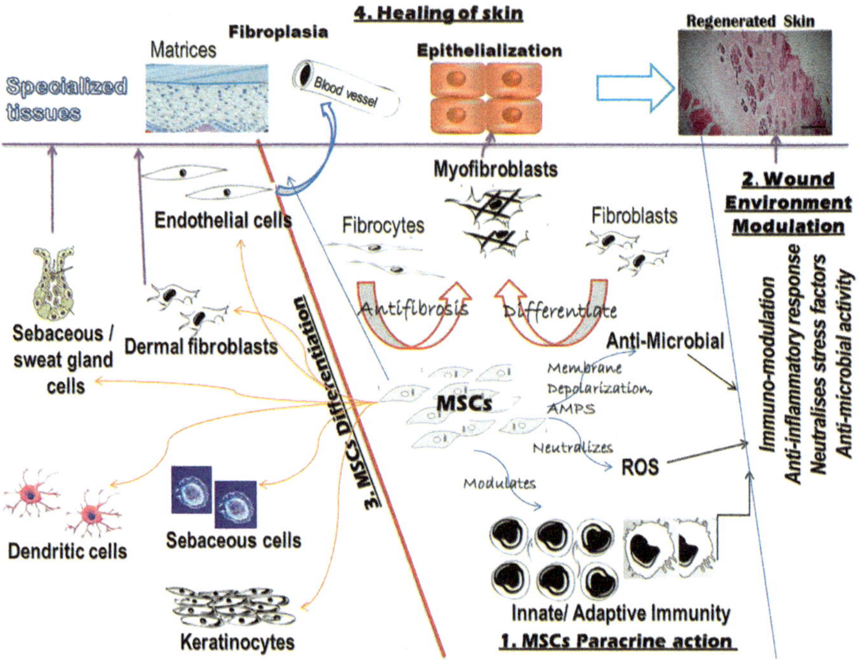

Fig. 11.1 MSCs potential therapeutic mechanisms in chronic wound healing. MSCs paracrine action (1) modulating wound environment in order to ensue normal healing pathways (2). MSCs differentiation into various cell lineages, although not fully confirmed (3). All these activities bring about the normal skin to regenerate (4)

or T-cell co-stimulatory molecules mostly fail to express on their surface (Krampera et al. 2006; Carrade et al. 2012; Bçrziòð et al. 2018; Enciso et al. 2018). MSCs express various factors or receptors that show antioxidative effects and/or prevent hypoxia-induced reperfusion injury, in addition to the anti-inflammatory action. Equine MSCs block reactive oxygen species (ROS) and prevent inflammation through STC-1 and TSG-6 receptors (antagonist of IL-1), respectively (Angelone et al. 2017). Equine MSCs expression of IL1-Ra potentially prevents inflammation and also hypoxia-induced tissue injury (Angelone et al. 2017). MSCs exosomes too carry immunomodulatory properties (Kordelas et al. 2014).

MSCs modulate both innate as well as acquired immunity (Kang et al. 2008; Kim et al. 2010; Lee et al. 2011; Wang et al. 2014; Wheat et al. 2017). The immunomodulatory actions are ensured through certain factors like haemoxygenase 1 (HO-1), inducible nitric oxide synthase (iNOS), transforming growth factor β1 (TGF-β1), prostaglandin E2 (PGE2), interleukin-68 (IL-68) and hepatocyte growth factor (HGF) (Lee et al. 2011; Carrade et al. 2012; Bearden et al. 2017; Chow et al. 2017; Parys et al. 2017) or through the direct cell–cell interaction via vascular cell adhesion molecule-1 (VCAM), intracellular adhesion molecule-1 (ICAM) and toll-like receptors (TLRs) (Tomchuck et al. 2008; DelaRosa and Lombardo 2010;

Akiyama et al. 2012). MSCs tend to polarize macrophage M1 to M2 phenotype through the cyclooxygenase-2-prostaglandin E2 (COX-2-PGE2) pathway. Mast cell degranulation and TNF-α secretion are suppressed. These cells induce anti-inflammatory cytokines and simultaneously reduce pro-inflammatory cytokines (Golchin et al. 2019). Furthermore, these cells express CD46, CD55 and CD59 surface markers that reduce complement system activation (Le Blanc and Davies 2015). MSCs are able to phagocytize but do not show direct cytotoxic or direct humoral effects due to their inability to secrete relevant enzymes like granzymes and perforins (Tso et al. 2010). Irrespective of the source (allogeneic and autologous MSCs), these cells suppress proliferation of the T-cell and their IFN-γ production (Colbath et al. 2017). Thus, the ready-to-use availability of MSCs ensures their immediate application in clinical conditions. MSCs apart from immunomodulatory effects may enhance myofibroblast differentiation of fibroblasts with the TGF-β1 expression (Masur et al. 1996; Wynn 2007), although these cells may also differentiate into the myofibroblasts under specific conditions (Zhao et al. 2009; De Boeck et al. 2010).

Inflammatory mediators tend to enhance MSCs anti-inflammatory/immunomodulatory activities but might decrease their secretory properties. In vitro studies mostly show that inflammatory mediators stimulate the immunomodulatory action of MSCs, although variably (Liu et al. 2006; Poncelet et al. 2007; Russell et al. 2016). Such variabilities with respect to the species and tissue types arise (Carrade and Borjesson 2013). In vivo studies involving osteoarthritis show that the inflammatory environment reduces MSCs migration and matrix secretory potential (Barrachina et al. 2016; Zayed et al. 2016; Reesink et al. 2017). Thus, further studies, well controlled and correlated between in vitro and in vivo environments, are desired.

11.2.2 MSCs Anti-Microbial Role

Irrational use of antimicrobials has become the cause of concern for development of superbugs. This great deal of antimicrobial resistance has put the use of antibiotics questionable and a great challenge ahead. This applies to chronic skin wounds as skin harbours microbial load (Zhao et al. 2013) and cattle mastitis. As MSCs secrete antimicrobial peptides, they may play a significant role to alleviate the need of antibiotics.

MSCs conditioned media (MSCs-CM) contains a plethora of antimicrobial agents that depolarize the gram-positive (*S. aureus*) and gram-negative (*E. coli*) membranes and are comparable to the antibiotics like polymyxin B and nisin, respectively. However, bactericidal activity of MSCs-CM in comparison to these antibiotics is limited. Various antimicrobial peptides (AMPs) like cystatin C, elafin, lipocalin and cathelicidin that show microbicidal effects too are detected whose effect gets obtunded in presence of specific antibodies (Harman et al. 2017). Dog study demonstrates MSCs-CM to secrete antimicrobial peptides like apolipoprotein B and D; amyloid-β peptide, cathepsin B, Protein S100-A4, galectin 3, CLEC3A, granulin, transferrin (Bujňáková et al. 2020). Equine MSCs from varied sources like bone marrow, endometrium and adipose tissue express lipocalin-2 (Cortés-Araya et al. 2018). It is worth mentioning here that both these equine studies fail to show MSCs β-defensin 1 expression as seen in MSCs from other species. Further, an advanced equine study demonstrates that MSCs-CM hampers growth of only some of the bacterial species while simultaneously significant reduction in the growth of methicillin-resistant *S. aureus* (MRSA) is seen as compared to the antibiotics. It is attributed in part to the cysteine protease activity of cathepsins being validated through protease inhibitors (Marx et al. 2020). All these in vitro results are supported by in vivo studies in dogs (Johnson et al. 2017), goats (Costa et al. 2019) and cattle (Lange-Consiglio et al. 2019; Peralta et al. 2020). Intravenous transplantation of MSCs evaluated in endotoxaemia (induced experimentally) appear safe, although its effectiveness remains to be seen (Kilcoyne et al. 2021).

11.3 Wound Healing

Wound, a break in the skin continuity, is a common clinical complaint reported in veterinary practice. Some of the wounds make connections to the mucosal linings with the resultant fistula formation. Wounds of various sizes and shapes, induced with different etiological agents are reported. Based on the aetiology and underlying systemic disease condition, wound management followed by subsequent healing and finally prognosis varies (Edlich et al. 1988). On the basis of the extent of damage, wounds may be superficial or deeper with the extent determining the time of healing. In clinical settings, wounds are classified as acute or chronic (Monaco and Lawrence 2003; Whitney 2005). In normal healthy individuals, acute wounds like surgical incision, accidental trauma and superficial burns undergo normal homeostasis. Regenerative type of healing tends to occur in injured skin. Various cells including the stem cells and/or secretome of the native or progenitor cells undertake all the phases of healing and tissue heals within the expected time (Monaco and Lawrence 2003). As such the clinical management in acute wounds is mainly aimed at ensuring the normal healing process to continue unabated and to facilitate the wound-healing processes in order to reduce morbidity. Contrarily, chronic wounds like infected wounds, ulcers, burns and fistulae undergo deranged wound healing with the failure of one or all the phases of the wound healing (Maxson et al. 2012). In chronic wounds factors like tissue hypoxia, infection, extensive necrosis, continuous exudation and inappropriate inflammatory reaction besides, some underlying systemic diseases, derange and/or exacerbate the normal healing processes (Guo and DiPietro 2010).

11.3.1 Skin Wounds and Their Healing Complicacies in Veterinary Medicine

Wounds that fail to heal normally and timely are generally classified as chronic wounds (Day 1993; Ghamsari et al. 1996). In veterinary practice, various types of chronic wounds are encountered ranging from epidermal wounds, infected wounds, ulcers and fistula. Further, wounds seen in patients suffering from underlying systemic diseases like diabetes, equine metabolic syndrome (EMS) and autoimmune diseases, and those in geriatric animals also fail to heal timely.

Wounds involving complete loss of epidermis and penetrating to some part of dermis show fibroproliferative response resulting in a scar formation. Scarred skin, a fibrous tissue, is devoid of all the functional dermal appendages viz., sweat glands, hair follicles, sebaceous glands and sensory nerve receptors (Martin 1997). Fibrous tissue has lower tensile strength which deranges skin physical appearance and affects its normal functioning (Clark 1994). Current cosmetic surgery and other therapeutic interventions fail to yield desirable results in such conditions (Mustoe et al. 2006). Infected epidermal wounds undergo dehiscence and subsequently delay healing. A prolonged neutrophil reactivity in such a wound creates environment of higher collagenases and elastases that degrades the secreted matrix and pro-healing factors like platelet-derived growth factor (PDGF) and transforming growth factor-β (TGF-β) (Diegelmann and Evans 2004; Edwards et al. 2004; Schönfelder et al. 2005; Jackson et al. 2012; Maxson et al. 2012).

In certain systemic ailments like diabetes, equine metabolic syndrome (EMS) and autoimmune diseases, the wound healing becomes sluggish and even sometimes deranged (Guo and DiPietro 2010). These systemic diseases constitute impaired white blood cell (lymphocyte) function and reduced blood circulation, in addition to the lesser nerve sensation, leading to impaired wound healing (Tiwary et al. 2011). Furthermore, mesenchymal stem cells (MSCs) function in conditions like EMS may be impaired as well (Marycz et al. 2016; Nawrocka et al. 2017). In geriatric patients, a clear mechanism behind the chronicity of wounds is currently poorly understood, although some recent insights show that the progenitor tissue cells like MSCs in aged animals undergo senescence with a little ability to proliferate and differentiate (Han et al. 2010; Gould et al. 2015). Aging is also associated with delayed macrophage and T-cell infiltration, reduced angiogenesis, epithelialization and matrix response (Ashcroft et al. 1997, 2002; Ballas and Davidson 2001; Han et al. 2010). Furthermore, hormonal effects and microvascular response are also impaired in geriatric patients (Ashcroft et al. 1997; Bentov and Reed 2014). Additionally, mitochondrial dysfunction and oxidative stress and higher microbial load might further complicate the geriatric wound healing (Taylor and James 2005; Gardner et al. 2013). Some lab animal-burn studies demonstrate diminished expression of hypoxia-inducible factor-1 (HIF-1) together with the presence of fewer bone marrow-derived angiogenic cells, resulting in delayed healing. The associated cells show reduced response and 'homing' to the burn area (Zhang et al. 2011; Du et al. 2013).

In cattle teat, surgical affections (that have public health and economic constraints) are reportedly higher as compared to the other locations of mammary glands (Nabil et al. 2018; Dar et al. 2022). Cattle teat has unilateral longitudinal blood circulation, and milk being a rich source of nutrients acts a nidus for the microbial growth with available milk fat further hampering the wound closure. This predisposes animals to mastitis with a potential zoonotic infection and economic consequences (Ghamsari et al. 1996). This makes early and desirable healing of teat fistula the most challenging task. Dog anal furunculosis, a chronic inflammatory disease, progresses to perianal fistulas (Day 1993). This inflammatory reaction might arise due to the inappropriate immune responses against the resident intestinal or commensal organisms of skin. This condition has histopathological resemblance to that of human fistulizing Crohn's disease (Ferrer et al. 2016). The disease relapses frequently and currently lacks any long-lasting effective therapy.

Apart from the typical chronic wounds, some of the wounds may develop excessive granulation tissues and thus associated complications. In equine, wounds below the carpus/hock joint resist to heal even in the healthiest horse and tend to develop exuberant granulation tissue (EGT) (Khashjoori et al. 2019). A wide range of predisposing factors are proposed for EGT which range from environmental, biochemical, immunologic and to the genetic ones (Theoret and Wilmink 2016). Some of the factors, however, are well established. One of the important factors attributed to EGT formation is the poor blood supply and subsequent limited oxygen availability to the lower limbs. This might occur due to the considerable distance between trunk and dependent limbs in horse, post wound thrombosis (50% to 60% microvasculature obstruction), loss of tissue and peripheral wound oedema (Theoret et al. 2001; Carter et al. 2003; Jørgensen et al. 2017). A decrease in blood supply decreases the appropriate growth factor/cytokines availability with resultant upregulation of pro-fibrotic mediators. This diminishes innate normal antibacterial response, rate of re-epithelialization and wound contraction, leading to the EGT (Wilmink et al. 2010; Bischofberger et al. 2011; Tracey et al. 2014). Only about 24% of wounds in horses heal by primary closure with the current standard treatment methods (Wilmink et al. 2010). Horse lower limb wounds are comparable to human keloid lesions, although with some reservations, and therapeutic solution to the EGT wounds might suffice therapy for the humans as well (Carter et al. 2003).

It is therefore imperative to develop treatment modality that could evade such lacunas in wound healing and help in early regeneration of these wounds. Wound healing through the prism of regenerative medicine involving stem cells is therefore the need of the hour.

11.3.2 In Vivo MSC Wound-Healing Studies

To develop evidence-based MSCs therapeutics in wounds, various stages of drug development including the in vitro and in vivo experimental and clinical trials are conducted. There are numerous lab animal studies that focus on MSCs therapeutic effects. Considering the importance of farm and pet animals in veterinary practice, studies on these animals are reviewed and discussed and tabulated in Tables 11.1 and 11.2.

MSCs from variable sources including the bone marrow (BM), adipose tissue (AD), peripheral blood (PB), Wharton's jelly (Wj), cord blood (CB) and embryonic stem cell (ESC) are utilized in wound-healing studies. It is worth mentioning that MSCs characterization in most of these in vivo studies is either unavailable or the characterization lacks standard ISCT criteria. The lack of data or non-uniform characterization therefore does not guarantee typical MSCs studies. Apart from autologous cells, allogeneic and xenogeneic cells too are utilized. MSCs are utilized at variable concentrations ranging from 0.1 million to 100 million (Kim et al. 2013; Mund et al. 2020). Some of the studies utilize cells only while others incorporate scaffolds like carboxymethylcellulose gel (Iacono et al. 2016; Lanci et al. 2019), hyaluronic acid (Martinello et al. 2018) or decellularized amniotic membrane (Khashjoori et al. 2019) as well. One of the studies utilizes vascular endothelial growth factor- (VEGF) and fibroblast growth factor- (FGF) transfected cells (Zakirova et al. 2019), while other study compares the healing potential of normoxia-cultured with that of hypoxia-cultured cells (Textor et al. 2017). The biomaterial implantation is made through variable routes (Spaas et al. 2013). Mostly, the effect of single-cell dose is evaluated but some of the studies evaluate the effect of repeated cell doses in wounds (Carstanjen et al. 2006; Zubin et al. 2015; Iacono et al. 2016; Rafee et al. 2017; Lanci et al. 2019). E*x vivo study in horses* demonstrates MSCs and their conditioned media to promote dermal wound repair through mobilization and enhanced pro-healing expression of fibroblasts (Bussche et al. 2015).

11.3.2.1 In Vivo Experimental Studies

Sheep and goats are mainly utilized for experimental evidence of wound-healing potential of MSCs, while dog and horse undergo both the experimental as well as the clinical applications as detailed in Table 11.1. Preferable choice of sheep and goat as model animals for wound study is due to their significant available backspace and their overall limited neurological development (Martinello et al. 2018). Despite the fact, the literature on MSC applications in experimental wound models in these species is currently limited (Table 11.1).

Various animal models involve multiple wounds to evaluate MSCs wound-healing potential. This helps to keep both control and experimental wounds in a single animal. All these studies create full-thickness square wounds barring a single study that studies linear wounds (Azari et al. 2011). The wound size ranges from as small as 0.5×2.0 cm^2 (Mund et al. 2020) to as large as 4.0×4.0 cm^2 (Martinello et al. 2018). The control animals are treated with commercial creams or left untreated. The animal number in these studies range from $n = 2$ (Mund et al. 2020) to $n = 29$ (Mackenzie and Flake 2001). Evaluation is conducted for a variable period ranging from 1 week (Kim et al. 2013) to 13 months (Mackenzie and Flake 2001). The healing is evaluated through clinical examination, histopathology, immunohistochemistry and/or through gene expression analysis. The cells are demonstrated to home to the site of tissue damage (Mackenzie and Flake 2001) and their viability remains for longer periods varying from 7 days to 33 days (Mackenzie and Flake 2001; Azari et al. 2011; Mund et al. 2020). None of these studies report any adverse reaction to MSCs transplantation.

11.3.2.1.1 Sheep Model

In a single available acute second-intention wound-healing model, peripheral blood MSCs (injected around wound margins and implanted at wound centre along with

Table 11.1 MSCs preclinical experimental studies on wound healing in various veterinary species

Model type/ animal number/study period	Biomaterial/cell dose	MSCs characterization	Overall result	Reference
Sheep full-thickness wound model				
Randomized placebo-controlled study ($n = 6$)/ 42 days	Allogeneic PB-MSCs (one million injected in wound margins and one million loaded with hyaluronic acid and implanted at wound centre)	–	Improved result	Martinello et al. (2018)
Goat full-thickness wound model				
Randomized placebo-controlled study ($n = 5$)/ 5 weeks	Allogeneic WJ-MSCs (nine million around wound and in wound bed)	–	Improved result	Azari et al. (2008)
Randomized placebo-controlled study ($n = 4$)/ 14 days	Allogeneic WJ-MSCs (three million cells injected intradermally around the wound at three injection sites and subcuticular in wound bed at three injection sites)	αSMA expression, positive AP activity CD34– and CD44+	Improved result	Azari et al. (2011)
Dog full-thickness wound model				
Randomized placebo-controlled study ($n = 10$)/ 7 days	Allogeneic BM-MSCs (10 thousand, 0.1 million, one million and ten million intradermally injected into wound bed and around margins)	CD9+ and CD44+, CD34– and CD45–	Improved result	Kim et al. (2013)
Randomized placebo-controlled study ($n = 6$)/ 14 days	BM-MSCs derived exosomes (2×10^6 derived 1 mL/wound)	CD105+, CD90+ and CD73+	Improved result	El-Tookhy et al. (2017)
Randomized placebo-controlled study ($n = 6$)/ 14 days	Exosomes from cryopreserved BM-MSCs (0.5×10^6) 1 mL/wound, carboxy methyl cellulose (CMC) gel, applied 4 times, with 1 day interval	CD105+, CD90+ and CD73+	Improved result	Bahr et al. (2021)
Equine lower limb or thoracic wound models				
Randomized placebo-controlled study ($n = 6$)/ 6 weeks	UCB-MSCs (15–20 million per wound, each wound treated with hypoxic- and normoxic-preconditioned cells. Some wounds were treated with above cell types laden in autologous fibrin gel	–	Improved result	Textor et al. (2017)

(continued)

Table 11.1 (continued)

Model type/animal number/study period	Biomaterial/cell dose	MSCs characterization	Overall result	Reference
Randomized placebo-controlled study ($n = 4$)/30 days	Allogeneic AD-MSCs, decellularized amniotic membrane, 1% silver sulfadiazine	–	Improved result	Khashjoori et al. (2019)
Randomized placebo-controlled study ($n = 2$)/33 days	Allogeneic cord blood-derived MSCs (1.00×10^8), intravenous	–	Improved result	Mund et al. (2020)
Randomized controlled study ($n = 12$)/28 days	Cord blood MSCs ($1.51–2.46 \times 10^8$), 50% HypoThermosol FRS, intravenous (injected 12 h later)	CD20+, CD44+, CD90+, and CD4, CD8, CD11, CD73 (low expression)	Improved result	Mund et al. (2021)

MSCs mesenchymal stem cells, *AD* adipose tissue derived, *BM* bone marrow derived, *UCB* umbilical cord blood derived, *PB* peripheral blood derived, *WJ* Wharton's jelly derived, *n* total number of animals in all the groups of the study; Improved result: Indicates improved outcome with MSCs transplantation even if only one parameter improves; Study period: Indicates maximum time period for which study is conducted

the hyaluronic acid) enhance wound healing evaluated through clinical examination, histopathology and expression analysis. Cell-treated wounds show an overall reduction in inflammation and expression of CD3+ and CD20+ cells. At day 15, MSCs promote a greater degree of matrix proliferation and neovascularization, re-epithelialization and increased contraction leading to wound closure while control wounds are still discernible. Further, hair keratine (hKER) and collagen type I (Col1α1) expression increases significantly from day 15 through to day 42, leading to a higher degree of dense mature cutaneous adnexa as compared to the control (Martinello et al. 2018). The reduction in inflammation in wounds might occur through inhibition of the immune cells especially B lymphocytes (CD3+ and CD20+). Higher granulation tissue in the early period of wound-healing phase is confirmed through a higher expression of Ki67. Matrix and vessel formation that increases through the study after cell treatment could occur through upregulation of growth factors (EGF, TGF-β1 and stromal-derived growth factor-1α) (Chow et al. 2017; Parys et al. 2017). The matrix typical to skin (collagen type I) is higher as compared to the control at all the phases (days 14 and 42) in cell-treated wounds. It shows that early remodeling also occurs with the cells. Thus, MSCs may enhance and improve second-intention wound healing in sheep.

11.3.2.1.2 Goat Model

In goats, MSCs therapeutic effect is evaluated in two types of acute wound models viz., square (Azari et al. 2008) and linear (Azari et al. 2011) wounds. The second-intention healing wounds are created on either side of the thorax keeping treatment and control in the same animal. The cell concentrations implanted vary with the former wounds implanted @ nine million and the latter @ three million cells. In the former study, evaluation is made through morphometry, immunohistochemical staining (α-smooth muscle actin) and in the latter study, only histology is conducted. In the linear wound model, cells survive in wounds up to day 12. Re-epithelialization in cell-treated wounds is complete by earliest at day 7 but still incomplete in control wounds. By day 12, reduced inflammation, thinner granulation tissue and minimal scar tissue formation in the cell-treated group are reported (Azari et al. 2011). In the square wound model, a significantly higher percent of wound healing characterized by epithelialization and wound contraction in cell-treated wounds as compared to control at week 5 occurs (Azari et al. 2008).

Thus, MSCs in sheep and goats wounds appear to enhance and improve healing. However, being limited in relation to the dose titration, route of implantation and advanced evaluation to confirm typical regeneration, in addition to the role of other biomaterials like scaffolds, further studies are desired.

11.3.2.1.3 Dog Model

In dogs, a single acute cutaneous wound model study demonstrates that MSCs irrespective of the concentration enhance wound healing. The cell concentration employed ranges from 10 thousand through 0.1 million and one million to ten million. Healing evaluation is made through histopathology, immunohistochemistry and gene expression analysis. The cells are detected at wound base up to day 14 and not in epidermis, although the time gap shows decrease in such numbers from day 3 to day 14. At day 7, cell-treated wounds show epithelial regeneration characterized by smaller epithelial gaps with higher proliferating cell nuclear antigen (PCNA), angiogenesis and collagen synthesis as compared to the control. Although collagen deposition is higher with smaller epithelial gap in higher cell concentration-treated group, it is statistically non-significant. Higher PCNA confirms the proliferation of keratinocytes in MSC-treated wounds (Kim et al. 2013). As there has been non-significant difference in the results with respect to the cell concentration, there might be requirement of the minimal cell concentration of one million per cm^2 of wound for a significant therapeutic effect to occur (Falanga 2005). Exosomes harvested from different concentrations of MSCs also enhance wound healing without scarring in experimentally incised dog wounds (3 bilateral wounds, 2 cm) (El-Tookhy et al. 2017) and (2 bilateral wounds, 2 cm) (Bahr et al. 2021). The wound size reduces right from day 3rd to 14th. In these studies, a single dose of the exosomes derived from the same cell concentration is applied although the latter study implants it in multiple doses. Exosomes may show pro-angiogenic effects including the angiogenesis and tubular formation, besides activation of wound-healing pathways and secretion of growth factor and collagen synthesis (Shabbir et al. 2015; Zhang et al. 2015).

Table 11.2 MSCs clinical studies on wound healing in dog and horse and perianal fistula in dog

Study design/ period	Cell source/cell dose/ route	MSC characterization	Result(s)	References
Dog skin wound studies				
Case report ($n = 2$)/ 15 days	Xenogeneic WJ-MSCs, polyvinyl alcohol	–	Improved result	Ribeiro et al. (2014)
Case report ($n = 1$)/ 50 days	Allogeneic BM-MSCs post caudal superficial epigastric axial pattern flap application/five million doses 23 days apart/local implantation on wound margins	–	–	Nanjappa et al. (2014)
Case report ($n = 1$)/ 480 days	PRP (at day 1 and every 48–72 h by spraying and every 4–6 days by injecting), AD-MSCs (05 applications at day 11, 17, 23, 31, 41/two million/local spraying	–	Improved result	Zubin et al. (2015)
Case report ($n = 6$)/ 7 days	Allogeneic BM-MSCs/ 2.5 million (two doses 5–7 days apart)/local injection	–	Improved result	Rafee et al. (2017)
Case series ($n = 6$)/ 6 weeks	Allogeneic BM-MSCs/ one million (2–3 doses 2 weeks apart)/local injection	–	Improved result	Bhatt et al. (2018)
Case report ($n = 1$)/ 25 days	Allogeneic AD-MSCs/ ten million/local intradermal injection [wounds treated with AD-MSCs; neck (control); chest (treated conservatively)]	CD90+ and MHC-I+ CD45−, CD34− and MHC-II-	Improved result	Enciso et al. (2020b)
Case series ($n = 24$)/ 19 days	Allogeneic AD-MSCs/ 30 million (single dose in wounds $\leq$10 cm^2 day 3 after wounding and double for wounds $\geq$10 cm^2 10 days after wounding/local injection around wounds	Trilineage differentiation	Improved result	Enciso et al. (2020a)
Case report ($n = 1$)/ 30 days	Allogeneic MSCs/two million/ local injection in wound margins	–	Improved result	Trindade and Trindade (2021)

(continued)

Table 11.2 (continued)

Study design/ period	Cell source/cell dose/ route	MSC characterization	Result(s)	References
Dog bite wound study				
Case report ($n = 1$)/ 7 days	Allogeneic AD-MSCs/ three million/topical application [Back wounds treated with on matrix containing three million transfected MSCs matrix; groin treated with matrix and combination of Ad5-VEGF165 and Ad5-FGF2 in amounts equivalent to those used to transduce the MSCs	CD10+, CD71+, CD44+, CD105+, stro-1+, Thy-1+, CD34−, CD45−	Improved result	Zakirova et al. (2019)
Dog perianal fistula				
Case report ($n = 6$)/ 6 months	Human embryonic stem cell-derived MSCs + cyclosporine/ commercial preparation/ local implantation	CD73+, CD90+, CD105+, CD13+, CD29, CD54+, CD44+ and CD166+ and CD31, CD34− and CD45−	Improved result	Ferrer et al. (2016)
Equine decubital ulcer studies				
Case report (multiple sores) ($n = 1$)/ 7 months	Amniotic fluid-derived MSCs (five million MSCs applied twice a week for 4 consecutive times for 15 days), PRP	Trilineage differentiation	Improved result	Iacono et al. (2012)
Controlled case series/ ($n = 7$)	Carboxymethylcellulose seeded heterologous amnion-derived MSCs (0.5×10^6 cells/cm^2 applied twice a week for 4 consecutive times)	CD44+ and CD90+ CD34− Trilineage differentiation	Improved result	Iacono et al. (2016)
Horse dermal wounds				
Case report ($n = 4$)/ 8 weeks	PB-MSCs (0.5×10^6 local and 0.125×10^6 intravenous)	CD34+ and CD117+ CD90 and CD105+	Improved result	Spaas et al. (2013)
Case report ($n = 1$)/ 39 days	Allogeneic WJ-MSCs (applied four times using a carboxymethylcellulose (CMC) gel in autologous plasma (4 doses of 5×10^6 at 4 days interval)	CD44+ and CD90+, trilineage differentiation	Improved result	Lanci et al. (2019)

11.3.2.1.4 Horse Model

MSCs therapeutic evaluation in three experimental studies on distal limb wounds (Textor et al. 2017; Khashjoori et al. 2019; Mund et al. 2020) and one thorax wound (Mund et al. 2020) is evaluated. One of the distal limb wound studies involves fore limb wounds (Textor et al. 2017) and other involves hind limb wounds (Khashjoori et al. 2019). In case of thorax and limb wounds, intravenously transplanted allogeneic parenteral cord blood MSCs show preferential homing and engraftment to wounds with the persistence of the cells for 33 days following wound creation, without any clinical adverse reactions (Mund et al. 2020). Same authors report that the cell-treated limb wounds in the first 2 weeks are closed by a less pronounced lag phase with no improvement in histological features. By day 28, the cell-treated wounds heal with 75% contraction and 25% epithelialization in contrast to the control wounds that approximately have 60% contraction and 40% epithelialization. Cytokine profile is a bit conflicting as decreased anti-inflammatory and pro-fibrotic cytokines are also recorded (Mund et al. 2021). The decreased anti-inflammatory and pro-fibrotic cytokines could be due to the first-pass effect. MSCs post intravenous transplantation might get entrapped in the lungs where anti-inflammatory cytokines are secreted and are absorbed from the pulmonary vasculature and distributed systemically. Finally, the effect at the local injury sites might be seen (Kean et al. 2013; White et al. 2016).

Apart from the intravenous transplantation in the above studies, local transplantation of MSCs seeded over decellularized amniotic membrane on hind limb distal wounds also leads to the significant promotion of wound healing. This occurs in relation to the geometric analysis of total wound area, epithelialization and contraction as compared to the other groups at day 30 (Khashjoori et al. 2019). Similarly, in the fore limb distal wound model, umbilical cord MSCs cultured under normoxia or hypoxia are implanted either alone or along with the fibrin gel. Healing is evaluated through histology, thermography, gene expression and wound surface area. Cell (irrespective of prior preconditioning) improves histological scoring and wound area. Histology shows MSC-treated wounds heal with tissue that is more pro-healing than pro-inflammatory. An enhanced TGF-β and cycloxygenase-2 (COX-2) gene expression at 1 week occurs in the cell-treated wounds (Textor et al. 2017). Continued enhancement of TGF-β1 in wounds leads to exuberant granulation tissue formation (Theoret et al. 2002) and epithelialization inhibition (Haber et al. 2003) in horses. MSCs paracrine secretion (especially microsomal RNA) that suppresses TGFβ/SMAD pathway results in reduced myofibroblast function (Fang et al. 2016). However, its early enhancement in wounds (up to first week) as shown in this study may have a pro-healing effect. This has been reported for pony wounds that have intense bursts of TGF-β1 in which wounds heal faster as compared to the horse (van den Boom et al. 2002). Furthermore, it is imperative to evaluate its antifibrotic isoform ($TGF\beta_3$) in MSC-treated wounds as has been done in other studies (Lichtman et al. 2016).

From all these experimental studies, it appears that MSCs enhance wound healing but variable periods exist for complete wound healing to occur. Besides, the route of transplantation shows considerable effect on the healing potential of the MSCs. The preconditioning, whether normoxic or hypoxic, may not have any effect on wound-healing potential of MSCs. The cell concentration fails to have a significant effect on the wound healing but slight differences in some of the assessment factors occur. Dose frequency remains to be evaluated. Variable and limited evaluation parameters make the comparison among studies inappropriate and thus, evaluation parameters must be uniformly sufficient.

11.3.2.2 Clinical Studies on Wound Healing

MSCs are evaluated clinically in case reports in dogs, cats and horses (Table 11.2). Only a single controlled dog study is available that groups clinical cases with respect to the treatment protocol (Enciso et al. 2020b). Few case reports have kept control wounds alongside the treated wounds (Spaas et al. 2013; Iacono et al. 2016; Zakirova et al. 2019; Enciso et al. 2020b). Maximum of n = 24 cases in dogs (Enciso et al. 2020b) and n = 7 cases in horses (Iacono et al. 2016) are incorporated. The effect of MSCs in wound along back/thorax and/or limbs in dog, and on lower limbs in horse are studied. In all these studies, intradermal injection of allogeneic MSCs is made. In a single study on horse additional intravenous transplantation of MSCs is made (Spaas et al. 2013). The cell concentration ranges from two million (Iacono et al. 2012) to 30 million (Enciso et al. 2020b) in dog and 0.5 million (Spaas et al. 2013; Iacono et al. 2016) to ten million (Carstanjen et al. 2006) in horse studies. The evaluation period varies from 5–7 days (Rafee et al. 2017) to 30 months (Zubin et al. 2015) in dog and 2 weeks (Carstanjen et al. 2006) to 7 months (Iacono et al. 2012) in horse.

11.3.2.2.1 Dog Clinical Studies

In case report, MSCs therapeutic effects are evaluated in acute wounds (Zubin et al. 2015; Bhatt et al. 2018; Enciso et al. 2020a, b) or in the chronic wounds (Ribeiro et al. 2014; Nanjappa et al. 2014; Rafee et al. 2017; Bhatt et al. 2018; Trindade and Trindade 2021) as detailed in Table 11.2. In addition, MSCs are utilized in bite wounds (Zakirova et al. 2019) and perianal fistula (Ferrer et al. 2016) cases.

In acute wounds, cell-treated wounds show almost normal skin with thin epidermis and orthokeratosis. Cell-treated wounds show no sign of dermal inflammation, and multiple hair follicles are in different stages of activity in contrast to the non-AD-MSC-treated control wounds. The epidermis of the control wound is characterized by epidermal hyperplasia, intracellular oedema, hypergranulosis and ortho- and para-keratotic hyperkeratosis. The dermis undergoes fibroplasia of all layers (Enciso et al. 2020b). In another study, an extensive wound with Achilles injury corrected surgically and is followed by PRP/MSCs application reveals that complete closure of the wound occurs 3 months after the start of the regenerative therapy. The wound remains free from signs of infection or necrosis during and following the regenerative therapy. The lameness of the hind limbs and the oedema of the Achilles's tendon disappear 20 days after intralesional application of MSCs and PRP. Pigmentation of the skin appears after about 2 months (Zubin et al. 2015).

In the case of non-healing bite wounds, MSCs injected 2 weeks later show edge epithelization from 6–7 to 10 mm on the seventh day. There are no signs of necrosis or inflammation throughout the entire healing period. The wound on dog back is completely healed after 1 month and the wound in the groin heals within 1.5 months with normal non-hypertrophic scar formation (Zakirova et al. 2019). Bite wounds tend to develop infection, but in the present study wounds remain clear of infection and could have been effectuated through combined use of MSCs and other antimicrobial agents.

Apart from acute wounds, MSCs also appear promising in relation to chronic wounds. The improvement with MSCs application is evident by the faster rate of coverage of the defect by epithelization around wound margins, significant reduction in swelling and inflammation and enhanced granulation tissue formation (Nanjappa et al. 2014; Rafee et al. 2017). In a detailed dog study, acute and chronic wounds exhibit significantly higher contraction and re-epithelialization post AD-MSCs transplantation as compared to the control wounds. Histopathology reveals that cell transplantation reduces inflammatory cells and enhances multiple hair follicles on day 7. Further, the cell-transplanted wounds have upregulated expression of granulocyte macrophage colony-stimulating factor (GM-CSF), citing possible cell recruitment towards the wounds (Enciso et al. 2020a).

Dog perianal fistula treated with human embryonic stem cell-derived MSCs (hESC-MSCs) injected locally improves within 3 months. Post cell treatment, reduced serum levels of IL-2 and IL-6 are recorded with a significant reduction in required cyclosporine dose. Few cases show recurrence after 6 months (Ferrer et al. 2016). Thus, further studies with multiple doses may offer some relief.

11.3.2.2.2 Equine Clinical Studies

In equine, MSCs therapeutic effect is evaluated in decubital ulcers (Iacono et al. 2012), chronic dermal wounds (Spaas et al. 2013), deep sore wounds (Iacono et al. 2016) and non-healing skin wound (Lanci et al. 2019) (Table 11.2). MSCs transplantation in palate defects successfully organizes in a manner similar to that in native soft palate tissue (Carstanjen et al. 2006).

In all the tabulated clinical studies, progressive early and effective wound healing upon MSCs implantation is shown. MSC-treated wounds show almost normal skin with thin epidermis and orthokeratosis. In chronic dermal wounds of metatarsus, MSCs injected locally and through intravenous routes lead to positive visual and clinical outcomes. Granulation tissue is formed within 4 weeks of the cell application, significantly earlier as compared to the control. The scar formation is delayed by more than 2 months as compared to the control (Spaas et al. 2013). In deep sore wounds of the carpus, fetlock and hock, carboxymethylcellulose (CMC) seeded heterologous amnion-derived MSCs lead to the faster dermal wound healing as compared to control (Iacono et al. 2016). Decubital septic ulcers of foal treated with MSCs and PRP improve earlier and effectively as compared to those implanted with PRP or aloe vera gel only. At 7 months, an ulcer treated with aloe gel is still incompletely healed (Iacono et al. 2012). In a non-healing skin wound, allogeneic Wharton's jelly (WJ) MSCs (applied four times) using a CMC in autologous plasma are implanted. Initial wound area reduced 4 days after the last application of WJ-MSCs. The wound regression rate is +74%. No local or systemic side effects are registered after WJ-MSCs application and no evident exuberant scar is observed after wound healing. At discharge, the mean wound area is 0.38 ± 0.01 cm^2 and the total regression rate is +80%. Five days later, the wound is completely healed (Lanci et al. 2019). In all the equine experimental and clinical studies (especially of the distal limb wounds), EGT formation is missing and thus a positive outcome can be expected with the MSCs transplantation.

Apart from these wounds, clinical conditions like teat fistula are worth mentioning. As described above, the need for teat fistula, especially transverse recurrent fistulae with fibrosis are cause of concern. Currently, no clinical study could be seen that employs MSCs in teat fistula. However, in our own preliminary clinical study, MSCs appear promising as fistulae were better healed with soft pliable tissue and the recurrence rate of fistulae was much lower than the control (Dar et al. 2022). Therefore further studies must be evaluated in detail to arrive at evidence based medicine.

Most of the above studies are uncontrolled clinical studies and as such spontaneous healing cannot be ruled out. Further wide variation in cell culture method, cell type, characterization, concentration, passage number and administration method exists among these studies. Besides, the route and type of cell implantation have been made either locally or parenterally. In a few studies, scaffolds like hyaluronic acid, CMC or fibrin gel have been utilized with the aim to reduce exposure of cells to toxic environments and ensure their slow release for long-lasting effects (Gugjoo et al. 2019, 2020a, b). Furthermore, the details on the mechanisms involved in therapy are less cleared. Furthermore, the animal numbers are too low to evaluate data statistically. Above all, the evaluation period remains mostly limited.

11.4 Atopic Dermatitis

Atopic dermatitis (AD) is one of the most frequently reported dog dermatopathy with a significant resemblance to that of the human (Hoffman and Dow 2016; Jiménez and Guerrero 2017). It is estimated to affect around 10–15% of dogs around the world (Hillier and Griffin 2001). AD is a multifactorial disorder attributed to genetic reasons (the particular breed, polymorphisms at specific gene) that make animals more susceptible to allergies (Roque et al. 2012; Bizikova et al. 2015). Acute dog AD is dominated by T helper 2 (Th2) cells resulting in the release of cytokines like interleukins (IL)-4, −5, −6, −13 and − 31 (Mineshige et al. 2018). Additionally, keratinocytes released thymic stromal lymphopoietin (TSLP) induces maturation/activation of the cells like dendritic cells (DCs) and mast cells (Klukowska-Rötzler et al. 2013; Olivry et al. 2016). Such a flurry of cytokines produces itching, inflammatory condition and compromise in skin (Reis et al. 2021). The chronic AD has mixture of immune mediator cells including the Th1, Th2, Th17 and Th22 producing interferon-gamma (IFNγ), IL-6, IL12 and granulocyte macrophage colony-stimulating factor (GM-CSF) (Daltro et al. 2020). In AD, skin barrier weakens (predisposing animals to increased allergen penetration) and also affects skin antimicrobial production (Roque et al. 2012; Bizikova et al. 2015; Jiménez and Guerrero 2017).

To evaluate MSCs therapeutic effect in dog allergic atopic dermatitis, one of the early preliminary studies that inject allogeneic AD-MSCs (1.3 million per kg body weight) intravenously in $n = 5$ cases fails to get satisfactory results (Hall et al. 2010). Contrarily, recent studies show much improvement in clinical signs with allogeneic AD-MSCs. Allogeneic AD-MSCs (1.5 million per kg body weight) transplanted intravenously decrease clinical scores of pruritus (VAS) and dermatitis (CADESI-04) to a significant extent as early as week one and improved further up to a month post-treatment (Table 11.3). The clinical improvement is demonstrated to be stable for the last follow-up of 6 months (Villatoro et al. 2018). Another study that implants multiple doses (6 doses at weekly intervals) of allogeneic AD-MSCs (0.5 million) intramuscularly too shows improvement in CADLI and pruritus index scores without any adverse effect on liver or blood parameters (Enciso et al. 2019). A more recent study that categorized dog AD clinical cases as mild, moderate and severe demonstrates that three doses of allogeneic AD-MSCs (two million per kg body weight) transplanted intravenously at 21 days interval improve epidermal thickness in the moderate and severe cases with all the physiological, haematological and biochemical features remaining within a normal range (Ramos et al. 2020). Comparing low-dose allogeneic AD-MSCs (0.5 million per kg b wt.) to that of the high dose (five million per kg b. wt.) implanted subcutaneously, animals receiving the latter cell concentrations showed improvement in clinical symptoms at day 30 post cell implantation. Thus, cell dose and frequency may have a significant effect on the outcome of the atopic dermatitis as higher cell dose shows better clinical outcome and repeated doses further improves the skin condition. It is worth mentioning that the potential prognostic markers like miRNA-483 may be developed for evaluation of therapeutic efficacy (Kaur et al. 2021). All these results demand further studies that compare cell dose, route and time of implantation, besides pathophysiology of the condition for developing the effective role of the cells.

The above studies demonstrate MSCs potential role in skin affections. This might be due to their ability to give rise to epidermal keratinocytes, epithelial cells (follicular), sebaceous gland cells and dendritic cells (Krause et al. 2001; Wu et al. 2007; Borena et al. 2009), and release proangiogenic factors like vascular endothelial growth factor and angiopoietin-1 (Wu et al. 2007). Further, MSCs might enhance wound healing by resolution of inflammation (Yang et al. 2018), suppress degradation of collagenous matrix (Jeon et al. 2010), enhance angiogenesis through differentiation into endothelial cells (Kim et al. 2012) and accelerate wound closure through transdifferentiation into myofibroblasts (Jackson et al. 2012; Rodriguez-Menocal et al. 2015). Further, MSCs might produce antifibrotic factors to prevent hypertrophic scarring (Foubert et al. 2017). However, such a role of MSCs largely remains to be confirmed in veterinary species.

11.5 Equine Laminitis

Equine laminitis is a complex multifactorial disease with a wide variety of contributing factors including inflammatory, metabolic disorders and endothelial or vascular dysfunction (Katz and Bailey 2012). Initiation of inflammation and secretion of metalloproteinases attributes to the basement membrane degradation leading to complete mechanical collapse of the affected foot (Engiles 2010). Currently, unconfirmed etiological basis of laminitis makes its definitive treatment impossible and thus palliative treatment is being offered (Pollitt 2010). One of the significant reasons for limited proliferation of hoof epithelium in laminitis could be the loss of hoof epidermal stem cells (p63+) (Carter et al. 2011). Thus, MSCs due to their proliferation, differentiation and paracrine effects (immunomodulation) may potentially be an effective treatment in laminitis. One of the studies demonstrates that MSCs like stem cells known as progenitor cells of the coronary corium of the equine feet exist that may have a potential role in equine laminitis (Marycz et al. 2021).

Limited available clinical studies on equine laminitis show MSCs to act as a beneficial adjunctive to the traditional managemental therapy (Table 11.4). In order to utilize MSCs effectively, digital implantation of allogeneic MSCs perfuses better through intravenous route than epidermal implantation (Sole et al. 2012). Among intravascular routes, intra-arterial infusion might symmetrically and homogeneously distribute MSCs in the distal limb, while intravenous regional limb perfusion (IVRLP) results in an asymmetrical and heterogeneous distribution. IVRLP offers variable and limited retention of MSCs and their distribution to the hoof. MSCs may not be implanted through coronary bands as they fail to migrate to the hoof (Spriet et al. 2013).

For better results, MSCs may be transplanted within the first 30 days of laminitis before lamellar wedge (hyperplastic undifferentiated lamella) formation (Morrison 2011; Dryden et al. 2013). Autologous or allogeneic MSCs from the umbilical cord (Morrison 2011; Dryden et al. 2013), bone marrow (Dryden et al. 2013) and adipose tissue (Angelone et al. 2017; Malyuk et al. 2021) are utilized at variable doses ranging from 15 million to 30 million. The success rate ranges from 70% (Morrison 2011) to 83% (Dryden et al. 2013). Laminitic animals treated within 1 month and after 3 months had a 100% and $< 50\%$ success rate, respectively (Dryden et al. 2013). A study with longer follow-up demonstrates digital palmer vein injected AD-MSCs (3 doses at monthly intervals) suspended in 15 mL PRP potentially improve vascularization, and structure and function of laminitic hoof. At 1 year, $n = 7$ out of $n = 9$ are active and the rest $n = 2$ develop recurrent laminitis. After 1–3 years of treatment, $n = 2$ survives out of $n = 7$ as those lost die due to the irrelevant cause (colic; $n = 3$) or are euthanized (recurrent laminitis; $n = 2$), while rest $n = 2$ are unavailable for the follow-up (Angelone et al. 2017).

Thus, MSCs may be the potential therapeutic candidates for equine laminitis. However, further studies that offer long-term follow-up with extensive detail on digital bone and hoof regeneration, besides the cell uniformity in relation to cell passage number, concentration, dose and route of implantation are desired. Further, early and late transplantations may also be compared in large animal numbers to confirm the effectiveness of MSCs in acute and chronic laminitis cases.

11.6 Mammary Gland

Cattle mammary gland affections like mastitis and udder fibrosis lead to a major economic loss to the dairy industry. Such affections not only pose a challenge to a veterinarian but also have public health issues in the form of milk infected with microbes and antibiotic residues. No specific and effective treatment is currently available to control mastitis, and udder and teat fibrosis (Dar et al. 2022). Normally mammary gland homeostasis is maintained by the existing resident stem cells (Martignani et al. 2009, 2014), although their characterization remains to be fully conducted in bovines (Martignani et al. 2014; Gugjoo et al. 2022). MSCs may suffice the current lack of mammary stem cells.

MSCs are demonstrated to exist in milk besides other body tissues and foetal membranes (Pipino et al. 2018). These cells can potentially show paracrine effects of tissue healing and control of infection as discussed above. In vitro cattle studies demonstrate that mammary gland epithelial cell secretory (β-casein and κ casein) potential increases (Zhao et al. 2017) and their apoptosis decreases (Wei et al. 2020) upon co-culture with cattle UC-MSCs. Thus, MSCs may prove useful for mastitis control and in increasing milk production of the udder. In vivo experimental (Peralta et al. 2020) or clinical (Lange-Consiglio et al. 2019; Ting et al. 2020) mastitis studies demonstrate that MSCs or their secretome reduce somatic cell count (SCC) and pH value and titrable acidity in the milk. However, the exact mechanisms like antimicrobial effect, immunomodulation and tissue regeneration remain to be evaluated in these studies. A single goat in vivo study shows a potential role of MSCs in reconstituting the mammary gland function through the prevention of inflammatory cell infiltration and thereby reducing the fibrosis (Costa et al. 2019). Thus, further studies are highly desired in this area.

11.7 MSCs Posology

Variable concentrations of MSCs have been utilized under in vivo experimental and clinical studies. All these studies show improvement in skin affections with MSCs. Currently, considerable consensus lacks on the utilization of effective cell concentration. In wounds, MSCs therapeutic concentration as per the wound area and by taking the underlying disease into the consideration can offer a suitable way to standardize the cell concentration. In various skin affections including wounds, MSCs have been implanted locally through topical, intradermal, subcutaneous or through the intravenous and intramuscular routes (Iacono et al. 2012; Zubin et al. 2015; Duscher et al. 2016; Braid et al. 2018; Zakirova et al. 2019; Mankuzhy et al. 2021). MSCs transplantation through all these routes offer some sort of benefit although not without lacunas. In parenteral routes, cells are required in large doses as they get trapped in the lungs with an additional threat of the hypercoagulable state leading to disseminated intravascular coagulation (Tatsumi et al. 2013; Liao et al. 2017). However, being less invasive, these routes may provide relief from further painful procedures. Among various parenteral routes, intramuscular transplantation may keep MSCs available for longer duration as compared to the intravenous route

Table 11.3 MSCs clinical studies on atopic dermatitis in dog

Study design/period	Biomaterial	Route	Evaluation criteria	Overall result	References
Atopic dermatitis					
Case report (pilot study), $n = 5$/ 90 days	Autologous AD-MSCs (1.3×10^6/kg b. wt.)	Intravenous	Canine atopic dermatitis extent and severity score and visual pruritus score	No improvement	Hall et al. (2010)
Case report (controlled) ($n = 22$)/6 month	Allogeneic AD-MSCs (1.5×10^6/kg b. wt.)	Intravenous	Clinical signs, haematological and biochemistry profiles, and AD severity (Canine Atopic Dermatitis Extent and Severity Index, version 4 (CADESI-04), visual analogue scale	Improved result	Villatoro et al. (2018)
Case report (pilot study), $n = 12$/ 6 weeks	Allogeneic AD-MSCs (6 doses at weeks interval of 0.5×10^6)	Intramuscular	Clinical scores and blood and liver profile	Improved result	Enciso et al. (2019)
Controlled clinical case series ($n = 16$)/ 82 days	Allogeneic AD-MSCs (3 doses at days 10, 31 and D52 of 2×10^6/kg b. wt.)	Intravenous	VAS, CADESI-04, histopathology haematological and biochemical evaluation	Improved result	Ramos et al. (2020)
Controlled clinical study/n=/90 days	Allogeneic AD-MSCs: Low dose (5×10^5 per kg b. wt.); high dose (5×10^6 per kg b wt.)	Subcutaneous	Pruritus visual analog scales (PVAS), CADESI-04, serum biochemistry	Improved result	Kaur et al. (2021)

MSCs mesenchymal stem cells, *AD* adipose tissue derived, *n* total number of animals in all the groups of the study; Improved result: Indicates improved outcome with MSCs transplantation even if only one parameter improves; Study period: Indicates maximum time period for which study is conducted

Table 11.4 MSCs clinical studies on chronic in horse

Study design/ period	Biomaterial	Evaluation criteria	Overall result	References
Chronic laminitis				
Case report (n = 12)/ 14 months	Allogeneic UC-MSCs (3–4 doses of 20–25 × 10^6 as retrograde venous digital perfusion)	–	Improved result	Morrison (2011)
Case report (30)/ 4 months	Allogeneic UC-MSCs or BM-MSCs (4 treatments at monthly interval of 20–30 × 10^6 as retrograde venous digital perfusion, using the palmar/ plantar digital vein)	Radiographically	Improved result	Dryden et al. (2013)
Case report (n = 9)/ 1 year	Allogeneic AD-MSCs (first dose) and autologous AD-MSCs (two doses). 15 × 10^6 in 15 mL PRP (implanted thrice at 1 month interval (digital palmer vein)	Clinical observations and venography	Improved result	Angelone et al. (2017)
Case report (n = 1)/ 1 year	Autologous AD-MSCs (12 × 10^6) thrice in palmar medial digital artery, autologous platelet-rich plasma	Clinical observations and radiography	Improved result	Malyuk et al. (2021)

MSCs mesenchymal stem cells, *AD* adipose tissue derived, *UC* umbilical cord derived, *n* total number of animals in all the groups of the study; Improved result: Indicates improved outcome with MSC transplantation even if only one parameter improves; Study period: Indicates maximum time period for which study is conducted

(Braid et al. 2018). In wounds, the local topical delivery is the simplest way to offer the therapy; however, the cell loss from draining and exposure to the local toxic environment may require further improvement in their delivery (Gugjoo et al. 2019, 2020a, b). As such, bioscaffold-based delivery mode of stem cells is currently being evaluated (Garg et al. 2014; Zare et al. 2015; Lanci et al. 2019). Peripheral intradermal injection which has the benefits of getting local environment to MSCs together with less invasiveness may be currently considered as a suitable mode of MSCs delivery in wounds. In the case of other affections like AD and laminitis, the parenteral route appears effective. In the case of laminitis, intra-arterial route appears better than intravenous especially in relation to the cell distribution. In mastitis, MSCs or their conditioned medium may better be infused through direct intramammary route; however, other routes may also be evaluated. The cell dose, frequency of dosing and its appropriate timing post affection, however, needs further deliberation to harvest maximum regenerative benefits of MSCs.

11.8 Conclusion(s)

Among various skin affections, wounds are most frequently reported. Chronic wounds pose a challenge to the clinician due to the lack of appropriate therapy. Clinical management of acute wounds should be aimed at ensuring the normal healing process to continue unabated and at enhanced pace to reduce morbidity. Other clinical cases commonly reported are, atopic dermatitis, equine laminitis, mastitis and udder and teat fibrosis. Stem cells especially MSCs have been evaluated under in vitro and in vivo studies. In vitro MSCs characterization remains a challenge in veterinary sciences as the surface markers expression variability occurs across species and tissue types. MSCs appear promising in both acute and chronic wounds. However, the clinical evaluation is made on dogs and horses only. The mechanisms involved in MSCs therapeutic effects in veterinary species are very limited. In case of atopic dermatitis, MSCs appear promising but details again fall short of the clinical standard. In laminitis MSCs as adjunct appear promising but lack of sufficient details over the exact stage of MSCs transplantation and its typical therapeutic effects make it little less understood. Limited available in vitro stem cell studies on mammary gland show promising role in mastitis. However, the lack of understanding on specific effects of MSCs on the microbes and their role in presence of tissue matrices need to be understood. MSCs in all these affections although show promising role but are not extensive enough to be considered suitable for drawing the conclusion. Therefore, uniform studies that evaluate uniform cell concentration, frequency and duration along with other biomaterials in a sufficiently large animal population are desired. The in vivo studies especially clinical ones need to be case-controlled with sufficient animal numbers to draw any conclusion for evidence-based medicine.

References

Akiyama K, Chen C, Wang DD, Xu X, Qu C, Yamaza T, Cai T, Chen WJ, Sun L, Shi S (2012) Mesenchymal stem- cell-induced immune-regulation involves FAS-ligand-/FAS-mediated T cell apoptosis. Cell Stem Cell 10(5):544–555

Angelone M, Conti V, Biacca C, Battaglia B, Pecorari L, Piana F, Gnudi G, Leonardi F, Ramoni R, Basini G, Dotti S, Renzi S, Ferrari M, Grolli S (2017) The contribution of adipose tissue-derived mesenchymal stem cells and platelet-rich plasma to the treatment of chronic equine laminitis: a proof of concept. Int J Mol Sci 18:2122

Ashcroft GS, Horan MA, Herrick SE, Tarnuzzer RW, Schultz GS, Ferguson MW (1997) Age-related differences in the temporal and spatial regulation of matrix metalloproteinases (MMPs) in normal skin and acute cutaneous wounds of healthy humans. Cell Tissue Res 290: 581–591

Ashcroft GS, Mills SJ, Ashworth JJ (2002) Ageing and wound healing. Biogerontology 3:337–345

Azari O, Babaei H, Molaei MM, Nematollahi-Mahani SN, Layasi T (2008) The use of Wharton's jelly-derived mesenchymal stem cells to accelerate second-intention cutaneous wound healing in goat. Iran J Vet Surg 3(3):15–26

Azari O, Babaei H, Derakhshanfar A, Nematollahi-Mahani SN, Poursahebi R, Moshrefi M (2011) Effects of transplanted mesenchymal stem cells isolated from Wharton's jelly of caprine

umbilical cord on cutaneous wound healing; histopathological evaluation. Vet Res Commun 35(4):211–222
Bahr MM, Amer MS, Abo-EL-Sooud K, Abdallah AN, Shehab GG, El-Tookhy OS (2021) Proficiency of carboxymethylcellulose as a cryoprotectant. Clinical and histological evaluation of cryopreserved heterogenous mesenchymal stem cell-exosomal hydrogel on critical size skin wounds in dogs. Int J Hematol Oncol Stem Cell Res 15(3):178–191
Ballas CB, Davidson JM (2001) Delayed wound healing in aged rats is associated with increased collagen gel remodeling and contraction by skin fibroblasts, not with differences in apoptotic or myofibroblast cell populations. Wound Repair Regen 9:223–237
Barrachina AR, Remachaa A, Romeroa FJ, Fortier LA (2016) Effect of inflammatory environment on equine bone marrow derived mesenchymal stem cells immunogenicity and immunomodulatory properties. Vet Immunol Immunopathol 171:57–65
Bçrziòð U, Matise-VanHoutana I, Pçtersone I, Dûrîtis I, Òikuïðins S, Bogdanova-Jâtniece A, Kâlis M, Svirskis D, Skrastiòa D, Ezerta A, Kozlovska T (2018) Characterisation and in vivo safety of canine adipose-derived stem cells. Proc Latv Acad Sci 72(3):160–171
Bearden RN, Huggins SS, Cummings KJ, Smith R, Gregory CA, Saunders WB (2017) In-vitro characterization of canine multipotent stromal cells isolated from synovium, bone marrow, and adipose tissue: a donor-matched comparative study. Stem Cell Res Ther 8:218
Bentov I, Reed MJ (2014) Anesthesia, microcirculation, and wound repair in aging. Anesthesiology 120:760–772
Bhatt J, Bhat AR, Singh AP, Indu B, Amarpal. (2018) Application of mesenchymal stem cells for surgical and chronic wound healing in dogs. Indian J Vet Surg 39(1):51–52
Bischofberger AS, Dart CM, Perkins NR, Dart AJ (2011) A preliminary study on the effect of manuka honey on second-intention healing of contaminated wounds on the distal aspect of the forelimbs of horses. Vet Surg 40:898–902
Bizikova P, Pucheu-Haston CM, Eisenschenk MNC, Marsella R, Nuttall T, Review SD (2015) Role of genetics and the environment in the pathogenesis of canine atopic dermatitis. Vet Dermatol 26(2):1–10
Borena BM, Pawde AM, Amarpal I, Aithal HP, Kinjavedkar P, Singh R, Kumar D (2009) Autologous bone marrow-derived cells for healing excisional dermal wounds of rabbits. Vet Rec 165(19):563–568
Braid LR, Wood CA, Wiese DM, Ford BN (2018) Intramuscular administration potentiates extended dwell time of mesenchymal stromal cells compared to other routes. Cytotherapy 20(2):232–244
Bujňáková D, Cuvalov A, Ižek MC, Humenik F, Salzet M, Ižková DC (2020) Canine bone marrow mesenchymal stem cell conditioned media affect bacterial growth, biofilm-associated Staphylococcus aureus and AHL-dependent quorum sensing. Microorgan 8:1478
Bussche L, Harman RM, Syracuse BA, Plante EL, Lu Y, Curtis TM, Van de Walle GR (2015) Microencapsulated equine mesenchymal stromal cells promote cutaneous wound healing in vitro. Stem Cell Res Ther 6:66
Cabezas J, Rojas D, Navarrete F, Ortiz R, Rivera G, Saravia F, Rodriguez-Alvarez L, Castro FO (2018) Equine mesenchymal stem cells derived from endometrial or adipose tissue share significant biological properties, but have distinctive pattern of surface markers and migration. Theriogenology 106:93–102
Carrade DD, Borjesson DL (2013) Immunomodulation by mesenchymal stem cells in veterinary species. Comp Med 63(3):207–217
Carrade DD, Lame MW, Kent MS, Clark KC, Walker NJ, Borjesson DL (2012) Comparative analysis of the immunomodulatory properties of equine adult-derived mesenchymal stem cells. Cell Med 4:1–11
Carstanjen B, Desbois C, Hekmati M, Behr L (2006) Successful engraftment of cultured autologous mesenchymal stem cells in a surgically repaired soft palate defect in an adult horse. Can J Vet Res 70:143–147
Carter CA, Jolly DG, Worden CE, Hendren DG, Kane CJM (2003) Platelet-rich plasma gel promotes differentiation and regeneration during equine wound healing. Exp Mol Pathol 74: 244–255

Carter RA, Engiles JB, Megee SO, Senoo M, Galantino-Homer HL (2011) Decreased expression of p63, a regulator of epidermal stem cells, in the chronic laminitic equine hoof. Equine Vet J 43(5):543–551

Chow L, Johnson V, Coy J, Regan D, Dow S (2017) Mechanisms of immune suppression utilized by canine adipose and bone marrow-derived mesenchymal stem cells. Stem Cells Dev 26(5): 374–389

Clark RAF (1994) Overview and general considerations of wound repair. In: Molecular and cellular biology of wound repair. Springer, Boston, MA, pp 3–33

Colbath AC, Dow SW, Phillips JN, McIlwraith CW, Goodrich LR (2017) Autologous and allogeneic equine mesenchymal stem cells exhibit equivalent immunomodulatory properties in vitro. Stem Cells Dev 26(7):503–511

Cortés-Araya Y, Amilon K, Rink BE, Black G, Lisowski Z, Donadeu FX, Esteves CL (2018) Comparison of antibacterial and immunological properties of mesenchymal stem/stromal cells from equine bone marrow, endometrium, and adipose tissue. Stem Cells Dev 27:1518–1525

Costa CRM, Feitosa MLT, Rocha AR, Bezerra DO, Leite YKC, Argolo Neto NP, Rodrigues HWS, Junior AS, Silva AS, Sarmento JLR, Silva LS, Carvalho MAM (2019) Adipose stem cells in reparative goat mastitis mammary gland. PLoS One 14(10):e0223751

Daltro S, Meira CS, Santos IP, dos Santos RR, Soares MBP (2020) Mesenchymal stem cells and atopic dermatitis: a review. Front Cell Dev Biol 8:326

Dar ER, Makhdoomi DM, Gugjoo MB, Shah SA, Ahmad SM, Shah RA, Ahmad SR, Parrah JD (2022) Cryopreserved allogeneic mesenchymal stem cells enhance wound repair in full thickness skin wound model and cattle clinical teat injuries. Current Res Translt Med. InPress

Day M (1993) Immunopathology of anal furunculosis in the dog. J Small Anim Pract 34(8): 381–388

De Boeck A, Narine K, De Neve W et al (2010) Resident and bone marrow-derived mesenchymal stem cells in head and neck squamous cell carcinoma. Oral Oncol 46:336–342

DelaRosa O, Lombardo E (2010) Modulation of adult mesenchymal stem cells activity by toll-like receptors: implications on therapeutic potential. Mediators Inflamm 2010:865601

Diegelmann RF, Evans MC (2004) Wound healing: an overview of acute, fibrotic and delayed healing. Front Biosci 9:283–289

Dominici M, Le Blanc K, Mueller I, Slaper-Cortenbach I, Marini F, Krause D, Deans R, Keating A, Prockop D, Horwitz E (2006) Minimal criteria for defining multipotent mesenchymal stromal cells. The International Society for Cellular Therapy position statement. Cytotherapy 8(4): 315–317

Dryden VC, Morrison S, Bras R, Morrell SA (2013) Using stem cells in clinical cases. J Equine Vet Sci 33:872–873

Du J, Liu L, Lay F, Wang Q, Dou C, Zhang X, Hosseini SM, Simon A, Rees DJ, Ahmed AK, Sebastian R, Sarkar K, Milner S, Marti GP, Semenza GL, Harmon JW (2013) Combination of HIF-1alpha gene transfection and HIF-1-activated bone marrow-derived angiogenic cell infusion improves burn wound healing in aged mice. Gene Ther 20:1070–1076

Duscher D, Barrera J, Wong VW, Maan ZN, Whittam ZN, Januszyk M, Gurtner GC (2016) Stem cells in wound healing: the future of regenerative medicine? A minireview. Gerontology 62: 216–225

Edlich R, Rodeheaver G, Morgan R, Berman D, Thacker J (1988) Principles of emergency wound management. Ann Emerg Med 17:1284–1302

Edwards JV, Howley P, Cohen IK (2004) In vitro inhibition of human neutrophil elastase by oleic acid albumin formulations from derivatized cotton wound dressings. Int J Pharm 284(1–2):1–12

El-Tookhy OS, Shamaa AA, Shehab GG, Abdallah AN, Azzam OM (2017) Histological evaluation of experimentally induced critical size defect skin wounds using exosomal solution of mesenchymal stem cells derived microvesicles. Int J Stem Cells 10(2):144–153

Enciso N, Ostronoff LLK, Mejías G, Leon LG, Fermin ML, Merino E, Fragio C, Avedillo L, Tejero C (2018) Stem cell factor supports migration in canine mesenchymal stem cells. Vet Res Commun 42(1):29–38

Enciso N, Amiel J, Pando J, Enciso J (2019) Multidose intramuscular allogeneic adipose stem cells decrease the severity of canine atopic dermatitis: a pilot study. Vet World 12(11):1747–1754

Enciso N, Avedillo L, Fermín ML, Fragío C, Tejero C (2020a) Cutaneous wound healing: canine allogeneic ASC therapy. Stem Cell Res Ther 11:261

Enciso N, Avedillo L, Fermín ML, Fragío C, Tejero C (2020b) Regenerative potential of allogeneic adipose tissue-derived mesenchymal cells in canine cutaneous wounds. Acta Vet Scand 62:13

Engiles JB (2010) Pathology of the distal phalanx in equine laminitis: more than just skin deep. Vet Clin N Am Equine Pract 26:155–165

Falanga V (2005) Wound healing and its impairment in the diabetic foot. Lancet 366(9498): 1736–1743

Fang S, Xu C, Zhang Y et al (2016) Umbilical cord-derived mesenchymal stem cell-derived exosomal microRNAs suppress myofibroblast differentiation by inhibiting the transforming growth factor-beta/SMAD2 pathway during wound healing. Stem Cells Transl Med 5:1425–1439

Ferrer L, Kimbrel EA, Lam A, Falk EB, Zewe C, Juopperi T, Lanza R, Hoffman A (2016) Treatment of perianal fistulas with human embryonic stem cell-derived mesenchymal stem cells: a canine model of human fistulizing Crohn's disease. Regen Med 11(1):33–43

Foubert P, Zafra D, Liu M, Rajoria R, Gutierrez D, Tenenhaus M, Fraser JK (2017) Autologous adipose-derived regenerative cell therapy modulates development of hypertrophic scarring in a red Duroc porcine model. Stem Cell Res Ther 8:261

Gardner SE, Hillis SL, Heilmann K, Segre JA, Grice EA (2013) The neuropathic diabetic foot ulcer microbiome is associated with clinical factors. Diabetes 62:923–930

Garg RK, Rennert RC, Duscher D, Sorkin M, Kosaraju R, Auerbach LJ, Lennon J, Chung MT, Paik K, Nimpf J, Rajadas J, Longaker MT, Gurtner GC (2014) Capillary force seeding of hydrogels for adipose-derived stem cell delivery in wounds. Stem Cells Transl Med 3:1079–1089

Ghamsari SM, Acorda JA, Taguchi K, Abe N, Yamada H (1996) Evaluation of wound healing of the teat with and without low level laser therapy in dairy cattle by laser Doppler flowmetry in comparison with histopathology, tensiometry and hydroxyproline analysis. Br Vet J 152:583–591

Golchin A, Farahany TZ, Khojasteh A, Soleimanifar F, Ardeshirylajimi A (2019) The clinical trials of mesenchymal stem cell therapy in skin diseases: an update and concise review. Curr Stem Cell Res Ther 14:22–33

Gould L, Abadir P, Brem H, Carter M, Conner-Kerr T, Davidson J, Schmader K (2015) Chronic wound repair and healing in older adults: current status and future research. J Am Ger Soc 63(3): 427–438

Gugjoo MB, Amarpal I (2018) Mesenchymal stem cell research in sheep: current status and future prospects. Small Rum Res 169:46–56

Gugjoo MB, Amarpal I (2020) Mesenchymal stem cell differentiation properties and available microenvironment. In: Gugjoo MB, Amarpal I (eds) Mesenchymal stem cell in veterinary sciences. Springer Nature, Berlin, pp 67–87

Gugjoo MB, Amarpal I, Chandra V, Wani MY, Dhama K, Sharma GT (2018a) Mesenchymal stem cell research in veterinary medicine. Curr Stem Cell Res Ther 13(8):645–657

Gugjoo MB, Amarpal I, Fazili MR, Shahe RA, Sharma GT (2018b) Mesenchymal stem cell: basic research and potential applications in cattle and buffalo. J Cell Physiol 234(6):8618–8635

Gugjoo MB, Amarpal I, Makhdoomi DM, Sharma GT (2019) Equine mesenchymal stem cells: properties, sources, characterization and potential therapeutic applications. J Equine Vet Sci 72: 16–27

Gugjoo MB, Amarpal I, Fazili MR, Shah RA, Mir MS, Sharma GT (2020a) Goat mesenchymal stem cell basic research and potential applications. Small Rum Res 183:106045

Gugjoo MB, Hussain S, Amarpal I, Shah RA, Dhama K (2020b) Mesenchymal stem cell-mediated immuno-modulatory and anti-inflammatory mechanisms in immune and allergic disorders. Rec Pat Inflamm Allergy Drug Discov 14(1):3–14

Gugjoo MB, Farooq F, Dar EJ, Ahmad SM, Shah SM, Amarpal, Parrah JD (2022) Adult stem cell research in light of the bovine mammary gland regenerative medicine. Curr Stem Cell Res Ther. InPress

Guo S, DiPietro LA (2010) Critical review in oral biology & medicine: factors affecting wound healing. J Dental Res 89(3):219–229

Haber M, Cao Z, Panjwani N et al (2003) Effects of growth factors (EGF, PDGF-BB and TGF-beta 1) on cultured equine epithelial cells and keratocytes: implications for wound healing. Vet Ophthalmol 6:211–217

Hall MN, Rosenkrantz WS, Hong JH, Griffin CE, Mendelsohn CM (2010) Evaluation of the potential use of adipose-derived mesenchymal stromal cells in the treatment of canine atopic dermatitis: a pilot study. Vet Ther 11:E1–E14

Han J, Liu JY, Swartz DD, Andreadis ST (2010) Molecular and functional effects of organismal ageing on smooth muscle cells derived from bone marrow mesenchymal stem cells. Cardiovasc Res 87:147–155

Harman RM, Yang S, He MK, Van de Walle GR (2017) Antimicrobial peptides secreted by equine mesenchymal stromal cells inhibit the growth of bacteria commonly found in skin wounds. Stem Cell Res Ther 8:157

Hillier A, Griffin CE (2001) The ACVD task force on canine atopic dermatitis (I): incidence and prevalence. Vet Immunol Immunopathol 81(3–4):147–151

Hoffman A, Dow S (2016) Concise review: stem cell trials using companion animal disease models. Stem Cells Express 34:1709–1729

Iacono E, Merlo B, Pirrone A, Saidijam M (2012) Effects of mesenchymal stem cells isolated from amniotic fluid and platelet-rich plasma gel on severe decubitus ulcers in a septic neonatal foal. Res Vet Sci 93(3):1439–1440

Iacono E, Lanci A, Merlo B, Ricci F, Pirrone A, Antonella C, Mariella J, Castegnetti C (2016) Effects of amniotic fluid mesenchymal stem cells in carboxymethyl cellulose gel on healing of spontaneous pressure sores: clinical outcome in seven hospitalized neonatal foals. Turk J Biol 40:484–492

Jackson W, Nesti L, Tuan R (2012) Mesenchymal stem cell therapy for attenuation of scar formation during wound healing. Stem Cell Res Ther 3(3):20

Jeon YK, Jang YH, Yoo DR, Kim SN, Lee SK, Nam MJ (2010) Mesenchymal stem cells' interaction with skin: wound-healing effect on fibroblast cells and skin tissue. Wound Repair Regen 18:655–661

Jiménez A, Guerrero F (2017) Células madre mesenquimales como nueva terapia en dermatología: conceptos básicos. Rev Clín Dermatol Vet 9:8–18

Johnson V, Webb T, Norman A, Coy J, Kurihara J, Regan D, Dow S (2017) Activated mesenchymal stem cells interact with antibiotics and host innate immune responses to control chronic bacterial infections. Sci Rep 7:9575

Jørgensen E, Bay L, Bjarnsholt T, Bundgaard L, Sørensen MA, Jacobsen S (2017) The occurrence of biofilm in an equine experimental wound model of healing by secondary intention. Vet Microbiol 204:90–95

Kang JW, Kang K-S, Koo HC, Park JR, Choi EW, Park YH (2008) Soluble factors-mediated immunomodulatory effects of canine adipose tissue-derived mesenchymal stem cells. Stem Cells Dev 17(4):681–693

Katz LM, Bailey SR (2012) A review of recent advances and current hypotheses on the pathogenesis of acute laminitis. Equine Vet J 44:752–761

Kaur G, Ramirez A, Xie C, Clark D, Dong C, Maki C, Ramos T, Izadyar F, Lopez Najera SO, Harb J, Hao J (2021) A double-blinded placebo-controlled evaluation of adipose-derived mesenchymal stem cells in treatment of canine atopic dermatitis. Vet Res Commun 46:251–260. https://doi.org/10.1007/s11259-021-09853-9

Kean TJ, Lin P, Caplan AI, Dennis JE, Kean J (2013) MSCs: delivery routes and engraftment, cell-targeting strategies, and immune modulation. Stem Cells Int 2013:732742

Khashjoori BK, Ghamsari SM, Dehghan MM, Mohajeri SF, Gholami H, Golshahi H (2019) Evaluation of the effects of adipose derived mesenchymal stem cells cultured on decellularized amniotic membrane in wound healing of distal part of the limbs in horse. Iran J Vet Med 13(1): 11–25

Kilcoyne I, Nieto JE, Watson JL, Galuppo LD, Borjesson DL (2021) Do allogeneic bone marrow derived mesenchymal stem cells diminish the inflammatory response to lipopolysaccharide infusion in horses? A pilot study. Vet Immunol Immunopathol 231:110146

Kim HS, Kim KH, Kim SH, Kim Y-S, Koo K-T, Kim T-I, Seol Y-J, Ku Y, Rhyu I-C, Chung C-P, Lee Y-M (2010) Immunomodulatory effect of canine periodontal ligament stem cells on allogenic and xenogenic peripheral blood mononuclear cells. J Periodont Implant Sci 40(6): 265–270

Kim S, Zhang H, Guo L, Kim J, Kim MH (2012) Amniotic mesenchymal stem cells enhance wound healing in diabetic NOD/SCID mice through high angiogenic and engraftment capabilities. PLoS One 7:1–11

Kim J-W, Lee J-H, Lyoo YS, Jung D-I, Park H-M (2013) The effects of topical mesenchymal stem cell transplantation in canine experimental cutaneous wounds. Vet Dermatol 24:242–253

Klukowska-Rötzler J, Chervet L, Müller EJ, Roosje P, Marti E, Janda J (2013) Expression of thymic stromal lymphopoietin in canine atopic dermatitis. Vet Dermatol 24(1):54–9.e9.0E14

Kordelas L, Rebmann V, Ludwig AK, Radtke S, Ruesing J, Doeppner TR, Epple M, Horn PA, Beelen DW, Giebel B (2014) MSC-derived exosomes: a novel tool to treat therapy-refractory graft-versus-host disease. Leukemia 28:970–973

Krampera M, Cosmi L, Angeli R, Liotta F, Andreini A, Santarlasci V, Mazzinghi B, Pizzolo G, Vinante F, Romagnani P, Maggi E, Romagnani S, Annunziato F (2006) Role for interferon-gamma in the immunomodulatory activity of human bone marrow mesenchymal stem cells. Stem Cells 24(2):386–398

Krause DS, Theise ND, Collector MI, Henegariu O, Hwang S, Gardner R, Neutzel S, Sharkis SJ (2001) Multi-organ, multi-lineage engraftment by a single bone marrow–derived stem cell. Cell 105:369–377

Lanci A, Merlo B, Mariella J, Castagnetti C, Iacono E (2019) Heterologous Wharton's jelly derived mesenchymal stem cells application on a large chronic skin wound in a 6-month-old filly. Front Vet Sci 6:9

Lange-Consiglio A, Gusmara C, Manfredi E, Idda A, Soggiu A, Greco V, Bonizzi L, Cremonesi F, Zecconi A (2019) Antimicrobial effects of conditioned medium from amniotic progenitor cells in vitro and in vivo: toward tissue regenerative therapies for bovine mastitis. Front Vet Sci 6:443

Le Blanc K, Davies LC (2015) Mesenchymal stromal cells and the innate immune response. Immunol Lett 168(2):140–146

Lee SH, Jeong SK, Ahn SK (2006) An update of the defensive barrier function of skin. Yonsei Med J 47(3):293–306

Lee WS, Suzuki Y, Graves SS, Iwata M, Venkataraman GM, Mielcarek M, Peterson LJ, Ikehara S, Torok-Storb B, Storb R (2011) Canine bone marrow-derived mesenchymal stromal cells suppress alloreactive lymphocyte proliferation in vitro but fail to enhance engraftment in canine bone marrow transplantation. Biol Blood Marrow Transplant 17(4):465–475

Liao L, Shi B, Chang H, Su X, Zhang L, Bi C, Shuai Y, Du X, Deng Z, Jin Y (2017) Heparin improves BMSC cell therapy: anticoagulant treatment by heparin improves the safety and therapeutic effect of bone marrow-derived mesenchymal stem cell cytotherapy. Theranost 7(1):106–116

Lichtman MK, Otero-Vinas M, Falanga V (2016) Transforming growth factor beta (TGF-beta) isoforms in wound healing and fibrosis. Wound Repair Regen 24:215–222

Liu H, Kemeny DM, Heng BC, Ouyang HW, Melendez AJ, Cao T (2006) The immunogenicity and immunomodulatory function of osteogenic cells differentiated from mesenchymal stem cells. J Immunol 176:2864–2871

Mackenzie TC, Flake AW (2001) Human mesenchymal stem cells persist, demonstrate site-specific multipotential differentiation, and are present in sites of wound healing and tissue regeneration after transplantation into fetal sheep. Blood Cells Mol Dis 27(3):601–604

Malyuk MO, Kharkevych YO, Brusko EP, Bokotko RR, Kovpak VV, Mazurkevych AY (2021) The effect of mesenchymal stem cells and platelet-rich plasma on the course of chronic laminitis

in horse: a clinical case. In: Veterinary science, technologies of animal husbandry and nature management, p 8
Mankuzhy PD, Ramesh ST, Thirupathi Y, Mohandas PS, Chandra V, Sharma GT (2021) The preclinical and clinical implications of fetal adnexa derived mesenchymal stromal cells in wound healing therapy. Wound Repair Regen 21:1–23
Martignani E, Eirew P, Eaves C, Baratta M (2009) Functional identification of bovine mammary epithelial stem/progenitor cells. Vet Res Commun 33:101–103
Martignani E, Cravero D, Miretti S, Accornero P, Baratta M (2014) Bovine mammary stem cells: new perspective for dairy science. Vet Q 34:52–58
Martin P (1997) Wound healing–aiming for perfect skin regeneration. Science 276(5309):75–81
Martinello T, Gomiero C, Perazzi A (2018) Allogeneic mesenchymal stem cells improve the wound healing process of sheep skin. BMC Vet Res 14:202
Marx C, Gardner S, Harman RM, de Walle GRV (2020) The mesenchymal stromal cell secretome impairs methicillin-resistant Staphylococcus aureus biofilms via cysteine protease activity in the equine model. Stem Cells Transl Med 9:746–757
Marycz K, Kornicka K, Basinska K, Czyrek A (2016) Equine metabolic syndrome affects viability, senescence, and stress factors of equine adipose-derived mesenchymal stromal stem cells: new insight into EqASCs isolated from EMS horses in the context of their aging. Oxid Med Cell Longev 2016:4710326
Marycz K, Pielok A, Kornicka-Garbowska K (2021) Equine hoof stem progenitor cells (HPC) CD29 + /nestin + /K15 + − a novel dermal/epidermal stem cell population with a potential critical role for laminitis treatment stem. Cell Rev Rep 17:1478–1485. https://doi.org/10.1007/s12015-021-10187-x
Masur SK, Dewal HS, Dinh TT et al (1996) Myofibroblasts differentiate from fibroblasts when plated at low density. Proc Natl Acad Sci USA, Cell Biol 93:4219–4223
Maxson S, Lopez EA, Yoo D, Danilkovitch-Miagkova A, Leroux MA (2012) Concise review: role of mesenchymal stem cells in wound repair. Stem Cells Transl Med 1(2):142–149
Mercati F, Pascucci L, Ceccarelli P, Dall'Aglio C, Pedini V, Gargiulo AM (2009) Expression of mesenchymal stem cell marker CD90 on dermal sheath cells of the anagen hair follicle in canine species. Eur J Histochem 53(3):159–166
Michler JK, Hillmann A, Savkovic V, Mülling CKW (2018) Horse hair follicles: a novel dermal stem cell source for equine regenerative medicine. Cytometry A 93(1):104–114
Mineshige T, Kamiie J, Sugahara G, Shirota K (2018) A study on periostin involvement in the pathophysiology of canine atopic skin. J Vet Med Sci 80(1):103–111
Monaco JL, Lawrence WT (2003) Acute wound healing an overview. Clin Plast Surg 30(1):1–12
Morrison S (2011) Successful use of allogenic umbilical cord-derived stem cells in nonresponsive chronic laminitic cases. J Equine Vet Sci 31:603
Mund SJK, Kawamura E, Awang-Junaidi AH, Campbell J, Wobeser B, MacPhee DJ, Honaramooz A, Barber S (2020) Homing and engraftment of intravenously administered equine cord blood-derived multipotent mesenchymal stromal cells to surgically created cutaneous wound in horses: a pilot project. Cell 9:1162
Mund SJK, MacPhee DJ, Campbell J, Honaramooz A, Wobeser B, Barber SM (2021) Macroscopic, histologic, and immunomodulatory response of limb wounds following intravenous allogeneic cord blood-derived multipotent mesenchymal stromal cell therapy in horses. Cell 10(11):2972
Mustoe TA, O'Shaughnessy K, Kloeters O (2006) Chronic wound pathogenesis and current treatment strategies: a unifying hypothesis. Plast Reconstr Surg 117(7S):35S–41S
Nabil MISK, Tarik MISK, EL-khamary A. (2018) A retrospective study of surgical affections of mammary glands in cattle and buffaloes and their management in the field. J Vet Med Sci 80 (10):1576–1583
Nanjappa MD, Ahmad RA, Amarpal I, Gugjoo MB, Pawde AM, Kinjavedkar P, Aithal HP, Ansari M, Chandra V, Sharma GT (2014) Caudal superficial epigastric axial pattern flap and stem cell therapy for the management of large wound on medial aspect of thigh in a dog. Adv Anim Vet Sci 2(3):188–191

Nawrocka D, Kornicka K, Smieszek A, Marycz K (2017) Spirulina platensis improves mitochondrial function impaired by elevated oxidative stress in adipose-derived mesenchymal stromal cells (ASCs) and intestinal epithelial cells (IECs), and enhances insulin sensitivity in equine metabolic syndrome (EMS) horses. Mar Drugs 15:237

Ojeh N, Pastar I, Tomic-Canic M, Stojadinovic O (2015) Stem cells in skin regeneration, wound healing, and their clinical applications. Int J Mol Sci 16(10):25476–25501

Olivry T, Mayhew D, Paps JS, Linder KE, Peredo C, Rajpal D, Hofland H, Cote-Sierra J (2016) Early activation of Th2/Th22 inflammatory and pruritogenic pathways in acute canine atopic dermatitis skin lesions. J Invest Dermatol 136(10):1961–1969

Parys M, Kruger JM, Yuzbasiyan-Gurkan V (2017) Evaluation of immunomodulatory properties of feline mesenchymal stem cells. Stem Cells Dev 26(10):776–785

Peralta OA, Carrasco C, Vieytes C, Tamayo MJ, Muñoz I, Sepulveda S, Tadich T, Duchens M, Melendez P, Mella A, Torres CG (2020) Safety and efficacy of a mesenchymal stem cell intramammary therapy in dairy cows with experimentally induced Staphylococcus aureus clinical mastitis. Sci Rep 2020(10):1–12

Pipino C, Mandatori D, Buccella F, Lanuti P, Preziuso A, Castellani F, Grotta L (2018) Identification and characterization of a stem cell-like population in bovine milk: a potential new source for regenerative medicine in veterinary. Stem Cells Dev 27(22):1587–1597

Pollitt CC (2010) The anatomy and physiology of the suspensory apparatus of the distal phalanx. Vet Clin N Am Equine Pract 26:29–49

Poncelet AJ, Vercruysse J, Saliez A, Gianello P (2007) Although pig allogeneic mesenchymal stem cells are not immunogenic in vitro, intracardiac injection elicits an immune response in vivo. Transplantation 83:783–790

Rafee MA, Amarpal I, Chandra V, Sharma GT, Kinjavdekar P (2017) Mesenchymal stem cell therapy for chronic wound healing in dogs. Intas Polivet 18:130–133

Ramos FO, Malard PF, Brunel HSS, Paludo GR, Castro MB, Silva PHS, Barreto-Vianna ARC, Lima EMM (2020) Canine atopic dermatitis attenuated by mesenchymal stem cells. J Adv Vet Anim Res 7(3):554–565

Reesink HL, Sutton RM, Shurer CR, Peterson RP, Tan JS, Su J, Paszek MJ, Nixon AJ (2017) Galectin-1 and galectin-3 expression in equine mesenchymal stromal cells (MSCs), synovial fibroblasts and chondrocytes, and the effect of inflammation on MSC motility. Stem Cell Res Ther 8:243

Reis BPZCD, Orge ID, Sampaio GLDA, Daltro SRT, Santos RRD, Meira CS, Soares MBP (2021) Mesenchymal stem cells in the context of canine atopic dermatitis: a review. Rev Bras Saúde Prod Anim, Salvador 22:e2122242021

Ribeiro J, Pereira T, Amorim I, Caseiro AR, Lopes MA, Lima J, Gartner A, Santos JD, Bártolo PJ, Rodrigues JM, Mauricio AC, Luís AL (2014) Cell therapy with human MSCs isolated from the umbilical cord Wharton jelly associated to a PVA membrane in the treatment of chronic skin wounds. Int J Med Sci 11(10):979–987

Rodriguez-Menocal L, Shareef S, Salgado M, Shabbir A, Van Badiavas E (2015) Role of whole bone marrow, whole bone marrow cultured cells, and mesenchymal stem cells in chronic wound healing. Stem Cell Res Ther 6:1–11

Roque JB, O'Leary CA, Duffy DL, Kyaw-Tanner M, Gharahkhani P, Vogelnest L et al (2012) Atopic dermatitis in West Highland white terriers is associated with a 1.3-Mb region on CFA 17. Immunogenetics 64(3):209–217

Russell KA, Chow NHC, Dukoff D, Gibson TWG, LaMarre J, Betts DH, Koch TG (2016) Characterization and immunomodulatory effects of canine adipose tissue- and bone marrow-derived mesenchymal stromal cells. PLoS One 11(12):e0167442

Russell KA, Garbin LC, Wong JM, Koch TG (2020) Mesenchymal stromal cells as potential antimicrobial for veterinary use—a comprehensive review. Front Microbiol 11:606404

Schönfelder U, Abel M, Wiegand C, Klemm D, Elsner P, Hipler U-C (2005) Influence of selected wound dressings on PMN elastase in chronic wound fluid and their antioxidative potential in vitro. Biomaterials 26(33):6664–6673

Shabbir A, Cox A, Rodriguez-Menocal L, Salgado M, Van Badiavas E (2015) Mesenchymal stem cell exosomes induce proliferation and migration of normal and chronic wound fibroblasts, and enhance angiogenesis in vitro. Stem Cells Dev 24:1635–1647

Sole A, Spriet M, Galuppo LD, Padgett KA, Borjesson DL, Wisner ER, Brosnan RJ, Vidal MA (2012) Scintigraphic evaluation of intra-arterial and intravenous regional limb perfusion of allogeneic bone marrow-derived mesenchymal stem cells in the normal equine distal limb using 99mTc-HMPAO. Equine Vet J 44:594–599

Spaas JH, Broeckx S, Van de Walle GR, Polettini M (2013) The effects of equine peripheral blood stem cells on cutaneous wound healing: a clinical evaluation in four horses. Clin Exp Dermatol 38:280–284

Spriet M, Trela JM, Padgett KA, Vidal MA (2013) Scintigraphic comparison of 4 different techniques of administration of mesenchymal stem cells to the equine foot. J Equine Vet Sci (Abstr) 33:867

Strioga M, Viswanathan S, Darinskas A, Slaby O, Michalek J (2012) Same or not the same? Comparison of adipose tissue-derived versus bone marrow-derived mesenchymal stem and stromal cells. Stem Cells Dev 21(14):2724–2752

Tan L, Liu X, Dou H, Hou Y (2020) Characteristics and regulation of mesenchymal stem cell plasticity by the microenvironment—specific factors involved in the regulation of MSC plasticity. Genes Dis 9:296–309. https://doi.org/10.1016/j.gendis.2020.10.006

Tatsumi K, Ohashi K, Matsubara Y, Ayako K, Takahiro O, Kakidachi H, Horii A, Kanegae K, Utoh R, Iwata T, Okano T (2013) Tissue factor triggers procoagulation in transplanted mesenchymal stem cells leading to thromboembolism. Biochem Biophys Res Commun 431:203–209

Taylor R, James T (2005) The role of oxidative stress in the development and persistence of pressure ulcers. In: Bader D, Bouten C, Colin D et al (eds) Pressure ulcer research. Springer, Berlin, Heidelberg, pp 205–232

Textor JA, Clark KC, Walker NJ, Aristizobal FA, Kol A, LeJeune SS, Bledsoe A, Davidyan A, Gray SN, Bohannon-Worsley LK, Woolard KD, Borjesson DL (2017) Allogeneic stem cells Alter gene expression and improve healing of distal limb wounds in horses. Stem Cells Transl Med 7(1):98–108

Theoret C, Wilmink JM (2016) Exuberant granulation tissue. In: Theoret C, Schumacher J (eds) Equine wound management. Wiley Online, Hoboken, NJ, pp 369–384

Theoret CL, Barber SM, Moyana TN, Gordon JR (2001) Expression of transforming growth factor β1, β3, and basic fibroblast growth factor in full-thickness skin wounds of equine limbs and thorax. Vet Surg 30:269–277

Theoret CL, Barber SM, Moyana TN et al (2002) Preliminary observations on expression of transforming growth factors beta1 and beta3 in equine full-thickness skin wounds healing normally or with exuberant granulation tissue. Vet Surg 31:266–273

Ting WJ, Shaw SW, Hii L-Y, Lin T-Y, Chang S-C, Liu K-Y, Shen P-C, Chen T-J, Peng S-Y (2020) Therapeutic effects of conditioned e DPBS from amniotic stem cells on lactating cow mastitis. Taiw J Obstet Gynecol 59:520e526

Tiwary R, Tripathi JN, Dwivedi KN (2011) Effect of medicinal plnt on wound healing in diabetics. Biomed Pharmacol 4(1):189–194

Tomchuck SL, Zwezdaryk KJ, Coffelt SB, Waterman RS, Danka ES, Scandurro AB (2008) Toll-like receptors on human mesenchymal stem cells drive their migration and immunomodulating responses. Stem Cells 26:99–107

Tracey AK, Alcott CJ, Schleining JA, Safayi S, Zaback PC, Hostetter JM, Reinertson EL (2014) The effects of topical oxygen therapy on equine distal limb dermal wound healing. Can Vet J 55: 1146–1152

Trindade ABH, Trindade JEBH (2021) Mesenchymal stem cell injection achieves closure of chronic wound in canine patient: a case report. Biomed J Sci Tech Res 34(1):26430–26432

Tso GH, Law HK, Tu W, Chan GC, Lau YL (2010) Phagocytosis of apoptotic cells modulates mesenchymal stem cells osteogenic differentiation to enhance IL-17 and RANKL expression on CD4+ T cells. Stem Cells 28(5):939–954 van den Boom R, Wilmink JM, O'Kane S et al (2002) Transforming growth factor-beta levels during second-intention healing are related to the different course of wound contraction in horses and ponies. Wound Repair Regen 10:188–194

Villatoro AJ, Hermida-Prieto M, Fernández V, Fariñas F, Alcoholado C, Rodríguez-García MI, Mariñas-Pardo L, Becerra J (2018) Allogeneic adipose-derived mesenchymal stem cell therapy in dogs with refractory atopic dermatitis: clinical efficacy and safety. Vet Rec 183:654. https://doi.org/10.1136/vr.104867

Wang Y, Chen X, Cao W, Shi Y (2014) Plasticity of mesenchymal stem cells in immunomodulation: pathological and therapeutic implications. Nat Immun 15(11):1009–1016

Wei S, Zhao Y, Zhang H, Wang L, Wu K, Luo C, Wang D, Xu L, Xiong Y (2020) Co-culture of umbilical cord mesenchymal stem cells and bovine mammary epithelial cells restrains apoptosis. Biomed J Sci Tech Res 25(3):19095

Wheat WH, Chow L, Kurihara JN, Regan DP, Coy JW, Webb TL, Dow SW (2017) Suppression of canine dendritic cell activation/maturation and inflammatory cytokine release by mesenchymal stem cells occurs through multiple distinct biochemical pathways. Stem Cells Dev 26(4): 249–262

White SV, Czisch CE, Han MH, Plant CD, Harvey AR, Plant GW (2016) Intravenous transplantation of mesenchymal progenitors distribute solely to the lungs and improve outcomes in cervical spinal cord injury. Stem Cells 34:1812–1825

Whitney JD (2005) Overview: acute and chronic wounds. Nurs Clin North Am 40(2):191–205

Wilmink JM, Herten JV, Weeren PR, Barneveld A (2010) Retrospective study of primary intention healing and sequestrum formation in horses compared to ponies under clinical circumstances. Eq Vet J 34:270–273

Wu Y, Chen L, Scott PG, Tredget EE (2007) Mesenchymal stem cells enhance wound healing through differentiation and angiogenesis. Stem Cells 25:2648–2659

Wynn TA (2007) Common and unique mechanisms regulate fibrosis in various fibroproliferative diseases. J Clin Invest 117:524–529

Yang H-M, Song W-J, Qiang L, Kim S-Y, Kim H-J, Ryu M-O, Ahn JO, Youn H-Y (2018) Canine mesenchymal stem cells treated with TNF-α and IFN-γ enhance anti-inflammatory effects through the COX-2/PGE2 pathway. Res Vet Sci 119:19–26

Zakirova EY, Shalimov DV, Garanina EE, Zhuravleva MN, Rutland CS, Rizvanov AA (2019) Use of biologically active 3D matrix for extensive skin defect treatment in veterinary practice: case report. Front Vet Sci 6:76

Zare S, Anjomshoa M, Kurd S, Chai J-K, Dahmardei M, Nilforoushzadeh MA, Rostamzadeh A (2015) Regenerative medicine: novel approach in burn wound healing. J Skin Stem Cell 2: e30351

Zayed MN, Schumacher J, Misk N, Dhar MS (2016) Effects of pro-inflammatory cytokines on chondrogenesis of equine mesenchymal stromal cells derived from bone marrow or synovial fluid. Vet J 217:26–32

Zhan X-S, El-Ashram S, Luo D-Z, Luo H-N, Wang B-Y, Chen S-F, Bai Y-S, Chen Z-S, Liu C-Y, Ji H-Q (2019) A comparative study of biological characteristics and transcriptome profiles of mesenchymal stem cells from different canine tissues. Int J Mol Sci 20:1485

Zhang X, Sarkar K, Rey S, Sebastian R, Andrikopoulou E, Marti GP, Fox-Talbot K, Semenza GL, Harmon JW (2011) Aging impairs the mobilization and homing of bone marrow derived angiogenic cells to burn wounds. J Mol Med 89:985–995

Zhang J, Guan J, Niu X, Hu G, Guo S, Li Q, Xie Z, Zhang C, Wang Y (2015) Exosomes released from human induced pluripotent stem cells-derived MSCs facilitate cutaneous wound healing by promoting collagen synthesis and angiogenesis. J Transl Med 13:49

Zhao HF, Chen J, Xu ZS et al (2009) Distribution and differentiation of mesenchymal stem cells in tumor tissue. Chin Med J (Engl) 122:712–715

Zhao G, Usui ML, Lippman SI, James GA, Stewart PS, Fleckman P, Olerud JE (2013) Biofilms and inflammation in chronic wounds. Adv Wound Care 2:389–399

Zhao L, Wang Z, Zhang J, Yang J, Gao XF, Wu B, Zhao G, Bao S, Hu S, Liu P, Li X (2017) Characterization of the single-cell derived bovine induced pluripotent stem cells. Tissue Cell 49(5):521–527

Zubin E, Conti V, Leonardi F, Zanichelli S, Ramoni R, Grolli S (2015) Regenerative therapy for the management of a large skin wound in a dog. Clin Case Rep 3(7):598–603

Mesenchymal Stem Cells Therapeutic Applications in Endocrine Disorders

12

Abstract

Endocrine diseases are now-a-days being increasingly reported with multiple factors that are considered to play a role which remains to be understood decisively. In animals, selective grading and selection of elite germplasm, and production-enhancing managemental techniques, besides, the endocrine-disrupting chemicals (EDCs) are some of the causes of endocrine disorders. In order to understand these growing problems and develop therapeutics, regenerative medicine utilizing mesenchymal stem cells (MSCs) is considered as one of the potential tools. Recently, insights in EDCs role in derangement of adipose tissue-derived MSCs and their correlation in metabolic diseases like equine metabolic syndrome (EMS) has been established. In vitro studies demonstrating the role of some of the chemicals in MSCs normalization may potentially help in understanding the disease and development of its treatment. MSCs innate resistance to oxidative stress and ability to ameliorate stress makes them potential anti-stress candidates. Their ability to differentiate into endocrine cells including insulin-producing cells (IPCs) and to secrete pro-healing factors including the immunomodulatory/anti-inflammatory, anti-apoptotic and regulation of cellular pathways makes them potentially suitable therapeutic agents in endocrine disorders. The current chapter focuses on the MSCs role in oxidative and some of the endocrine disorders in veterinary medicine.

Keywords

Mesenchymal stem cell · Diabetes mellitus · Oxidative stress · Metabolic syndrome

M. B. Gugjoo, *Therapeutic Applications of Mesenchymal Stem Cells in Veterinary Medicine*, https://doi.org/10.1007/978-981-19-3277-9_12

12.1 Introduction

Endocrine disorders in veterinary species are characterized as hormonal overproduction (cat hyperthyroidism and dog hyperadrenocorticism) or deficient or absent hormone secretion (hypothyroidism, diabetes mellitus type 1, primary hypoparathyroidism and primary hypoadrenocorticism). Some of the endocrine disorders arise as a result of altered tissue responsiveness to hormones such as diabetes mellitus (DM) type 2 and metabolic syndrome (MS) (Kooistra et al. 2009). Furthermore, the endocrine tissue may not be responsive to the feedback mechanisms leading to inappropriate hormone release. Overproduction of hormone from endocrine tissues may also result from ailments of secondary tissue sources like renal disease-induced hyperphosphatemia leading to parathyroid hyperplasia. Autoimmune diseases may also be one of the causes of endocrine diseases (DeFronzo 1997; Hani et al. 1999; Eizirik and Mandrup-Poulsen 2001). Thus, there may be primary or the secondary causes of the endocrine disorders depending upon the aetiological agent affecting the tissue type.

Farm animals like ruminants and horses are mostly being reared to boost the economy, while pet animals like dogs and cats are being reared as family members. As such these animals are reproduced through selective breeding for a genetically superior genome with attractive traits, and high production and reproduction potential. This selective reproduction and thereby higher production of farm animals puts selection pressure that leads to oxidative stress (Jóźwik et al. 2012). Stress-induced oxidative damage accelerates cell ageing and elicits various body disorders affecting metabolic and endocrine systems, among others. Stress influences the process of stem and progenitor cell differentiation via the central nervous system like hypothalamus pituitary axis. Stem cells of the pituitary gland and of the peripheral effector organs like adrenal gland have active developmental signalling pathways. Stress induces migration of stem cells explaining mingling of cellular components and aberrant expression of peptide receptors. This leads to autonomous changes in the endocrine tissues progressing to the endocrine disorders (Bornstein et al. 2018).

Domestic and pet animal species have variable occurrence of these endocrine disorders. Among various endocrine disorders, DM and MS are often life-threatening and tend to show resemblance to that of the human disease. However, other disorders of these glands are not less commonly reported, especially in pets or horses. Currently, there is lack of any definitive therapeutics that could address the actual inciting cause. As such their understanding and treatment through the prism of regenerative medicine employing mesenchymal stem cells (MSCs) is being evaluated. Regenerative medicine may offer hope in such cases although MSCs applications remain to be evaluated in veterinary species. Besides, the systemic affections that develop as sequelae of these endocrine disorders like diabetic neuropathy, nephropathy or retinopathy also remains to be evaluated. In this chapter main focus remains on DM and MS as currently MSCs are evaluated in relation to these two endocrine disorders to some extent in veterinary species like dog and horse.

12.2 Endocrine Disrupting Chemicals and MSCs

There is a growing concern that endocrine-disrupting chemicals (EDCs) available as environmental pollutants (size from 100 pm to 1 nm) are able to affect endocrine functions. Their structural similarity with endogenous hormones makes it possible to disrupt endocrine function in animals. This may occur either through their interaction with hormone transporter proteins, affecting hormone metabolism or in some cases blocking their effects. EDCs include both chemical compounds like organochlorine pesticides, phthalates, etc. as well as the heavy metals like cadmium, lead and mercury (Schantz and Widholm 2001). Recent insights in the field show that EDCs also promote mesenchymal stem cell changes deranging their proliferation, differentiation, trophic action and immunomodulation. This has been correlated with an increased MSCs adipogenic differentiation and decreased osteogenic differentiation, activation of the pro-inflammatory cytokines and increased oxidative stress (Bateman et al. 2017). EDCs may promote MSCs adipogenic differentiation through the peroxisome proliferator-activated receptor gamma (PPARγ) ligand (Pakzad et al. 2013). As increased MSCs adipogenesis is associated with their reduced RUNX2, a transcription factor regulating osteogenic gene expression-driven osteogenesis might be hampered (Ricciardi et al. 2012; Hayrapetyan et al. 2015). The long-term oxidative stress tends to affect MSCs proliferation, promotes their adipogenesis and reduces osteogenesis, in addition to the alteration in their immunomodulation potential (Bustos et al. 2014; Kizilay Mancini et al. 2015; Denu and Hematti 2016; Gu et al. 2016). Oxidative stress and change in ligand signalling may also promote MSCs epigenetic changes with resultant inappropriate MSCs biology (Bateman et al. 2017). A chronic exposure to EDCs incites low-level inflammation that in contrast to severe inflammation promotes MSCs pro-inflammatory mediators and immune response induction (Ma et al. 2014). EDCs induce oxidative stress through production of reactive oxygen species (ROS) in cells. Mitochondrial ROS and oxidation of its lipids promote autophagy, which is considered important for MSCs self-renewal, differentiation and pluripotency. Autophagy level effectuates MSCs-regulated CD4+ T cell immunosuppression through secretion of TGF-β1 (Gao et al. 2016). However, excessive ROS and oxidative stress may impair autophagy leading to reduced self-renewal and promotion of early senescence of the stem cells (Denu and Hematti 2016). Under hypoxia, autophagy-regulated apoptosis of BM-MSCs occurs via AMP-activated protein kinase/mammalian target of rapamycin (AMPK/mTOR) pathway (Zhang et al. 2016). Thus, quite a significant correlation of EDCs and the endocrine disorders especially through modulation of MSCs seems to occur. However, this literature is currently very limited and as such in-depth insights are highly desired in relation to the EDCs and their role in MSCs deranged functionality. This shall provide proof-of-concept for the human medicine as well to tackle endocrine disorders.

12.3 Endocrine Disorders and MSCs

MSCs or their secretome are considered to have potential to relieve endocrine sufferings, although studies are very limited to provide evidence-based medicine (Fig. 12.1). This is attributed to their ability to home particular tissue/organ and transdifferentiate into specific cells like insulin-producing cells (IPCs) upon availability of the suitable environment. The cells potentially secrete pro-healing factors including the anti-inflammatory and/or immunomodulatory factors (Davis et al. 2012; Gugjoo et al. 2020). MSCs inhibit secretion of pro-inflammatory factors like IL6 and TNFα and simultaneously guide macrophage 1 (M1) pro-inflammatory phenotype towards anti-inflammatory M2 (Xie et al. 2016; Zhao et al. 2018). Further, β-cell dysfunction can be alleviated by MSCs through reversal of dedifferentiation being mediated through secretion of IL-1R antagonist (Huang et al. 2021). MSCs secretome may restore insulin receptor substrate 1 (IRS-1) and protein kinase B (Akt) phosphorylation and promote glucose transporter 4 (GLUT4) expression and membrane translocation in muscles. Such a secretome may also raise glycogen synthesis related protein p-GSK3β expression levels, thereby increasing the liver glycogen storage for glucose homeostasis. Furthermore, the secretome may inhibit β-cell apoptosis and restore insulin secretion (Sun et al. 2018). MSCs secretome may prevent obesity-induced insulin resistance through attenuation of dyslipidaemia and inhibition of the fat cell hypertrophy (Zhao et al. 2018). One of the interesting

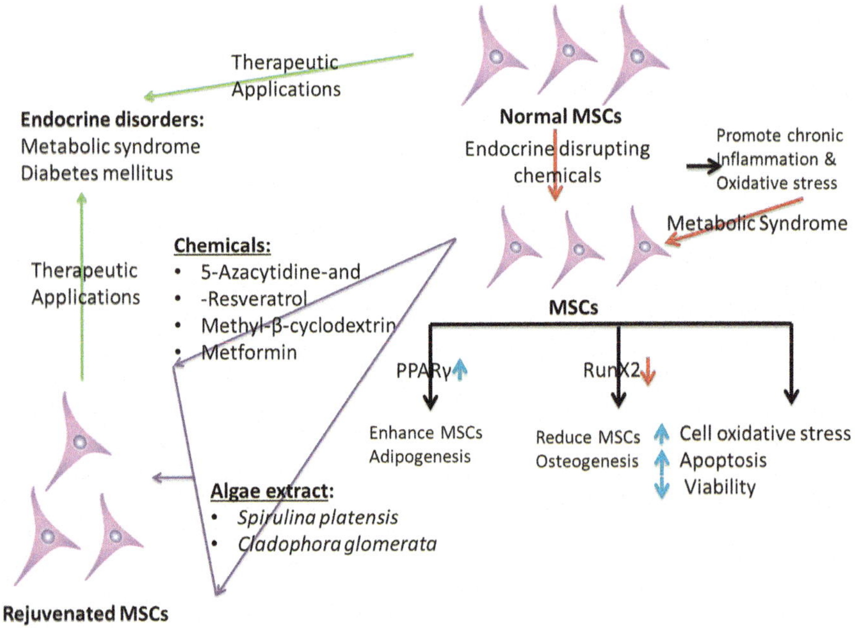

Fig. 12.1 Pictorial representation of effects of the metabolic syndrome on mesenchymal stem cells (MSCs), their rejuvenation and potential applications in endocrine disorders

findings is that MSCs may itself have a role to play in development of insulin resistance in aged ones as one of the lab animal studies demonstrates that BM-MSCs secretome contains miR-29b-3p. Its absorption by adipocytes, cardiomyocytes and hepatocytes results in insulin resistance. Thus targeting upregulated miR-29b-3p may potentially help in amelioration of insulin resistance (Su et al. 2019).

MSCs are tolerant to oxidative stress and might play an important role to evade oxidative stress possibly through their antioxidant action (Brandl et al. 2011). These cells may prevent lipid peroxidation through hemeoxygenase-1 (HO-1) production which arrests superoxide anion (O_2^-). Additionally, MSCs may also decrease reactive oxygen species (ROS) like H_2O_2 and oxygen-free radicals through glutathione production (Vanella et al. 2012; Calió et al. 2014; Ayatollahi et al. 2014). One of the major reasons for cell ageing is cell replication-induced telomere shortening. The telomere shortening may be prevented by MSCs through their catalytic activity of the subunit human telomerase reverse transcriptase (hTERT) (Estrada et al. 2013).

12.4 Metabolic Syndrome

Metabolic syndrome (MS), an endocrine disorder, is commonly seen in horses and dogs among various veterinary species. MS in animals may have characteristics resembling that of the human counterpart, although some differences especially in dogs are seen (Verkest 2014). Equine metabolic syndrome (EMS) shows pathological obesity, hyperinsulinemia, hyperglycaemia and altered hepatic function and predisposes animals to laminitis (Frank et al. 2010). Dog metabolic syndrome also shows insulin resistance, hyperlipidaemia (non-atherogenic) and hypertension but unlikely the hyperglycaemia seen in human and horse (Verkest 2014). Currently, aetiological factors responsible for metabolic syndrome are unknown. Obesity may predispose animals to the metabolic syndrome as it acts as a reserve of pro-inflammatory mediators that might produce adverse effects locally or systemically (Frank et al. 2010). An association of the EMS with the EDCs in horses is also demonstrated (Durward-Akhurst et al. 2019). Recently, the endocrine relationship between hepatic stellate cells (HSCs) and AD-MSCs is considered to have a critical relationship in the development of metabolic syndrome, although the physiological crosstalk between these two cells remains elusive (Marycz 2021). In general, EMS is characterized by ROS accumulation and decreased superoxide dismutase (SOD) activity, besides the mitochondrial deterioration (Marycz et al. 2016a, b, c). Currently, no standard treatment exists that could be considered as a definitive treatment. Mostly, dietary management and exercise is being recommended as antidiabetic drug metformin does not achieve effective concentration upon oral intake (Hustace et al. 2009) and anti-obesity drug pioglitazone fails to provide insulin-sensitizing effects in horse (Wearn et al. 2011).

Metabolic syndrome exhibits numerous affections and one among those is the deranged activity of adipose tissue-derived MSCs (Kornicka et al. 2018a; Bourebaba et al. 2019; Smieszek et al. 2019; Weiss et al. 2020), epithelial stem cells and pancreatic islet-like cells (Kornicka et al. 2018b) (Fig. 12.1). AD-MSCs derived from EMS-affected horses appear senescent and apoptotic with reduced viability and ability to differentiate (Marycz et al. 2016a, b, c). Their osteogenic (bone BMP-2 and Coll-1 downregulation, and lower RUNX2, OCN and ALP comparable to control) and chondrogenic (decreased vimentin, decorin and Sox9) differentiation potentials are reduced. This occurs due to the accumulation of excessive ROS with an endoplasmic reticulum stress (Marycz et al. 2016a, 2018). Various chemicals appear to be useful to rejuvenate such senescent and apoptotic MSCs to a more viable and mitochondrial stress free. These chemicals include 5-Azacytidine-and-Resveratrol (Kornicka et al. 2018b), metformin (Smieszek et al. 2019) and methyl–βcyclodextrin (Szydlarska et al. 2018). Additionally, extracts from algae like *Spirulina platensis* (Nawrocka et al. 2017) and *Cladophora glomerata* (Marycz et al. 2017; Bourebaba et al. 2019) also improve affected MSCs functionality. EMS affected horse AD-MSCs viability, reduced oxidative stress and apoptosis, besides the chondrogenic differentiation may be improved through water extract of freshwater alga *Cladophora glomerata* (Marycz et al. 2017; Bourebaba et al. 2019). *Cladophora glomerata* water extract enriched during a biosorption process in Cr (III) trivalent chromium and chromium picolinate improves mitochondrial dynamics, reduced endoplasmic reticulum stress and decrease in mitochondrial biogenesis-related gene (PDK4) expression in affected cells. Such an exposure additionally induces anti-oxidative protection from SOD activity (Marycz et al. 2017). Apart from direct exposure to 5-Azacytidine-and-Resveratrol, normal AD-MSCs microvesicles activated with these chemicals may also improve EMS-affected MSCs viability and increase their life span (Weiss et al. 2020). Furthermore, enhanced adipogenesis of MSCs encouraged by EDCs and seen in metabolic syndrome may be inhibited by interleukin (IL)-15. MSCs adipogenic differentiation may be inhibited by IL-15 through activation of the signal transducers and activators of transcription (STAT) pathway as demonstrated in cattle BM-MSCs. It may also attenuate insulin-induced phosphoinositide 3-kinase (PI3K)/Akt signalling activation (Shi et al. 2019). Thus, to improve metabolic syndrome in animals, allogeneic MSCs from normal animals or those treated with the above chemicals may effectuate improvement in metabolic syndrome.

A single case report of EMS transplanted with 3 weekly doses of 5-Azacytidine-and-Resveratrol treated autologous AD-MSCs (30 million, intravenous) shows reduced blood levels of liver enzymes including glutamate dehydrogenase (GLDH), gamma-glutamyltransferase (GGT), lactate dehydrogenase (LDH) and aspartate transaminase (AST). It is worth mentioning that the standard recommended protocol of dietary management and exercise is simultaneously followed (Marycz et al. 2019). Thus, EMS-affected animal derived AD-MSCs that tend to deteriorate may be rejuvenated with various chemicals and may be used clinically. Extensive studies for actual safety and efficacy are highly desired as EMS is the complex and multifactorial disorder. Further, MSCs sources, passage, suitable time and route of transplantation, concentration, number of doses, besides their actual mechanism of action need to be evaluated. Furthermore, it is imperative to determine effect of EMS on the MSCs from sources other than adipose tissue.

12.5 MSCs and Diabetes Mellitus (DM)

Diabetes, a group of metabolic diseases, is characterized by chronic hyperglycaemia resulting from defects in insulin secretion and/or function. DM type 1 may commonly result from an abnormal autoimmune response (T-cell) leading to the β-cells destruction, while DM type 2 is characterized by insulin resistance from peripheral organs resulting from the abnormal metabolic activity (insufficient production and/or activity) (DeFronzo 1997; Hani et al. 1999; Eizirik and Mandrup-Poulsen 2001). β-cell dysfunction occurs with glucotoxicity and lipotoxicity. This includes their dysregulation of autophagy-related mechanisms that induces inflammatory and oxidative stress and finally leads to either type of DM (Rother 2007; Lee et al. 2019). DM occurs in almost all the animal species; however, its clinical significance is mainly attributed to dog and cat, and to some extent to horse, goat and cattle. Sheep has been utilized as a model animal for human diabetes mellitus (Dickinson et al. 1991). Clinically, DM type II occurs in cat (O'Brien 2002) and horse (Durham et al. 2009), while DM type I is reported in dog (Davison et al. 2008), goat (Braun et al. 2008) and cattle (Clark 2003). In cattle, genetic predisposition has not been confirmed while infectious aetiology like bovine viral diarrhoea and foot and mouth disease may be one of the reasons (Tajima et al. 1992, 1999; Taniyama et al. 1999). In dogs, the main aetiological reason for DM type 1 (more appropriately termed as insulin-dependent DM) is pancreatitis (Alejandro et al. 1988; Hoenig 2002) or autoimmune-mediated β-cell destruction (Atkins et al. 1979). In horse, pituitary neoplasia and adrenal cortical hyperplasia are considered to be the main causes of the DM (Stogdale 1986).

12.5.1 In Vitro MSC Studies

Various large animal in vitro studies confirm MSCs ability to differentiate into insulin-producing cells (IPCs). MSCs differentiation into the particular cell type may be ensured through culture addition of a cocktail of chemicals and/or scaffolds required. Such a differentiation cocktail is developed based on the evolutionary insulin-producing cell developmental biology. Cattle UC-MSCs are induced with H-DMEM, HGF, mercaptoethanol and niacinamide supplementation to differentiate into IPCs-expressing pancreatic and duodenal homeobox 1 (*PDX-1*) besides insulin (Xiong et al. 2014). Similarly, cattle placental MSCs under nicotinamide, extendin-4, glucose and poly-D-Lysine differentiate into IPCs-expressing insulin, glucagon, Paired box protein *(Pax)-4*, *Pax-6*, Homeobox protein *Nkx6.1* and Forkhead box protein (*Fox*) (Peng et al. 2017). Use of chemicals (DMEM-HG, β-mercaptoethanol), and nicotinamide (for the initial 8 day period) in MSCs culture or further addition of exendin-4, B27, nonessential amino acids, l-glutamine and/or basic fibroblast growth factor enhances insulin aggregates. However, the latter combination appears to lead to significantly bigger insulin aggregates (Camara et al. 2021).

MSCs induced into IPCs may improve with the extendin-4 (Rodprasert et al. 2021). Extendin-4 induces pancreatic endoderm marker (*Pdx-1*) and further addition of nicotinamide and β-mercaptoethanol synergistically induces this transcription factor. *Pdx-1* is essential for early pancreatic development and β-cells formation, besides acts as an insulin secretion activator. Extendin-4 additionally upregulates *Nkx2.2* transcription factor, involved in early pancreatic development and *MafA* transcripts that control and activate insulin expression and secretion. The combined use of nicotinamide and β-mercaptoethanol induces some of the transcription factors (*Nkx2.2*, *Isl-1* and *MafA*) in the induced MSCs. Contrarily, nicotinamide and β-mercaptoethanol and exendin-4 synergistically induce *Pdx-1*, *Isl-1* and *MafA* from induced MSCs. Extendin-4 therefore may have an essential role to play in the early development of IPCs (*Pdx-1* and *Nkx2.2*) and late-stage insulin secretion (*MafA*) from MSCs and shows synergism with the nicotinamide and β-mercaptoethanol (Camara et al. 2021).

Apart from the chemicals/hormones, extracellular matrix and culture technique may affect MSCs ability to differentiate into insulin-producing cells. Dog AD-MSCs fail to differentiate into insulin-producing cells (IPCs) in a 2D culture while 3D culture that employs human recombinant peptide effectuates dog AD-MSCs differentiation into IPCs. Insulin production of these cells is confirmed both under ex vivo and in vivo lab animal studies (Teshima et al. 2020). MSCs IPCs production potential also varies with their source. Dog BM-MSCs require 3D culture techniques (hanging drop or Matrigel-embedded culture) for generating and maintaining the colony-like structure of IPCs contrary to the AD-MSCs that show efficient colony-like structure of IPCs under low attachment 2D culture. In hanging drop culture technique, dog BM-MSCs show higher *Pdx1* and lower pancreatic beta-cell markers (*Nkx-6.1*, *Isl-1*, *Glut-2*, and *Insulin*) while in Matrigel-embedded culture, mRNA expression of pancreatic-relating markers (*Glucagon* and *Glp1r*) does not occur. Furthermore, dog BM-MSCs overexpression of pancreatic commitment regulator (*pdx1*) fails to generate IPCs in 2D but such cell induction may significantly increase in 3D culture. In low attachment (2D) culture, dog AD-MSCs significant upregulation of pancreatic beta-cell markers (*Nkx-6.1*, *Isl-1*, *Maf-A*, *Glut-2*, and *Insulin*) occurs. AD-MSCs efficiency also significantly improves with the manipulation of Notch signalling (Rodprasert et al. 2021). Thus, culture technique and signalling manipulation may improve MSCs IPCs production ability although their specific source like adipose tissue may be more amenable towards IPCs commitment. There might be differences in MSCs gene regulatory pathways across the source and species based on the priming under in vivo conditions prior to harvesting. In dog CD73 marker expression may remain low as compared to the human MSCs and such differences may have a role to play in variable expression. In order to regulate such pathways, extrinsic manipulation of the environment like further addition of melatonin may efficiently induce MSCs into IPCs. Melatonin binds to its receptors on AD-MSCs and activates transforming growth factor β (TGF-β) pathway to improve their viability (Li et al. 2021). In an in vitro insulin resistance model (utilizing differentiated 3T3-L1 murine adipocytes), dog AD-MSCs conditioned medium augments mRNA and protein expression levels of IRS-1 and GLUT4 (insulin indicators) and may occur through fibroblast growth factor-1 (Kim et al. 2018). Also dog AD-MSCs ameliorate murine acute pancreatitis, which is considered to be one of the main causes for dog DM (Kim et al. 2016). Ex vivo studies show promising results although with some ifs and buts, it is therefore imperative to have further insights into the cellular process and finally into the in vivo safety, feasibility and utility as currently the therapeutics remains a big challenge.

12.5.2 In Vivo Studies

12.5.2.1 Experimental Studies

In vivo MSCs studies for DM in veterinary species are exclusively conducted in dog with improved results (Table 12.1), albeit with some concerns. MSCs therapeutic effects are evaluated in experimental and clinical insulin-dependent DM (IDDM). Experimental randomized controlled diabetes models are induced with alloxan and streptozotocin and treated with xenogeneic-differentiated BM-MSCs (Gabr et al. 2018) or autologous BM-MSCs (Zhu et al. 2011). Transplantation of five million xenogeneic (human) differentiated cells (in a TheraCyte capsules under the rectus sheath) ensures euglycemia in $n = 4$ dogs confirmed by the glucose tolerance curves. In $n = 2$ dogs, blood glucose levels reduce but fail to reach normal level. Serum of all dogs ($n = 6$) shows human insulin and C-peptide but negligible concentration of dog insulin, depicting functioning of transplanted xenogeneic cells. Further, removal of transplanted capsules containing cells leads to the immediate return of diabetes. TheraCyte capsule potentially prevents immune response against the transplanted cells. Considering xenogeneic MSCs insulin induction, chances of immune response

Table 12.1 MSCs preclinical experimental studies on diabetes in dog

Study design/ period	Biomaterial	Evaluation criteria	Results	References
Dog diabetes model				
Randomized placebo-controlled study ($n = 30$)/ 16 weeks	Autologous BM-MSCs expressing human insulin (50×10^6 intrahepatic)	Fluorescent microscopy, radioimmunoassay (RIA), immunohistochemistry	Improved outcome	Zhu et al. (2011)
Randomized placebo-controlled study ($n = 10$)/ 18 months	Xenogeneic BM-MSCs (human, 5×10^6 in TheraCyte capsule placed below rectus sheath)	Histology, immuno-histochemistry, PCR	Improved outcome	Gabr et al. (2018)

MSCs mesenchymal stem cells, *BM* bone marrow derived, *n* total number of animals in all the groups of the study; Improved outcome: Indicates improved outcome with MSCs transplantation even if only one parameter improves; Study period: Indicates maximum time period for which study is conducted increase. As transplanted animals show cellular response and simultaneously lack humoral response, it could have been prevented by the adjuvant use of azathioprine. The main concern in the study that remains is the reduced insulin response observed after some period. This may be attributed to the decreased circulation to the capsule by its fibrosis (Gabr et al. 2018). In order to avoid such lacunas, a two-step procedure has been performed. Initial implantation of the capsule and ensuring its sufficient vascularization is followed by cell loading that can survive and function properly (Sörenby et al. 2008). To waive off such concerns, direct transplantation of autologous MSCs into the highly vascularized area like liver through hepatic artery is also performed. In dog study, the effect of insulin-transfected autologous BM-MSCs show comparable effect on glucose as that of the non-transfected cells (@50 million) but higher than the control. This glucose tolerance, however, may not be comparable to the normal islet-like cells. As C-peptide is detected in transfected cell-transplanted animals, β cell regeneration can be excluded (Zhu et al. 2011). Thus, MSCs may play some role in amelioration of diabetes, but β-cell regeneration currently remains the challenge.

12.5.2.2 Clinical Studies

In limited dog clinical IDDM studies, MSCs positive role in relation to clinical improvement is demonstrated (Table 12.2). Multiple intravenous doses (4–5) of

Table 12.2 MSC's clinical studies on diabetes in dog

Study design/period	Cell source	Cell dose	Evaluation criteria	Results	References
Insulin-dependent diabetes mellitus model					
Case report ($n = 4$)/ 6 months	Cultured pancreatic islet cells co-aggregated with high numbers AD-MSCs (Neo-Islet cells) (2×10^5/kg b wt intraperitoneal)		ELISA, blood examination	Improved outcome	Gooch et al. (2019)
Case report ($n = 4$)/ 5 months	Allogeneic AD-MSCs (3–5 times monthly intravenously)		Clinical evaluation, fasting C-peptide monitoring, blood and urine analysis	Improved outcome	Rhew et al. (2021)

MSCs mesenchymal stem cells, *AD* adipose tissue derived, *n* total number of animals in all the groups of the study; Improved outcome: Indicates improved outcome with MSCs transplantation even if only one parameter improves; Study period: Indicates maximum time period for which study is conducted

allogeneic AD-MSCs enhances vitality of the 50% of the animals ($n = 4$). Clinical improvement exhibited in polyuria, polydipsia, fructosamine and HbA1c, although with concurrent increase in C-peptide. Further, none of the animals appear to have cystitis, a common problem in DM (Rhew et al. 2021). To enhance efficacy and prevent rejection of islet cells, intraperitoneal transplantation of AD-MSCs and islet-like cells (termed as Neo-Islet cells) is made. These cells engraft, re-differentiate and produce insulin without being immune rejected in $n = 4$. In $n = 2$, reduced immune response in the form of anti-immune complex antibodies occurs post Neo-Islet cell transplantation. The insulin dose requirement reduces by 50% with a significant reduction in serum glucose concentration (Gooch et al. 2019). All these studies favour MSCs therapeutic role in IDDM in dogs but actual regeneration of islet cells and maintenance of their efficacy for a long period remains a challenge. The clinical studies have been made on limited animals and lack control, besides complete amelioration of symptoms remains to be seen. Further studies are therefore desired that evaluate safety, efficacy and thereby determine posology of the stem cell therapy. This may be possible through uniform in vivo studies involving specific MSCs, their dose, route and repetition. Furthermore, MSCs engraftment, therapeutic role and their viability check remain to be seen for their evidence-based definitive clinical applications.

12.6 Conclusion(s)

Endocrine diseases develop from varied aetiologies and are mostly managed through amelioration of symptoms. In order to understand endocrine diseases and develop their definitive therapeutics, MSCs are being evaluated. These cells may itself be the reason behind the endocrine disorders due to their changes in ageing and/or through the EDCs. However, these cells are resistant to the stress and rejuvenate with various chemicals and thus can potentially be used for therapeutic applications. The cells are able to differentiate into IPCs although such a differentiation remains to be seen under in vivo conditions. MSCs may potentially ameliorate clinical signs in some of the conditions like metabolic syndrome and DM in animals. The limited studies need to be further progressed towards goal-oriented work. In vitro studies that harness their pro-healing potential in endocrine diseases may be conducted, and on that basis, MSCs in vivo applications may be performed. The studies that evaluate the effects of MSCs source, passage number, concentration, together with their route of implantation and above all their mechanism of action are required. Uniform studies on oxidative stress and endocrine diseases that have sufficient animal number and are conducted for longer periods are to be be made to overcome the loopholes of currently available research/ clinical studies.

References

Alejandro R, Feldman EC, Shienvold FL, Mintz DH (1988) Advances in canine diabetes mellitus research: etiopathology and results of islet transplantation. J Am Vet Med Assoc 193:1050–1055

Atkins CE, Hill JR, Johnson RK (1979) Diabetes mellitus in the juvenile dog: a report of four cases. J Am Vet Med Assoc 175:362–368

Ayatollahi M, Hesami Z, Jamshidzadeh A, Gramizadeh B (2014) Antioxidant effects of bone marrow mesenchymal stem cell against carbon tetrachloride-induced oxidative damage in rat livers. Int J Organ Transplant Med 5:166–173

Bateman ME, Strong AL, McLachlan JA, Burow ME, Bunnell BA (2017) The effects of endocrine disruptors on adipogenesis and osteogenesis in mesenchymal stem cells: a review. Front Endocrinol 7:171

Bornstein SR, Steenblock C, Chrousos GP, Schally AV, Beuschlein F, Kline G, Krone NP, Licinio J, Wong ML, Ullmann E, Ruiz-Babot G, Boehm BO, Behrens A, Brennand A, Santambrogio A, Berger I, Werdermann M, Sancho R, Linkermann A, Lenders JW, Eisenhofer G, Andoniadou CL (2018) Stress-inducible-stem cells: a new view on endocrine, metabolic and mental disease? Mol Psychiatry 24:2–9

Bourebaba L, Michalak I, Baouche M, Kucharczyk K, Marycz K (2019) Cladophora glomerata methanolic extract promotes chondrogenic gene expression and cartilage phenotype differentiation in equine adipose-derived mesenchymal stromal stem cells affected by metabolic syndrome. Stem Cell Res Ther 10:392

Brandl A, Meyer M, Bechmann V, Nerlich M, Angele P (2011) Oxidative stress induces senescence in human mesenchymal stem cells. Exp Cell Res 317:1541–1547

Braun U, Gansohr B, Seidel M, Dumelin J, Wenger B, Schade B, Pospischil A (2008) Diabetes mellitus type 1 in a goat. Schweiz Arch Tierheilkd 150(12):608–612

Bustos ML, Huleihel L, Kapetanaki MG, Lino-Cardenas CL, Mroz L, Ellis BM et al (2014) Aging mesenchymal stem cells fail to protect because of impaired migration and antiinflammatory response. Am J Respir Crit Care Med 189(7):787–798

Calió ML, Marinho DS, Ko GM, Ribeiro RR, Carbonel AF, Oyama LMM, Ormanji T, Guirao P, Calió PL, Reis LA, de Jesus Simões M, Lisbôa-Nascimento T, Ferreira AT, Bertoncini CR (2014) Transplantation of bone marrow mesenchymal stem cells decreases oxidative stress, apoptosis, and hippocampal damage in brain of a spontaneous stroke model. Free Radic Biol Med 70:141–154

Camara BOS, Ocarino NM, Bertassoli BM, Malm C, Araújo FR, Reis AMS, Jorge EC, Alves EGL, Serakides R (2021) Differentiation of canine adipose mesenchymal stem cells into insulin-producing cells: comparison of different culture medium compositions. Domest Anim Endocrinol 74:106572

Clark Z (2003) Diabetes mellitus in a 6-month-old Charolais heifer calf. Can Vet J 44(11):921

Davis NE, Hamilton D, Fontaine MJ (2012) Harnessing the immunomodulatory and tissue repair properties of mesenchymal stem cells to restore β cell function. Curr Diab Rep 12:612–622

Davison LJ, Weenink SM, Christie MR et al (2008) Autoantibodies to GAD65 and IA-2 in canine diabetes mellitus. Vet Immunol Immunopathol 126:83–90

DeFronzo RA (1997) Pathogenesis of type 2 diabetes: metabolic and molecular implications for identifying diabetes genes. Diabetes Rev 5:177–269

Denu RA, Hematti P (2016) Effects of oxidative stress on mesenchymal stem cell biology. Oxidative Med Cell Longev 2016:2989076

Dickinson JE, Meyer BA, Chmielowiec S, Palmer SM (1991) Streptozocin-induced diabetes mellitus in the pregnant ewe. Am J Obstet Gynecol 65(6):1673–1677

Durham AE, Hughes KJ, Cottle HJ, Rendle DI, Boston RC (2009) Type 2 diabetes mellitus with pancreatic beta cell dysfunction in 3 horses confirmed with minimal model analysis. Equine Vet J 41(9):924–929

Durward-Akhurst SA, Schultz NE, Norton EM, Rendahl AK, Besselink H, Behnisch PA, Brouwer A, Geor RJ, Mickelson JR, McCue ME (2019) Associations between endocrine disrupting chemicals and equine metabolic syndrome phenotypes. Chemosphere 218:652–661

Eizirik DL, Mandrup-Poulsen T (2001) A choice of death. The signal transduction of immunemediated b-cell apoptosis. Diabetologia 44:2115–2133

Estrada JC, Torres Y, Benguría A, Dopazo A, Roche E, Carrera-Quintanar L, Pérez RA, Enríquez JA, Torres R, Ramírez JC, Samper E, Bernad A (2013) Human mesenchymal stem cell-replicative senescence and oxidative stress are closely linked to aneuploidy. Cell Death Discov 4:e691

Frank N, Geor RJ, Bailey SR, Durham AE, Johnson PJ (2010) American College of Veterinary Internal Medicine. J Vet Intern Med 24(3):467–475

Gabr MM, Zakaria MM, Refaie AF, Ismail AM, Khater SM, Ashamallah SA, Azzam MM, Ghoneim MA (2018) Insulin-producing cells from adult human bone marrow mesenchymal stromal cells could control chemically induced diabetes in dogs: a preliminary study. Cell Transplant 27(6):937–947

Gao L, Cen S, Wang P, Xie Z, Liu Z, Deng W, Su H, Wu X, Wang S, Li J, Ouyang Y, Wu Y, Shen H (2016) Autophagy improves the immunosuppression of CD4+ T cells by mesenchymal stem cells through transforming growth factor-β1. Stem Cells Transl Med 5(11):1496–1505

Gooch A, Zhang P, Hu Z, Son NL, Avila N, Fischer J, Roberts G, Sellon R, Westenfelder C (2019) Interim report on the effective intraperitoneal therapy of insulin-dependent diabetes mellitus in pet dogs using "Neo-Islets," aggregates of adipose stem and pancreatic islet cells (INAD 012-776). PLoS One 14(9):e0218688

Gu Y, Li T, Ding Y, Sun L, Tu T, Zhu W et al (2016) Changes in mesenchymal stem cells following long-term culture in vitro. Mol Med Rep 13(6):5207–5215

Gugjoo MB, Amarpal, Fazili MR, Shah RA, Mir MS, Sharma GT (2020) Goat mesenchymal stem cell basic research and potential applications. Small Rum Res 183:106045

Hani EH, Stoffers DA, Chevre JC, Durand E, Stanojevic V, Dina C, Habener JF, Froguel P (1999) Defective mutations in the insulin promoter factor-1 (IPF-1) gene in late-onset type 2 diabetes mellitus. J Clin Invest 104:R41–R48

Hayrapetyan A, Jansen JA, van den Beucken JJ (2015) Signaling pathways involved in osteogenesis and their application for bone regenerative medicine. Tissue Eng Part B Rev 21(1):75–87

Hoenig M (2002) Comparative aspects of diabetes mellitus in dogs and cats. Mol Cell Endocrinol 197:221–229

Huang Q, Huang Y, Liu J (2021) Mesenchymal stem cells: an excellent candidate for the treatment of diabetes mellitus. Int J Endocrinol 2021:9938658

Hustace JL, Firshman AM, Mata JE (2009) Pharmacokinetics and bioavailability of metformin in horses. Am J Vet Res 70(5):665–668

Jóźwik A, Krzyżewski J, Strzałkowska N, Poławska E, Bagnicka E, Wierzbicka A, Niemczuk K, Lipińska P, Horbańczuk JO (2012) Relations between the oxidative status, mastitis, milkquality, and disorders of reproductive functions in dairy cows—a review. Anim Sci Paper Rep 30:297–307

Kim H-W, Song W-J, Li Q, Han S-M, Jeon K-O, Park S-C, Ryu M-O, Chae H-K, Kyeong K, Youn H-Y (2016) Canine adipose tissue-derived mesenchymal stem cells ameliorate severe acute pancreatitis by regulating T cells in rats. J Vet Sci 17(4):539–548

Kim H-J, Li Q, Song W-J, Yang H-M, Kim S-Y, Park S-C, Ahn J-O, Youn H-Y (2018) Fibroblast growth factor-1 as a mediator of paracrine effects of canine adipose tissue derived mesenchymal stem cells on in vitro-induced insulin resistance models. BMC Vet Res 14:351

Kizilay Mancini O, Shum-Tim D, Stochaj U, Correa JA, Colmegna I (2015) Age, atherosclerosis and type 2 diabetes reduce human mesenchymal stromal cell-mediated T-cell suppression. Stem Cell Res Ther 6:140

Kooistra HS, Galac S, Buijtels JJCW, Meij MBP (2009) Endocrine diseases in animals. Horm Res 71(1):144–147

Kornicka K, Smieszek A, Węgrzyn AS, Röcken M, Marycz K (2018a) Immunomodulatory properties of adipose-derived stem cells treated with 5-azacytydine and resveratrol on peripheral blood mononuclear cells and macrophages in metabolic syndrome animals. J Clin Med 7:383

Kornicka K, Smieszek A, Szłapka-Kosarzewska J, Irwin Houston JM, Roecken M, Marycz K (2018b) Characterization of apoptosis, autophagy and oxidative stress in pancreatic islets cells and intestinal epithelial cells isolated from equine metabolic syndrome (EMS) horses. Int J Mol Sci 19:3068

Lee YH, Kim J, Park K, Lee MS (2019) β-Cell autophagy: mechanism and role in β-cell dysfunction. Mol Metab 27s:S92–S103

Li B, Cheng X, Aierken A, Du J, He W, Zhang M, Tan N, Kou Z, Peng S, Jia W, Tang H, Hua J (2021) Melatonin promotes the therapeutic effect of mesenchymal stem cells on type 2 diabetes mellitus by regulating TGF-β pathway. Front Cell Dev Biol 9:799571

Ma S, Xie N, Li W, Yuan B, Shi Y, Wang Y (2014) Immunobiology of mesenchymal stem cells. Cell Death Differ 21(2):216–225

Marycz K (2021) The hepatic stellate cells (HSCs) and adipose derived stem progenitor cells (ASCs): are they critical multitargeted endogenous metabolic modulators in syndrome X and EMS? Letter to editor. Stem Cell Rev Rep 17(5):1931–1933. https://doi.org/10.1007/s12015-021-10233-8

Marycz K, Kornicka K, Basinska K, Czyrek A (2016a) Equine metabolic syndrome affects viability, senescence, and stress factors of equine adipose-derived mesenchymal stromal stem cells: new insight into EqASCs isolated from EMS horses in the context of their aging. Oxidative Med Cell Longev 2016:4710326. https://doi.org/10.1155/2016/4710326

Marycz K, Kornicka K, Grzesiak J, Śmieszek A, Szłapka J (2016b) Macroautophagy and selective mitophagy ameliorate chondrogenic differentiation potential in adipose stem cells of equine metabolic syndrome: new findings in the field of progenitor cells differentiation. Oxidative Med Cell Longev 2016:1–18. https://doi.org/10.1155/2016/3718468

Marycz K, Kornicka K, Markedziak M, Golonka P, Nicpoń J (2016c) Equine metabolic syndrome impairs adipose stem cells osteogenic differentiation by predominance of autophagy over selective mitophagy. J Cell Mol Med 20:2384–2404

Marycz K, Michalak I, Kocherova L, Mar˛edziak M, Weiss C. (2017) The Cladophora glomerata enriched by biosorption process in Cr(III) improves viability, and reduces oxidative stress and apoptosis in equine metabolic syndrome derived adipose mesenchymal stromal stem cells (ASCs) and their extracellular vesicles (MV's). Mar Drugs 15:385

Marycz K, Weiss C, Śmieszek A, Kornicka K (2018) Evaluation of oxidative stress and mitophagy during adipogenic differentiation of adipose-derived stem cells isolated from equine metabolic syndrome (EMS) horses. Stem Cells Int 2018:5340756

Marycz K, Szłapka-Kosarzewska J, Geburek F, Kornicka-Garbowska K (2019) Systemic administration of rejuvenated adipose-derived mesenchymal stem cells improves liver metabolism in equine metabolic syndrome (EMS)—new approach in veterinary regenerative medicine. Stem Cell Rev Rep 15(6):842–850

Nawrocka D, Kornicka K, Smieszek A, Marycz K (2017) Spirulina platensis improves mitochondrial function impaired by elevated oxidative stress in adipose-derived mesenchymal stromal cells (ASCs) and intestinal epithelial cells (IECs), and enhances insulin sensitivity in equine metabolic syndrome (EMS) horses. Mar Drugs 15:237

O'Brien TD (2002) Pathogenesis of feline diabetes mellitus. Mol Cell Endocrinol 197:213–219

Pakzad M, Fouladdel S, Nili-Ahmadabadi A, Pourkhalili N, Baeeri M, Azizi E et al (2013) Sublethal exposures of diazinon alters glucose homostasis in Wistar rats: biochemical and molecular evidences of oxidative stress in adipose tissues. Pestic Biochem Physiol 105(1): 57–61

Peng SY, Chou CW, Kuo YH, Shen PC, Shaw SS (2017) Potential differentiation of islet-like cells from pregnant cow-derived placental stem cells. Taiwan J Obstet Gynecol 56(3):306–311

Rhew S-Y, Park S-M, Li Q, An J-H, Chae H-K, Lee J-H, Ahn J-O, Song W-J, Youn H-Y (2021) Efficacy and safety of allogenic canine adipose tissue-derived mesenchymal stem cell therapy for insulin-dependent diabetes mellitus in four dogs: a pilot study. J Vet Med Sci 83(4):592–600

Ricciardi M, Malpeli G, Bifari F, Bassi G, Pacelli L, NwaboKamdje AH et al (2012) Comparison of epithelial differentiation and immune regulatory properties of mesenchymal stromal cells derived from human lung and bone marrow. PLoS One 7(5):e35639

Rodprasert W, Nantavisai S, Pathanachai K, Pavasant P, Osathanon T, Sawangmake C (2021) Tailored generation of insulin producing cells from canine mesenchymal stem cells derived from bone marrow and adipose tissue. Sci Rep 11:12409

Rother K (2007) Diabetes treatment-bridging the divide. N Engl J Med 356:1499–1501

Schantz SL, Widholm JJ (2001) Cognitive effects of endocrine-disrupting chemicals in animals. Environ Health Perspect 109(12):1197–1206

Shi M, Li Z, Miao Z, Guo Y, Yi L (2019) Interleukin-15 inhibits adipogenic differentiation of cattle bone marrow-derived mesenchymal stem cells via regulating the crosstalk between signal transducer and activator of transcription 5A and Akt signalling. Biochem Biophys Res Commun 517(2):346–352

Smieszek A, Kornicka K, Szłapka-Kosarzewska J, Androvic P, Valihrach L, Langerova L, Rohlova E, Kubista M, Marycz K (2019) Metformin increases proliferative activity and viability of multipotent stromal stem cells isolated from adipose tissue derived from horses with equine metabolic syndrome. Cell 8:80

Sörenby AK, Kumagai-Braesch M, Sharma A, Hultenby KR, Wernerson AM, Tibell AB (2008) Preimplantation of an immunoprotective device can lower the curative dose of islets to that of free islet transplantation: studies in a rodent model. Transplantation 86(2):364–366

Stogdale L (1986) Definition of diabetes mellitus. Cornell Vet 76(2):156–174

Su T, Xiao Y, Xiao Y, Guo Q, Li C, Huang Y et al (2019) Bone marrow mesenchymal stem cells-derived exosomal MiR-29b-3p regulates aging-associated insulin resistance. ACS Nano 13: 2450–2462

Sun Y, Shi H, Yin S, Ji C, Zhang X, Zhang B et al (2018) Human mesenchymal stem cell derived exosomes alleviate type 2 diabetes mellitus by reversing peripheral insulin resistance and relieving β-cell destruction. ACS Nano 12:7613–7628

Szydlarska J, Weiss C, Marycz K (2018) The effect of methyl-β-cyclodextrin on apoptosis, proliferative activity, and oxidative stress in adipose-derived mesenchymal stromal cells of horses suffering from metabolic syndrome (EMS). Molecules 23:287

Tajima M, Yazawa T, Hagiwara K, Kurosawa T, Takahashi K (1992) Diabetes mellitus in cattle infected with bovine viral diarrhoea mucosal disease virus. J Vet Med Assoc 39:616–620. https://doi.org/10.1111/j.1439-0442.1992.tb00225.x

Tajima M, Yuasa M, Kawanabe M, Taniyama H, Yamato O, Maede Y (1999) Possible causes of diabetes mellitus in cattle infected with bovine viral diarrhoea virus. J Vet Med 46:207–215. https://doi.org/10.1046/j.1439-0450.1999.00230.x

Taniyama H, Hirayama K, Kagawa Y et al (1999) Immunohistochemical demonstration of bovine viral diarrhoea virus antigen in the pancreatic islet cells of cattle with insulin-dependent diabetes mellitus. J Comp Pathol 121:149–157. https://doi.org/10.1053/jcpa.1999.0312

Teshima T, Okamoto K, Dairaku K, Nagashima T, Michishita M, Suzuki R, Matsumoto H, Koyama H (2020) Generation of insulin-producing cells from canine adipose tissue-derived mesenchymal stem cells. Stem Cells Int 2020:8841865

Vanella L, Sanford C Jr, Kim DH, Abraham NG, Ebraheim N (2012) Oxidative stress and heme oxygenase-1 regulated human mesenchymal stem cells differentiation. Int J Hyperth 2012:1

Verkest KR (2014) Is the metabolic syndrome a useful clinical concept in dogs? A review of the evidence. Vet J 199(1):24–30

Wearn JM, Crisman MV, Davis JL, Geor RJ, Hodgson DR, Suagee JK, Ashraf-Khorassani M, McCutcheon LJ (2011) Pharmacokinetics of pioglitazone after multiple oral dose administration in horses. J Vet Pharmacol Ther 34(3):252–258

Weiss C, Kornicka-Grabowska K, Mularczyk M, Siwinska N, Marycz K (2020) Extracellular microvesicles (MV's) isolated from 5-azacytidine-and-resveratrol-treated cells improve viability and ameliorate endoplasmic reticulum stress in metabolic syndrome derived mesenchymal stem cells. Stem Cell Rev Rep 16:1343–1355

Xie Z, Hao H, Tong C, Cheng Y, Liu J, Pang Y, Si Y, Guo Y, Zang L, Mu Y, Han W (2016) Human umbilical cord-derived mesenchymal stem cells elicit macrophages into an anti-inflammatory phenotype to alleviate insulin resistance in type 2 diabetic rats. Stem Cells 34(3):627–639

Xiong H, Bai C, Wu S, Gao Y, Lu T, Hu Q, Guan W, Ma Y (2014) Biological characterization of mesenchymal stem cells from bovine umbilical cord. Anim Cells Syst 18(1):59–67. https://doi.org/10.1080/19768354.2014.880370

Zhang Z, Yang M, Wang Y, Wang L, Jin Z, Ding L, Hu T, Jiang W, Gao G, Yang J, Lu B, Cao F, Hu T (2016) Autophagy regulates the apoptosis of bone marrow-derived mesenchymal stem cells under hypoxic condition via AMP-activated protein kinase/mammalian target of rapamycin pathway. Cell Biol Int 40(6):671–685

Zhao H, Shang Q, Pan Z, Bai Y, Li Z, Zhang H et al (2018) Exosomes from adipose-derived stem cells attenuate adipose inflammation and obesity through polarizing M2 macrophages and beiging in white adipose tissue. Diabetes 67:235–247

Zhu S, Lu Y, Zhu J, Xu J, Huang H, Zhu M, Chen Y, Zhou Y, Fan X, Wang Z (2011) Effects of intrahepatic bone-derived mesenchymal stem cells autotransplantation on the diabetic Beagle dogs. J Surg Res 168(2):213–223

Mesenchymal Stem Cells Therapeutic Applications in Eye and Adnexa Ailments 13

Abstract

Ocular affections pose a major challenge to ophthalmologists in veterinary and human medicine. Eye being quite a sensitive and delicate organ, the functional and structural recovery of which has to be complete to maintain sight. To ensure complete recovery of the eye and its sight in clinically protracted cases, ocular affections are being studied through the prism of regenerative medicine employing stem cells. In comparison to humans, literature on veterinary ocular regenerative medicine is very limited. Due to the limited availability of limbal or corneal stem cells and the lack of their typical characterization protocols in animals, alternative mesenchymal stem cells (MSCs) are being evaluated. MSCs share some of the characteristic features with the limbal and corneal stem cells. As such their role is being evaluated in numerous eye affections with a promising outcome, although the literature on veterinary species is currently very limited. In dog, cat and horse, MSCs role is evaluated in a few clinical affections of cornea and retina and/or dog dry eye conditions involving conjunctiva and cornea. The current chapter discusses MSCs in relation to the ocular stem cells and shows their potential therapeutic mechanism and clinical applications.

Keywords

Corneal stem cells · Limbal stem cells · Mesenchymal stem cells · Corneal ulcers · Keratitis · Keratoconjunctivitis

M. B. Gugjoo, *Therapeutic Applications of Mesenchymal Stem Cells in Veterinary Medicine*, https://doi.org/10.1007/978-981-19-3277-9_13

13.1 Introduction

Eye, a specialized light-sensitive sensory structure, is sensitive even to the mildest of the insults arising externally or systemically. Like humans, ocular affections are not uncommon in animal species and tend to occur equally on either side (Athar et al. 2018). Impaired vision of any degree has adverse effects on physiological and productive traits, besides the associated morbidity in animals (Moschos 2014). Incidence of ocular affections in animals varies with the species, which is being determined by their population and geographical location. Ocular affections are commonly reported in equines (horse and mule) followed in order by cats, birds, dogs, sheep, cattle and at last the goat (Athar et al. 2018). Such a higher incidence in horses is also reported in other studies but is shown to be comparable to that of the dog (Tamilmahan et al. 2013; Pandey et al. 2018). Horses have large-sized eyes and are placed laterally making it more susceptible to damage (Clode 2011). Horse being an important draught animal in the area and its performance is dependent on the effectiveness of the vision. Their early reporting to the referral hospitals could be accounted for their higher incidence (Athar et al. 2018). Dogs are commonly kept as pets and thus its early clinical reporting including that for the affections of the eye is expected to give higher incidence. Eye affections may be categorized based on the structural involvement although some of the affections may have an indirect effect on the functionality of its other structures. There are wide varieties of ocular affections in animals including corneal ulcer, corneal dystrophy or dryness of cornea; cataract or subluxation of lens; uveitis of uveal tract and retinal detachment or glaucoma, among others. Corneal ulcers/wounds are one of the most common affections in horses (Andrew and Willis 2005; Wada et al. 2010) and dogs (Gelatt and Brooks 2011), although other affections like glaucoma and cataract are not reported lesser either. In addition to the global affections, associated structures like lacrimal and meibomian glands, and conjunctival goblet cells may also be affected leading to kerato-conjunctivitis sicca (KCS), commonly reported in dogs (Williams and Tighe 2018) and horses (Schwarz et al. 2008). Some of the diseases like dog glaucoma and KCS resemble that of human, making animal studies potentially translational to human (Cislo-Pakuluk and Marycz 2017; Sgrignolia et al. 2019).

13.2 Why Mesenchymal Stem Cell (MSC) Therapy?

Eye affections tend to be more severe and serious considering the importance of each and every structure for the normal vision. Thus slightest of the eye affection cannot be ignored and is taken seriously. Eye affections are either inherited (crystalline dystrophy and fibrosis of cornea or retinal affections) (Pascolini and Mariotti 2012; Robaei and Watson 2014; Wiik et al. 2015) or acquired from trauma or through systemic diseases (diabetes, immune-mediated, and/or infections), in addition to that develop with age (Williams 2008; Pate et al. 2012; Cislo-Pakuluk and Marycz 2017; Davis et al. 2019; Saldinger et al. 2020). In all these affections, an initial inflammatory response may be followed by the structural changes progressing towards an

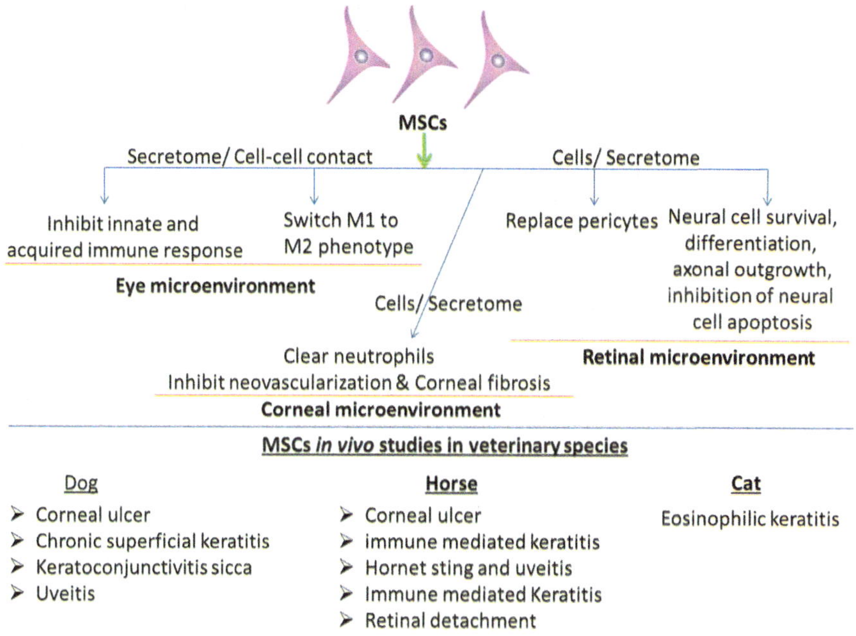

Fig. 13.1 Mesenchymal stem cells (MSCs) therapeutic effects and their potential applications in eye and its adnexa

impaired vision or blindness. In ophthalmology, therapeutics are aimed at restoring both the structural and functional properties for an accurate vision. Affections of cornea, lens or the retina are hard to treat to the fullest possible function with the present therapeutic strategies. Transplantation techniques are common in human ophthalmology with mixed but overall improved results. In animal practice, corneal and lens transplants are rare in comparison to humans. These transplantation techniques in animals are not well standardized, have a shortage of donor and corneal banks, besides being costly affair (Gelatt and Brooks 2011). Thus, a therapeutic option is desired that can be economical and is available as a ready-to-use.

Stem cells especially mesenchymal stem cells (MSCs) offer a suitable option to provide an all-in-one solution for diverse ailments of varied eye structures due to their characteristic features (Fig. 13.1). These cells can be harvested from numerous body sources and from foetal membranes (Gugjoo et al. 2020). Eye structures like cornea, limbus and retina too are not devoid of the stem cells as is demonstrated in animals and human (Funderburgh et al. 2005; Reh and Fischer 2006; Branch et al. 2012; Sancak et al. 2014; Kafarnik et al. 2020). The characterization of these stem cells, however, is not as extensive as that of the other tissues/organs. In comparison to the corneal and/or limbal stem cells, retinal stem cell studies are very much limited. Limbal/corneal stem cells have limited established marker identification for their effective isolation and characterization, especially in animals. Further, their only source remains limbus or cornea, and as such the cells are very limited in availability (Gonzalez et al. 2018). Human limbal epithelial and corneal stem cells share many characteristic features of mesenchymal stem cells set by the International Society for Cellular Therapy (Branch et al. 2012; Veréb et al. 2016). In comparison to limbal stem cells, human corneal stem cells appear much closer to bone marrow MSCs (Veréb et al. 2016). MSCs are identified near the epithelial stem cells at the limbus (Shukla et al. 2020). Area-specific localization of stem cells of corneal and limbal is well established in human, although such studies are very much limited in animals (Patruno et al. 2018). Dog stem cells derived from the limbus and central cornea demonstrate their immune-privileged character. These cells have an ability to express CD90, CD73, CD105, N-cadherin and Pax6 and simultaneously lack haematopoietic (CD34) expression. The cells differentiate into trilineage (osteoblasts, chondrocytes and adipocytes) and keratocyte-like cells. The keratocyte-like cells express keratocan, lumican and ALDH1A3, although with a downregulated Pax6 and N-cadherin expression (Kafarnik et al. 2020). The dog limbal stem cell secretome shows a good concentration of MCP-1 (mediating role in angiogenesis), IL-8 (elimination of pathogens through neutrophil induction and enhances endothelial proliferation), vascular endothelial growth factor (VEGF; vasculogenesis and angiogenesis) and IL-10 (anti-inflammatory effect) (Villatoro et al. 2020). In horses, corneal limbal stem cells share a similar immunostaining pattern (p63+, CK14+, CK3−) as that of the horse corneal epithelial stem cells. However, the cells isolated from central cornea (corneal stem cells) fail to proliferate (Moriyama et al. 2014). In bovine, less than 4% of the corneal stromal cells appear to be progenitor cells expressing Pax6 (Funderburgh et al. 2005) and may be mostly located at the limbus and central cornea expressing stem cell markers like CK14/CK5 (Chen et al. 2010). All these studies depict the presence of innate eye stem cells with characteristics resembling that of the mesenchymal stem cells. Due to the lack of well-defined markers along with the lack of sufficient sources for these stem cells, their counterpart MSCs are being therapeutically evaluated for the ocular affections.

MSCs, with numerous sources, bear characteristics like self-renewal, multipotency, transdifferentiation, in addition to being immune-privileged with the ability to immunomodulate the local environment. Among various set criteria, secretory profile and ability to differentiate show variability among the MSCs derived from various sources and across the species (Gugjoo et al. 2019, 2020; Voga et al. 2021). Based on the available microenvironment, these cells secrete various trophic factors (soluble factors and/or microvesicles) and antimicrobial proteins that potentiate them as therapeutic candidates for eye affections (Fig. 13.1) (Gugjoo et al. 2019, 2020). Microvesicles/exosomes encapsulate the proteins, lipids and miRNAs that can have important therapeutic effects through reduction in inflammation, reduction of cellular damage and apoptosis (Yu et al. 2016; Mead and Tomarev 2017; He et al. 2018). MSCs ability to transfer functional mitochondria to the damaged cells may additionally provide therapeutic benefits. This may occur through tunnelling nanotubes (Feng et al. 2019), gap junctions (Sinclair et al. 2016) or through the exosomes (Gugjoo et al. 2020).

MSCs immunomodulate the local environment either through direct cell–cell contact and/or through secretome. The immunomodulation may occur through secretome containing nitric oxide (NO), indoleamine 2,3-dioxygenase (IDO), transforming growth factor (TGF-β1), interleukins (ILs)-6, -10, tumour necrosis factor-stimulated gene 6 (TSG6), prostaglandin E2 (PGE2), thrombospondin type 1 (TSP1) and hepatocyte growth factor (HGF) (Wang et al. 2018; Gugjoo et al. 2019, 2020). Differential expression of the immunomodulatory factors especially TGF-β1, NO, IDO and IL-6 occurs as per the MSCs source (Lee et al. 2011; Carrade et al. 2012; Cassano et al. 2018) and species involved (Carrade et al. 2012). MSCs or their secretome polarizes pro-inflammatory macrophage M1 to pro-healing macrophage M2 phenotype and modulates the immune cells (Zhang et al. 2014; Dostert et al. 2017). These cells may activate corneal stem cells and promote macrophage switch through the tumour necrosis factor (TNF)-stimulated gene-6 (TSG-6) (Di et al. 2017). MSCs may hamper INF-γ expression of CD4+ T-cells in equine recurrent uveitis possibly through the cell–cell contact and prostaglandin signalling (Saldinger et al. 2020). However, these mechanisms may not be seen in all the species and with all the MSCs.

MSCs may secrete neurotrophic factors like ciliary neurotrophic factor (CNTF), hepatocyte growth factor (HGF), brain-derived neurotrophic factor (BDNF), glial cell-derived neurotrophic factor (GDNF), platelet-derived growth factor (PDGF), nerve growth factor (NGF), neurotrophin-3, 4/5 (NT-3, 4/5), insulin-like growth factor 1 (IGF1), basic fibroblast growth factor (FGF2), epidermal growth factor (EGF), PEDF and erythropoietin (EPO) (Mead et al. 2015; Fontanilla et al. 2016; Noverina et al. 2019; Usategui-Martín et al. 2020). Such secretory action may have therapeutic effects in retinal damages and might lead to the neural cell survival, differentiation, axonal outgrowth and inhibition of neural cell apoptosis (Konala et al. 2016; Salehi et al. 2017). These cells may also replace pericytes in damaged vasculature of the retina due to their comparable morpho-functionality (Mendel et al. 2013; Ezquer et al. 2016; Caplan 2017).

Corneal affections tend to increase its vascularity and opacity and thereby impair the vision. In vitro corneal scratch assays demonstrate equine BM-MSCs, and human UC-MSCs potentially decrease the scratch area in corneal fibroblast cultures (Sherman et al. 2017; Kacham et al. 2021). MSCs conditioned medium contains a higher concentration of TGF-β1, epithelial marker CK8/18 and ocular developmental markers, Pax6 and WNT-7A (Sherman et al. 2017; Kacham et al. 2021). MSCs may clear neutrophils from the cornea to ensure its early healing (Shang et al. 2020). MSCs and their conditioned medium/microvesicles may downregulate the expression of VEGF and matrix metalloproteinase-9 (MMP-9) as per the local environment required in some of the ocular affections like corneal wounds (Ghazaryan et al. 2016; Adak et al. 2021). Microvesicles containing the microRNA (miRNA) may reduce corneal fibrosis (Shojaati et al. 2019) and neovascularization (Pan et al. 2021). Neovascularization inhibition in the cornea may be mediated through miR-340-5p that binds and represses the expressions of prospero homeobox 1 (PROX1) and epithelial membrane protein 2 (EMP2) (Pan et al. 2021). Furthermore, increase in the expression of potent anti-angiogenic factor, thrombospondin-1 (TSP-1) and the reduction in inflammation-related proangiogenic factor MMP-2 might prevent angiogenesis in cornea (Parys et al. 2017). Thus, MSCs have potential to waive off the adversity of the ocular affections, although currently such studies are very limited in animals.

13.3 In Vivo Studies

Numerous lab animal studies favour MSCs efficacy in treating eye affections including the cornea (Yao et al. 2012), retina (Zwart et al. 2009; Tzameret et al. 2015) and uvea (Bai et al. 2018). However, there are very limited studies on MSCs therapeutic applications in veterinary species (Fig. 13.1). MSCs therapeutic potential is evaluated in clinical affections of cornea like keratitis, ulcers, etc. (Spaas et al. 2011; Marfe et al. 2012; Arantes-Tsuzuki et al. 2019; Falcao et al. 2020; Pereira et al. 2020), adnexa-like keratoconjunctivitis sicca (KCS) (Villatoro et al. 2015; Bittencourt et al. 2016; Sgrignolia et al. 2019), uveal tissue-like total uveitis (Shupyk et al. 2020) and retina (Marfe et al. 2012) (Tables 13.1 and 13.2), although potential goes beyond to other affections of these structures and the other structures and their affections like glaucoma and uveitis (Mannino et al. 2021). In this section, the studies that employ MSCs in ocular affections of the common veterinary species are being discussed.

13.3.1 Corneal Ailments

Corneal damage is common as it faces the external environment directly. The affections like ulcers and keratitis arise through trauma, infection and/or immune response (Greenberg 2017). A complicated corneal ulcer with an active protease activity may progress towards the loss of vision (Cequier et al. 2021). To address refractory corneal affections whether immune-mediated or else, aim remains to prevent infection and enhance early healing with minimal scar formation (Marfe et al. 2012). MSCs are evaluated in corneal ulcer in dogs (Arantes-Tsuzuki et al. 2019; Falcao et al. 2020) and horses (Marfe et al. 2012) (Table 13.1).

13.3.1.1 Corneal Ulcers

Corneal ulcer in a diabetic dog (poodle) refractory to the conventional treatment treated topically by two doses of allogeneic AD-MSCs (@ three million 48 h apart) shows an improvement in signs by day 7. By 69th day clinical signs of eye lids, conjunctiva and cornea together with the fundus reduce together with a negative fluorescein test (Arantes-Tsuzuki et al. 2019). In case series study ($n = 26$), subconjunctival transplantation of allogeneic AD-MSCs improves deep ulcers that remain refractory to the conventional treatment. By day 14, $n = 22$ cases (84.6%) show complete ulcer healing. $n = 3$ animals remain refractory to any healing signs while $n = 1$ remains to be evaluated due to the severity of the case. These non-healed ulcers are managed by surgery. It is worth mentioning that the discomfort level varies among the cases and may directly have bearing on the new lesion or further deepening of the already existing condition. The differences in results might be

Table 13.1 MSCs clinical studies on corneal affections in horse, dog and cat

Clinical case	Study design/period	Biomaterial	Evaluation criteria	Overall result	References
Horse corneal ulcers and retinal detachment	Case report [$n = 5$: Ulcer ($n = 1$), keratitis ($n = 1$), hornet sting ($n = 1$), uveitis ($n = 1$), bilateral retinal detachment ($n = 1$)]/6 months	Allogeneic PB-MSCs (ophthalmic artery and locally)	Clinical examination, ultrasound	Improved outcome	Marfe et al. (2012)
Equine immune-mediated keratitis (IMMK)	Case report ($n = 4$)/18 months	Autologous BM-MSCs (3–5 doses of 15×10^6 at 3–4 weeks apart in 3 cases while only single dose in one case, all subconjunctively)	Clinical examination	Improved outcome	Davis et al. (2019)
Dog recurrent corneal ulcer	Case report ($n = 1$)/69 days	Allogeneic AD-MSCs [3×10^6 (2 doses at 48 days interval topically]	Clinical examination	Improved outcome	Arantes-Tsuzuki et al. (2019)
Dog deep corneal ulcers	Case series ($n = 26$)/14 days	Allogeneic AD-MSCs (2.5×10^5 multiple installations)	Clinical examination	Improved outcome	Falcao et al. (2020)
Dog chronic superficial keratitis	Comparative study ($n = 8$)/110 days	Allogeneic AD-MSCs (1×10^6, 2 doses at monthly interval, injected subconjunctival and lower lateral perilimbal region	Clinical examination	Improved outcome	Pereira et al. (2020)
Refractory feline eosinophilic keratitis (FEK)	Case report ($n = 5$)/11 months	Allogeneic AD-MSCs (2 doses of 2×10^6 implanted subconjunctivally, 2 months apart)	Clinical examination	Improved outcome	Villatoro et al. (2018)

MSCs mesenchymal stem cells, *AD* adipose tissue, *BM* bone marrow derived, *PB* peripheral blood derived, *n* total number of animals in all the groups of the study; Improved result: Indicates improved outcome with MSC transplantation even if only one parameter improves; Study period: Indicates maximum time period for which study is conducted

Table 13.2 MSCs clinical studies on Keratoconjunctivitis in dog

Study design/ period	Biomaterial	Evaluation criteria	Overall result	References
Keratoconjunctivitis in dog				
Case report ($n = 12$)/ 9 months	Allogeneic AD-MSCs (5×10^6 around lacrimal glands & 3×10^6 around third eyelid)	Schirmer tear test, ocular surface evaluation	Improved outcome	Villatoro et al. (2015)
Case report ($n = 15$)/ 12 months	Allogeneic AD-MSCs (1×10^6 lacrimal glands dorsal and third eyelid)	Schirmer tear test, ocular surface evaluation	Improved outcome	Bittencourt et al. (2016)
Case report ($n = 3$)/ 6 months	AD-MSCs (10×10^6 in main lacrimal gland and 5×10^6 in accessory lacrimal gland)	Ocular examination, tear production	Improved outcome	Gimenez et al. (2017)
Case report ($n = 22$)/ 6 months	Allogeneic AD-MSCs (4 doses of 1×10^6 into conjunctival sac after every 7 days)	Immunohistochemistry and immunocytochemistry	Improved outcome	Sgrignolia et al. (2019)
Case-controlled studies (28)/ 180 days	Allogeneic AD-MSCs (10×10^6 for body weight $\leq$10 kg, 20×10^6 for body weight 10–20 kg, 30×10^6 for body weight 20–30 kg) intravenously	Schirmer tear test, ocular surface evaluation	Improved outcome	Hermida-Prieto et al. (2021)

MSCs mesenchymal stem cells, *AD* adipose tissue, *n* total number of animals in all the groups of the study; Improved result: Indicates improved outcome with MSCs transplantation even if only one parameter improves; Study period: Indicates maximum time period for which study is conducted

attributed to the differences in cell transplantation and variable case severity (Falcao et al. 2020). Studies that are case controlled with placebo maintained are highly desired. MSCs comparative studies against the current therapeutic options and mechanisms involved need to be understood. These studies are limited enough to draw any conclusion.

13.3.1.2 Dog Chronic Superficial Keratitis

In dog chronic superficial keratitis, MSCs transplanted subconjunctivally and at lower lateral perilimbal region on 0 and 30th day intervals lead to the progressive improvement in corneal lesions. Although up to a month, no major observable improvement occurs but with time (at 110th day follow-up), the cell implanted cornea shows significant regression in neovascularization and inflammatory infiltrate comparable to that with the conventional treatment (topical Prednisolone, 1%). Around 36% area of the total cornea becomes non-neovascularized. Pigmentation remains unchanged which would even prevent MSCs actions. It is worth mentioning that topical administration of routine medications is avoided in MSCs-treated cases.

MSCs transplantation appears safe considering the normal fluorescein, Schirmer and tonometry tests (Pereira et al. 2020).

13.3.1.3 Horse Mixed Eye Lesions

In horse case series involving a single case each of refractory corneal ulcer, immune-mediated keratitis, hornet sting and uveitis, transplantation of MSCs leads to an improvement. Corneal ulcers treated with allogeneic PB-MSCs (into ophthalmic artery and local instillation) show reduced inflammation and lacrimation at day 15 and heal within 3 months. In hornet sting, corneal ulcer significantly reduces after a fortnight. In uveitis-induced corneal ulcer, local and intravenous treatment of PB-MSCs heals corneal ulcer completely by 2-month period. Similarly in corneal keratitis, PB-MSCs injected in ophthalmic artery and applied locally led to an improvement by fortnight without any relapse up to a month's period (Marfe et al. 2012). Thus, MSCs appear promising; however, placebo-controlled studies with a significant number of cases are desired to confirm their utility in clinical cases of corneal ulcers and keratitis.

13.3.1.4 Immune-Mediated Keratitis (IMMK)

Immune-mediated keratitis (IMK) like eosinophilic keratitis is commonly reported in cats (Villatoro et al. 2018), horses (Lassaline-Utter et al. 2014) and dogs (De Geyer and Raymond-Letron 2010). In IMMK, current treatment options fail to prevent recurrence and thus stem cells are being evaluated.

In case of equine IMK, subconjunctival injection of autologous BM-MSCs (@15 million; 3–5 injection at 3–4 weeks interval in $n = 3$ cases and a single dose in $n = 1$) provides a positive response in 75% cases ($n = 3$ out of $n = 4$). Increase in corneal clarity and reduction in neovascularization and surface irregularity occur. The improvement in these cases remains unaffected by the frequency of cell doses (Davis et al. 2019). A clinical refractory feline eosinophilic keratitis (FEK) ($n = 5$) treated locally with allogeneic AD-MSCs (two million, two doses 2 months apart) lead to an improved ocular signs at week 4 in 40% ($n = 2$) cases. This improvement occurs before the second dose. A progressive decrease in ocular signs like corneal plaque, corneal vascularization and hyperaemia occurs. The cornea gets eliminated from eosinophils and mast cells at 2-month follow-up. In other animals, complete remission of clinical signs and cytology occurs at a delayed period (6 months) after two doses of MSCs. It is worth mentioning that no other systemic or local treatment is received in the follow-up period (Villatoro et al. 2018). MSCs repeated transplantation may be required although a single dose may work in some cases. Thus, grading of the clinical case is mandatory to determine the frequency of MSCs transplantation. MSCs may prevent immune-mediated keratitis through immunomodulation and antibacterial actions and can potentially prevent fibrous tissue formation. Studies on such properties of MSCs under in vivo studies are desired as the current study does not provide any impetus. Long-term studies that focus on MSCs mechanism of action, their route and dosing frequency as per the condition and severity are highly desired.

13.3.2 Keratoconjunctivitis Sicca (KCS)

KCS is an immune-mediated disease that has many associated aetiological factors. Its prevalence in dogs ranges from 4 to 20% (Barabino and Dana 2004) while in humans a slightly higher percentage is reported (Abelson et al. 2009). Dogs may preferentially act as model animals for humans as both share developmental origin (Sgrignolia et al. 2019). Currently, KCS is managed by lifelong required palliative treatment, besides its efficacy and safety are variable (Villatoro et al. 2015). In order to develop an effective treatment protocol, stem cell transplantation is evaluated through various routes including the conjunctival (Sgrignolia et al. 2019), lacrimal and third eyelid gland (Park et al. 2013; Villatoro et al. 2015; Bittencourt et al. 2016). Besides, intravenous transplantation of allogeneic AD-MSCs of different concentrations as per the weight is also made (Hermida-Prieto et al. 2021) (Table 13.2). Allogeneic AD-MSCs delivered into the lacrimal and third eyelid gland of dogs are well tolerated even after repeated MSC transplantations without severe consequences (Park et al. 2013). These cells may survive for at least 2 weeks post periocular transplantation (Wood et al. 2012) and are able to improve clinical signs. Conjunctival absorption may increase goblet cell density (Sgrignolia et al. 2019). All these routes are amenable although direct lacrimal gland transplantation may offer maximum cell availability.

Variable doses of allogeneic MSCs are transplanted @ five million in lacrimal gland and three million around third eyelid (Villatoro et al. 2015), ten million in main lacrimal gland and five million around accessory lacrimal glands (Gimenez et al. 2017) and one million in both lacrimal gland and third eyelid (Bittencourt et al. 2016; Sgrignolia et al. 2019). As KCS usually occurs in aged dogs, it is better to use allogeneic cells to prevent cell harvesting from aged patients that may undergo morbidity with little therapeutic effect. MSCs local transplantation significantly improves the clinical signs. This includes tear quantity and quality with reduced density of polymorphonuclear inflammatory cells in the palpebral conjunctiva (Villatoro et al. 2015; Bittencourt et al. 2016; Gimenez et al. 2017). The immunological markers (CD4, IL-1, IL-6 and TNFα) in the palpebral conjunctiva and third eyelid gland also reduce up to 6 months period (Sgrignolia et al. 2019). The clinical remission does not occur up to 9 months (Villatoro et al. 2015) or a year (Bittencourt et al. 2016) although one of the reports show remission at 6 months (Sgrignolia et al. 2019). Complete clinical remission may vary with the severity of the KCS. In mild-to-moderate cases, complete clinical remission may be observed while in severe KCS it may not be completely cured or the single dose of the cells may not be sufficient (Bittencourt et al. 2016). The recurrence of clinical signs may possibly be due to the reactivation of immune-mediated aggression to the lacrimal glands or conjunctival goblet cells (Sgrignolia et al. 2019). Apart from the local applications, MSCs transplanted intravenously also appear feasible as clinical signs improve as early as 15 days in comparison to the control. Such a route may additionally help in resolving concurrent ailments like cutaneous atopy as demonstrated in this study (Hermida-Prieto et al. 2021). In severe KCS, chronicity of the signs concurrently develops structural changes and for evaluation of therapeutic effects on resolution of structural changes, longer duration studies are required. MSCs appear quite promising in KCS and their therapeutic effects may largely be mediated through inflammation suppression and by restoring activity of conjunctival goblet cells and tear production as is also been reported in humans (Lee et al. 2015; Møller-Hansen et al. 2021). However, the long-lasting effects may have to be ensured for which repeated dosing may be needed and effects are required to be evaluated for longer periods. Thus, further extensive studies that confirm its efficacy followed by actual posology to provide evidence-based medicine are desired.

13.3.3 Retinal and Optic Nerve Disorders

MSCs may protect retinal ganglion cells and stimulate axonal regeneration in the optic nerve. This may potentially occur through their trophic action (secretion of neurotrophic factors or by activation of endogenous cells) (Mead et al. 2015; Zhang et al. 2015). In veterinary species, no typical MSCs-enhanced optic nerve regeneration study could be located. However, MSCs may play a potential role to enhance optic nerve regeneration and thus studies are desired.

13.3.3.1 Retinal Detachment

Horse retinal detachments may occur due to recurrent uveitis or injury, although congenital ocular disorders may also predispose horses towards it. Vision loss both in complete or partial retinal detachment is seen (Strobel et al. 2007). A single clinical study of retinal detachment evaluates MSCs therapeutic role in a 2-month-old case report of equine bilateral retinal detachment treated conventionally for 2 months without any positive effect on eyesight. Post peripheral blood MSCs intravenous injection, optic nerve regains function with an improved visual acuity by third month. By the sixth month period, the animal jumps and crosses some poles. Ultrasonography confirms clinical remission in the left eye but not in the right eye that shows only slight improvement (Marfe et al. 2012). Differences in pathophysiology together with insufficient cell concentration in the right eye might have led to differences in results. Further studies are therefore desired for better understanding.

13.3.3.2 Glaucoma

Glaucoma, rise in intraocular pressure (IOP) that threatens sight through retinal neuropathy, occurs with multiple aetiologies and is one of the most difficult ophthalmological ailments encountered in veterinary practice (Gelatt and MacKay 2004a, b). In glaucoma, various life-long medications are required to prevent such an increase in pressure, although gradual changes keep on occurring leading to terminal vision loss (Roubeix et al. 2015). Currently, no stem cell study on glaucoma is conducted in animals to evaluate the therapeutic benefits. However, various lab animal studies show MSCs promising therapeutic effects. The lab animal model developed through cauterization of three episcleral veins (EVC) and transplanted with MSCs into the anterior chamber injection shows localization of the cells to the ciliary processes and the trabecular meshwork with improvement in the intraocular pressure (IOP) (Roubeix et al. 2015). In another lab animal glaucoma model (optic nerve crush), BM-MSCs exosomes show significant neuroprotective and neuritogenic effects. Intravitreal injection of exosomes promotes significant survival of retinal ganglion cells (RGCs) and their axonal regeneration through prevention of RGC axonal loss occurs partially (Mead and Tomarev 2017). Furthermore, in a retinal ischemia model, MSCs extracellular vesicles injected into vitreous humour significantly enhance functional recovery, and decrease neuroinflammation and apoptosis. These vesicles may be located in retinal neurons, RGCs and microglia (Mathew et al. 2019). These diverse aetiological experimental models provide proof-of-therapeutic principles of MSCs or their extracellular vesicles. Thus, potential neuroprotective and regenerative effects in retinal disorders including the glaucoma may occur with these biomaterials and are desired to be evaluated in animals.

In addition to these structures, MSCs may show some therapeutic effect on some other conditions like total uveitis and rare conditions like neurodegenerative disease. In a single study on the dog total uveitis, MSCs transplantation improves outcome through the regenerative effects and reduction of inflammation (Shupyk et al. 2020). A neurodegenerative disease that occurs with the null mutation in soluble lysosomal enzyme tripeptidyl peptidase-1 (TPP1) gene in dachshund dog breed exhibits a disorder similar to human CLN2 disease. In this disorder, most neurological functions including vision progressively decline (Whiting et al. 2015; Tracy et al. 2016). At an early stage, intravitreal transplantation of single dose of TPPI-transduced BM-MSCs (either 2–3 million or 10–13 million) substantially inhibits ophthalmological changes including that of the retinal functional deficits and its structural changes. The results obtained, however, may be dose-dependent and thus additional doses may be required for improved results (Tracy et al. 2016). These preliminary studies are very limited and lack details of MSCs therapeutic effects. Further detailed studies are therefore desired to evaluate the therapeutic efficacy of MSCs in these eye affections.

13.4 Conclusion(s)

Ocular affections are common and tend to make animals blind unless treated early and effectively. Many ocular ailments lack effective treatment options. There are very limited studies for veterinary species on eye stem cells and/or MSCs for ocular affections. Limbal and corneal stem cells have mainly been characterized in dogs and cattle under in vitro studies although without any in vivo applications. Mesenchymal stem cells (MSCs) bear stem cell properties and can be harvested from numerous sources. Limited literature currently exists on their potential applications in clinical cases of corneal ulcers, chronic superficial keratitis, immune-mediated keratitis (IMMK), retinal detachment and KCS. These studies show that MSCs can be applied through local as well as the parenteral routes. Transplantation of MSCs leads to clinical remission though the overall clinical improvement is determined by the severity. Most of these studies are case-uncontrolled and lack placebo control. A very limited animal number and shorter duration studies make results non-conclusive. MSCs therapeutic mechanisms in eye affections in these large animals remain to be studied. Therefore, studies may initially be performed under uniform experimental models and focused towards MSCs therapeutic mechanisms. Further, comparative studies based on cell source, concentration, dose, route and frequency are desired to confirm MSCs safety and efficacy to establish evidence-based medicine with proper MSCs posology. In addition to the above-mentioned ocular affections, MSCs may be evaluated for other affections like glaucoma, retinal and optic disorders.

References

Abelson MB, Ousler GW, Maffei C (2009) Dry eye in 2008. Curr Opin Ophthalmol 20(4):282–286

Adak S, Magdalene D, Deshmukh S, Das D, Jaganathan BG (2021) A review on mesenchymal stem cells for treatment of retinal diseases. Stem Cell Rev Rep 6:1–20

Andrew SA, Willis AM (2005) Diseases of the cornea and sclera. In: Gilger BC (ed) Equine ophthalmology. Elsevier Saunders, St. Louis, pp 157–251

Arantes-Tsuzuki PDM, Mazzonetto PC, Lo Turco EG (2019) Treatment for canine corneal ulcer using adipose tissue derived mesenchymal stem cell therapy—case report. Preprints 120185

Athar H, Parrah JD, Makhdoomi DM, Dar SH, Fazili MR, Dar M (2018) A retrospective study of eye affections in animals of Kashmir Valley. Haryana Vet 57(2):234–238

Bai L, Shao H, Wang H, Zhang Z, Su C, Dong L, Yu B, Chen X, Li X, Zhang X (2018) Effects of mesenchymal stem cell-derived exosomes on experimental autoimmune uveitis. Sci Rep 8(1): 9889

Barabino S, Dana MR (2004) Animal models of dry eye: a critical assessment of opportunities and limitations. Invest Ophthalmol Vis Sci 45(6):1641–1646

Bittencourt MKW, Barros MA, Martins JFP, Vasconcellos JPC, Morais BP, Pompeia C, Bittencourt MD, Evangelho KS, Kerkis I, Wenceslau CV (2016) Allogeneic mesenchymal stem cell transplantation in dogs with keratoconjunctivitis sicca. Cell Med 8:63–77

Branch MJ, Hashmani K, Dhillon P, Jones DR, Dua HS, Hopkinson A (2012) Mesenchymal stem cells in the human corneal limbal stroma. Invest Ophthalmol Vis Sci 53:5109–5116

Caplan AI (2017) New MSC: MSCs as pericytes are sentinels and gatekeepers. J Orthop Res 35(6): 1151–1159

Carrade DD, Lame MW, Kent MS, Clark KC, Walker NJ, Borjesson DL (2012) Comparative analysis of the immunomodulatory properties of equine adult-derived mesenchymal stem cells. Cell Med 4(1):1–11

Cassano JM, Fortier LA, Hicks RB, Harman RM, Van de Walle GR (2018) Equine mesenchymal stromal cells from different tissue sources display comparable immune-related gene expression profiles in response to interferon gamma (IFN)-γ. Vet Immunol Immunopathol 202:25–30

Cequier A, Sanz C, Rodellar C, Barrachina L (2021) The usefulness of mesenchymal stem cells beyond the musculoskeletal system in horses. Animals 11:931

Chen B, Mi S, Wright B, Connon CJ (2010) Investigation of K14/K5 as a stem cell marker in the limbal region of the bovine cornea. PLoS One 5(10):e13192

Cislo-Pakuluk A, Marycz K (2017) A promising tool in retina regeneration: current perspectives and challenges when using mesenchymal progenitor stem cells in veterinary and human ophthalmological applications. Stem Cell Rev and Rep 13:598–602

Clode AB (2011) Diseases and surgery of the cornea. In: Gilger BC (ed) Equine ophthalmology. Wiley, Hoboken, NJ, pp 181–266. ISBN 9781437708462

Davis AB, Schnabel LV, Gilger BC (2019) Subconjunctival bone marrow-derived mesenchymal stem cell therapy as a novel treatment alternative for equine immune-mediated keratitis: a case series. Vet Ophthalmol 22(5):674–682

De Geyer G, Raymond-Letron I (2010) Kératopathie calcique en bandelette chez le chat: à propos de deux cas. Pratique Médicale et Chirurgicale de l'Animal de Compagnie, Spécial Cancérologie/Académie Vétérinaire de France 45:33–39

Di G, Du X, Qi X, Zhao X, Duan H, Li S, Xie L, Zhou Q (2017) Mesenchymal stem cells promote diabetic corneal epithelial wound healing through TSG-6–dependent stem cell activation and macrophage switch. Invest Opthalmol Vis Sci 58(10):4344

Dostert G, Mesure B, Menu P, Velot É (2017) How do mesenchymal stem cells influence or are influenced by microenvironment through extracellular vesicles communication? Front Cell Dev Biol 5:6

Ezquer M, Urzua CA, Montecino S, Leal K, Conget P, Ezquer F (2016) Intravitreal administration of multipotent mesenchymal stromal cells triggers a cytoprotective microenvironment in the retina of diabetic mice. Stem Cell Res Ther 7:1–17

Falcao MSA, Brunel HS, Peixer MAS, Dallago BSL, Costa FF, Queiroz LM, Campbell P, Malard PF (2020) Effect of allogeneic mesenchymal stem cells (MSCs) on corneal wound healing in dogs. J Tradit Complement Med 10:440–445

Feng Y, Zhu R, Shen J, Wu J, Lu W, Zhang J, Zhang J, Liu K (2019) Human bone marrow mesenchymal stem cells rescue endothelial cells experiencing chemotherapy stress by mitochondrial transfer via tunneling nanotubes. Stem Cells Dev 28(10):674–682

Fontanilla CV, Gu HY, Liu QP, Zhu TZ, Zhou CW, Johnstone BH, March KL, Pascuzzi RM, Farlow MR, Du YS (2016) Adipose-derived stem cell conditioned media extends survival time of a mouse model of amyotrophic lateral sclerosis. Sci Rep 5:16953

Funderburgh ML, Du Y, Mann MM, SundarRaj N, Funderburgh JL (2005) PAX6 expression identifies progenitor cells for corneal keratocytes. FASEB J 19:1371–1373

Gelatt K, Brooks DE (2011) Penetrating corneal grafts. Chapter 8: surgery of the cornea and sclera. In: Gelatt K, Gelatt JP (eds) Veterinary ophthalmic surgery. Saunders Elsevier, Philadelphia, p 223

Gelatt KN, MacKay EO (2004a) Prevalence of the breed-related glaucomas in pure-bred dogs in North America. Vet Ophthalmol 7:97–111

Gelatt KN, MacKay EO (2004b) Secondary glaucomas in the dog in North America. Vet Ophthalmol 7:245–259

Ghazaryan E, Zhang Y, He Y, Liu X, Li Y, Xie J, Su G (2016) Mesenchymal stem cells in corneal neovascularization: comparison of different application routes. Mol Med 14(4):3104–3112

Gimenez E, Bayon A, Gil-Chinchilla J, Garcia-Bernal D, Atucha N, Talavera J (2017) Therapeutic potential of adipose-tissue derived mesenchymal stem cells in the treatment of canine keratoconjunctivitis sicca: preliminary clinical observations. Paper presented at: Annual Scientific Meeting of the European College of Veterinary Ophthalmologists (ECVO-2017), Estoril, Portugal

Gonzalez G, Sasamoto Y, Ksander BR, Frank MH, Frank NY (2018) Limbal stem cells: identity, developmental origin and therapeutic potential. Wiley Interdiscip Rev Dev Biol 7(2). https://doi.org/10.1002/wdev.303

Greenberg S (2017) Ulcerative keratitis and keratomalacia in horses: medical management and surgical considerations. Equine Vet Educ 31:179–182

Gugjoo MB, Amarpal, Makhdoomi DM, Sharma GT (2019) Equine mesenchyaml stem cells: properties, sources, characterization and potential therapeutic applications. J Equine Vet Sci 72:16–27. https://doi.org/10.1016/j.jevs.2018.10.007

Gugjoo MB, Hussain S, Amarpal SRA, Dhama K (2020) Mesenchymal stem cell mediated Immuno-modulatory and anti-inflammatory mechanisms in immune and allergic disorders. Recent Pat Inflamm Allergy Drug Discov 14(1):3–14. https://doi.org/10.2174/1872213X14666200130100236

He GH, Zhang W, Ma YX, Yang J, Chen L, Song J, Chen S (2018) Mesenchymal stem cells-derived exosomes ameliorate blue light stimulation in retinal pigment epithelium cells and retinal laser injury by VEGF-dependent mechanism. Int J Ophthalmol 11:559–566

Hermida-Prieto M, García-Castro J, Mariñas-Pardo L (2021) Systemic treatment of immune mediated keratoconjunctivitis sicca with allogeneic stem cells improves the Schirmer tear test score in a canine spontaneous model of disease. J Clin Med 10:5981

Kacham S, Bhure TS, Eswaramoorthy SD, Naik G, Rath SN, Parcha SR, Basu S, Sangwan VS, Shukla S (2021) Human umbilical cord-derived mesenchymal stem cells promote corneal epithelial repair in vitro. Cell 10:1254

Kafarnik C, McClellan A, Dziasko M, Daniels JT, Guest DJ (2020) Canine corneal stromal have multipotent mesenchymal stromal cell properties in vitro. Stem Cells Dev 29(7):425–439

Konala VBR, Mamidi MK, Bhonde R, Das AK, Pochampally R, Pal R (2016) The current landscape of the mesenchymal stromal cell secretome: a new paradigm for cell-free regeneration. Cytotherapy 18(1):13–24

Lassaline-Utter M, Miller C, Wotman KL (2014) Eosinophilic keratitis in 46 eyes of 27 horses in the mid-Atlantic United States (2008-2012). Vet Ophthalmol 17:311–320

Lee WS, Suzuki Y, Graves SS, Iwata M, Venkataraman GM, Mielcarek M et al (2011) Canine bone marrow-derived mesenchymal stromal cells suppress alloreactive lymphocyte proliferation in vitro but fail to enhance engraftment in canine bone marrow transplantation. Biol Blood Marrow Transplant 17(4):465–475

Lee MJ, Ko AY, Ko JH, Lee HJ, Kim MK, Wee WR, Khwarg SI, Oh JY (2015) Mesenchymal stem/stromal cells protect the ocular surface by suppressing inflammation in an experimental dry eye. Mol Ther 23(1):139–146

Mannino G, Russo C, Longo A, Anfuso CD, Lupo G, Lo Furno D, Giuffrida R, Giurdanella G (2021) Potential therapeutic applications of mesenchymal stem cells for the treatment of eye diseases. World J Stem Cells 13(6):632–644

Marfe G, Massaro-Giordano M, Ranalli M, Cozzoli E, Di Stefano C, Malafoglia V, Polettini M, Gambacurta A (2012) Blood derived stem cells: an ameliorative therapy in veterinary ophthalmology. J Cell Physiol 227:1250–1256

Mathew B, Ravindran S, Liu X, Torres L, Chennakesavalu M, Huang C-C, Feng L, Zelka R, Lopez J, Sharma M, Roth S (2019) Mesenchymal stem cell-derived extracellular vesicles and retinal ischemia-reperfusion. Biomaterials 197:146–160

Mead B, Tomarev S (2017) Bone marrow-derived mesenchymal stem cells-derived exosomes promote survival of retinal ganglion cells through miRNA-dependent mechanisms. Stem Cells Transl Med 6(4):1273–1285

Mead B, Berry M, Logan A, Scott RAH, Leadbeater W, Scheven BA (2015) Stem cell treatment of degenerative eye disease. Stem Cell Res 14:243–257

Mendel TA, Clabough EBD, Kao DS, Demidova-Rice TN, Durham JT, Zotter BC, Seaman SA, Cronk SM, Rakoczy EP, Katz AJ, Herman IM, Peirce SM, Yates PA (2013) Pericytes derived from adipose-derived stem cells protect against retinal vasculopathy. PLoS One 8(5):e65691

Møller-Hansen M, Larsen A-C, Toft PB, Lynggaard CD, Schwartz C, Bruunsgaard H, Haack-Sørensen M, Ekblond A, Kastrup J, Heegaard S (2021) Safety and feasibility of mesenchymal stem cell therapy in patients with aqueous deficient dry eye disease. Ocul Surf 19:43–52

Moriyama H, Kasashima Y, Kuwano A, Wada S (2014) Anatomical location and culture of equine corneal epithelial stem cells. Vet Ophthalmol 17(2):106–112

Moschos MM (2014) Physiology and psychology of vision and its disorders: a review. Med Hypothesis Discov Innov Ophthalmol 3(3):83–90

Noverina R, Widowati W, Ayuningtyas W, Kurniawan D, Afifah E, Laksmitawati DR, Rinendyaputri R, Rilianawati R, Faried A, Bachtiar I, Wirakusumah FF (2019) Growth factors profile in conditioned medium human adipose tissue-derived mesenchymal stem cells (CM-hATMSCs). Clin Nutr Exp 2019(24):34–44

Pan J, Luo X, Zhao S, Li J, Jiang Z (2021) miR-340-5p mediates the therapeutic effect of mesenchymal stem cells on corneal neovascularization. Graefes Arch Clin Exp Ophthalmol 260(2):497–507. https://doi.org/10.1007/s00417-021-05394-8

Pandey P, Shahi A, Kumar D, Shukla MK (2018) Incidence of eye affections in dogs. Indian J Vet Sci Biotechnol 13(4):65–67

Park SA, Reilly CM, Wood JA, Chung DJ, Carrade DD, Deremer SL, Seraphin RL, Clark KC, Zwingenberger A, Borjesson DL, Hayashi K, Russell P, Murphy CJ (2013) Safety and immunomodulatory effects of allogeneic canine adipose-derived mesenchymal stromal cells transplanted into the region of the lacrimal gland, the gland of the third eyelid and the knee joint. Cytotherapy 15(12):1498–1510

Parys M, Kruger JM, Yuzbasiyan-Gurkan V (2017) Evaluation of immunomodulatory properties of feline mesenchymal stem cells. Stem Cells Dev 26(10):776–785

Pascolini D, Mariotti SP (2012) Global estimates of visual impairment: 2010. Br J Ophthalmol 96: 614–618

Pate DO, Clode AB, Olivry T et al (2012) Immunohistochemical and immunopathologic characterization of superficial stromal immune-mediated keratitis in horses. Am J Vet Res 73:1067–1073

Patruno M, Perazzi A, Martinello T, Gomiero C, Maccatrozzo L, Iacopetti I (2018) Investigations of the corneal epithelium in veterinary medicine: state of the art on corneal stem cells found in different mammalian species and their putative application. Res Vet Sci 118:502–507

Pereira AL, Bittencourt MW, Barros MA, Malago R, Martins JFP, Morais B, Montiani-Ferreira F, Jose de Vasconcellos PC (2020) Subconjunctival use of allogeneic mesenchymal stem cells to treat chronic superficial keratitis in German Shepherd dogs: pilot study. Res Square. https://doi.org/10.21203/rs.3.rs-51716/v1

Reh TA, Fischer AJ (2006) Retinal stem cells. Methods Enzymol 419:52–73

Robaei D, Watson S (2014) Corneal blindness: a global problem. Clin Exp Ophthalmol 42:213–214

Roubeix C, Godefroy D, Mias C, Sapienza A, Riancho L, Degardin J, Fradot V, Ivkovic I, Picaud S, Sennlaub F, Denoyer A, Rostene W, Sahel JA, Parsadaniantz SM, Brignole-Baudouin F, BaudouinC. (2015) Intraocular pressure reduction and neuroprotection conferred by bone marrow-derived mesenchymal stem cells in an animal model of glaucoma. Stem Cell Res Ther 6(1):177

Saldinger LK, Nelson SG, Bellone RR, Lassaline M, Mack M, Walker NJ, Borjesson DL (2020) Horses with equine recurrent uveitis have an activated CD4+ T- cell phenotype that can be modulated by mesenchymal stem cells in vitro. Vet Ophthalmol 23(1):160–170

Salehi H, Amirpour N, Razavi S, Esfandiari E, Zavar R (2017) Overview of retinal differentiation potential of mesenchymal stem cells: a promising approach for retinal cell therapy. Ann Anat 210:52–63

Sancak IG, Ozen A, Pinarli FA, Tiryaki M, Ceylan A, Acar U, Delibasi T (2014) Limbal stem cells in dogs and cats their identification culture and differentiation into keratinocytes. Kafkas universitesi Veteriner Fakültesi Dergisi 20(6):909–914

Schwarz BC, Sallmutter T, Nell B (2008) Keratoconjunctivitis sicca attributable to parasympathetic facial nerve dysfunction associated with hypothyroidism in a horse. J Am Vet Med Assoc 233(11):1761–1766

Sgrignolia MR, Silvaa DA, Nascimentoa FF, Sgrignolia DAM, Naia GA, da Silvac MG, de Barrosd MA, Bittencourtd MKW, de Moraisd BP, Dinalloe HR, Fogliae BTD, Cabrerae WB, Farese EC, Andrade SF (2019) Reduction in the inflammatory markers CD4, IL-1, IL-6 and TNFα in dogs with keratoconjunctivitis sicca treated topically with mesenchymal stem cells. Stem Cell Res 39: 101525

Shang Q, Chu Y, Li Y, Han Y, Yu D, Liu R, Zheng Z, Song L, Fang J, Li X, Cao L, Gong Z, Zhang L, Chen Y, Wang Y, Shao C, Shi Y (2020) Adipose-derived mesenchymal stromal cells promote corneal wound healing by accelerating the clearance of neutrophils in cornea. Cell Death Dis 11(8):707

Sherman AB, Gilger BC, Berglund AK, Schnabel LV (2017) Effect of bone marrow-derived mesenchymal stem cells and stem cell supernatant on equine corneal wound healing in vitro. Stem Cell Res Ther 8:120

Shojaati G, Khandaker I, Funderburgh ML, Mann MM, Basu R, Stolz DB, Geary ML, Santos AD, Deng SX, Funderburgh JL (2019) Mesenchymal stem cells reduce corneal fibrosis and inflammation via extracellular vesicle-mediated delivery of miRNA. Stem Cells Transl Med 8(11): 1192–1201

Shukla S, Shanbhag SS, Tavakkoli F, Varma S, Singh V, Basu S (2020) Limbal epithelial and mesenchymal stem cell therapy for corneal regeneration, current eye research 2019. Curr Eye Res 45(3):265–277

Shupyk OV, Bokotko RR, Savchuk TL, Danilov VB, Kladnytska LV, Kharkevych YO, Pasnichenko OS, Blahyi RS, Hraborenko NI, Krystyniak YM (2020) Effectiveness of mesenchymal stem cells in uveitis in dogs, depending on the method of their administration. Sci Tech Bull State Sci Res Control Inst Vet Med Prod Fodder Addit Inst Anim Biol 21(2):219–229

Sinclair KA, Yerkovich ST, Hopkins PMA, Chambers DC (2016) Characterization of intercellular communication and mitochondrial donation by mesenchymal stromal cells derived from the human lung. Stem Cell Res Ther 7(1):91

Spaas JH, Gambacurta A, Polettini M, Broeckx S, Van Hoeck F, De Schauwer C, Van De Walle GR, Van Soom A (2011) Purification and expansion of stem cells from equine peripheral blood, with clinical applications. Vlaams Diergeneeskd Tijdschr 80:129–135

Strobel BW, Wilkie DA, Gilger BC (2007) Retinal detachment in horses: 40 cases (1998–2005). Vet Ophthalmol 10(6):380–385

Tamilmahan P, Zama MMS, Pathak R, Muneeswaran NS, Karthik K (2013) A retrospective study of ocular occurrence in domestic animals: 799 cases. Vet World 6(5):274–276

Tracy CJ, Whiting REH, Pearce JW, Williamson BG, Vansteenkiste DP, Gillespie LE, Leilani Castaner J, Bryan JN, Coates JR, Jensen CA, Katz ML (2016) Intravitreal implantation of TPP1-transduced stem cells delays retinal degeneration in canine CLN2 neuronal ceroid lipofuscinosis. Exp Eye Res 152:77–87

Tzameret A, Sher I, Belkin M et al (2015) Epiretinal transplantation of human bone marrow mesenchymal stem cells rescues retinal and vision function in a rat model of retinal degeneration. Stem Cell Res 15(2):387–394

Usategui-Martín R, Puertas-Neyra K, García-Gutiérrez MT, Fuentes M, Pastor JC, Fernandez-Bueno I (2020) Human mesenchymal stem cell secretome exhibits a neuroprotective effect over in vitro retinal photoreceptor degeneration. Mol Ther Methods Clin Dev 2020(17):1155–1166

Veréb Z, Póliska S, Albert R, Olstad OK, Boratkó A, Csortos C, Moe MC, Facskó A, Petrovski G (2016) Role of human corneal stroma-derived mesenchymal-like stem cells in corneal immunity and wound healing. Sci Rep 6:26227

Villatoro AJ, Fernández V, Claros S, Rico-Llanos GA, Becerra J, Andrades JA (2015) Use of adipose-derived mesenchymal stem cells in keratoconjunctivitis sicca in a canine model. BioMed Res Int 2015:527926

Villatoro AJ, Claros S, Fernández V, Alcoholado C, Fariñas F, Moreno A, Becerra J, Andrades JA (2018) Safety and efficacy of the mesenchymal stem cell in feline eosinophilic keratitis treatment. BMC Vet Res 14:116

Villatoro AJ, Alcoholado C, Martín-Astorga MC, Rico G, Fernández V, Becerra J (2020) Characterization of the secretory profile and exosomes of limbal stem cells in the canine species. PLoS One 15(12):e0244327

Voga M, Kovac V, Majdic G (2021) Comparison of canine and feline adipose-derived mesenchymal stem cells/medicinal signaling cells with regard to cell surface marker expression, viability, proliferation, and differentiation potential. Front Vet Sci 7:610240

Wada S, Hobo S, Niwa H (2010) Ulcerative keratitis in thoroughbred racehorses in Japan from 1997 to 2008. Vet Ophthalmol 13(2):99–105

Wang M, Yuan Q, Xie L (2018) Mesenchymal stem cell-based immunomodulation: properties and clinical application. Stem Cells Int 2018:3057624

Whiting RE, Pearce JW, Castaner LJ, Jensen CA, Katz RJ, Gilliam DH, Katz ML (2015) Multifocal retinopathy in Dachshunds with CLN2 neuronal ceroid lipofuscinosis. Exp Eye Res 134: 123e132

Wiik A, Ropstad E, Ekesten B, Karlstam L, Wade C, Lingaas F (2015) Progressive retinal atrophy in Shetland sheepdog is associated with a mutation in theCNGA1gene. Anim Genet 46(5): 515–521

Williams DL (2008) Immunopathogenesis of keratoconjunctivitis sicca in the dog. Vet Clin North Am Small Anim Pract 38(2):251–268

Williams DL, Tighe A (2018) Immunohistochemical evaluation of lymphocyte populations in the nictitans glands of normal dogs and dogs with keratoconjunctivitis sicca. Open Vet J 8:47–52

Wood JA, Chung DJ, Park SA, Zwingenberger AL, Reilly CM, Ly I, Walker NJ, Vernau W, Hayashi K, Wisner ER, Cannon MS, Kass PH, Cherry SR, Borjesson DL, Russell P, Murphy CJ (2012) Periocular and intra-articular injection of canine adipose derived mesenchymal stem cells: an in vivo imaging and migration study. J Ocul Pharmacol Ther 28(3):307–317

Yao L, Li Z-r, Su W-r, Li Y-p, Lin M-l, Zhang W-x, Liu Y, Wan Q, Liang D (2012) Role of mesenchymal stem cells on cornea wound healing induced by acute alkali burn. PLoS One 7(2): e30842

Yu B, Shao H, Su C, Jiang Y, Chen X, Bai L, Zhang Y, Li Q, Zhang X, Li X (2016) Exosomes derived from MSCs ameliorate retinal laser injury partially by inhibition of MCP-1. Sci Rep 6: 1–12

Zhang B, Yin Y, Lai RC, Tan SS, Choo ABH, Lim SK (2014) Mesenchymal stem cells secrete immunologically active exosomes. Stem Cells Dev 23(11):1233–1244

Zhang L, Coulson-Thomas VJ, Ferreira TG et al (2015) Mesenchymal stem cells for treating ocular surface diseases. BMC Ophthalmol 15:155. https://doi.org/10.1186/s12886-015-0138-4

Zwart I, Hill A, Al-Allaf F et al (2009) Umbilical cord blood mesenchymal stromal cells are neuroprotective and promote regeneration in a rat optic tract model. Exp Neurol 216(2):439–448

The manufacturer's authorised representative in the EU is Springer Nature Customer Service Centre GmbH, Europaplatz 3, 69115 Heidelberg, Germany. If you have any concerns regarding our products, please contact ProductSafety@springernature.com

Printed and bound by CPI Group (UK) Ltd, Croydon, CR0 4YY
01/10/2026
02225860-0014